DATE DUE

VOLUME FIVE HUNDRED AND SEVENTY FOUR

METHODS IN ENZYMOLOGY

Enzymes of Epigenetics, Part B

METHODS IN ENZYMOLOGY

Editors-in-Chief

ANNA MARIE PYLE
Departments of Molecular, Cellular and Developmental
Biology and Department of Chemistry
Investigator, Howard Hughes Medical Institute
Yale University

DAVID W. CHRISTIANSON
Roy and Diana Vagelos Laboratories
Department of Chemistry
University of Pennsylvania
Philadelphia, PA

Founding Editors

SIDNEY P. COLOWICK and NATHAN O. KAPLAN

VOLUME FIVE HUNDRED AND SEVENTY FOUR

METHODS IN
ENZYMOLOGY

Enzymes of Epigenetics, Part B

Edited by

RONEN MARMORSTEIN

*Abramson Family Cancer Research Institute
Perelman School of Medicine at the University of
Pennsylvania, Philadelphia, PA, United States*

AMSTERDAM • BOSTON • HEIDELBERG • LONDON
NEW YORK • OXFORD • PARIS • SAN DIEGO
SAN FRANCISCO • SINGAPORE • SYDNEY • TOKYO

Academic Press is an imprint of Elsevier

Academic Press is an imprint of Elsevier
50 Hampshire Street, 5th Floor, Cambridge, MA 02139, USA
525 B Street, Suite 1800, San Diego, CA 92101-4495, USA
The Boulevard, Langford Lane, Kidlington, Oxford OX5 1GB, UK
125 London Wall, London, EC2Y 5AS, UK

First edition 2016

Copyright © 2016 Elsevier Inc. All rights reserved.

No part of this publication may be reproduced or transmitted in any form or by any means, electronic or mechanical, including photocopying, recording, or any information storage and retrieval system, without permission in writing from the publisher. Details on how to seek permission, further information about the Publisher's permissions policies and our arrangements with organizations such as the Copyright Clearance Center and the Copyright Licensing Agency, can be found at our website: www.elsevier.com/permissions.

This book and the individual contributions contained in it are protected under copyright by the Publisher (other than as may be noted herein).

Notices
Knowledge and best practice in this field are constantly changing. As new research and experience broaden our understanding, changes in research methods, professional practices, or medical treatment may become necessary.

Practitioners and researchers must always rely on their own experience and knowledge in evaluating and using any information, methods, compounds, or experiments described herein. In using such information or methods they should be mindful of their own safety and the safety of others, including parties for whom they have a professional responsibility.

To the fullest extent of the law, neither the Publisher nor the authors, contributors, or editors, assume any liability for any injury and/or damage to persons or property as a matter of products liability, negligence or otherwise, or from any use or operation of any methods, products, instructions, or ideas contained in the material herein.

ISBN: 978-0-12-805381-2
ISSN: 0076-6879

For information on all Academic Press publications visit our website at https://www.elsevier.com/

Publisher: Zoe Kruze
Acquisition Editor: Zoe Kruze
Editorial Project Manager: Sarah Lay
Production Project Manager: Magesh Kumar Mahalingam
Cover Designer: Greg Harris

Typeset by SPi Global, India

CONTENTS

Contributors	xi
Preface	xv

Part I
Epigenetic Technologies

1. Identification and Quantification of Histone PTMs Using High-Resolution Mass Spectrometry 3
K.R. Karch, S. Sidoli, and B.A. Garcia

1.	Introduction	4
2.	Histone Extraction from Cells	6
3.	Bottom-Up Mass Spectrometry	9
4.	Offline Fractionation of Histone Species	18
5.	Middle-Down Mass Spectrometry	19
6.	Top-Down Mass Spectrometry	24
	References	28

2. Substrate Specificity Profiling of Histone-Modifying Enzymes by Peptide Microarray 31
E.M. Cornett, B.M. Dickson, R.M. Vaughan, S. Krishnan, R.C. Trievel, B.D. Strahl, and S.B. Rothbart

1.	Introduction	32
2.	Assay Optimization	34
3.	Assay Methodology	37
4.	Enzyme Specificity Profiling by Microarray	42
5.	Summary and Perspectives	49
	Acknowledgments	50
	References	50

3. ArrayNinja: An Open Source Platform for Unified Planning and Analysis of Microarray Experiments 53
B.M. Dickson, E.M. Cornett, Z. Ramjan, and S.B. Rothbart

1.	Introduction	54
2.	ArrayNinja	55
3.	Planning Custom Microarrays with ArrayNinja	61

4.	Data Analysis Features of ArrayNinja	64
5.	Benchmarking ArrayNinja Against ImageQuant TL	69
6.	Methodological Details	72
7.	Limitations, Assumptions, Other Features, and Future Development	74
8.	Summary	76
	Acknowledgments	76
	References	76

4. Chemical Biology Approaches for Characterization of Epigenetic Regulators   79

D. Barsyte-Lovejoy, M.M. Szewczyk, P. Prinos, E. Lima-Fernandes, S. Ackloo, and C.H. Arrowsmith

1.	Introduction	80
2.	Validation of Chemical Probes for Use in Cell-Based Experiments	81
3.	Inhibitor Enabled Discovery	92
4.	Conclusions	99
	Acknowledgments	100
	References	100

5. Mapping Lysine Acetyltransferase–Ligand Interactions by Activity-Based Capture   105

D.C. Montgomery and J.L. Meier

1.	Introduction	106
2.	Technical Aspects	110
3.	Discussion	118
	References	121

6. Investigating Histone Acetylation Stoichiometry and Turnover Rate   125

J. Fan, J. Baeza, and J.M. Denu

1.	Introduction	126
2.	Labeling and Methods for Sample Preparation	128
3.	Sample Analysis	132
4.	Data Analysis and Kinetic Modeling	140
5.	Discussion and Perspective	145
	Acknowledgments	146
	References	146

7. Rapid Semisynthesis of Acetylated and Sumoylated Histone Analogs   149
A. Dhall, C.E. Weller, and C. Chatterjee

1. Introduction   150
2. Materials and Methods   152
3. Semisynthesis of Sumoylated Histone H4   153
4. Preparation of Acetylated Histone H3 Analogs   156
5. Generation of Designer MNs   159
6. Summary and Conclusions   163
Acknowledgments   164
References   164

8. An IF–FISH Approach for Covisualization of Gene Loci and Nuclear Architecture in Fission Yeast   167
K.-D. Kim, O. Iwasaki, and K. Noma

1. Introduction   168
2. Case Studies in the Application of the IF–FISH Approach   168
3. Supplies   171
4. Protocol   174
5. Notes   178
Acknowledgments   179
References   179

Part II
Small Molecule Epigenetic Regulators

9. Biology, Chemistry, and Pharmacology of Sirtuins   183
A. Bedalov, S. Chowdhury, and J.A. Simon

1. Introduction   184
2. Sirtuins and Metabolism   187
3. Sirtuins and Regulation of Cellular NAD^+ Levels   189
4. Sirtuin Functions   189
5. Sirtuins and Metabolic Disorders   190
6. Sirtuins and Cancer   191
7. Sirtuin Activity Assays   194
8. Identification of First-Generation Sirtuin Inhibitors   197
9. Second-Generation Splitomicin Inhibitors   200
10. Other Sirtuin Inhibitors   201
11. Concluding Remarks   205
References   205

10. Synthesis and Assay of SIRT1-Activating Compounds   213
H. Dai, J.L. Ellis, D.A. Sinclair, and B.P. Hubbard

1. Introduction   214
2. Materials   215
3. Methods   217
4. Notes   240
Acknowledgments   242
References   242

11. Synthesis and Assays of Inhibitors of Methyltransferases   245
X.-C. Cai, K. Kapilashrami, and M. Luo

1. Introduction to Methyltransferases   246
2. Designing and Synthesizing Inhibitors of Methyltransferases   249
3. Evaluating Methyltransferase Inhibitors   283
4. Conclusion   296
References   298

Part III
Epigenetics and Biological Connections

12. Exploring the Dynamic Relationship Between Cellular Metabolism and Chromatin Structure Using SILAC-Mass Spec and ChIP-Sequencing   311
P. Mews and S.L. Berger

1. Introduction   312
2. Analyzing the Turnover Dynamics of Histone Modifications   314
3. Genome-Wide Mapping of Histone Modifications   320
4. Summary   327
References   327

13. Current Proteomic Methods to Investigate the Dynamics of Histone Turnover in the Central Nervous System   331
L.A. Farrelly, B.D. Dill, H. Molina, M.R. Birtwistle, and I. Maze

1. Introduction   332
2. Early Methods to Study Histone Turnover in Brain   334
3. Current Proteomic Methods to Study Histone Turnover in Brain   337
4. Retrospective Birth Dating of Histones in Human Postmortem Brain   342
5. Conclusion   347

6. Methodology: Preparing Chromatin from Neurons for Mass
   Spectrometry Analysis of Histone Variants and Turnover   348
   Acknowledgments   352
   References   352

14. ChIP-Sequencing to Map the Epigenome of Senescent Cells Using Benzonase Endonuclease   355
T.S. Rai and P.D. Adams

1. Introduction   356
2. Buffer Compositions   358
3. Protocol   359
4. Conclusion   362
   References   363

15. Exploiting Chromatin Biology to Understand Immunology   365
J.L. Johnson and G. Vahedi

1. Introduction: Design Principle of Immune Responses   366
2. Epigenome: Our Software   366
3. Mapping DNA Methylation   367
4. Mapping Histone Modifications by Chromatin Immunoprecipitation
   and Sequencing   368
5. Limitations of ChIP-seq   369
6. MNase-seq for Nucleosome Positioning   370
7. DNase-seq and ATAC-seq for DNA Accessibility   371
8. Analysis of NGS Data   372
9. Painting an Enhancer Landscape   376
10. Chromatin Biology to Understand Gene Regulation in Immune Cells   377
11. Role of Intrinsic and Extrinsic Signals on Enhancer Formation   378
12. Conclusions   379
    Acknowledgments   380
    References   380

Author Index   *385*
Subject Index   *413*

CONTRIBUTORS

S. Ackloo
Structural Genomics Consortium, University of Toronto, Toronto, ON, Canada

P.D. Adams
CR-UK Beatson Labs, Institute of Cancer Sciences, University of Glasgow, Glasgow, United Kingdom

C.H. Arrowsmith
Structural Genomics Consortium; Princess Margaret Cancer Centre, University of Toronto, Toronto, ON, Canada

J. Baeza
School of Medicine and Public Health, Wisconsin Institute for Discovery, University of Wisconsin, Madison, WI, United States

D. Barsyte-Lovejoy
Structural Genomics Consortium, University of Toronto, Toronto, ON, Canada

A. Bedalov
Fred Hutchinson Cancer Research Center, Seattle, WA, United States

S.L. Berger
Perelman School of Medicine, University of Pennsylvania, Philadelphia, PA, United States

M.R. Birtwistle
Icahn School of Medicine at Mount Sinai, New York, NY, United States

X.-C. Cai
Memorial Sloan Kettering Cancer Center, New York, NY, United States

C. Chatterjee
University of Washington, Seattle, WA, United States

S. Chowdhury
Fred Hutchinson Cancer Research Center, Seattle, WA, United States

E.M. Cornett
Center for Epigenetics, Van Andel Research Institute, Grand Rapids, MI, United States

H. Dai
Sirtuin DPU, GlaxoSmithKline (GSK), Collegeville, PA, United States

J.M. Denu
School of Medicine and Public Health, Wisconsin Institute for Discovery, University of Wisconsin; Morgridge Institute for Research, Madison, WI, United States

A. Dhall
University of Washington, Seattle, WA, United States

B.M. Dickson
Center for Epigenetics, Van Andel Research Institute, Grand Rapids, MI, United States

B.D. Dill
The Rockefeller University Proteomics Resource Center, The Rockefeller University, New York, NY, United States

J.L. Ellis
Sirtuin DPU, GlaxoSmithKline (GSK), Collegeville, PA, United States

J. Fan
School of Medicine and Public Health, Wisconsin Institute for Discovery, University of Wisconsin, Madison, WI, United States

L.A. Farrelly
Icahn School of Medicine at Mount Sinai, New York, NY, United States

B.A. Garcia
Perelman School of Medicine, University of Pennsylvania, Philadelphia, PA, United States

B.P. Hubbard
University of Alberta, Edmonton, AB, Canada

O. Iwasaki
The Wistar Institute, Philadelphia, PA, United States

J.L. Johnson
Institute for Immunology, Pereleman School of Medicine, University of Pennsylvania, Philadelphia, PA, United States

K. Kapilashrami
Memorial Sloan Kettering Cancer Center, New York, NY, United States

K.R. Karch
Perelman School of Medicine, University of Pennsylvania, Philadelphia, PA, United States

K.-D. Kim
The Wistar Institute, Philadelphia, PA, United States

S. Krishnan
University of Michigan, Ann Arbor, MI, United States

E. Lima-Fernandes
Structural Genomics Consortium, University of Toronto, Toronto, ON, Canada

M. Luo
Memorial Sloan Kettering Cancer Center, New York, NY, United States

I. Maze
Icahn School of Medicine at Mount Sinai, New York, NY, United States

J.L. Meier
Chemical Biology Laboratory, National Cancer Institute, Frederick, MD, United States

P. Mews
Perelman School of Medicine, University of Pennsylvania, Philadelphia, PA, United States

H. Molina
The Rockefeller University Proteomics Resource Center, The Rockefeller University, New York, NY, United States

D.C. Montgomery
Chemical Biology Laboratory, National Cancer Institute, Frederick, MD, United States

K. Noma
The Wistar Institute, Philadelphia, PA, United States

P. Prinos
Structural Genomics Consortium, University of Toronto, Toronto, ON, Canada

T.S. Rai
Institute of Biomedical and Environmental Health Research, University of the West of Scotland, Paisley, United Kingdom

Z. Ramjan
Center for Epigenetics, Van Andel Research Institute, Grand Rapids, MI, United States

S.B. Rothbart
Center for Epigenetics, Van Andel Research Institute, Grand Rapids, MI, United States

S. Sidoli
Perelman School of Medicine, University of Pennsylvania, Philadelphia, PA, United States

J.A. Simon
Fred Hutchinson Cancer Research Center, Seattle, WA, United States

D.A. Sinclair
Glenn Labs for the Biological Mechanisms of Aging, Harvard Medical School, Boston, MA, United States; The University of New South Wales, Sydney, NSW, Australia

B.D. Strahl
University of North Carolina at Chapel Hill, Chapel Hill, NC, United States

M.M. Szewczyk
Structural Genomics Consortium, University of Toronto, Toronto, ON, Canada

R.C. Trievel
University of Michigan, Ann Arbor, MI, United States

G. Vahedi
Institute for Immunology, Pereleman School of Medicine, University of Pennsylvania, Philadelphia, PA, United States

R.M. Vaughan
Center for Epigenetics, Van Andel Research Institute, Grand Rapids, MI, United States

C.E. Weller
University of Washington, Seattle, WA, United States

PREFACE

These two volumes of *Methods of Enzymology* cover the rapidly developing field of Epigenetics. The central dogma of molecular biology, first proposed by Francis Crick in 1956, provided a framework for understanding the transfer of genetic information, which flows from DNA to RNA to protein. However, in the early 1980s, it was discovered that methylation of the DNA can change its function, hinting that genetic information transfer could be altered in other ways. In the 1990s, it was discovered that gene function could indeed be altered in many other ways and the modern field of Epigenetics was born. Epigenetics, a term first coined by Conrad Waddington in 1942 as "the branch of biology which studies the causal interactions between genes and their products, which bring the phenotype into being," is now known as the study of heritable changes in gene expression due to internal or environmental signals that results in the change of cellular function or physiological phenotype, but that is not caused by changes in the genetic information. An example of epigenetic regulation is cell differentiation, whereby cells of an organism with identical genetic information, such as cells of the nose, eyes, and hair, carry out different cellular functions.

At the heart of epigenetics is the regulation of chromatin, the packaged form of DNA. The building blocks of chromatin are nucleosome core particles containing about 146 bp of DNA wrapped around an octamer of histone proteins, two copies each of histones H2A, H2B, H3, and H4. Chromatin mediates all DNA-templated events, including DNA transcription, replication, and repair, and is regulated by many proteins and noncoding RNAs. The proteins that regulate chromatin include posttranslational modification (PTM) "writer" enzymes, "eraser" enzymes that remove these PTMs, and "reader" proteins that bind chemically modified histones or DNA. PTMs of DNA include methylation at the 5 position of cytosine and various oxidation states of 5-methyl-cytosine. PTMs of histones include acetylation, methylation, and ubiquitination on lysine residues; methylation and citrullination of arginine residues; and phosphorylation of threonine and serine residues. ATP-dependent chromatin-remodeling enzymes that function to reposition nucleosomes within chromatin; histone chaperone proteins that insert or eviction of histone variants in and out of chromatin, respectively; and noncoding RNA molecules also contribute to chromatin regulation.

Chromatin-regulatory proteins work together to mediate epigenetic regulation, with ramifications for cellular function and physiological phenotype, and over the last decade, it has become apparent that the dysfunction of epigenetic regulators can drive many diseases including metabolic and neurodegenerative disorders and various cancers. Over the last decade, we have seen significant progress on understanding the molecular mechanisms underlying epigenetic regulation leading to new insights into cellular function and physiological phenotype, the development of new technologies, and the development of small-molecule probes to study these biological processes and small epigenetic drugs that are currently in clinical trials to treat disease.

The remarkable progress in the field of epigenetic research that has occurred over the last two decades is highlighted in these two volumes of *Methods of Enzymology*, entitled Enzymes of Epigenetics. In Volume 1, Chapters 1 through 5 cover *Chromatin Structure and Histones*. This includes strategies for in vitro chromatin assembly (Chapter 1) and the assembly of protein–chromatin complexes (Chapter 2) for biochemical, biophysical, and structural studies, preparation of recombinant centromeric nucleosomes with and without nonhistone centromere proteins (Chapter 3), functional characterization of histone deposition by histone chaperones (Chapter 4), and methods to study nucleosome sliding by ATP-dependent chromatin-remodeling enzymes (Chapter 5).

In Volume 1, Chapters 6 through 13 cover *Posttranslational Histone Modification Enzymes and Complexes*. This includes in vitro activity assays for MYST histone acetyltransferases and adaptation for high-throughput inhibitor screening (Chapter 6), and the preparation, biochemical analysis, and structure determination of the classical (Chapter 7) and sirtuin (Chapter 8) histone lysine deacetylases, SET domain histone methyltransferases (Chapter 9), LSD1/KDM1A (Chapter 10), and JmjC (Chapter 12). Also included are chapters on LSD1 histone demethylase assays and inhibition (Chapter 11) and preparation and analysis of native chromatin-modifying complexes (Chapter 13).

In Volume 1, Chapters 14 and 15 cover *Histone Modification Binders* and include the preparation, biochemical analysis, and structure determination of the acetyl-lysine reader bromodomains (Chapter 14) and the readers of the methyllysine mark (Chapter 15).

In Volume 1, Chapters 16 through 20 cover *DNA Modifications and Nucleic Acid Regulators*. This includes quantification of oxidized 5-methylcytosine bases and TET enzyme activity (Chapter 16), characterization of how

DNA modifications affect DNA binding by C_2H_2 zinc finger proteins (Chapter 17), regulation of chromosome ends by the riboprotein telomerase complex (Chapter 18), detection and analysis of long noncoding RNAs (Chapter 19), and identification of centromeric RNAs involved in histone dynamics in vivo (Chapter 20).

In Volume 2, Chapters 1 through 8 cover *Epigenetic Technologies*. This includes identification and quantification of histone PTMs using high-resolution mass spectrometry (Chapter 1), substrate specificity profiling of histone-modifying enzymes using peptide microarray (Chapter 2), an open-source platform to analyze such microarrays (Chapter 3), chemical biology approaches for characterization of epigenetic regulators (Chapter 4), mapping lysine acetyltransferase–ligand interactions by activity-based capture (Chapter 5), methods for investigating histone acetylation stoichiometry and turnover rate (Chapter 6), semisynthesis of acetylated and sumoylated histone analogs (Chapter 7), and an IF-FISH approach for covisualization of gene loci and nuclear architecture in fission yeast (Chapter 8).

In Volume 2, Chapters 9 through 11 cover *Small-Molecule Epigenetic Regulators* and include the preparation and analysis of sirtuin deacetylase inhibitors (Chapter 9) and activators (Chapter 10) and methyltransferase inhibitors (Chapter 11).

In Volume 2, Chapters 12 through 15 cover *Epigenetics and Biological Connections* and include exploring the dynamic relationship between cellular metabolism and chromatin structure using SILAC-mass spec and ChIP-sequencing (Chapter 12), proteomic methods to investigate the dynamics of histone turnover in the central nervous system (Chapter 13), ChIP-seq techniques to map the epigenome of senescent cells (Chapter 14), and exploiting chromatin biology to understand immunology (Chapter 15).

I expect that these volumes will be a useful resource for investigators in the epigenetics field as well as those outside of the epigenetics field who would like to incorporate epigenetics into their own research programs.

RONEN MARMORSTEIN
Abramson Family Cancer Research Institute,
Perelman School of Medicine at the University of Pennsylvania,
Philadelphia, PA, United States

PART I

Epigenetic Technologies

CHAPTER ONE

Identification and Quantification of Histone PTMs Using High-Resolution Mass Spectrometry

K.R. Karch, S. Sidoli, B.A. Garcia[1]

Perelman School of Medicine, University of Pennsylvania, Philadelphia, PA, United States
[1]Corresponding author: e-mail address: bgarci@mail.med.upenn.edu

Contents

1. Introduction	4
2. Histone Extraction from Cells	6
2.1 Materials and Buffer Recipes	6
2.2 Cell Harvest	7
2.3 Nuclei Isolation	7
2.4 Acid Extraction	8
3. Bottom-Up Mass Spectrometry	9
3.1 Materials and Buffer Recipes	10
3.2 Derivatization and Digestion	10
3.3 Desalting	11
3.4 Online RP-HPLC and MS Acquisition	12
3.5 Data Analysis	14
4. Offline Fractionation of Histone Species	18
4.1 Materials and Buffer Recipes	18
4.2 Histone Variant Purification	18
5. Middle-Down Mass Spectrometry	19
5.1 Materials and Buffer Recipes	20
5.2 Digestion	21
5.3 WCX-HILIC and MS	21
5.4 Data Analysis	22
6. Top-Down Mass Spectrometry	24
6.1 Materials and Buffer Recipes	25
6.2 Top-Down MS Using Direct Infusion	25
6.3 Data Analysis	26
References	28

Abstract

DNA is organized into nucleosomes, composed of 147 base pairs of DNA wrapped around an octamer of histone proteins including H2A, H2B, H3, and H4. Histones are critical regulators of many nuclear processes, including transcription, DNA damage repair, and higher order chromatin structure. Much of their function is mediated through extensive and dynamic posttranslational modification (PTM) by nuclear enzymes. Histone PTMs are thought to form a code, where combinations of PTMs are responsible for specific biological functions. Here, we present protocols to identify and quantify histone PTMs using nanoflow liquid chromatography coupled to mass spectrometry (MS). We first describe how to purify histones and prepare them for MS. We then describe three MS platforms for histone PTM analysis, including bottom-up, middle-down, and top-down approaches, and explain the relative benefits and pitfalls of each approach. We also include tips to increase the throughput of large experiments.

1. INTRODUCTION

DNA must be highly organized and tightly regulated within the nucleus to maintain proper gene expression. The cell accomplishes this task by organizing DNA into a protein–DNA complex called chromatin. Within chromatin, DNA is contained in nucleosomes, which are composed of 147 base pairs of DNA wrapped around an octamer of histone proteins with two copies of each core histone—H2A, H2B, H3, and H4 (Luger, Mäder, Richmond, Sargent, & Richmond, 1997). Linker histone H1 can bind the free DNA that exists between nucleosomes. Due to their intimate association with DNA, histones are major regulators of chromatin structure and function. Histone proteins are extensively and dynamically posttranslationally modified on specific residues by a myriad of enzymes in the nucleus. These posttranslational modifications (PTMs) mediate histone function by directly altering the chemistry of the surrounding chromatin or through the action of other proteins that can bind these modifications. A growing body of research supports the hypothesis that PTMs form a "histone code" and can act in tandem to illicit a specific biological response (Jenuwein & Allis, 2001). Histone PTM profiles are critical to maintain nuclear stability, and aberrant regulation of histone PTMs is implicated in many diseases including cancer. As such, the ability to identify and quantify histone PTMs in biological systems is vital for understanding nuclear processes and how disease states may arise.

Histone PTM analysis had traditionally been accomplished using antibody-based approaches such as Western blots, chromatin immunoprecipitation, and deep sequencing. These methods have been instrumental in elucidating the roles of many histone PTMs but suffer from several critical drawbacks (Britton, Gonzales-Cope, Zee, & Garcia, 2011). For example, many antibodies are not entirely site specific and can cross-react with similar modifications on different residues. A similar issue is epitope occlusion, where a PTM on a nearby residue can block interaction with an antibody. Perhaps the biggest drawback is that these methods require previous knowledge of the modification and are therefore unable to identify novel PTMs (Rothbart et al., 2015).

Mass spectrometry (MS) is an unbiased and quantitative method to comprehensively analyze histone PTM profiles. One of the greatest advantages of MS is that it can identify novel PTMs and can also measure the cooccurrence of PTMs on a single peptide. As such, MS has emerged as a critical tool for characterization of histone modifications. There are three major MS approaches, namely, bottom-up, middle-down, and top-down MS, each of which is useful for specific applications (Fig. 1). Bottom-up MS involves digestion of a protein sample into small peptides (5–15 amino acid residues) followed by online separation by reversed phase chromatography coupled to tandem MS. This method is very robust and sensitive. One major drawback, however, is that the cooccurrence between PTMs located on different peptides cannot be measured. Top-down MS, on the other hand, is performed by directly analyzing intact proteins and, as such, preserves complete connectivity between PTMs. However, this method is much less sensitive than bottom-up MS and thus has much larger sample

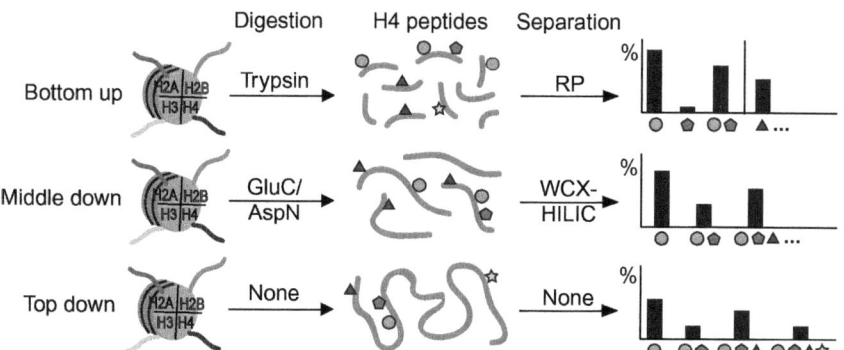

Figure 1 Workflows for bottom-up, middle-down, and top-down histone PTM analysis by high-resolution tandem MS. In bottom-up MS, the relative abundances and PTM cooccurrences can be monitored for PTMs contained within a single tryptic peptide. Longer peptides are generated in middle-down MS, allowing for better connectivity than bottom-up MS. In top-down MS, full connectivity is preserved, allowing for identification of complete protein isoforms. (See the color plate.)

requirement. Furthermore, the data analysis is much more challenging due to the large complexity of the tandem mass spectra, which sometimes results in the impossibility of discriminating proteoforms when cofragmented. Using chip-based infusion, rather than injection with a syringe, can reduce the sample requirement for top-down experiments that do not include online chromatographic separation. Middle-down MS offers a compromise between these two methods and is performed by digesting proteins into large peptides (about 30–60 amino acid residues). Analyzing large peptides allows for connectivity of many PTM locations, but offers better sensitivity and simpler data analysis than the top-down approach.

Bottom-up MS is the most commonly used MS approach for histone PTM analysis as it is technically more facile than the other approaches and does not require specialized equipment such as a 2D HPLC or chip-based electrospray ionization. Furthermore, small peptides generated in bottom-up MS are fragmented by collision-induced dissociation (CID) or high-energy collision dissociation (HCD), which is available in most commercial instruments. Large peptides or intact histone proteins, however, result in high charge state analytes when electrospray ionized and therefore do not fragment well with CID or HCD. Electron transfer dissociation (ETD) fragmentation is highly efficient for highly charged peptides or proteins and it is thus the fragmentation technique of choice for middle-down and top-down approaches.

In this chapter, we present the protocol to isolate histone proteins from cells and prepare the protein for MS analysis. We also outline how to perform bottom-up, middle-down, and top-down MS to identify and characterize histone PTMs. Tips to increase the throughput of experiments are also included.

2. HISTONE EXTRACTION FROM CELLS

Histones are among the most basic proteins in the cell, and as such, can be easily purified using an acid extraction. Here, we describe how to first isolate nuclei from cells and then perform an acid extraction to obtain purified histones. The whole protocol takes 1 day.

2.1 Materials and Buffer Recipes
1. 0.25% Trypsin
2. Phosphate-buffered saline (PBS): 137 mM NaCl, 2.7 mM KCl, 10 mM Na$_2$HPO$_4$·2H$_2$O, 2 mM KH$_2$PO$_4$

3. Nuclei isolation buffer (NIB): 15 mM Tris–HCl (pH 7.5), 60 mM KCl, 15 mM NaCl, 5 mM MgCl$_2$, 1 mM CaCl$_2$, 250 mM sucrose
4. 1 M DTT
5. 200 mM AEBSF
6. 2.5 µM Microcystin
7. 5 M Sodium butyrate

2.2 Cell Harvest

2.2.1 Cell Harvest: Tissue Samples

1. Rinse tissue sample with cold PBS (4°C). This can be stored at −80°C after flash freezing or used immediately.
2. Use a razor blade to cut the tissue into small pieces, as small as possible.
3. Record the approximate volume of the tissue sample.

2.2.2 Cell Harvest: Cell Cultures

1. For adherent cells, detach the cells by covering the plate with a thin layer of trypsin 0.25% for 5 min at room temperature. Move cells to a centrifuge tube and spin at 300 rcf for 5 min. For cells grown in suspension, centrifuge cells at 300 rcf for 5 min. Aspirate the trypsin or pour it off into bleach (10% final concentration).
2. Wash with PBS (10 × volume). Centrifuge cells at 600 rcf for 5 min and remove PBS.
3. Repeat step 2. Pellets can be stored at −80°C or used immediately.

2.3 Nuclei Isolation

1. If cell pellets were frozen, thaw them on ice.
2. Estimate the amount of NIB that will be needed, which is approximately 50 × the total volume of cell pellets. Chill the buffer on ice.
3. Add inhibitors to the NIB to the final concentrations of 1 mM DTT, 500 µM AEBSF, 5 nM microcystin, and 10 mM sodium butyrate. For 50 mL buffer, add 50 µL of 1 M DTT, 125 µL of 200 mM AEBSF, 100 µL of 2.5 µM microcystin, and 100 µL of 5 M sodium butyrate. Inhibitors will degrade over time, so prepare NIB + inhibitors freshly for each experiment and store inhibitor stocks at −20°C.
4. Move 25% of the NIB + inhibitors to a new beaker and add 10% NP-40 alternative to a final concentration of 0.3%. For 50 mL, add 1.5 mL of 10% NP-40 alternative.
5. Resuspend cell pellets in 10 × volume of NIB with NP-40 and homogenize by gentle pipetting (cultured cells) or douncing (tissue samples).

6. Incubate on ice for 5 min to lyse outer cell membranes. Nuclei isolation efficiency can be approximated using Trypan Blue staining.
7. Centrifuge at 4°C for 5 min at 1000 rcf. The supernatant contains the cytoplasm and can be reserved if desired. Otherwise, discard the supernatant. The pellet contains nuclei and should be smaller than the original cell pellet.
8. Wash the pellet by resuspending it in NIB + inhibitors without NP-40 alternative (10 × volume of cell pellet).
9. Centrifuge cells at 4°C at 1000 rcf for 5 min.
10. Repeat steps 8 and 9 until no NP-40 remains. Usually, two washes are sufficient. NP-40 forms bubbles when resuspending and so lack of bubbles indicates successful removal of NP-40.
11. Nuclei can be stored in NIB + inhibitors + 5% glycerol at −80°C after freezing in liquid nitrogen or used immediately for acid extraction.

2.4 Acid Extraction

1. Gently vortex the nuclei pellet and slowly add 0.4 N H_2SO_4 (5 × volume of pellet, ie, add 5 mL H_2SO_4 to a 1 mL pellet).
2. Incubate at 4°C with intermittent mixing or on a rotator for 1 h up to overnight. We recommend incubation of 2 h for pellets larger than 500 μL or 4 h for pellets smaller than 500 μL. Longer incubation can result in extraction of other basic proteins besides histones.
3. Centrifuge extracts at 3400 rcf for 5 min. The supernatant contains the histone proteins and the pellet contains other proteins.
4. Transfer the supernatant to a new tube.
5. Repeat steps 3 and 4 to remove any traces of the pellet.
6. Gently add 100% TCA to the supernatant to a final concentration of 20% (ie, add 1/4 volume of the supernatant).
7. Incubate on ice for at least 1 h without agitation. For most samples, 1 h is sufficient. We recommend overnight incubation for small samples (<50 μL pellet).
8. Centrifuge at 3400 rcf for 5 min. The histones will form a film around the bottom of the tube. Other proteins and nonprotein material will form a white pellet at the bottom of the tube, which cannot be solubilized.
9. Carefully aspirate the supernatant, avoiding the protein film on the side of the tube.

10. Wash the protein with acetone + 0.1% HCl by pipetting gently down the side of the tube. Use a glass pipettor as acetone will dissolve plastic pipette tips.
11. Centrifuge at 3400 rcf for 5 min. Aspirate acetone.
12. Repeat steps 10 and 11 using 100% acetone two times.
13. Allow remaining acetone to evaporate by leaving the tubes open for 30 min up to overnight.
14. Resuspend the histone film in ddH$_2$O. The volume will depend on the size of the tube and pellet, but generally 100 μL is sufficient for pellets in a 1.5 mL microcentrifuge tube.
15. Centrifuge at 3400 rcf for 2 min.
16. Move supernatant to new tube, being careful to avoid the pellet, which can be discarded.
17. Measure the protein concentration using a Bradford assay or another method. Histones can be stored in ddH$_2$O at −80°C. If samples are dilute, concentrate them in a vacuum centrifuge.
18. If doing bottom-up MS, continue with Section 3. If doing middle-down or top-down MS, continue with Section 4 then 5 or 6, respectively.

3. BOTTOM-UP MASS SPECTROMETRY

Bottom-up MS is the most commonly used MS platform for proteomics. In bottom-up MS, proteins are digested into small peptides (5–15 amino acid residues) with trypsin, which are then separated with online reversed phase high-performance liquid chromatography (RP-HPLC) and analyzed via tandem MS (MS/MS). Histones are among the most basic proteins in the cell and contain a large number of lysine and arginine residues. Therefore, digestion with trypsin results in peptides that are too small to be retained by RP chromatography. To overcome this issue, histone proteins can be chemically derivatized on the ξ-amino groups of unmodified or monomethylated lysine residues (Garcia et al., 2007). This derivatization prevents trypsin cleavage after lysine residues, allowing the enzyme to cleave only after arginine residues thus generating longer peptides. We recommend using propionic anhydride as the derivatization reagent due to its high efficiency (Sidoli et al., 2015). After digestion with trypsin, a second round of derivatization is performed to modify the amino groups of the newly generated N-termini. This increases the hydrophobicity of the peptides and allows for better interaction with RP columns. Propionylation and trypsinization can be performed either in microcentrifuge tubes or in 96-well plates to reduce sample preparation time if a large number of samples are being processed.

3.1 Materials and Buffer Recipes

1. 100 mM ammonium bicarbonate, pH 8.5
2. Ammonium hydroxide (NH_4OH)
3. Propionic anhydride
4. Isopropanol
5. pH paper
6. Trypsin
7. Glacial acetic acid
8. Vacuum centrifuge
9. C18 Disc (3M Empore)
10. Methanol (MeOH)
11. Wash buffer: 0.1% acetic acid in water
12. Elution buffer: 75% acetonitrile, 5% acetic acid, 20% water
13. Buffer A: 0.1% formic acid in water (all MS grade solvents)
14. Buffer B: 0.1% formic acid in 75% acetonitrile/25% water (all MS grade solvents)

3.2 Derivatization and Digestion

1. Dry samples down in a vacuum centrifuge and resuspend in 20 μL ammonium bicarbonate, pH 8.5. Ensure that the pH of the samples is between 7 and 9 using pH paper. If they are too basic, add some glacial acetic acid. If they are too acidic, add some powdered ammonium bicarbonate. If using the plate format, transfer samples to a 96-well plate.
2. Prepare the propionylation reagent by combining propionic anhydride and isopropanol in a 1:3 ratio. The propionic anhydride in the reagent will begin to dissociate into propionic acid so it is important to use it immediately after preparation. If processing samples on a plate, make new reagent for each plate. If processing samples in microcentrifuge tubes, make new reagent every three to five samples.
3. Add 10 μL of the propionylation reagent to the sample and vortex briefly. The pH will drop to 4–6 due to the propionic acid.

4. Immediately add 3–7 μL NH$_4$OH to bring the pH up to ~8, checking with pH paper. Do not allow the sample to go above pH 10 as propionylation of other residues such as serine can occur at high pH.
5. Incubate the samples at 37°C for 15 min.
6. Dry down samples to less than 5 μL in a vacuum centrifuge.
7. Repeat steps 1–6 one more time.
8. Resuspend samples in 50–100 μL of ammonium bicarbonate.
9. Add trypsin in a 1:20 protease:histone ratio and incubate at 37°C for at least 6 h up to overnight.
10. Quench trypsin by freezing at −80°C or adding glacial acetic acid to lower pH to ~4.
11. Dry samples to 20 μL or less. If dried to less than 20 μL, add more ammonium bicarbonate to a final volume of 20 μL.
12. Repeat steps 2–6 twice.

3.3 Desalting

Desalting is performed on home-made C$_{18}$ columns called stage-tips.

1. Cut off the last centimeter of a P1000 tip to make the opening bigger. Use this pipet tip to punch out a piece of C$_{18}$ material from the disc. If desalting more than 25 μg of protein, use two discs. Ensure that there is no space between the discs.
2. Use a piece of fused silica capillary (or any other long, thin item) to firmly push the C$_{18}$ material out of the P1000 pipet tip into a P200 pipet tip. Ensure that the C$_{18}$ disc is firmly positioned in the tip, but do not push hard enough to puncture the material.
3. Place the column in a microcentrifuge adaptor inside a 1.5- or 2-mL microcentrifuge tube.
4. Activate the resin by adding 50 μL methanol to the column. Push the methanol through using compressed air or by spinning in a centrifuge at $500 \times g$ for 30–60 s.
5. Repeat step 4 once. After this point, keep the disc wet at all times.
6. Equilibrate the column by adding 200 μL washing buffer to the column and push the solution through as described in step 5. Note that the collection tube may need to be emptied.
7. Repeat step 6 once.
8. Add wash buffer to samples to a final volume of approximately 200 μL. The pH should be below 4.0 (check with pH paper). If needed, adjust the pH using glacial acetic acid.

9. Add the sample to the stage-tip and centrifuge for 2–5 min at $200 \times g$ until the sample passes through the column.
10. Wash the column by adding 50 μL wash buffer to the stage-tip and centrifuge at $500 \times g$ for 30–60 s until the buffer passes through.
11. Repeat step 10 once.
12. Remove the collection tube and replace with a clean 1.5-mL microcentrifuge tube.
13. Add 75 μL elution buffer to the column and centrifuge at $200 \times g$ for 2–5 min.
14. Repeat step 13 once.
15. Dry down sample in a vacuum centrifuge to less than 5 μL. Desalted samples can be stored at $-80°C$.
16. Resuspend samples to approximately 1 μg/μL in buffer A for MS analysis.

3.4 Online RP-HPLC and MS Acquisition

We present protocols to perform bottom-up MS using nanoelectrospray ionization (nano-ESI) on a hybrid LTQ-Orbitrap (ie, Thermo Orbitrap Elite).

MS data can be acquired in two different modes: data-dependent acquisition (DDA) or data-independent acquisition (DIA). DDA is the traditional method used in histone PTM analysis, where a high-resolution full MS scan is acquired followed by low-resolution MS/MS acquisition of the topN (eg, 10) most abundant peaks from the full MS scan. Peptide abundance is quantified by integrating the area of the extracted ion chromatogram (XIC) at the MS1 level. One issue with DDA is that it cannot accurately quantify coeluting isobaric peptides (different peptides that have the same mass) because they cannot be discriminated at the MS1 level. To overcome this limitation, coeluting isobaric peptides must be targeted across their elution profile and quantified based on the fragment ion intensities.

DIA is performed by acquiring a high-resolution full MS scan followed by a series of high-resolution MS/MS scans covering a larger m/z window (ie, 50) that step across the desired m/z range (Gillet et al., 2012). All ions present in the m/z window get fragmented and detected together, and so the size of the window and complexity of the sample will dictate the quality of the MS/MS spectra and consequently identification. DIA methodology has been optimized for histone PTM analysis also performing low-resolution MS/MS scans, and 50 m/z windows were found to be a good balance for allowing sufficient identification and cycle speed (Sidoli, Simithy, Karch, Kulej, & Garcia, 2015). Peptide identifications should be obtained before the DIA experiment using a spectral library generated with DDA data. In other words, DIA is not the recommended acquisition method to identify unknown peptides. On the other hand, one major benefit of DIA is that MS/MS spectra are obtained for all peptides across the entire elution peak. Thus, DIA provides higher confidence in determining the correct chromatographic peak for a given peptide, as it allows for XIC for both precursor and fragment ions, which will appear as coeluting peak profiles. The abundance of the peptide is still quantified by calculating the area under the curve for the parent ion, as the precursor ion signal is always more intense than any fragment mass ion. Moreover, DIA eliminates the need for targeting of coeluting isobaric species, as all analytes are already fragmented at each MS scan cycle and allows for future data mining. DIA methods have been gaining popularity due to these advantages.

1. Fit the HPLC with a C_{18} column (75 μM inner diameter, 10–20 cm in length), either purchased commercially or packed in-house.
2. Load 1–2 μg histone peptides onto the column with an autosampler.
3. Program the HPLC gradient: 0–30% buffer B in 30 min, 30–100% buffer B in 5 min, 100% buffer B for 8 min. For HPLC systems that do not automatically equilibrate the column before sample loading, include equilibration steps in your gradient: 100–0% buffer B in 1 min and 0% buffer B for 10 min. Set the flow rate to 250–300 nL/min.
4. Program a method to acquire and record MS data.
    a. DDA: Composed of three segments. All full MS scans are obtained in the Orbitrap and all MS/MS scans are obtained in the ion trap.
        i. Segment 1: 14 min. One full MS scan followed by CID MS/MS of the top seven most abundant ions (in DDA mode).
        ii. Segment 2: 27 min. One full MS scan followed by CID fragmentation of isobaric species [for, eg, human and mouse H3 9–17aa 1 acetyl (528.296 m/z), H3 18–26aa 1 acetyl (570.841 m/z), H4 4–17 1 acetyl (768.947 m/z), H4 4–17 2 acetyls (761.939 m/z), H4 4–17aa 3 acetyls (754.930 m/z)] and DDA of the top five most abundant ions.
        iii. Segment 3: 19 min. One full MS scan followed by DDA of the top 10 most abundant ions.
        iv. *Note*: The time and length of each segment will vary depending on the exact elution times of the targeted species, which can vary between columns. For first time users, run one sample to

determine the elution time for the targeted species and modify the method, if needed, to allow for accurate targeting of the desired species.

    b. DIA: Set up method according to Table 1.

3.5 Data Analysis

Histone PTM analysis is a computationally challenging process. Given that histones are highly modified proteins, each peptide can contain several PTM

Table 1 Parameters for DIA MS Acquisition of Histone Samples

Scan #	Scan Type	Detector	Scan Range	MS2 Scan Range
1	MS1	Orbitrap	300–1100	N/A
2	MS2	Ion trap	300–350	120–1500
3	MS2	Ion trap	350–400	120–1500
4	MS2	Ion trap	400–450	120–1500
5	MS2	Ion trap	450–500	130–1500
6	MS2	Ion trap	500–550	140–1500
7	MS2	Ion trap	550–600	155–1500
8	MS2	Ion trap	600–650	170–1500
9	MS2	Ion trap	650–700	185–1500
10	MS1	Orbitrap	300–1100	N/A
11	MS2	Ion trap	700–750	195–1500
12	MS2	Ion trap	750–800	210–1500
13	MS2	Ion trap	800–850	225–1500
14	MS2	Ion trap	850–900	240–1500
15	MS2	Ion trap	900–950	250–1500
16	MS2	Ion trap	950–1000	265–1500
17	MS2	Ion trap	1000–1050	280–1500
18	MS2	Ion trap	1050–1100	295–1500

Note: The full MS1 scan is performed twice within the same duty cycle to allow for a more resolved definition of the precursor peak profile. This is not necessary for MS2 ions, as these are commonly not used for quantification, but only for increasing the confidence of the selected chromatographic peak and, in specific cases, to discriminate isobaric forms. The differences in MS2 scan range, in particular for the low mass range, are due to the intrinsic limitations of the ion trap, which cannot hold for scanning fragment ions smaller than $\sim$1/3 of the isolated precursor mass.

acceptor sites, resulting in a large number of possible modified forms for a given peptide. For example, the H3 9–17 peptide can contain mono-, di-, or trimethylation (me1/me2/me3) on K9 and K14 as well as acetylation (ac) on K9, for a total of 10 possible forms of the peptide. Some isobaric peptides are also generated by the derivatization process, further complicating analysis. For example, the H3 9–17 peptide has two sets of isobaric peptides: unmodified and K9me1K14ac ($[M+2H]^{2+} = 535.304$ m/z) and K9me3K14ac and K9me2 ($[M+2H]^{2+} = 521.306$ m/z. Many of these isobaric species arise because the mass of a propionyl group is the same as the mass of an acetyl group and monomethyl group (56.026 Da). Fortunately, many of these peptides elute at different times and therefore do not require MS/MS targeting across their elution profiles in DDA experiments (ie, K9me3K14ac and K9acK14ac). This is because the modifications impart different hydrophobicities, causing the peptides to elute at different times (Fig. 2).

Some isobaric species, however, coelute and cannot be discriminated based on retention time (RT). A prime example is the H4 4–17 peptide that contains four lysine residues that can be acetylated, resulting in many isobaric species. The diacetylated peptide is the most complicated example, as there are six isobaric forms that coelute. These peptides can be differentiated by collecting MS/MS spectra across their elution and using the elution profiles of unique fragment ions to define the ion chromatogram (Fig. 3A).

In both DDA and DIA experiments, the relative abundances of histone peptides must be calculated by extracting the raw peak area of all modified and unmodified forms for a given peptide sequence. The relative abundance is obtained by dividing the total area of a single peptide isoform (including all charge states) by the total area of that peptide sequence in all of its modified forms. We developed a software tool to automate this type of data analysis for both DDA and DIA experiments (Yuan et al., 2015).

3.5.1 Software-Based Peak Area Extraction and Abundance Calculation
EpiProfile is a freely available Matlab-based automated tool to quantify histone PTM profiles in DDA experiments (Yuan et al., 2015). The software reads raw data and provides a table of quantified histone peptides, layouts (MS1 elution profiles), and annotated MS/MS spectra used for identification. EpiProfile can quantify coeluting isobaric peptides (if they were targeted in the MS method) as well as isotopically labeled peptides (ie, labeled by SILAC).

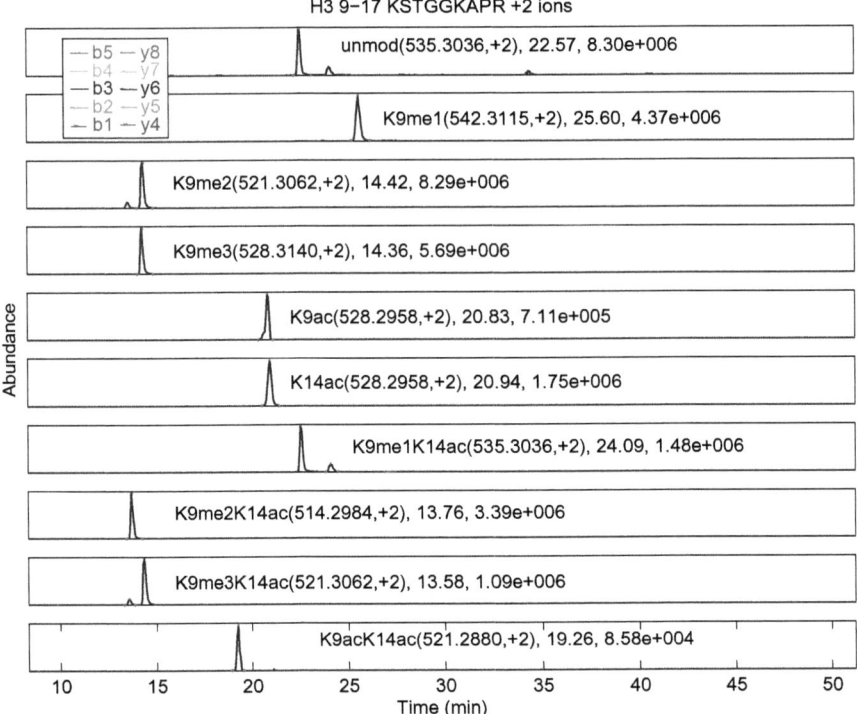

Figure 2 Example layout for H3 9–17 peptide provided by EpiProfile. Each row represents an extracted ion chromatogram (XIC) for a specific modified form of the peptide. The script next to each peak provides the modification state, m/z value, charge state, rentention time, and intensity, respectively. The XIC of fragment ions are also illustrated in colors, however, they cannot be easily visualized because they overlap with the XIC trace. Note that some isobaric peptides (ie, K9meK14ac and K9acK14ac) have nearly the same mass but elute at different times, while others (ie, K9ac and K14ac) have overlapping XICs.

EpiProfile uses previous knowledge about relative elution times to aid in quantification and identification of histone peptides. It uses this information, as well as mass information, to perform XIC of each unique peptide. EpiProfile then calculates the area under the curve of the XIC, which is then used to calculate the relative abundance of each peptide.

1. Open the "params.txt" file in EpiProfile and specify the file path containing your raw files. Other options can be specified as well, such as isotopic labeling and mass tolerance.
2. Open Matlab and specify the file path containing EpiProfile.
3. Enter "EpiProfile" in the command window to start the program.

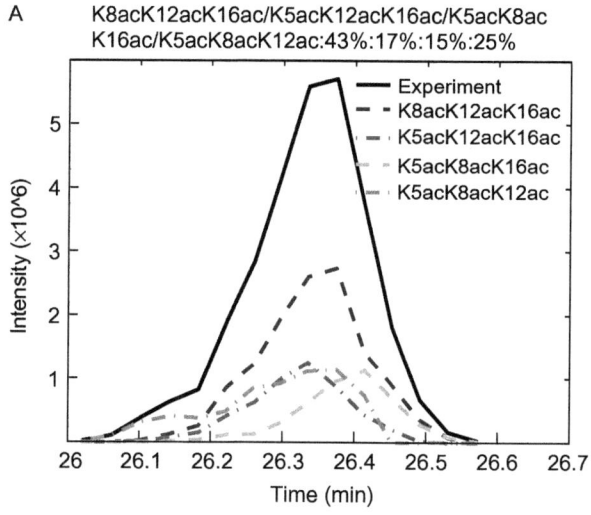

	1,Sample 1			2,Sample2		
Peptide	RT(min)	Area	Ratio	RT(min)	Area	Ratio
KSTGGKAPR(H3 9-17)						
unmod	22.57	8.58E+08	0.245241	22.22	4.21E+08	0.248842
K9me1	25.6	1.04E+09	0.296224	25.29	3.75E+08	0.221391
K9me2	14.42	5.76E+08	0.164828	15.09	3.75E+08	0.221797
K9me3	14.36	3.45E+08	0.098683	14.97	2.44E+08	0.144006
K9ac	20.83	5.84E+07	0.016689	20.85	1.67E+07	0.009887
K14ac	20.94	1.71E+08	0.0488	20.97	6.25E+07	0.036953
K9me1K14ac	24.09	1.93E+08	0.055091	23.99	6.23E+07	0.036792
K9me2K14ac	13.76	1.85E+08	0.052866	13.47	9.81E+07	0.057953
K9me3K14ac	13.58	6.87E+07	0.01964	13.29	3.52E+07	0.020818
K9acK14ac	19.26	6.78E+06	0.001938	19.27	2.64E+06	0.001561

Figure 3 EpiProfile allows for quantification of histone PTMs, including isobaric species. (A) Fragment ion XICs for H4 4–17 peptide containing three acetyl groups. These fragment ion XICs are used for quantification because the parent ion XICs overlap. (B) Example of EpiProfile quantification output from "histone_ratios.xls" file. The retention time, area under the XIC and relative abundance (ratio) for each peptide in each sample are listed. (See the color plate.)

4. The program will run and the results will be stored in the same file path that was specified in the "params.txt" file. Several result files and folders will be provided:
   a. An excel file called "histone_ratios.xls." This file contains the RT, area, and relative abundance of each peptide for all raw files (Fig. 3B).
   b. A folder called "histone_layouts." This contains all of the XICs used for area calculations, including all of the XICs for fragment ions (ie,

Fig. 2). The "details" folder in this directory contains XICs for coeluting isobaric species (ie, Fig. 3A).

5. The layout files can be used to manually validate that the program chose the correct peak to quantify.

   Note: EpiProfile, although it is the software we recommend for histone analysis, is not the only one available. Manual quantification can be performed using the Xcalibur QualBrowser (for Thermo instruments), which is mostly used to visualize raw files. The mass of the monoisotopic peak can be entered to obtain the XIC, and the area under the curve function automatically calculates the peak area. Moreover, the free software Skyline can be adopted for the purpose (MacLean et al., 2010). Skyline is optimized to extract precursor and fragment XIC upon a precompiled list of peptides, and thus it is definitely more automated than the manual quantification procedure. However, it does not include unique features of EpiProfile such as RT-based peak calling and automatic calculation of ratios for coeluting isobaric peptides.

4. OFFLINE FRACTIONATION OF HISTONE SPECIES

4.1 Materials and Buffer Recipes

1. Offline buffer A: 5% acetonitrile, 0.2% trifluoroacetic acid (TFA) in water (all HPLC-grade)
2. Offline buffer B: 95% acetonitrile, 0.188% TFA in water (all HPLC-grade)
3. 5 μm C_{18} column (size depending on application)

4.2 Histone Variant Purification

1. Attach a C_{18} column to the offline HPLC and set the flow rate according to the size of the column. Generally, a flow rate of 0.2 mL/min is used for a 2.1 mm column, 0.8 mL/min for a 4.6 mm column, and 2.5 mL/min for a 10 mm column. Allow the column to equilibrate in buffer A for 10× column volumes.
2. Load the sample. Generally, 100–200 μg histone should be loaded for a 4.6 mm column.
3. Run the following gradient: 30–60% buffer B over 100 min, 60–100% buffer B over 20 min, 100–30% buffer B over 10 min. The elution profile is shown in Fig. 4.

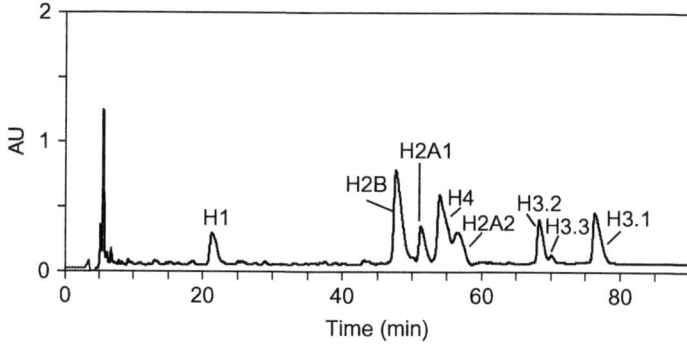

Figure 4 Chromatogram for histone variant purification. This chromatogram was obtained by injected 200 μg of acid-extracted histone from HeLa cells on a 4.6 mm C_{18} column. Note that other cell types may contain varying abundances of histone variants, but the elution times should remain the same. The identity of each peak is provided.

4. Set up a fraction collector to collect samples in 1 min intervals from 15 to 80 min.
5. Transfer the desired fractions to centrifuge tubes and dry to completion in a vacuum centrifuge to remove organic solvent and TFA. Fractions corresponding to the same histone can be pooled before drying. Dried histones can be stored at room temperature. If reconstituted, store at −80°C.

5. MIDDLE-DOWN MASS SPECTROMETRY

Middle-down MS is a different proteomics strategy valuable for quantifying combinatorial histone PTMs. In middle-down MS, proteases that generate long peptides are employed to allow for greater connectivity between PTM sites than bottom-up MS. It is important to use a protease that does not cleave in the tail domain as this is where most PTMs are catalyzed. Endoproteinase GluC (in bicarbonate buffers) is recommended for histone H3, H4, and canonical H2A, as it generates N-terminal peptides of 50, 53, and 41 amino acid residues, respectively. Endoproteinase AspN is a valid alternative, as it cleaves all canonical histones providing middle-down sized N-terminal peptides; specifically, 71 amino acid residues for H2A, 76 for H3, 24 for H4, and 51 for H2B. Because the peptides are longer and trypsin is not being used, derivatization of the protein is unnecessary. The high charge states occupied by histone tail peptides are incompatible with CID fragmentation. ETD fragmentation, on the other hand, generally results in high coverage of the histone tail peptides.

Middle-down histone tail peptides do not separate well by RP chromatography. Weak cation exchange hydrophilic interaction liquid chromatography (WCX-HILIC) is currently the stationary phase of choice for this application, as it employs a hydrophilic stationary phase containing negatively charged residues (ie, glutamic acid), which is the ideal binding pocket for basic hydrophilic polypeptides (Jung et al., 2013; Sidoli et al., 2014; Young et al., 2009). The HILIC separation occurs by decreasing the amount of organic buffer in the solvent so that more hydrophobic species elute first. The WCX separation is accomplished by incorporating a pH gradient into the mobile phase. As the pH decreases during the gradient, the resin becomes increasingly protonated, which removes the charge on the resin and abolishes the cation exchange interaction.

One drawback of middle-down MS compared to bottom-up MS is reduced sensitivity. Larger peptides are electrospray ionized in multiple charge states, which reduces the signal for any given charge state compared to smaller peptides. Similarly, larger peptides can have a larger number of modified forms, which dilutes the signal for any single form. This limitation can be partially alleviated by using larger amounts of starting material. Fractionation of histones before MS analysis is the most effective method to increase sensitivity because it reduces the complexity of the sample, and thus allows for loading more material of a single histone variant. Another caveat of middle-down MS is that the data analysis for middle-down experiments is more complicated and less automated compared to bottom-up MS, although few software tools have been developed for the purpose (DiMaggio, Young, Baliban, Garcia, & Floudas, 2009; Sidoli et al., 2014).

5.1 Materials and Buffer Recipes

1. 100 mM ammonium acetate, pH 4.0
2. 10 mM Tris–HCl, pH 7.5
3. GluC
4. AspN
5. Offline buffer A: 5% acetonitrile, 0.2% TFA in water (all HPLC-grade)
6. Offline buffer B: 95% acetonitrile, 0.188% TFA in water (all HPLC-grade)
7. Online buffer A: 75% acetonitrile, 20 mM propionic acid, pH 6.0 (generated using ammonium hydroxide)
8. Online buffer B: 15% acetonitrile, 0.2% formic acid

9. WCX-HILIC column: 75 μm ID, 15 cm length, 1.7 μm (diameter) 1000 Å (porosity) particle size is recommended. The 3 μm (diameter) 1500 Å (porosity) particle size can also be used

5.2 Digestion

5.2.1 GluC

1. Resuspend sample in 100 mM ammonium acetate to a final concentration of 0.5 μg/μL.
2. Dilute the GluC enzyme to 0.2 μg/μL in the same buffer.
3. Add GluC to a final concentration of 1:10 GluC:histone.
4. Incubate for 6 h at room temperature.
5. Block digestion by adding 1% formic acid.

5.2.2 AspN (Alternative to GluC)

1. Resuspend enzyme to 40 ng/μL in 10 mM Tris–HCl, pH 7.5.
2. Resuspend sample in the same buffer and add enzyme at 1:100 enzyme: histone by weight.
3. Digest at 37°C for 6 h.
4. Block digestion by adding 1% formic acid.

5.3 WCX-HILIC and MS

Middle-down has been optimized by different research laboratories (Jung et al., 2013; Sidoli et al., 2014; Young et al., 2009). Currently, the most automated platform includes an RP trap column for sample loading and a WCX-HILIC analytical column for the gradient (Fig. 5). This setup allows

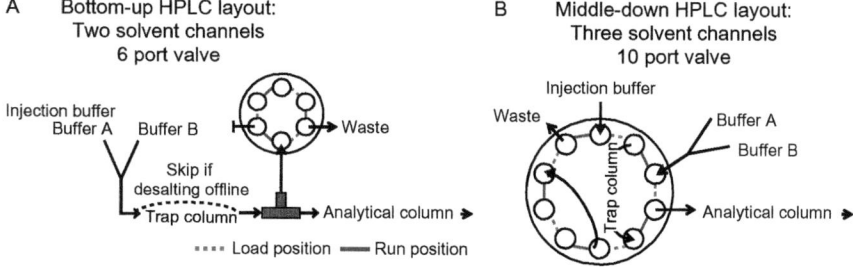

Figure 5 Valve layouts for bottom-up and middle-down HPLC-MS/MS analysis. (A) Bottom-up HPLC layout. The trap column can be used to desalt online. (B) Middle-down HPLC layout. Three solvent channels are needed: two are used to deliver the WCX-HILIC gradient, and one is used to load and desalt the sample on the trap column.

for sample loading in aqueous conditions and separation using WCX-HILIC (Sidoli et al., 2014).
1. Program the HPLC method as follows: from 0% to 55% buffer B in 1 min, from 55% to 85% B in 160 min, and from 85% to 100% in 5 min. If the HPLC is not programmed for automated column equilibration before sample loading then include this part in the method: switch the valve in position load (Fig. 5B), from 100% to 0% B in 1 min and isocratic flow at 0% B for 10 min. The flow rate of the analysis should be 250–300 nL/min.
2. Program the MS acquisition method to perform MS/MS DDA of the 5–8 most abundant precursor masses. Do not apply dynamic exclusion, as this increases the number of isobaric forms quantified. The full MS scan range should be 450–750 m/z, as this is the region of the most intense charge states for histone polypeptides. If only histone H3 is analyzed, the window can be narrowed to 660–720 m/z, in order to include only charge state 8^+.
3. Program the MS/MS acquisition to be performed with ETD at a resolution of ~30,000. The reaction time should be around 20 ms for polypeptides with 8–10 charges. Include three microscans to improve the quality of the MS/MS spectra acquired, as ETD spectra are overall less reproducible than CID.
4. Load ~2 μg of sample onto the HPLC trap column. The sample can be loaded as is from the digestion step (Section 5.2.1 or 5.2.2).
5. Run the HPLC-MS/MS method as programmed. Since the sample is loaded onto a trap column it does not require prior desalting, as the salts will be eluted during the trap loading and thus they will not be sprayed into the MS.

5.4 Data Analysis

Identification and quantification of middle-down peptides is a more challenging process than bottom-up. This is because each precursor mass might correspond to hundreds of isobarically modified peptides. Currently, the most comprehensive algorithm to quantify histone peptides employs mixed integer linear optimization (DiMaggio et al., 2009). However, this software is currently not in a user-friendly format and it is not flexible through unexpected modification sites. Thus, we developed a workflow that includes database searching and quantification based on virtual histograms. Briefly, the obtained LC-MS/MS result file is searched using Mascot (Matrix

Science), which provides identification and spectrum intensities, and then filtered with the freely available isoScale slim (http://middle-down.github.io/Software) (Sidoli et al., 2014).

1. Collect all raw files and submit them to a deconvolution tool. MS/MS spectra ions should be all singly charged before Mascot database searching. We recommend Xtract as deconvolution algorithm if Thermo Scientific instrument is used (eg, LTQ-Orbitrap), although it is important to highlight that utilizing Thermo instruments is not mandatory.

2. Perform database searching using a sequence database containing only histones; large databases increase dramatically searching time. Search parameters should be as following: MS mass tolerance: 2.2 Da, to include possible errors of the deconvolution algorithm in isotopic recognition. MS/MS mass tolerance: 0.01 Da. Enzyme: GluC (or AspN) with no missed cleavages. Sample preparation does not generate static modifications on the peptide, while variable modifications can be chosen as desired.

    Note: Recommended variable modifications are mono- and dimethylation (KR), trimethylation (K), acetylation (K) and, optionally, phosphorylation (ST).

3. Export Mascot results in .csv, including the following information to the file: all Query level information, all the default information (these last are already selected by default).

4. Import the .csv file in isoScale slim. Set the tolerance for the search (recommended: 30 ppm) and the type of fragmentation adopted (in this case ETD). The result table contains only confidently assigned combinatorial PTMs. A peptide is defined as confident if all the modifications are uniquely validated by site determining ions, which are ions that unambiguously confirm the localization of a modification site. For instance, in order to verify that a methylation is on H3K27 the software requires at least one ion proving that the modification is not either on H3R26 or H3K36. The output table contains the list of peptides that passed the site determining ions validation and their MS/MS ion intensity. isoScale also quantifies cofragmented isobaric species, as soon as all these species pass the confidence threshold, in a similar manner as EpiProfile performs for bottom-up peptides (Section 3.5.1).

5. The output table contains duplicates (peptides with the same sequence and PTM combination). Remove them by using the "Remove duplicates" option in Excel.

Note: From the relative abundance of the combinatorial marks, it is possible to extract the relative abundance of single marks simply by summing all relative abundances of peptides that contain the given PTM. To estimate which histone marks tend to coexist with each other with high or low frequency, it is possible to calculate the interplay score (Jung et al., 2013; Schwämmle, Aspalter, Sidoli, & Jensen, 2014). This score is calculated as:

$$I_{xy} = \log_2\left(F_{xy}/\left(F_x \times F_y\right)\right)$$

where I_{xy} is the interplay score between the marks X and Y, F_{xy} is the coexistence frequency (or relative abundance) of the two marks, and F_x or F_y are the frequencies of the single marks in the dataset. Basically, F_{xy} is the observed coexistence frequency, while $(F_x \star F_y)$ is the theoretical coexistence frequency, calculated based on the relative abundance of single PTMs. This calculation provides a score of how much two marks tend to coexist or be mutually exclusive on the same histone protein. Positive values indicate tendency to coexist higher than if the two marks were completely independent from each other, while negative values indicate the opposite. This score has been used to predict crosstalk between histone marks (Schwämmle et al., 2014).

6. TOP-DOWN MASS SPECTROMETRY

Top-down MS is performed by fragmenting intact protein using ETD fragmentation. The biggest advantage of top-down MS over the other methods is that it provides a global view of the intact protein sequence. However, this advantage is accompanied by the caveat that it is also the least sensitive method because the histone proteins will occupy the maximal number of charge states and modified forms as compared to bottom-up and middle-down strategies. Furthermore, the data analysis is not only more difficult, but also currently prohibitive in specific cases. For instance, while the bottom-up sample preparation can produce peptides with up to six isobaric forms, an intact histone has potentially hundreds of thousands of combinatorial forms with the same identical precursor mass. The exponential number can be explained considering the many modifiable sites (eg, K and R) and the permutations of isobaric modified forms (eg, me3 is equivalent to me1me2, which is equivalent to me1me1me1, which is equivalent to the same combination on other sites, etc.). Therefore, a single MS/MS spectrum might contain an impressive number of isobaric proteoforms, of which at the moment we can only scrape its surface.

Generally, fractionation is performed before conducting a top-down experiment to reduce the complexity of the sample. Histone species can be fractioned as described in Section 4. Usually, histone proteins are directly infused into the MS in top-down analysis, but online RP separation can also be conducted if desired (Contrepois, Ezan, Mann, & Fenaille, 2010; Eliuk, Maltby, Panning, & Burlingame, 2010; Tian et al., 2010). Direct infusion using a syringe requires a large amount of sample (about 20–30 μg for 10 min injection), so using a chip-based infusion system, such as an Advion Nanomate, is strongly recommended, as the amount of sample required can be as low as 5 μg for a 30 min injection. In these systems, sample is picked up by a small tip and delivered to a chip containing nano-ESI nozzles. Voltage is applied to the chip to enable ionization into the MS. Very small sample volumes can be stably sprayed for long periods of time using this system (10 μL of sample lasts for about 30–40 min).

6.1 Materials and Buffer Recipes
1. Sample buffer: 75% acetonitrile/25% of 0.8% formic acid adjusted to pH 2.5 with TFA
2. Chip-based nano-ESI autosampler such as Advion Nanomate

6.2 Top-Down MS Using Direct Infusion
The acquisition method in top-down MS will vary depending on the exact application. Most top-down experiments aim to characterize a specific modified form of histone protein. In this case,
1. If sample contains salt, desalt according to Section 3.3 using C_8 resin in place of C_{18} resin. One disc of C_8 resin can bind approximately 10 μg of intact histone protein. Use multiple discs if the sample contains more than 10 μg of protein. If offline fractionation was used, there is no need to desalt the sample.
2. Resuspend histone protein in sample buffer to a concentration of approximately 1 μg/μL.
3. Infuse sample in a chip-based direct infusion system and adjust parameters to achieve stable spray. Generally, stable spray can be achieved using a capillary temperature of 150°C, voltage between 1.7 and 2.5 kV, and a gas pressure between 0.3 and 0.4 psi. Capillary temperatures above 170°C can damage the chip and are not recommended. These

parameters will need to be adjusted for each run before data acquisition and may also need adjustment as data is being acquired. A sudden complete loss of signal usually indicates that the nozzle has become clogged, in which case another nozzle should be selected.
4. Run an MS acquisition method to collect data. The method will depend largely on the application. To analyze a specific modified form of a histone protein, set up an MS1 scan that spans the m/z range of that form but excludes other species (ie, Fig. 6A). Perform a data-dependent MS/MS on the top isotope of the distribution with a large number of microscans (10–20 should provide a clean spectra depending on the quality of signal and spray). Generally, a 3–5 min acquisition time is adequate to obtain a high-quality MS/MS spectrum that can be identified during data analysis. If characterization of more than one species is desired, set up the method to take MS1 scans corresponding to the other species followed by data-dependent MS2 scans.

6.3 Data Analysis

Top-down data require deconvolution before analysis, where each multiply charged ion is reduced to its singly charged monoisotopic mass. This process facilitates data analysis as most programs cannot analyze MS/MS spectra containing highly charged fragment ions. The deconvoluted data can then be searched using Mascot (Matrix Science) software.
1. Convert files into mzxml format using any of a number of available programs for this purpose.
2. Deconvolute data using any of a number of available programs for this purpose. Examples include the Xtract module from Thermo Xcalibur software and MS-Deconv (freely available at http://bix.ucsd.edu/projects/msdeconv/).
3. Use proper software to search the deconvoluted data. The traditional Mascot license can analyze intact molecules up to 16 kDa, making it suitable for intact histone proteins. Other software tools can be used, such as ProSightPC (Thermo) or the freely available MS-Align+ (Liu et al., 2012). Use a database containing all of the histone protein sequences, including variants. Select the modifications of interest as variable PTMs.

Note: Manual validation of the results will be necessary, as all softwares are not optimized for heavily modified proteins such as histones. In order to achieve a confident localization for a given PTM, it is necessary to have specific fragment ions between two possible modifiable

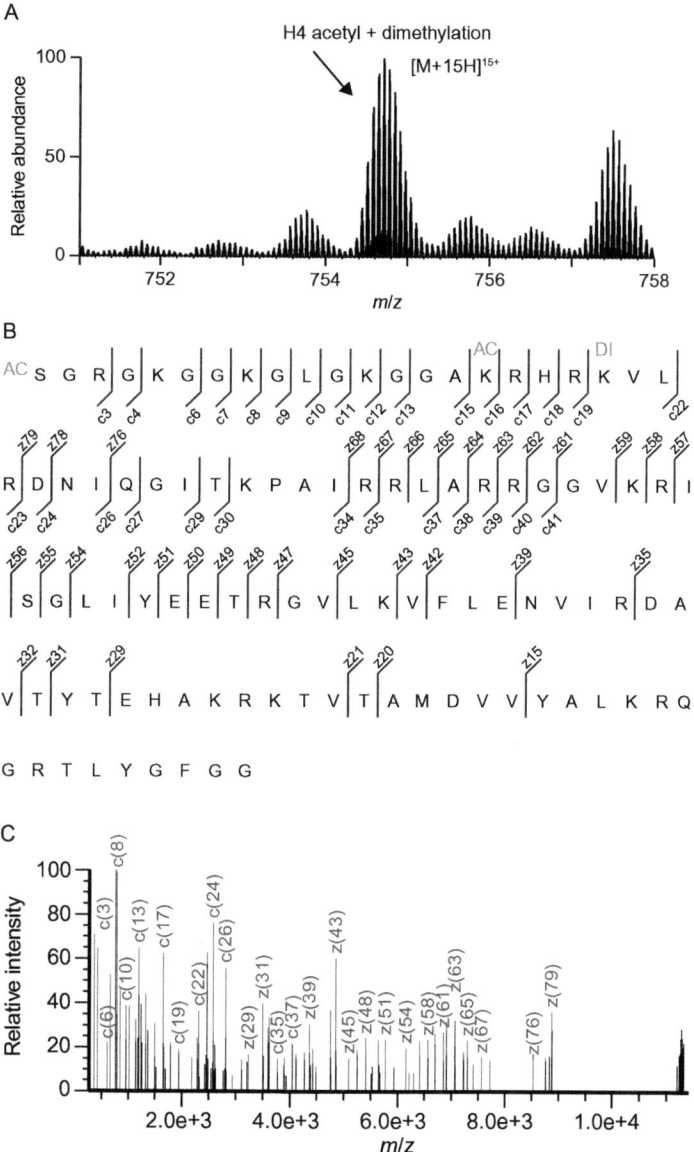

Figure 6 Example data obtained in a top-down experiment for intact H4 containing 1 dimethyl and 2 acetyl groups. (A) Full MS scan prior to isolation and fragmentation of intact H4 containing an acetyl and dimethyl group. The isolation window is set so that the species of interest is the most abundant ion in the scan so that it will be selected for data-dependent MS2 acquisition. (B) Fragment ion coverage of the H4 protein containing N-terminal acetylation (AC), K16ac (AC), and K20me2 (DI). There are fragment ions flanking each modification, allowing for confident identification of the PTMs. (C) Deconvoluted MS/MS spectra. Each identified fragment ion is indicated in gray and the most abundant fragment ions are labeled.

sites, confirming on which amino acid residue the modification is localized (Fig. 6B and C). Software will provide the "most probable" localization of the PTMs, but they rarely provide a score describing whether another localization for a given modification is equally probable. isoScale slim is applicable for this purpose, as previously illustrated for middle-down (Section 5.4). However, it demands result files from Mascot.

REFERENCES

Britton, L.-M. P., Gonzales-Cope, M., Zee, B. M., & Garcia, B. A. (2011). Breaking the histone code with quantitative mass spectrometry. *Expert Review of Proteomics, 8*(5), 631–643. http://doi.org/10.1586/epr.11.47.

Contrepois, K., Ezan, E., Mann, C., & Fenaille, F. (2010). Ultra-high performance liquid chromatography–mass spectrometry for the fast profiling of histone post-translational modifications. *Journal of Proteome Research, 9*(10), 5501–5509. http://doi.org/10.1021/pr100497a.

DiMaggio, P. A., Young, N. L., Baliban, R. C., Garcia, B. A., & Floudas, C. A. (2009). A mixed integer linear optimization framework for the identification and quantification of targeted post-translational modifications of highly modified proteins using multiplexed electron transfer dissociation tandem mass spectrometry. *Molecular & Cellular Proteomics, 8*(11), 2527–2543. http://doi.org/10.1074/mcp.M900144-MCP200.

Eliuk, S. M., Maltby, D., Panning, B., & Burlingame, A. L. (2010). High resolution electron transfer dissociation studies of unfractionated intact histones from murine embryonic stem cells using on-line capillary LC separation. *Molecular & Cellular Proteomics, 9*(5), 824–837. http://doi.org/10.1074/mcp.M900569-MCP200.

Garcia, B. A., Mollah, S., Ueberheide, B. M., Busby, S. A., Muratore, T. L., Shabanowitz, J., & Hunt, D. F. (2007). Chemical derivatization of histones for facilitated analysis by mass spectrometry. *Nature Protocols, 2*(4), 933–938. http://doi.org/10.1038/nprot.2007.106.

Gillet, L. C., Navarro, P., Tate, S., Röst, H., Selevsek, N., Reiter, L., ... Aebersold, R. (2012). Targeted data extraction of the MS/MS spectra generated by data-independent acquisition: A new concept for consistent and accurate proteome analysis. *Molecular & Cellular Proteomics, 11*(6). http://doi.org/10.1074/mcp.O111.016717. O111.016717.

Jenuwein, T., & Allis, C. D. (2001). Translating the histone code. *Science, 293*(5532), 1074–1080. http://doi.org/10.1126/science.1063127.

Jung, H. R., Sidoli, S., Haldbo, S., Sprenger, R. R., Schwämmle, V., Pasini, D., ...Jensen, O. N. (2013). Precision mapping of coexisting modifications in histone H3 tails from embryonic stem cells by ETD-MS/MS. *Analytical Chemistry, 85*(17), 8232–8239. http://doi.org/10.1021/ac401299w.

Liu, X., Sirotkin, Y., Shen, Y., Anderson, G., Tsai, Y. S., Ting, Y. S., ... Pevzner, P. A. (2012). Protein identification using top-down spectra. *Molecular & Cellular Proteomics, 11*(6). M111.008524. http://doi.org/10.1074/mcp.M111.008524.

Luger, K., Mäder, A. W., Richmond, R. K., Sargent, D. F., & Richmond, T. J. (1997). Crystal structure of the nucleosome core particle at 2.8 Å resolution. *Nature, 389*(6648), 251–260. http://doi.org/10.1038/38444.

MacLean, B., Tomazela, D. M., Shulman, N., Chambers, M., Finney, G. L., Frewen, B., ... MacCoss, M. J. (2010). Skyline: An open source document editor for creating and analyzing targeted proteomics experiments. *Bioinformatics, 26*(7), 966–968. http://doi.org/10.1093/bioinformatics/btq054.

Rothbart, S. B., Dickson, B. M., Raab, J. R., Grzybowski, A. T., Krajewski, K., Guo, A. H., ... Strahl, B. D. (2015). An interactive database for the assessment of histone antibody specificity. *Molecular Cell, 59*(3), 502–511. http://doi.org/10.1016/j.molcel.2015.06.022.

Schwämmle, V., Aspalter, C.-M., Sidoli, S., & Jensen, O. N. (2014). Large scale analysis of co-existing post-translational modifications in histone tails reveals global fine structure of cross-talk. *Molecular & Cellular Proteomics, 13*(7), 1855–1865. http://doi.org/10.1074/mcp.O113.036335.

Sidoli, S., Schwämmle, V., Ruminowicz, C., Hansen, T. A., Wu, X., Helin, K., & Jensen, O. N. (2014). Middle-down hybrid chromatography/tandem mass spectrometry workflow for characterization of combinatorial post-translational modifications in histones. *Proteomics, 14*(19), 2200–2211. http://doi.org/10.1002/pmic.201400084.

Sidoli, S., Simithy, J., Karch, K. R., Kulej, K., & Garcia, B. A. (2015). Low resolution data-independent acquisition in an LTQ-Orbitrap allows for simplified and fully untargeted analysis of histone modifications. *Analytical Chemistry, 87*(22), 11448–11454. http://doi.org/10.1021/acs.analchem.5b03009.

Sidoli, S., Yuan, Z.-F., Lin, S., Karch, K., Wang, X., Bhanu, N., ... Garcia, B. A. (2015). Drawbacks in the use of unconventional hydrophobic anhydrides for histone derivatization in bottom-up proteomics PTM analysis. *Proteomics, 15*(9), 1459–1469. http://doi.org/10.1002/pmic.201400483.

Tian, Z., Zhao, R., Tolić, N., Moore, R. J., Stenoien, D. L., Robinson, E. W., ... Paša-Tolić, L. (2010). Two-dimensional liquid chromatography system for online top-down mass spectrometry. *Proteomics, 10*(20), 3610–3620. http://doi.org/10.1002/pmic.201000367.

Young, N. L., DiMaggio, P. A., Plazas-Mayorca, M. D., Baliban, R. C., Floudas, C. A., & Garcia, B. A. (2009). High throughput characterization of combinatorial histone codes. *Molecular & Cellular Proteomics, 8*(10), 2266–2284. http://doi.org/10.1074/mcp.M900238-MCP200.

Yuan, Z.-F., Lin, S., Molden, R. C., Cao, X.-J., Bhanu, N. V., Wang, X., ... Garcia, B. A. (2015). EpiProfile quantifies histone peptides with modifications by extracting retention time and intensity in high-resolution mass spectra. *Molecular & Cellular Proteomics, 14*(6), 1696–1707. http://doi.org/10.1074/mcp.M114.046011.

CHAPTER TWO

Substrate Specificity Profiling of Histone-Modifying Enzymes by Peptide Microarray

E.M. Cornett*, B.M. Dickson*, R.M. Vaughan*, S. Krishnan[†],
R.C. Trievel[†], B.D. Strahl[‡], S.B. Rothbart*,[1]

*Center for Epigenetics, Van Andel Research Institute, Grand Rapids, MI, United States
[†]University of Michigan, Ann Arbor, MI, United States
[‡]University of North Carolina at Chapel Hill, Chapel Hill, NC, United States
[1]Corresponding author: e-mail address: scott.rothbart@vai.org

Contents

1. Introduction 32
2. Assay Optimization 34
   2.1 Design of Custom Print Formats 34
   2.2 Detection Reagent Considerations 36
3. Assay Methodology 37
   3.1 Microarray-Based Lysine Methyltransferase (KMT) Assays with G9a 38
   3.2 Solution-Based KMT Assays with G9a 40
   3.3 Microarray-Based Lysine Demethylase (KDM) Assays with JMJD2A 40
4. Enzyme Specificity Profiling by Microarray 42
   4.1 High-Throughput Profiling of G9a KMT Activity 42
   4.2 High-Throughput Profiling of JMJD2A KDM Activity 46
5. Summary and Perspectives 49
Acknowledgments 50
References 50

Abstract

The dynamic addition and removal of covalent posttranslational modifications (PTMs) on histone proteins serves as a major mechanism regulating chromatin-templated biological processes in eukaryotic genomes. Histone PTMs and their combinations function by directly altering the physical structure of chromatin and as rheostats for effector protein interactions. In this chapter, we detail microarray-based methods for analyzing the substrate specificity of lysine methyltransferase and demethylase enzymes on immobilized synthetic histone peptides. Consistent with the "histone code" hypothesis, we reveal a strong influence of adjacent and, surprisingly, distant histone PTMs on the ability of histone-modifying enzymes to methylate or demethylate their substrates. This platform will greatly facilitate future investigations into histone substrate specificity and mechanisms of PTM signaling that regulate the catalytic properties of histone-modifying enzymes.

1. INTRODUCTION

Eukaryotic genomes are tightly packaged in cell nuclei by their ability to associate with histone proteins. These histone-DNA complexes are first organized into nucleosomes, which are further folded into higher-order chromatin fibers that are poorly understood (Khorasanizadeh, 2004; Kornberg & Lorch, 1999; Luger, Mader, Richmond, Sargent, & Richmond, 1997). A major focus of modern biomedical research has been to understand how genetic information is accessed in the context of chromatin to control DNA-templated processes like gene transcription, replication, and repair, and how deregulation of these processes contributes to the initiation and progression of human disease (Detrich, 1986; Maze, Noh, & Allis, 2013; Portela & Esteller, 2010).

Posttranslational modifications (PTMs) on histone proteins have emerged as key epigenetic regulators of genome accessibility and function (Kouzarides, 2007). Fundamental breakthroughs in our understanding of chromatin regulation have been made through the identification of protein machineries that add (write), remove (erase), and bind (read) these marks (Rothbart & Strahl, 2014). Fueled by technological advances in mass spectrometry-based proteomics, more than 20 unique histone PTMs have been identified at upward of 80 different histone residues, many of which cluster in the unstructured N- and C-terminal tail domains that protrude from the nucleosome core (Huang, Sabari, Garcia, Allis, & Zhao, 2014; Zhao & Garcia, 2015).

In 2000, the concept of a "histone code" emerged as a hypothesis to stimulate new thinking about how histone PTMs might function in a combinatorial manner to dynamically regulate chromatin interactions of histone reader proteins (Strahl & Allis, 2000). In addition, it was postulated that much like histone PTM readers, the enzymes that write and erase these marks would themselves be influenced by preexisting PTM patterns. While significant effort has been placed on identifying and characterizing enzymes and effector proteins responsible for writing, erasing, and reading histone marks (Fig. 1), deciphering regulatory mechanisms of combinatorial PTM patterning has proven challenging, due in part to the sheer complexity of the histone PTM landscape.

To address this issue, we developed a high-density histone peptide microarray platform to enable rapid and high-throughput biochemical

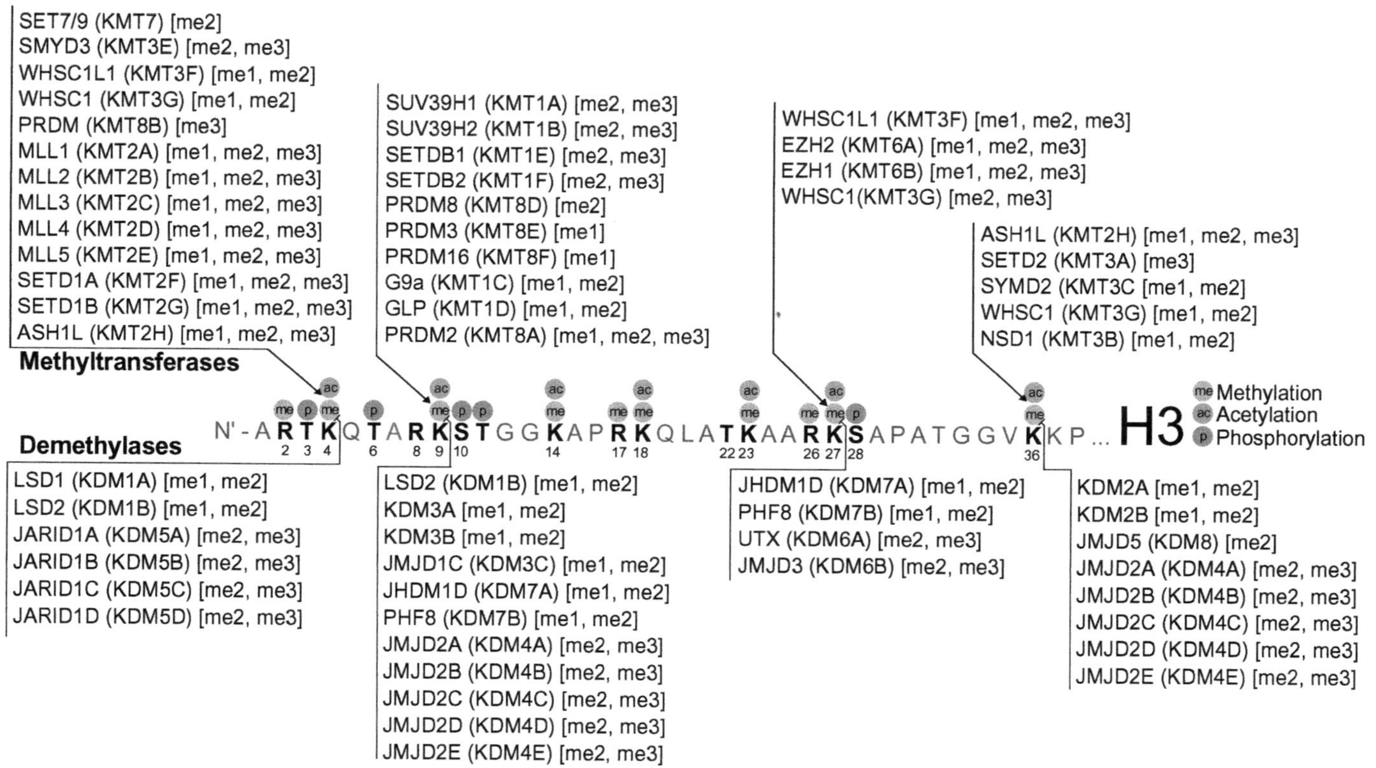

SET7/9 (KMT7) [me2]
SMYD3 (KMT3E) [me2, me3]
WHSC1L1 (KMT3F) [me1, me2]
WHSC1 (KMT3G) [me1, me2]
PRDM (KMT8B) [me3]
MLL1 (KMT2A) [me1, me2, me3]
MLL2 (KMT2B) [me1, me2, me3]
MLL3 (KMT2C) [me1, me2, me3]
MLL4 (KMT2D) [me1, me2, me3]
MLL5 (KMT2E) [me1, me2, me3]
SETD1A (KMT2F) [me1, me2, me3]
SETD1B (KMT2G) [me1, me2, me3]
ASH1L (KMT2H) [me1, me2, me3]

SUV39H1 (KMT1A) [me2, me3]
SUV39H2 (KMT1B) [me2, me3]
SETDB1 (KMT1E) [me2, me3]
SETDB2 (KMT1F) [me2, me3]
PRDM8 (KMT8D) [me2]
PRDM3 (KMT8E) [me1]
PRDM16 (KMT8F) [me1]
G9a (KMT1C) [me1, me2]
GLP (KMT1D) [me1, me2]
PRDM2 (KMT8A) [me1, me2, me3]

WHSC1L1 (KMT3F) [me1, me2, me3]
EZH2 (KMT6A) [me1, me2, me3]
EZH1 (KMT6B) [me1, me2, me3]
WHSC1 (KMT3G) [me2, me3]

ASH1L (KMT2H) [me1, me2, me3]
SETD2 (KMT3A) [me3]
SYMD2 (KMT3C) [me1, me2]
WHSC1 (KMT3G) [me1, me2]
NSD1 (KMT3B) [me1, me2]

Methyltransferases

N'- A R T K Q T A R K S T G G K A P R K Q L A T K A A R K S A P A T G G V K K P ... **H3**
    2 3 4   6 8 9 10   14   17 18   22 23   26 27 28   36

me — Methylation
ac — Acetylation
p — Phosphorylation

Demethylases

LSD1 (KDM1A) [me1, me2]
LSD2 (KDM1B) [me1, me2]
JARID1A (KDM5A) [me2, me3]
JARID1B (KDM5B) [me2, me3]
JARID1C (KDM5C) [me2, me3]
JARID1D (KDM5D) [me2, me3]

LSD2 (KDM1B) [me1, me2]
KDM3A [me1, me2]
KDM3B [me1, me2]
JMJD1C (KDM3C) [me1, me2]
JHDM1D (KDM7A) [me1, me2]
PHF8 (KDM7B) [me1, me2]
JMJD2A (KDM4A) [me2, me3]
JMJD2B (KDM4B) [me2, me3]
JMJD2C (KDM4C) [me2, me3]
JMJD2D (KDM4D) [me2, me3]
JMJD2E (KDM4E) [me2, me3]

JHDM1D (KDM7A) [me1, me2]
PHF8 (KDM7B) [me1, me2]
UTX (KDM6A) [me2, me3]
JMJD3 (KDM6B) [me2, me3]

KDM2A [me1, me2]
KDM2B [me1, me2]
JMJD5 (KDM8) [me2]
JMJD2A (KDM4A) [me2, me3]
JMJD2B (KDM4B) [me2, me3]
JMJD2C (KDM4C) [me2, me3]
JMJD2D (KDM4D) [me2, me3]
JMJD2E (KDM4E) [me2, me3]

Fig. 1 The dynamic regulation of lysine methylation on histone H3. Shown are major sites of methylation (me), acetylation (ac), and phosphorylation (p) on the N-terminal tail domain of histone H3. Known writers (methyltransferases; KMTs) and erasers (demethylases; KDMs) of lysine methylation are clustered by major histone substrate residue(s). Methylation products and substrates (mono-, me1; di-, me2; tri-, me3) of KMT and KDM reactions, respectively, are listed. Enzyme identification reflects both conventional and generic (Allis et al., 2007) nomenclature. (See the color plate.)

characterization of histone PTM-specific antibodies and readers in the context of complex histone PTM patterns (Fuchs, Krajewski, Baker, Miller, & Strahl, 2011; Rothbart et al., 2015; Rothbart, Krajewski, Strahl, & Fuchs, 2012). Briefly, synthetic biotinylated peptides, posttranslationally modified with up to eight physiologically relevant combinations of lysine acetylation and methylation (mono-, di-, and trimethyl), arginine methylation (mono, symmetric dimethyl, and asymmetric dimethyl) and citrullination, and serine/threonine phosphorylation, are deposited on streptavidin-coated glass slides that can then be used to examine aspects of protein function like binding and enzymatic activity.

Methods for the synthesis of combinatorially modified biotinylated histone peptides, microarray fabrication using these peptides, and the characterization of histone readers and antibodies with histone peptide microarrays were previously described (Rothbart et al., 2012). In this chapter, we now detail the utility of this same peptide microarray platform, which is commercially available through Epicypher and Millipore, for high-throughput substrate specificity profiling of histone lysine methyltransferases (KMTs) and demethylases (KDMs). We further reveal the influence of neighboring, and surprisingly distant, PTMs on the catalytic properties of histone-modifying enzymes.

2. ASSAY OPTIMIZATION

2.1 Design of Custom Print Formats

A number of variables should be considered when designing an enzyme assay for microarray screening. For instance, incubation times and enzyme concentrations that have been optimized for solution-based assays may not translate to a microarray format, particularly since substrate concentrations of an immobilized peptide or protein can be limiting by several orders of magnitude. To optimize assay conditions, including variables of time, buffer composition, enzyme concentration, and detection reagent, we used our recently developed open source software package, ArrayNinja (see chapter "ArrayNinja: An Open Source Platform for Unified Planning and Analysis of Microarray Experiments" by Dickson et al.). ArrayNinja unifies the planning and analysis of microarray experiments to facilitate streamlined microarray customization and data processing. Shown in Fig. 2A is a format we have found useful for enzyme assay optimization that partitions 48 subarrays on a single slide by hydrophobic wax. Detailed procedures for wax

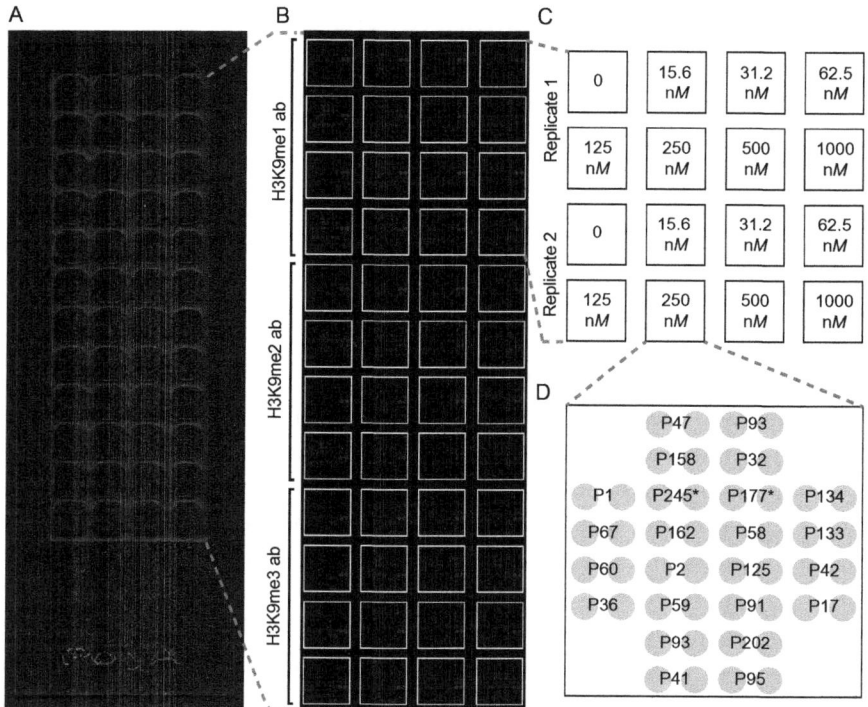

Fig. 2 Optimization of a G9a KMT assay on microarrays. (A) A 48-well optimization array partitioned by hydrophobic wax. (B and C) Scanned image of an optimization array assay following a 2-h incubation with seven G9a concentrations, a no enzyme control (0), and three detection antibodies (H3K9me1, EpiCypher #13-0014; H3K9me2, Abcam #1220; H3K9me3, Active Motif #39161), all in duplicate. (D) Print layout of individual arrays partitioned on 48-well optimization slides. Each array consists of 24 unique histone peptides printed in duplicate. Peptide identifications numbers correspond to those used previously (Rothbart et al., 2015). P245 and P177 are $H3_{(1-32)}$ and $H3_{(1-43)}$, respectively. Biotin-4-Fluorescein (not shown) is printed in the corners of each array as a spotting control and a landmark to aid in alignment for data analysis.

deposition using a stamping device are described elsewhere (Partyka, Wang, Zhao, Cao, & Haab, 2014). These wax boundaries produce a well around each subarray of 64 unique features (275 μm spot diameter and 100 μm between spots) and allows for hybridization of 6 μL reaction volumes. Representative slides for the optimization of a KMT assay with G9a are shown, where we were able to test seven enzyme concentrations and three detection antibodies (ie, antibodies that recognize the orders of H3K9 methylation that are generated by G9a), all in duplicate, on a single slide (Fig. 2B).

2.2 Detection Reagent Considerations

A critical optimization step when designing an assay for histone-modifying enzyme activity on microarrays is determining how catalysis will be monitored. Detection strategies most successful in our hands have relied on radioisotope incorporation or histone PTM-specific antibodies. Advantages and disadvantages of both radioisotopic and antibody readout of enzyme function are discussed in Sections 2.2.1 and 2.2.2, and methods for performing these assays are described in Section 3.

2.2.1 Radioisotopes

The activity of enzymes that write histone PTMs, including methyltransferases, acetyltransferases, and kinases can be monitored on microarrays by radioisotope incorporation. The use of radioisotopes for profiling the substrate specificity of the enzymes responsible for erasing histone PTMs presents a significant technical challenge for microarray analysis, as this would require the tedious synthesis of large peptide libraries that contain these isotopes. Radioactivity assays are advantageous when substrate specificity is unknown, and with enough chemical diversity in a substrate library, can be informative in identifying the target residue. However, the degree or order of modification (eg, mono-, di-, or trimethylation on lysine residues) cannot be defined with this detection strategy. Use of radioisotopes also enables a clear assessment of the influence neighboring PTMs have on enzymatic activity, since these direct labeling strategies are not confounded by the influence of PTMs on antibody-epitope recognition (Rothbart et al., 2015).

A major drawback of radioactivity assays, specifically methyltransferase and acetyltransferase assays, is the time required to collect data from an experiment. Tritium and ^{14}C are low-energy beta emitters, requiring upward of 1 month exposure time to X-ray film to visualize results. Autoradiography enhancing reagents do not appear to be compatible with this array format. While it may be possible to decrease exposure times with the aid of phosphor screens, this has not been empirically tested. Kinases are more amenable to radioisotope-based microarray assays, as they rely on the use of the high-energy beta emitter [γ-^{33}P] ATP as a cofactor (Rothbart et al., 2012), which is detectable by autoradiography following a brief exposure.

2.2.2 Antibodies

Histone PTM-specific antibodies are widely used chromatin biochemistry tools, and more than 1000 commercial antibodies are available that recognize the diverse modification states on histone proteins. There are a number of advantages to using antibodies as detection reagents for histone-modifying enzymes on microarrays. Unlike radioisotopic readouts, antibody detection methods enable site-specific analysis of a particular substrate residue of interest and can discriminate between the various states of methylation found on histones. In addition, a number of pan PTM antibody reagents have been developed, many of which react well with the intended modifications on histones (data not shown). However, a number of these pan reagents show epitope specificity problems, albeit less strict than conventional antibodies, resulting in unequal discrimination among PTMs when presented in diverse sequence contexts. Antibody hybridization procedures are straightforward, require little material, do not generate radioactive waste, and can be detected within several hours.

A significant caveat to using antibodies as detection reagents is the influence, both positive and negative, of neighboring PTMs. Inclusion of control detection experiments in the absence of modifying enzyme or cofactor can help assess the behavior of the antibody being used. In addition, we recently characterized the specificity of over 100 commercially available histone PTM antibodies by microarray, and we created a public-facing web portal, the Histone Antibody Specificity Database (www.histoneantibodies.com) (Rothbart et al., 2015). This online resource is a useful tool for aiding in the selection of antibodies suitable for enzyme profiling experiments.

3. ASSAY METHODOLOGY

Later we detail methods for assaying the methyltransferase activity of G9a and the demethylase activity of JMJD2A—both enzymes that write and erase H3K9 methylation, respectively—on our previously described peptide microarray platform (Rothbart et al., 2012). This print format benefits from the ability to assay >250 unique peptide substrates in a single assay. In addition, two identical subarrays are printed on each slide and can be separated by a silicon adhesive seal (eg, Epicypher #11-3001), by hydrophobic wax deposited with a slide imprinter (eg, Gel Company #WSP48-1), or by a PAP pen (eg, Sigma #Z377821). Subarray separation is useful for comparison of multiple conditions on the slide. For example, we commonly incubate the array in the absence or presence of enzyme on the same slide prior to antibody hybridization. We detail an antibody-based detection procedure, applicable to both classes of enzymes, and we also describe radioisotopic assays using the cofactor adenosyl-L-methionine, S-[methyl-^{3}H] (^{3}H-SAM) for histone methyltransferase assays on microarrays and in solution.

3.1 Microarray-Based Lysine Methyltransferase (KMT) Assays with G9a

Equipment and reagents
1. Tris–HCl, pH 8.8
2. MgCl$_2$
3. Recombinant human G9a, amino acids 913-1210 (Epicypher #15-1008)
4. S-adenosyl methionine (SAM)
5. Dithiothreitol (DTT)
6. Cold PBS, pH 7.6
7. Powdered bovine serum albumin (BSA)
8. Tween-20
9. Nontreated four-well dish (such as Thermo Scientific #267061)
10. 25 × 60 mm cover slips
11. Biotinylated histone peptide microarray, custom printed as previously described (Rothbart et al., 2012), or commercially available through Epicypher (#11-2001) or Millipore (#16-671).
12. Humidified microarray incubation chamber (such as VWR#97000-384).

Additional equipment and reagents specific to antibody detection
1. Histone PTM-specific primary antibody.
2. Fluorescent-labeled secondary antibody (such as Alexa Fluor® 647-conjugated anti-rabbit (Thermo #A-21244) or anti-mouse (Thermo #A-21235)).
3. Microarray scanner capable of scanning at ≤25 μm and equipped with lasers compatible with your secondary antibody conjugates.

Additional equipment and reagents specific to radioisotope detection
1. ^{3}H-SAM (Perkin Elmer NET155001MC)
2. X-ray film
3. Film developer

General procedure
1. Equilibrate a microarray slide in KMT reaction buffer (50 mM Tris–HCl, pH 8.8, 5 mM MgCl$_2$, 4 mM DTT) for 10 min at room temperature.

2. Incubate the slide in a humidified microarray incubation chamber with 200 µL[1] KMT reaction buffer containing 1 µM enzyme[2] and 60 µM SAM[3] under a coverslip[4] at 30°C for 2 h.
3. Wash the slide 2 × 5 min with PBS, pH 7.6 in the cold.
4. Dry the slide. We prefer a mini-centrifuge for this purpose, such as Sigma #Z674672. Alternatively, slides can be dried over a stream of filtered compressed air or by centrifugation in a 50-mL conical tube in a swinging-bucket rotor at 800 × g for 1 min.
5. Follow procedures below for detection with radioisotope or antibody.

Additional procedures and considerations for radioisotope detection

1. Step 3 above should be performed 5 × 5 min, and liquid waste should be discarded in accordance with your institution's radiation safety procedures.
2. For containment, the preferred drying method in step 4 above is centrifugation in a 50-mL conical tube.
3. Following step 4 above, expose the dry slide to X-ray film in an autoradiography cassette at −80°C for at least 1 month.
4. Bring the sealed autoradiography cassette to room temperature and process film through a developer.

Additional procedures and considerations for antibody detection

1. Prepare a dilution of primary antibody in hybridization buffer (PBS, pH 7.6, 5% BSA (w/v), 0.1% Tween-20). In general, begin with a dilution factor at the high end of the detection range by western blot.
2. Incubate the slide with 200 µL of diluted antibody under a coverslip for 1 h at 4°C in a humidified microarray incubation chamber.
3. Remove the coverslip and wash 3 × 5 min with cold PBS.
4. Prepare a dilution of fluorescent secondary antibody (1:5000–1:10,000 for Alexa Fluor® 647) in hybridization buffer and incubate the slide for 30 min in the dark with gentle rotation at 4°C.
5. Wash the slide 3 × 5 min with PBS in the dark at 4°C.
6. Dip the slide 5 × in 0.1 × PBS to remove excess salt.
7. Dry the slide by centrifugation or compressed air as described earlier.

---

[1] The volume requirement for a silicon adhesive seal is 350 and 500 µL for arrays separated by PAP pen.
[2] Enzyme concentration was optimized on 48-well slides at a range from 0 to 1 µM (Fig. 2B); 1 µM G9a was used for large-format arrays (Fig. 3A).
[3] For radioisotope assays, use 5 uCi ³H-SAM.
[4] 48-well optimization slides were incubated with 6 µL in each well without a coverslip.

8. Scan the slide using a microarray scanner at ≤ 25 μm with a laser appropriate for your fluorescent secondary antibody.
9. Slides can be stored at 4°C protected from light for several days without appreciable loss of fluorescent signal.

3.2 Solution-Based KMT Assays with G9a
Equipment and reagents
1. Tris–HCl, pH 8.8
2. $MgCl_2$
3. Recombinant human G9a, amino acids 913-1210 (Epicypher #15-1008)
4. DTT
5. ^{3}H-SAM (Perkin Elmer #NET155001MC)
6. H3$_{(1-20)}$ peptide (Epicypher #12-0001)
7. Trifluoroacetic acid (TFA)
8. $NaHCO_3$, pH 9.0
9. P81 phosphocellulose filter paper (Whatman)

General procedure
1. Dilute G9a to a final concentration of 25 nM in KMT reaction buffer (50 mM Tris–HCl, pH 8.8, 5 mM $MgCl_2$, 4 mM DTT) containing 0.5 μCi ^{3}H-SAM.
2. Start reactions by addition of substrate peptide at a final concentration of 10 μM.
3. Quench reactions by addition of TFA to a final concentration of 0.5% (w/v).
4. Spot equal amounts of quenched reactions on P81 filter paper and dry at room temperature for 10 min.
5. Wash filter papers in a single beaker 4×5 min with 50 mM $NaHCO_3$, pH 9.0.
6. Dry filter papers at room temperature for 20 min.
7. Place filter papers in scintillation vials and add an appropriate amount of scintillation fluid to submerge the paper.
8. Count tritium with a scintillation counter.

3.3 Microarray-Based Lysine Demethylase (KDM) Assays with JMJD2A
Equipment and reagents
1. HEPES, pH 7.5
2. NaCl

3. Recombinant human JMJD2A, amino acids 1-350 (Krishnan, Collazo, Ortiz-Tello, & Trievel, 2012; see chapter "Purification, Biochemical Analysis, and Structure Determination of JmjC Lysine Demethylases" by Krishnan and Trievel)
4. $(NH_4)_2Fe(SO_4)_2$
5. L-Ascorbic acid
6. 2-OG
7. Cold PBS, pH 7.6
8. Powdered BSA
9. Tween-20
10. Nontreated four-well dish (such as Thermo Scientific #267061)
11. Biotinylated histone peptide microarray, custom printed as previously described (Rothbart et al., 2012) or commercially available through Epicypher (#11-2001) or Millipore (#16-671)
12. Humidified microarray incubation chamber (such as VWR #97000-384)
13. Histone PTM-specific primary antibody
14. Fluorescent-labeled secondary antibody (such as Alexa Fluor® 647-conjugated anti-rabbit (Thermo #A-21244) or anti-mouse (Thermo #A-21235))
15. Microarray scanner capable of scanning at ≤25 μm and equipped with lasers compatible with your secondary antibody conjugates

General procedure
1. Equilibrate a microarray slide in KDM reaction buffer (50 mM HEPES, pH 7.5, 50 mM NaCl, 50 μM $(NH_4)_2Fe(SO_4)_2$, 1 mM L-ascorbic acid, 1 mM 2-OG) for 10 min at room temperature.
2. Incubate the slide in a humidified microarray incubation chamber with 500 μL KDM reaction buffer containing 0.313 μM enzyme[5] at 25 °C for 18 h.
3. Wash the slide 3 × 5 min with PBS, pH 7.6 in the cold.
4. Dry the slide. We prefer a mini-centrifuge for this purpose, such as Sigma #Z674672. Alternatively, slides can be dried by over a stream of filtered compressed air or by centrifugation in a 50-mL conical tube in a swinging-bucket rotor at 800 × g for 1 min.
5. Prepare a dilution of primary antibody in hybridization buffer (PBS, pH 7.6, 5% BSA (w/v), 0.1% Tween-20). In general, begin with a dilution factor at the high end of the detection range by western blot (1:1000–1:5000).

---

[5] Enzyme concentration was optimized on 48-well slides at a range from 0 to 625 nM (Fig. 4A); 313 nM JMJD2A was used for large-format arrays (Fig. 4B).

6. Incubate the slide with 500 μL of diluted antibody 1 h at 4°C in a humidified microarray incubation chamber.
7. Wash the slide 3 × 5 min with cold PBS.
8. Prepare a dilution of fluorescent secondary antibody (1:5000–1:10,000 for Alexa Fluor® 647) in hybridization buffer and incubate the slide for 30 min in the dark with gentle rotation at 4°C.
9. Wash the slide 3 × 5 min with PBS in the dark at 4°C.
10. Dip the slide 5 × in 0.1 × PBS to remove excess salt.
11. Dry the slide by centrifugation or compressed air as described earlier.
12. Scan the slide using a microarray scanner at ≤25 μm with a laser appropriate for your fluorescent secondary antibody.
13. Slides can be stored at 4°C protected from light for several days without appreciable loss of fluorescent signal.

4. ENZYME SPECIFICITY PROFILING BY MICROARRAY

Histone peptide microarrays have been a robust biochemical tool for the in vitro characterization of histone antibodies and readers in the context of complex histone PTM patterns (Fuchs et al., 2011; Rothbart et al., 2015, 2012). We therefore sought to expand the utility of this competitive assay platform for the interrogation of histone-modifying enzymes that act on peptide substrates. Here, we detail results obtained from successful screening of histone methyltransferase and demethylase activities using the methods described in Section 3.

4.1 High-Throughput Profiling of G9a KMT Activity

Since the seminal discovery of the first histone lysine methyltransferase (Rea et al., 2000), and the link between site-specific histone methylation and gene transcription (Lee, Teyssier, Strahl, & Stallcup, 2005), much research has focused on the role of histone methylation signaling through chromatin in human health and disease (Greer & Shi, 2012). Notably, over 50 known and predicted lysine methyltransferase enzymes have been identified (Petrossian & Clarke, 2011), many of which have histone substrates (Fig. 1). The lysine methyltransferase G9a (EKMT2/KMT1C) is a SET (Su(var)3-9-Enhancer of zeste-Trithorax) domain-containing protein identified as the major enzyme responsible for catalysis of mono- and dimethylation on H3K9 (Collins et al., 2005; Tachibana, Sugimoto, Fukushima, & Shinkai, 2001). G9a activity through euchromatic H3K9 methylation has primarily been associated with gene silencing

(Shankar et al., 2013). Several additional histone substrates of G9a have been identified, including H3K27 (Tachibana et al., 2001; Wu et al., 2011), H3K56 (Yu et al., 2012), and H1 (Tachibana et al., 2001; Trojer et al., 2009; Weiss et al., 2010). G9a also methylates a number of nonhistone proteins (Rathert et al., 2008; West et al., 2010).

Using assay conditions optimized on 48-well microarrays (Fig. 2), we subjected G9a to methyltransferase assays on a large array format. This platform allows for simultaneous monitoring of enzymatic activity on over 250 unique substrates in a competitive assay format, yielding over 2000 specific catalytic measurements in a single experiment. Following incubation with G9a, we first detected new methylation of H3K9 with an H3K9me2-specific antibody (Fig. 3A). Antibody reactivity in the absence of G9a was screened as a control. Consistent with previous observations (Rothbart et al., 2015), reactivity of this antibody with H3K9me2-containing peptides was both positively and negatively impacted by neighboring modifications. As discussed earlier, this behavior of histone PTM-specific antibodies is a potential disadvantage of using these reagents to detect enzymatic activity (Fig. 3B). In particular, phosphorylation of H3S10 completely inhibited this antibody from recognizing H3K9me2. Based on this observation, H3S10p peptides were excluded from subsequent analysis following detection with this antibody to circumvent false-negative conclusions.

Another caveat of detecting the products of enzyme reactions on arrays with site-specific antibodies is the potential for missing substrates (ie, residues being methylated other than H3K9 would not be detected by the antibody). We therefore developed a procedure to screen for G9a activity by monitoring radioisotope incorporation using ^{3}H-SAM as a cofactor (see method above). The majority of peptides detected by the H3K9me2 antibody after G9a incubation also incorporated tritiated methyl groups (Fig. 3A), consistent with the conclusion that the preferential histone target of G9a is H3K9 (Collins et al., 2005; Tachibana et al., 2001). Notably, a number of distinct methylation substrates in the sequence context of $H3_{(15-41)}$ were detected by radioactivity, consistent with the observation that G9a can also methylate H3K27 in vitro (Tachibana et al., 2001; Wu et al., 2011). We were unable to detect H3K56 methylation by G9a on our array platform.

Analysis of quantified signal intensities from both antibody and radioisotope detection methods revealed new insights regarding the behavior of G9a catalytic activity toward H3K9 in the presence of adjacent and distant histone PTMs (Fig. 3C). Modifications that enhanced and perturbed G9a

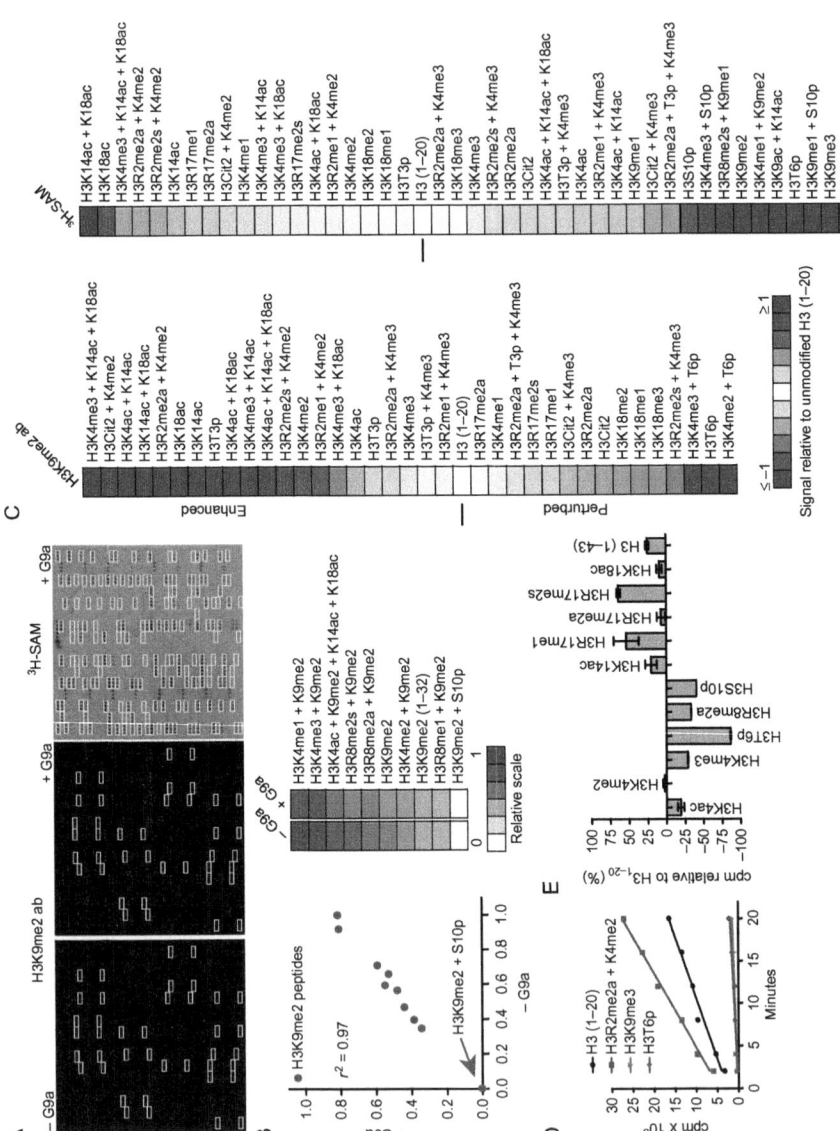

Fig. 3 See legend on opposite page.

activity toward $H3_{(1-20)}$ peptides were largely consistent between detection strategies. The unbiased nature of radioisotope incorporation allowed us to scrutinize the contribution of H3S10p to G9a activity, which, consistent with a previous observation (Rathert, Dhayalan, et al., 2008), was largely inhibited by this neighboring mark. In general, single PTMs toward the N-terminus of $H3_{(1-20)}$ inhibited G9a activity on H3K9, and PTMs toward the C-terminus of $H3_{(1-20)}$ enhanced G9a activity. In addition, several combinations of H3R2 modification (methylation and citrullination) with H3K4 methylation enhanced G9a activity. These latter results were surprising, connecting distant PTMs and their combinations to G9a activity and function. Understanding how these long-range PTMs contribute to G9a activity will be an exciting area of future study.

To validate the dynamics of combinatorial PTMs to G9a activity observed on microarrays, we performed in-solution filter-binding assays with ^{3}H-SAM (Fig. 3D). The rate of methyl incorporation using $H3_{(1-20)}$ peptide substrates was determined to be linear over the course of 20 min, and no activity was measured on H3K9me3 peptides. Consistent with single

Fig. 3 Analysis of G9a histone substrate specificity and the combinatorial PTM influence on KMT activity. G9a histone substrate specificity was profiled using the described antibody and radioisotopic detection strategies on large-format histone peptide microarrays. (A) Representative images of arrays detected with an H3K9me2 antibody (Abcam #1220) following hybridization in the absence or presence of 1 μM G9a for 2 h (*left*). *White boxes* demarcate peptides detected by H3K9me2 antibody in the absence of G9a. Image of autoradiography film exposed for 1.5 months following a 2-h array assay with 1 μM G9a and 5 μCi ^{3}H-SAM (*right*). For comparison, *yellow boxes* demarcate peptides detected with H3K9me2 antibody in the presence of G9a. (B) Scatter plot (*left*) and heat map (*right*) of H3K9me2 peptides detected by the above-mentioned H3K9me2 antibody in the absence or presence of G9a. Correlation coefficient (r^2) was calculated by linear regression analysis using GraphPad Prism v6. For heat maps, relative signal intensities are plotted using JavaTreeView (Saldanha, 2004) from 0 (*white*, no binding) to 1 (*blue*, strong binding). (C) Heat maps depicting the effects of combinatorial PTMs on the enzymatic activity of G9a from panel (A). Enhanced (1, *blue*) and occluded (-1, *red*) effects are depicted. Peptide signal intensities are presented relative to $H3_{(1-20)}$ (0, *white*) following detection with the above-mentioned H3K9me2 antibody (*left*) or autoradiography (*right*). (D) In-solution ^{3}H-SAM filter-binding assays monitoring G9a activity as a function of time on the listed histone peptide substrates. Data points are presented as counts per minute (cpm), each from three independent measurements. (E) In-solution ^{3}H-SAM filter-binding assays monitoring G9a activity following a 10-min incubation with the listed histone peptide substrates. Data points are presented as cpm relative to an $H3_{(1-20)}$ substrate. Error bars represent ±S.E.M. from three independent experiments. (See the color plate.)

time point measurements on the arrays and a previous observation (Rathert, Dhayalan, et al., 2008), H3T6p completely inhibited G9a catalysis (Fig. 3D). In addition, we validated an enhancement of G9a activity on $H3_{(1-20)}$ peptides modified with asymmetric dimethylation at arginine 2 (H3R2me2a) in combination with H3K4me2. Furthermore, single in-solution filter-binding assay measurements along this linear scale for a number of other histone PTMs was consistent with the changes observed on the arrays (Fig. 3C and E). Collectively, these results demonstrate the utility of histone peptide microarrays in capturing the dynamics of histone methylation in the context of the "histone code," and reveal combinatorial histone PTM patterns that may regulate the chromatin activity of G9a.

4.2 High-Throughput Profiling of JMJD2A KDM Activity

Methylation on lysine residues was long thought to be an irreversible modification until the discovery of the first lysine demethylase (KDM), LSD1 (lysine-specific demethylase 1), shown to remove mono- and dimethylation on histone H3K4 using its FAD-dependent amine oxidase activity (Shi et al., 2004). Since this landmark discovery, over 30 KDMs with activity specific for all of the major lysine methylation sites on histones have been identified (Fig. 1; Dimitrova, Turberfield, & Klose, 2015). This includes the discovery of an additional class of demethylase enzymes, the Jumonji demethylases, that coordinate iron and utilize α-ketoglutarate (2-OG) as a cofactor (Tsukada et al., 2006). JMJD2A (KMD4A), a Jumonji family KDM, was the first demethylase discovered to remove lysine trimethylation (Klose et al., 2006; Whetstine et al., 2006). JMJD2A is known to catalyze the removal of trimethylation at H3K9 and H3K36 and has been implicated in the regulation of gene expression, DNA damage signaling, DNA replication, and site-specific copy-number regulation (Black et al., 2013).

JMJD2A demethylase assays on histone peptide microarrays were first optimized using a 48-well array (Fig. 2). Using a single optimization slide, we simultaneously detected signal with three different antibodies to the states of H3K9 methylation following hybridization with six concentrations of JMJD2A, all in duplicate (Fig. 4A). Consistent with previous reports (Tsukada et al., 2006), we show that H3K9me3 peptides are reduced to lower order methyl states (K9me2 and K9me1) in the presence of JMJD2A. The H3K9me2 antibody used for these experiments cross reacts with H3K36me2 (Rothbart et al., 2015), and we were able to observe demethylation of H3K36me3-containing peptides at the higher JMJD2A

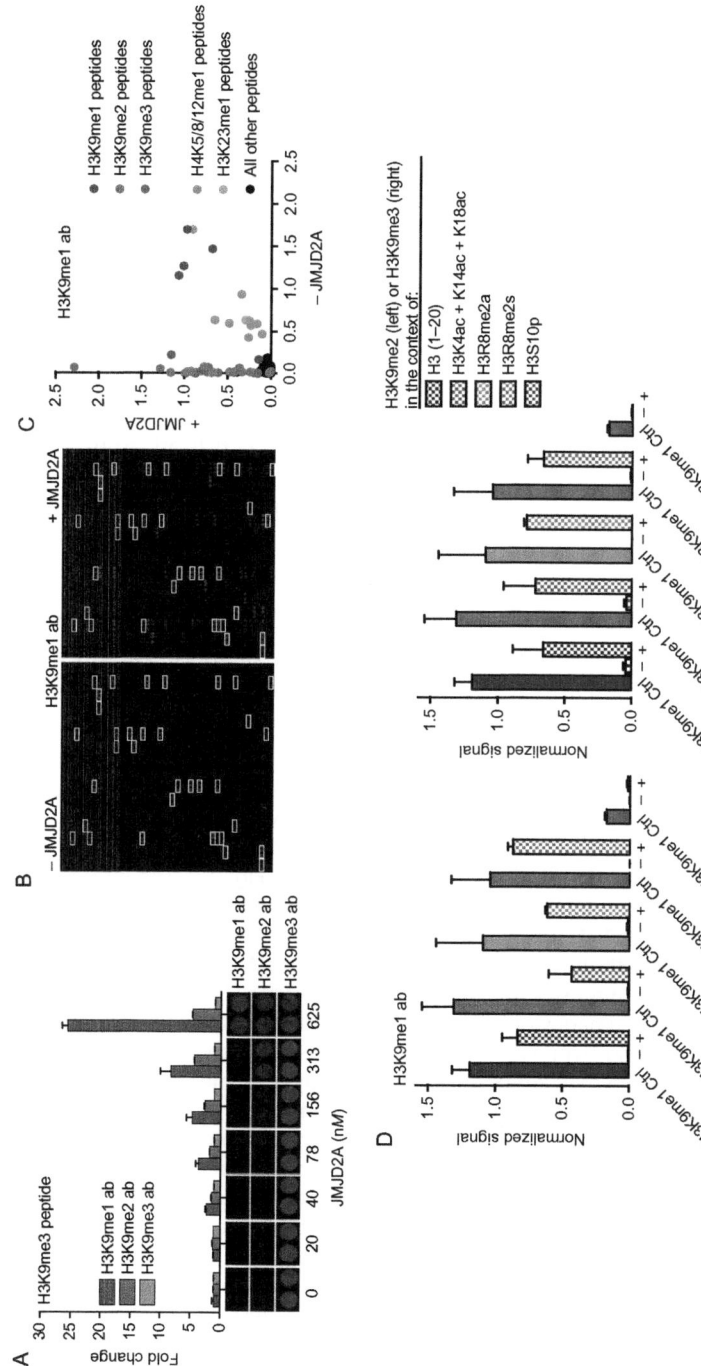

Fig. 4 See legend on next page.

concentrations used in these optimization experiments (data not shown). In general, detecting reaction products was a more robust readout for enzyme activity than monitoring the disappearance of substrate (Fig. 4A). This is perhaps due to the high sensitivity of primary antibodies and the amplification provided by secondary antibody conjugation.

Using assay conditions optimized on the 48-well microarray, we performed JMJD2A demethylase assays on a large-format array. Following incubation with JMJD2A, we monitored the appearance of H3K9me1 by antibody detection (Fig. 4B). Antibody reactivity in the absence of JMJD2A was tested as a control. Analysis of quantified signal intensities from both JMJD2A treated and untreated arrays revealed JMJD2A is able to demethylate the majority of H3K9me3- and H3K9me2-containing peptides, whereas the H3K9me1-containing peptides are not substrates (Fig. 4C). The H3K9me1 antibody used also cross-reacted with H3K23me1 and monomethylation of lysines 5, 8, and 12 on the H4 tail, and none of these PTMs were substrates of JMJD2A (Fig. 4C).

We next sought to determine how neighboring PTMs influenced the catalysis of JMJD2A toward H3K9me2 and H3K9me3. As discussed earlier, the behavior of histone PTM antibodies (positive and negative) when presented with combinatorially modified PTM epitopes is a factor that must be considered when using these reagents for enzyme assay detection. The arrays

Fig. 4 Analysis of JMJD2A histone substrate specificity and the combinatorial PTM influence on KDM activity. (A) Optimization of JMJD2A demethylase activity on H3$_{(1-20)}$K9me3 peptides in a 48-well microarray (see Fig. 2A). The indicated orders of H3K9 methylation were detected by antibody hybridization (H3K9me1, EpiCypher #13-0014; H3K9me2, Active Motif #39239; H3K9me3, Active Motif #39161) following 18 h incubation with the indicated concentrations of JMJD2A. Fold change is expressed relative to signal intensity in the absence of enzyme. Error bars represent ±S.E.M. (B) Representative images of arrays detected with an H3K9me1 antibody (EpiCypher #13-0014) following hybridization in the absence or presence of 313 n*M* JMJD2A for 18 h. *White boxes* demarcate peptides detected by H3K9me1 antibody in the absence of JMJD2A. (C) Scatter plot of all peptides on the large array detected with the above-mentioned H3K9me1 antibody following hybridization in the absence or presence of JMJD2A. Signal intensities are normalized to IgG control spots. (D) Bar graphs depicting the effects of combinatorial PTMs on the enzymatic activity of JMJD2A from panel (B). Signal intensities from H3K9me1 antibody detection are normalized to IgG control spots. Shown are signals for the indicated peptides that also contain H3K9me2 (*left*) or H3K9me3 (*right*), both in the absence (−) and presence (+) of JMJD2A. To control for H3K9me1 antibody specificity in the context of these additional PTMs, normalized signal intensities from these peptides that also contain H3K9me1 (H3K9me1 Ctrl) are plotted as a reference. (See the color plate.)

used for these experiments displayed five combinatorially modified $H3_{(1-20)}$ peptides that also harbored all three methylation states at H3K9 (Fig. 4D). Focusing on these peptides allowed us to properly control for the behavior of the H3K9me1 antibody toward the product of JMJD2A demethylation in the context of neighboring PTMs. We saw little effect of H3 polyacetylation or H3R8 dimethylation on the activity of JMJD2A toward H3K9me2 or H3K9me3 substrates. However, we show that like G9A (above), and consistent with a previous observation (Ng et al., 2007), JMJD2A activity on H3K9me2 and H3K9me3 is blocked by neighboring H3S10 phosphorylation (Fig. 4D). These results demonstrate the utility of histone peptide microarrays for profiling the substrate specificity of KDMs and for deciphering the impact of the "histone code" on JMJD2A activity.

5. SUMMARY AND PERSPECTIVES

Peptide microarrays have been a valuable tool for studying the influence of histone PTMs on antibody and reader protein recognition (Rothbart et al., 2012). The methods presented in this chapter describe the expanded utility of this tool for investigating how PTMs influence the enzymatic activity of the writers and erasers of these marks. While others have previously used array formats for analyzing substrate specificity of histone methyltransferases, libraries for these assays were designed primarily for motif mining through amino acid substitutions (Dhayalan, Kudithipudi, Rathert, & Jeltsch, 2011; Kudithipudi, Dhayalan, Kebede, & Jeltsch, 2012; Kudithipudi, Kusevic, Weirich, & Jeltsch, 2014; Kudithipudi, Lungu, Rathert, Happel, & Jeltsch, 2014; Rathert, Dhayalan, et al., 2008; Rathert, Zhang, Freund, Cheng, & Jeltsch, 2008; Smith, Settles, Hallows, Craven, & Denu, 2011). Our studies describe a robust high-throughput assay format optimized for both substrate specificity profiling and analysis of the impact that PTMs have on both KMT and KDM activity. It is now possible to investigate one of the key tenets of the original "histone code" hypothesis, that neighboring PTMs on histones impact the function of the enzymes that write and erase these marks. Indeed, we found that PTMs adjacent to the target amino acid both enhance and abrogate the KMT and KDM activity of both G9a and JMJD2A, respectively. Furthermore, we identified several distant PTMs that strongly perturb and enhance G9a catalytic function.

While peptide microarrays are a powerful technology for studying the activity of histone-modifying enzymes, there are several limitations that should be noted. First, not all histone-modifying enzymes are active on peptide substrates, limiting the enzymes for which this platform can be employed. This is an outstanding issue for all peptide-based approaches, and an important challenge for the field, as substrate profiling of these enzymes remains challenging. Another important consideration when using antibodies as the detection method is ensuring the proper peptide diversity is represented in the displayed library. For example, the peptide that corresponds to the product of catalysis in the context of neighboring PTMs should be represented to accurately control for antibody specificity.

Characterization of the impact that PTMs have on the catalytic activity of histone-modifying enzymes will be crucial in understanding the dynamic histone PTM language. It is important to note that combined, nearly 80 KMTs and KDMs have been identified. Additionally, the methods described above can serve as guidelines for adapting this microarray approach for histone-modifying enzymes that catalyze the addition or removal of modifications other than lysine methylation. Thus, microarray-based enzyme assays and similar approaches are poised to be widely used tools to aid in unraveling how the histone PTM landscape is created, maintained, and erased.

ACKNOWLEDGMENTS
This work was supported in part by grants from the National Institutes of Health to S.B.R. (CA181343) and B.D.S. (GM110058). B.D.S. also acknowledges support from the W.M. Keck Foundation and is a cofounder of EpiCypher.

REFERENCES
Allis, C. D., Berger, S. L., Cote, J., Dent, S., Jenuwien, T., Kouzarides, T., ... Zhang, Y. (2007). New nomenclature for chromatin-modifying enzymes. *Cell, 131*(4), 633–636.
Black, J. C., Manning, A. L., Van Rechem, C., Kim, J., Ladd, B., Cho, J., ... Whetstine, J. R. (2013). KDM4A lysine demethylase induces site-specific copy gain and rereplication of regions amplified in tumors. *Cell, 154*(3), 541–555.
Collins, R. E., Tachibana, M., Tamaru, H., Smith, K. M., Jia, D., Zhang, X., ... Cheng, X. (2005). In vitro and in vivo analyses of a Phe/Tyr switch controlling product specificity of histone lysine methyltransferases. *The Journal of Biological Chemistry, 280*(7), 5563–5570.
Detrich, H. W., 3rd. (1986). Isolation of sea urchin egg tubulin. *Methods in Enzymology, 134*, 128–138.
Dhayalan, A., Kudithipudi, S., Rathert, P., & Jeltsch, A. (2011). Specificity analysis-based identification of new methylation targets of the SET7/9 protein lysine methyltransferase. *Chemistry & Biology, 18*(1), 111–120.
Dimitrova, E., Turberfield, A. H., & Klose, R. J. (2015). Histone demethylases in chromatin biology and beyond. *EMBO Reports, 16*(12), 1620–1639.
Fuchs, S. M., Krajewski, K., Baker, R. W., Miller, V. L., & Strahl, B. D. (2011). Influence of combinatorial histone modifications on antibody and effector protein recognition. *Current Biology, 21*(1), 53–58.

Greer, E. L., & Shi, Y. (2012). Histone methylation: A dynamic mark in health, disease and inheritance. *Nature Reviews. Genetics*, *13*(5), 343–357.
Huang, H., Sabari, B. R., Garcia, B. A., Allis, C. D., & Zhao, Y. (2014). SnapShot: Histone modifications. *Cell*, *159*(2). 458–458.e451.
Khorasanizadeh, S. (2004). The nucleosome: From genomic organization to genomic regulation. *Cell*, *116*(2), 259–272.
Klose, R. J., Yamane, K., Bae, Y., Zhang, D., Erdjument-Bromage, H., Tempst, P., ... Zhang, Y. (2006). The transcriptional repressor JHDM3A demethylates trimethyl histone H3 lysine 9 and lysine 36. *Nature*, *442*(7100), 312–316.
Kornberg, R. D., & Lorch, Y. (1999). Twenty-five years of the nucleosome, fundamental particle of the eukaryote chromosome. *Cell*, *98*(3), 285–294.
Kouzarides, T. (2007). Chromatin modifications and their function. *Cell*, *128*(4), 693–705.
Krishnan, S., Collazo, E., Ortiz-Tello, P. A., & Trievel, R. C. (2012). Purification and assay protocols for obtaining highly active Jumonji C demethylases. *Analytical Biochemistry*, *420*(1), 48–53.
Kudithipudi, S., Dhayalan, A., Kebede, A. F., & Jeltsch, A. (2012). The SET8 H4K20 protein lysine methyltransferase has a long recognition sequence covering seven amino acid residues. *Biochimie*, *94*(11), 2212–2218.
Kudithipudi, S., Kusevic, D., Weirich, S., & Jeltsch, A. (2014). Specificity analysis of protein lysine methyltransferases using SPOT peptide arrays. *Journal of Visualized Experiments*, *93*, e52203.
Kudithipudi, S., Lungu, C., Rathert, P., Happel, N., & Jeltsch, A. (2014). Substrate specificity analysis and novel substrates of the protein lysine methyltransferase NSD1. *Chemistry & Biology*, *21*(2), 226–237.
Lee, D. Y., Teyssier, C., Strahl, B. D., & Stallcup, M. R. (2005). Role of protein methylation in regulation of transcription. *Endocrine Reviews*, *26*(2), 147–170.
Luger, K., Mader, A. W., Richmond, R. K., Sargent, D. F., & Richmond, T. J. (1997). Crystal structure of the nucleosome core particle at 2.8 A resolution. *Nature*, *389*(6648), 251–260.
Maze, I., Noh, K. M., & Allis, C. D. (2013). Histone regulation in the CNS: Basic principles of epigenetic plasticity. *Neuropsychopharmacology*, *38*(1), 3–22.
Ng, S. S., Kavanagh, K. L., McDonough, M. A., Butler, D., Pilka, E. S., Lienard, B. M., ... Oppermann, U. (2007). Crystal structures of histone demethylase JMJD2A reveal basis for substrate specificity. *Nature*, *448*(7149), 87–91.
Partyka, K., Wang, S., Zhao, P., Cao, B., & Haab, B. (2014). Array-based immunoassays with rolling-circle amplification detection. *Methods in Molecular Biology*, *1105*, 3–15.
Petrossian, T. C., & Clarke, S. G. (2011). Uncovering the human methyltransferasome. *Molecular & Cellular Proteomics*, *10*(1). M110.000976.
Portela, A., & Esteller, M. (2010). Epigenetic modifications and human disease. *Nature Biotechnology*, *28*(10), 1057–1068.
Rathert, P., Dhayalan, A., Murakami, M., Zhang, X., Tamas, R., Jurkowska, R., ... Jeltsch, A. (2008). Protein lysine methyltransferase G9a acts on non-histone targets. *Nature Chemical Biology*, *4*(6), 344–346.
Rathert, P., Zhang, X., Freund, C., Cheng, X., & Jeltsch, A. (2008). Analysis of the substrate specificity of the Dim-5 histone lysine methyltransferase using peptide arrays. *Chemistry & Biology*, *15*(1), 5–11.
Rea, S., Eisenhaber, F., O'Carroll, D., Strahl, B. D., Sun, Z. W., Schmid, M., ... Jenuwein, T. (2000). Regulation of chromatin structure by site-specific histone H3 methyltransferases. *Nature*, *406*(6796), 593–599.
Rothbart, S. B., Dickson, B. M., Raab, J. R., Grzybowski, A. T., Krajewski, K., Guo, A. H., ... Strahl, B. D. (2015). An interactive database for the assessment of histone antibody specificity. *Molecular Cell*, *59*(3), 502–511.

Rothbart, S. B., Krajewski, K., Strahl, B. D., & Fuchs, S. M. (2012). Peptide microarrays to interrogate the "histone code" *Methods in Enzymology*, *512*, 107–135.

Rothbart, S. B., & Strahl, B. D. (2014). Interpreting the language of histone and DNA modifications. *Biochimica et Biophysica Acta*, *1839*(8), 627–643.

Saldanha, A. J. (2004). Java Treeview—Extensible visualization of microarray data. *Bioinformatics*, *20*(17), 3246–3248.

Shankar, S. R., Bahirvani, A. G., Rao, V. K., Bharathy, N., Ow, J. R., & Taneja, R. (2013). G9a, a multipotent regulator of gene expression. *Epigenetics*, *8*(1), 16–22.

Shi, Y., Lan, F., Matson, C., Mulligan, P., Whetstine, J. R., Cole, P. A., ... Shi, Y. (2004). Histone demethylation mediated by the nuclear amine oxidase homolog LSD1. *Cell*, *119*(7), 941–953.

Smith, B. C., Settles, B., Hallows, W. C., Craven, M. W., & Denu, J. M. (2011). SIRT3 substrate specificity determined by peptide arrays and machine learning. *ACS Chemical Biology*, *6*(2), 146–157.

Strahl, B. D., & Allis, C. D. (2000). The language of covalent histone modifications. *Nature*, *403*(6765), 41–45.

Tachibana, M., Sugimoto, K., Fukushima, T., & Shinkai, Y. (2001). Set domain-containing protein, G9a, is a novel lysine-preferring mammalian histone methyltransferase with hyperactivity and specific selectivity to lysines 9 and 27 of histone H3. *The Journal of Biological Chemistry*, *276*(27), 25309–25317.

Trojer, P., Zhang, J., Yonezawa, M., Schmidt, A., Zheng, H., Jenuwein, T., & Reinberg, D. (2009). Dynamic histone H1 isotype 4 methylation and demethylation by histone lysine methyltransferase G9a/KMT1C and the Jumonji domain-containing JMJD2/KDM4 proteins. *The Journal of Biological Chemistry*, *284*(13), 8395–8405.

Tsukada, Y., Fang, J., Erdjument-Bromage, H., Warren, M. E., Borchers, C. H., Tempst, P., & Zhang, Y. (2006). Histone demethylation by a family of JmjC domain-containing proteins. *Nature*, *439*(7078), 811–816.

Weiss, T., Hergeth, S., Zeissler, U., Izzo, A., Tropberger, P., Zee, B. M., ... Schneider, R. (2010). Histone H1 variant-specific lysine methylation by G9a/KMT1C and Glp1/KMT1D. *Epigenetics & Chromatin*, *3*(1), 7.

West, L. E., Roy, S., Lachmi-Weiner, K., Hayashi, R., Shi, X., Appella, E., ... Gozani, O. (2010). The MBT repeats of L3MBTL1 link SET8-mediated p53 methylation at lysine 382 to target gene repression. *The Journal of Biological Chemistry*, *285*(48), 37725–37732.

Whetstine, J. R., Nottke, A., Lan, F., Huarte, M., Smolikov, S., Chen, Z., ... Shi, Y. (2006). Reversal of histone lysine trimethylation by the JMJD2 family of histone demethylases. *Cell*, *125*(3), 467–481.

Wu, H., Chen, X., Xiong, J., Li, Y., Li, H., Ding, X., ... Zhu, B. (2011). Histone methyltransferase G9a contributes to H3K27 methylation in vivo. *Cell Research*, *21*(2), 365–367.

Yu, Y., Song, C., Zhang, Q., DiMaggio, P. A., Garcia, B. A., York, A., ... Grunstein, M. (2012). Histone H3 lysine 56 methylation regulates DNA replication through its interaction with PCNA. *Molecular Cell*, *46*(1), 7–17.

Zhao, Y., & Garcia, B. A. (2015). Comprehensive catalog of currently documented histone modifications. *Cold Spring Harbor Perspectives in Biology*, *7*(9), a025064.

CHAPTER THREE

ArrayNinja: An Open Source Platform for Unified Planning and Analysis of Microarray Experiments

B.M. Dickson, E.M. Cornett, Z. Ramjan, S.B. Rothbart[1]

Center for Epigenetics, Van Andel Research Institute, Grand Rapids, MI, United States
[1]Corresponding author: e-mail address: scott.rothbart@vai.org

Contents

1. Introduction — 54
2. ArrayNinja — 55
   - 2.1 Obtaining and Running ArrayNinja — 57
   - 2.2 Overview of the ArrayNinja Database — 58
3. Planning Custom Microarrays with ArrayNinja — 61
4. Data Analysis Features of ArrayNinja — 64
   - 4.1 Microarray Image Preparation — 65
   - 4.2 Quantification of Microarray Data — 66
5. Benchmarking ArrayNinja Against ImageQuant TL — 69
   - 5.1 Homogeneous Noise — 69
   - 5.2 Inhomogeneous Noise — 70
   - 5.3 Implicit Assumption of Local Noise Correction — 71
6. Methodological Details — 72
   - 6.1 Default (Whole Spot) Quantification — 72
   - 6.2 Nonlocal Noise Thresholding — 72
   - 6.3 Local Noise Thresholding — 72
   - 6.4 Variegated Spot Morphology — 73
7. Limitations, Assumptions, Other Features, and Future Development — 74
   - 7.1 Limitations and Assumptions — 74
   - 7.2 Other Features — 75
   - 7.3 Future Development — 75
8. Summary — 76

Acknowledgments — 76

References — 76

Abstract

Microarray-based proteomic platforms have emerged as valuable tools for studying various aspects of protein function, particularly in the field of chromatin biochemistry. Microarray technology itself is largely unrestricted in regard to printable material and platform design, and efficient multidimensional optimization of assay parameters requires fluidity in the design and analysis of custom print layouts. This motivates the need for streamlined software infrastructure that facilitates the combined planning and analysis of custom microarray experiments. To this end, we have developed ArrayNinja as a portable, open source, and interactive application that unifies the planning and visualization of microarray experiments and provides maximum flexibility to end users. Array experiments can be planned, stored to a private database, and merged with the imaged results for a level of data interaction and centralization that is not currently attainable with available microarray informatics tools.

1. INTRODUCTION

Microarray technology is a pillar platform of functional proteomics, enabling massively parallel interrogation of macromolecular interactions in highly sensitive, reproducible, and miniaturized assay formats. Indeed, a variety of protein and peptide microarray platforms have been developed in the last decade for high-throughput analysis of protein–protein (Moore, Ajala, & Zhu, 2015), protein–peptide (Bock, Kudithipudi, et al., 2011; Bua et al., 2009; Fuchs, Krajewski, Baker, Miller, & Strahl, 2011; Garske et al., 2010; Kim et al., 2006; Nady, Min, Kareta, Chedin, & Arrowsmith, 2008; Rothbart, Krajewski, Strahl, & Fuchs, 2012), and protein–DNA (Hall et al., 2004; Hu et al., 2009) interactions. Additionally, peptide and protein microarrays have been useful for characterizing the selectivity of antibodies (Bock, Dhayalan, et al., 2011; Michaud et al., 2003; Rothbart et al., 2015; Rothbart, Lin, et al., 2012), and for profiling the substrate specificity of histone-modifying enzymes (Cornett et al., 2016; Dhayalan, Kudithipudi, Rathert, & Jeltsch, 2011; Kudithipudi, Kusevic, Weirich, & Jeltsch, 2014; Smith, Settles, Hallows, Craven, & Denu, 2011). The designs of these microarray platforms are as diverse as their applications, and taking advantage of the inherent flexibility of microarray technology requires an integrated workflow that couples the planning and analysis of microarray experiments.

Microarrays are fabricated by precision robotic printing of biomaterial (eg, synthetic peptides) onto a solid surface (eg, functionalized microscope glass slides). Currently, the design and construction of custom arrays requires a rigorous understanding of the printing mechanics (ie, how the arrayer maps positions in a source plate onto the array surface). For example, if the experimentalist wants to print a specific feature in a defined coordinate on the slide, then they need to know exactly where to load that feature on the source plate so that the action of the printer results in the correct spot positioning. Additionally, at the quantification and analysis stage, the experimentalist has to repeat this process to integrate the feature identities with the postexperiment imaged slide. This latter process can be facilitated by a GAL (GenePix Associated List) file, a tab-delimited list file generated by most microarray printers that contains spot coordinates and nondescript identifiers. An issue with GAL files is that they have no unified syntax. Frequently, a GAL file will fail to load in specific densitometry software, leaving the experimentalist to implement their own workaround. Such workarounds do exist, but they are arduous and present a bottleneck to the experimentalist (Rothbart, Krajewski, et al., 2012). Moreover, for every new slide layout, complete reengineering of the analysis workflow is required. Ironically, this aspect of microarray analysis is arguably the single most time-consuming step, yet is rarely discussed with detail in publications. In addition, the critical step of converting spot identifiers into human interpretable tags is not typically addressed.

To circumvent these issues, we have developed a portable, open source, and interactive application, ArrayNinja, that unifies the planning and visualization of microarray experiments and provides maximum flexibility to the experimentalist. Specifically, this platform is aimed at eroding the barriers between array layout creation, practical implementation, and the ensuing quantification and analysis of microarray data. Herein, we detail the installation, structure, user interface, and methodologies of ArrayNinja.

2. ArrayNinja

The objective of ArrayNinja is to connect the planning and analysis stages of microarray experiments in a seamless and version-less application, thus providing maximal fluidity to the cycle of experimentation. ArrayNinja achieves this by programmatically recreating the action of the microarrayer in order to automate the mapping of biomaterial from microarray slide to source plate, or vice versa. This provides a feature that is unique to ArrayNinja in that the experimentalist can plan a custom print layout and populate the corresponding source plate by interacting with a modeled representation of the printed slide. Entering substrate identifiers for the spots on the modeled slide automatically populates a table that represents the physical layout of the source plate required for the print run.

Source plates are collected in a private database running inside ArrayNinja, which allows rapid modification of slide layouts and makes the process of merging substrate identifiers (and human interpretable tags) with imaged data completely automatic. Once a slide is scanned, the image can be loaded and paired with a source plate and slide layout to automatically populate the image with feature identifiers. The image can be viewed simultaneously with the identifiers for the highest level of data interaction and control.

The ArrayNinja platform is free to use and can be edited to create a specialized application for even the most exotic printer. The current version of ArrayNinja supports Aushon 2470 and GeneMachines OmniGrid 100 arrayers, both with a 4×4 pin configuration. Once downloaded, the computer running ArrayNinja does not need to connect to the Internet. The database is a local one, living inside ArrayNinja. This means that the experimentalist does not need to sit at a particular computer where the printer or scanner software is installed to plan experiments or to quantify data. Additionally, ArrayNinja is not tied to a particular computer operating system nor is it tied to a particular brand. Lastly, ArrayNinja does not use GAL files to map identities of coordinates. The action of the printer is recreated within ArrayNinja so that there are no versioning issues with regard to reading GAL files from different printers of different eras. This seamless, version-less, license-free unification of planning and quantification is what makes ArrayNinja unique.

There are three variables that determine the print layout of a microarray: (1) the arrayer architecture, (2) the arrayer settings, and (3) the composition of the source plate(s) that will be used during printing. Given knowledge of points (2) and (3), a microarray layout can be precisely predicted by simulating the arrayer architecture. ArrayNinja simulates the motion of the arrayer and uses a database to associate, store, and retrieve arrayer settings and source plates. ArrayNinja can be modified to simulate any arrayer and is completely independent of the printed substrate and slide surface chemistry. As such, ArrayNinja is free of any company- or era-specific versioning issues. This is also what allows ArrayNinja to reverse-engineer a source plate from a microarray layout—which is entirely unique to ArrayNinja.

The ArrayNinja platform requires a running structured query language (SQL) database and web server, so distribution via a virtual machine (VM) has been selected to reduce complexity for the end user. Additionally, this allows the user to maintain a local and private database of experiment designs. We will discuss the use of the VM distribution later, but the tools can also be obtained as a Docker application via https://hub.docker.com/ by pulling bradleydickson/arrayninja. To facilitate the practical use of ArrayNinja, we have included a working example within the VM and Docker. The details of this example are described in Section 4.

It is worth noting that since ArrayNinja is a web server, installing the application on one accessible laboratory computer allows all privileged users access and use of ArrayNinja simultaneously without needing to install locally. While we discuss the VM here, a laboratory-wide version of ArrayNinja can be easily set up by running the Docker container (bradleydickson/arrayninja).

2.1 Obtaining and Running ArrayNinja

As a prerequisite, Oracle VM VirtualBox software should be correctly installed on the machine where ArrayNinja will run. VirtualBox is free and can be obtained from www.virtualbox.org. Once VirtualBox is installed, the ArrayNinja VM can be downloaded from http://research.vai.org/Tools/arrayninja.

Starting ArrayNinja

1. After downloading and uncompressing the VM image, start VirtualBox.
2. The ArrayNinja machine can be added to VirtualBox by clicking Machine → Add and selecting arrayninja.vbox from the uncompressed arrayninja folder.
3. To start ArrayNinja, select arrayninja from the list of machines in the VirtualBox interface and click the green "start" arrow. The VM will output a message stating that the ArrayNinja tools can be accessed by navigating your web browser to localhost:2080. We suggest Google Chrome or Firefox. To stop the VM after using ArrayNinja, use Machine → Close → Power Off.
4. Open your web browser and enter localhost:2080 in the URL bar. If you are using a remote computer, you would enter the IP address of that remote computer followed by :2080 to specify the correct port.

The ArrayNinja index page will load after navigating to localhost:2080, presenting a few options for the toolset to deploy. This index page also runs scripts to make certain the SQL database is initialized. Users can plan array experiments or quantify an experiment with ArrayNinja. Later we discuss each of these use cases after first discussing some details of the underlying SQL database.

2.2 Overview of the ArrayNinja Database

The ArrayNinja database depends on three types of tables: "substrate tables," "source plate tables," and "printer settings tables." We will first discuss how to browse through these tables and then introduce each table's structure. We chose phpMyAdmin as an interface for maintenance and browsing of tables. phpMyAdmin has an intuitive browser interface.

Table Management Interface

1. Enter localhost:2080/phpMyAdmin in your browser's URL bar. All of the ArrayNinja data are stored in the "arrayplates" database, listed in the left frame of the phpMyAdmin window (Fig. 1A).
2. Click on "arrayplates" to see the tables that belong to the database. Several tables are preloaded to provide working examples.

We will next outline each of the three distinct table structures and then look at how to load/create custom tables.

Fig. 1 Screenshots of the ArrayNinja database manager. (A) The phpMyAdmin interface and tables preloaded in the "arrayplates" database. (B) The preloaded "SubstrateTable."

SubstrateTable: The SubstrateTable (Fig. 1B) is the hub of information and must be maintained directly by users. This table contains all of the features that might be printed and links each of them to a unique numerical identifier. Additions, removals, ID changes, etc., all need to be reflected in this table.

The SubstrateTable is a running list of all features that can be printed on the microarray. To view the SubstrateTable, click on SubstrateTable in the list of tables displayed by phpMyAdmin (Fig. 1B). Column by column, the format is as follows:

ID = a unique numerical identifier for the substrate
Context = feature context (eg, H3 1-20)
Sequence = the amino acid/base/or SMILES sequence for the feature
TEXTid = simple text identifier (eg, H3)
Annotation = a human interpretable shorthand for the feature (eg, H3K9me2)

The preloaded SubstrateTable is required for the packaged data analysis example. For custom use, edit this table to incorporate unique features. Users can build upon this table or replace its contents but cannot change the table name.

We generally manage this table by keeping an up-to-date spreadsheet of all printable features. Such a table can be exported from any spreadsheet software as a comma separated values (c.s.v.) table and imported directly to the database via phpMyAdmin, as discussed later.

Source plate table: Each array layout requires a unique "source plate table," named to reflect an experiment or layout, that defines the 384-well plate(s) used during printing. ArrayNinja is configured to use up to five 384-well source plates for a single print. Each 384-well plate has 24 columns (A through X) and 16 rows. The source plate table reflects this with 24 columns and up to 80 rows. Each entry in this table is given by a feature ID and location in the table specifies location in the 384-well plate. In other words, the table position is taken literally for the source plate position. For new experiments, we populate source plate tables by interacting with the planning tool as described later. For experiments with existing source plates, c.s.v. tables for the source plates can be imported directly via phpMyAdmin as outlined later. A working example of a source plate table is loaded in the "abEX" table, where this name indicates use of the antibody example source plates. This table holds the source plate information that was used when printing the packaged example with an Aushon 2470 microarrayer. The table represents two 384-well plates, where the loaded feature identifier is recorded for each plate well. We discuss creation of this table later. These types of tables can either be loaded through the phpMyAdmin interface or created with the ArrayNinja planning tool.

Printer settings table: Each array layout requires a unique printer settings table, where the table name is defined by the source plate table name, as both are required to specify a layout. This table has 1 row and 11 columns that specify scanning resolution, spot diameter, spot spacing (edge-to-edge), number of "blocks" per source plate row (number of times a 4 × 4 pin cluster can fit on a plate row), number of plate rows printed/to be printed, replicates (counting from 0), top margin on the microscope slide, side margin on the microscope slide (right side for Aushon and left for OmniGrid), feature limit (long direction for Aushon, short direction for OmniGrid), number of subarrays on the slide, and the spacing between subarrays. The "abEXparameters" table holds the printer settings that were used when printing the packaged example with an Aushon 2470 microarrayer. These tables are created exclusively and automatically by the ArrayNinja planning tool, as discussed later.

Table import: Here, we detail a general procedure for importing a table, and we build the example around importing a new substrate table. This procedure works for any table and could be used to import an existing source plate. It is useful to create table copies, which preserve the table structure for importing new data, or serve as risk-free templates when practicing table manipulation.

Create an Empty Table Template
1. Click "arrayplates" in the list of databases (Fig. 1, left frame)
2. Click the checkbox next to SubstrateTable
3. Select "copy table with prefix" from the "with selected" dropdown box floating under the list of tables. This dropdown box is visible in Fig. 1A. A new page will load.
4. Enter SubstrateTable in the "From" field, and enter TestTable in the "To" field. An empty table has been created, and the table format is identical to the SubstrateTable.

In the following, we use the name "TestTable" as an example and go through the steps for importing new data.

Import Table Values: Example with TestTable
1. Select the arrayplates database.
2. Click "Empty" in the TestTable row, which is in the middle of the phpMyAdmin display. This action removes all entries in the TestTable.
3. Click TestTable to view the now empty table.

4. Click the "Import" tab near the top of the page.
5. Use the "choose file" button to find your substrate table in c.s.v. format, and click "Go." The c.s.v. format can be exported from any spreadsheet software. The required format for a SubstrateTable was specified earlier.

The new data from the imported c.s.v. file will populate the table. This procedure can be used to maintain any of the data used by ArrayNinja. This next example will look at importing a previously used source plate.

Import an Established Source Plate
1. Copy the table named "empty" into a table with your experiment name.
2. Import your source plate c.s.v. file into the new table.
3. Use the planning tool in the next section to first load your source plate and to set up your matching arrayer settings.
4. Save your work using the name specified in step 1 of this section.

The MySQL language and phpMyAdmin also offer a number of alternatives to maintain the tables. phpMyAdmin is a very robust platform with its own documentation (www.phpmyadmin.net). Tables may be directly imported from c.s.v. files and subsequently moved or renamed by interacting with the graphical phpMyAdmin interface.

3. PLANNING CUSTOM MICROARRAYS WITH ArrayNinja

The ability to easily customize microarray print formats enables experimentalists to fully capture the power of this technology. The ArrayNinja planning tool provides a simple interface for array design and source plate population to streamline custom microarray builds, and seamlessly connects the microarray fabrication step to data analysis.

Starting the Planning Tool
1. Access the planning tool by entering localhost:2080 in the browser's URL and clicking the appropriate planning option (arrayer specific). For the packaged example, select Aushon.
2. The tool first asks for source plate data, which is a table to be retrieved from the SQL database that defines the source plate for the printing job. Type the name of the source plate you want to edit in the "Load plate" text box, or if a new source plate is to be designed type "empty" and press enter to load the selected source plate. You must press the enter key before proceeding.

Once a source plate table is loaded, the array slide is drawn according to the printer settings that are associated with the source plate, and a series of printed spots appears (Fig. 2). The spots are uniquely color-coded to indicate

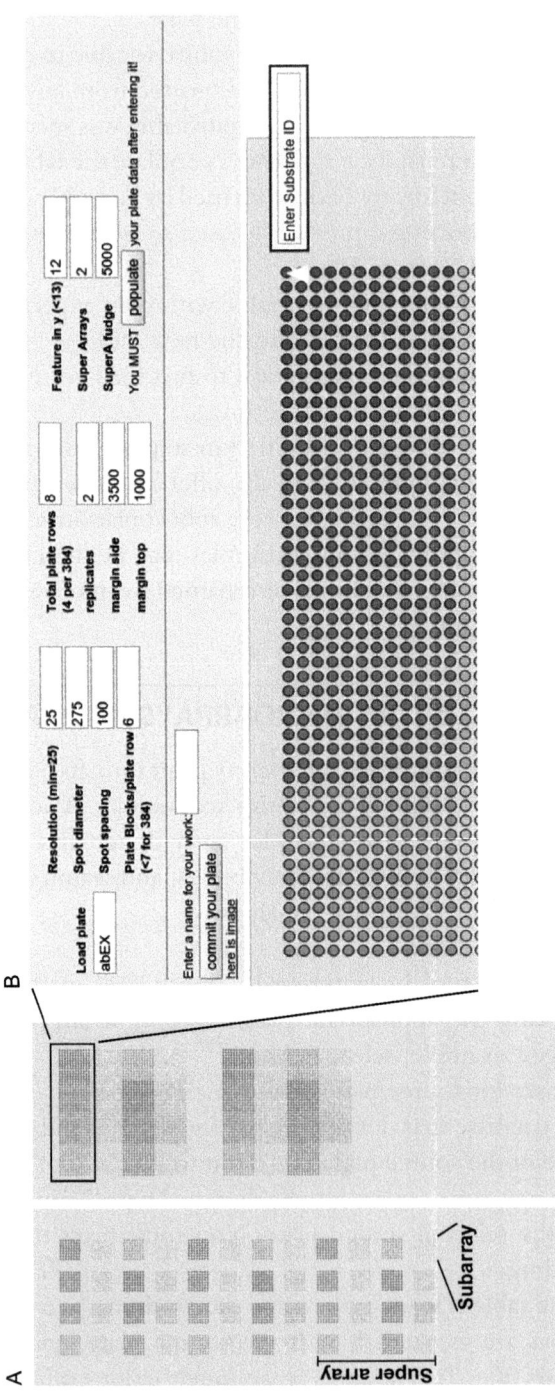

Fig. 2 Screenshots of the ArrayNinja planning tool. (A) A microarray layout consisting of 48 subarrays and three super arrays (*left*), and a microarray layout of two super arrays (*right*). (B) The ArrayNinja planning interface for designing array layouts, populating source plate tables, and naming experiments. (See the color plate.)

the pin responsible for printing that spot. A 4 × 4 pin configuration is hard-coded. Note that all distances are in micrometers.

Planning Tool Input Fields Example
1. Resolution: Sets the resolution at which the experiment will be scanned. Enter 25.
2. Spot diameter: This is specific to the pins loaded in the printhead. Enter 275.
3. Spot spacing: Distance between spots, edge-to-edge. Enter 100.
4. Plate blocks per plate row: A 4 × 4 pin configuration breaks a 384-well plate into a 6 × 4 set of unique "blocks." This parameter sets how many of the six possible columns the printer will extract. Enter 4.
5. Total plate rows: A 4 × 4 pin configuration breaks a 384-well plate into a set of 6 × 4 blocks, each plate having four "rows." This parameter sets the total number of rows to be extracted, thus controlling the required number of 384-well plates. Enter 8 (which means two 384-well plates).
6. Replicates: Number of spot replicates counting from zero. Counting from zero has been adopted from the Aushon print control software. ArrayNinja currently dictates that replicates will be placed horizontally across the slide. Ensure this is consistent with the actual printing. Enter 1.
7. Margin side: For Aushon, this sets the distance from the right edge of the slide. For OmniGrid, this sets the distance from the left of the slide. Enter 4400.
8. Margin top: Sets the distance from the top of the slide. Enter 3900.
9. Feature in y: For Aushon, this sets the number of unique spots that can be deposited down the slide in a column before starting a new column. For OmniGrid, this sets the number of unique spots in a row (excluding replicate counting). Enter 8.
10. Super arrays: Sets the number of repeated arrays. Fig. 2A (left) has 3 super arrays and 48 subarrays, while Fig. 2A (right) has 2 super arrays and 2 subarrays. This language, and counting from 1, has been adopted from the Aushon printer control software. Enter 3
11. SuperA fudge: Distance between the repeat arrays. Enter 0.

Having entered the above parameters, ArrayNinja produces the layout that we previously used to create 48-well microarrays to optimize histone methyltransferase and demethylase assays (Cornett et al., 2016). Many other possible print formats are permitted, limited only by the printer architecture and printhead format.

Population of a Source Plate

Once the desired layout is achieved, the underlying source plate must be populated. Ordinarily, this would require considerable mental gymnastics to deduce the mapping from slide to source plate for each spot. In ArrayNinja, substrate identities may be specified directly for each spot by interacting with the slide.

Finalize Slide Layout and Source Plate

1. Mouseover a spot. A text input field appears (Fig. 2B).
2. Without moving the mouse position, press backspace/delete to erase the existing entry.
3. Type the identifier for the substrate that should be printed on that spot.
4. Repeat for all unique spots that need to be defined.
5. Press the "Populate" button to prepare data structures for SQL.
6. Name the new experiment by typing in the text field, where prompted.
7. Press the "commit your plate" button.

The new source plate will be printed to the screen so that it may be quality-controlled or copied and pasted. We find it is helpful to print this source plate layout and use it at the bench to construct the actual source plate. Simultaneously, the source plate and print parameters have been logged in the SQL database. By navigating to phpMyAdmin, you can check that the new source plate has been created. Any unwanted tables may be deleted via phpMyAdmin, so there is no harm in practicing table operations. If a source plate was loaded and the planning tool is being used only to fix print settings, simply avoid changing spot identifiers in the preceding steps. The source plate and the new printer settings will be checked in to the SQL database, with the specified name, after pressing "Populate" and "Commit your plate."

Notice that there is a link stating "here is image" (Fig. 2B). Clicking this link will produce a slide mock-up image. Save this as an svg for making images or a reference template. To do this, use the browser's File → Save as dialog and enter a file name ending with the svg extension. The resulting file can be viewed in a web browser or image editing packages like Inkscape, Adobe Illustrator, etc.

4. DATA ANALYSIS FEATURES OF ArrayNinja

To facilitate the practical use of ArrayNinja, we packaged a working example in the VM (and Docker). The example experimental data (an image) can be viewed by typing localhost:2080/abEX.png in the browser's URL. This slide contains three "Super Arrays" of 2304 spots each, and each spot contains histone peptides. Each of the three blocks was incubated with a different histone antibody. The upper super array was hybridized with an H3K9me1 antibody, the middle block with an H3K9me2 antibody, and the lower block with an H3K9me3 antibody. We will refer to this example throughout the following overview of the ArrayNinja analysis suite, where we will quantify the top panel of the slide.

4.1 Microarray Image Preparation

The first step of analysis is to prepare an image for quantification. We suggest using ImageMagick (http://www.imagemagick.org/) to prepare images for quantification with ArrayNinja. This approach, detailed later, allows ArrayNinja to be used for a wide range of image sources without rewriting any analysis aspects of the program.

ImageMagick is a free, open source utility that provides precise control over image preparation. In Section 5, we quantify images from a Typhoon Trio scanner, which outputs grayscale images that present the brightest spots as the darkest pixels. This image must be inverted before it can be quantified with ArrayNinja. In addition, scanners that output separate image files for each laser channel scanned must be merged. Below is a suggested workflow to invert and merge a 2-color scan with ImageMagick.

1. Install ImageMagick
2. Invert images (if needed) with the command: convert INPUT.tif -negate OUTPUT.tif

Use the following series of commands to merge two scans.

1. convert -depth 16 SIGNAL_CHANNEL.TIF -clone 0 -channel GB -evaluate set 0 -delete 0 out.png 2 > error.file
2. convert -depth 16 CONTROL_CHANNEL.TIF -clone 0 -channel RB -evaluate set 0 -delete 0 outa.png 2 > error.file
3. convert -depth 16 CONTROL_CHANNEL.TIFF -clone 0 -channel RG -evaluate set 0 -delete 0 outB.png 2 > error.file
4. convert out.png outa.png outB.png -set colorspace RGB -combine merged.png

These commands generate an image "merged.png" that is now compatible with ArrayNinja. This workflow is the most versatile, allowing for quantification of microarray images acquired from any scanner that outputs a tif file. Even images acquired with X-ray film can be treated with ArrayNinja in this way. Simply invert the image and use the same image as SIGNAL_CHANNEL and CONTROL_CHANNEL in the above commands to generate the "merged.png."

Commands (1), (2), and (3) above generate a red, green, and (arbitrary) blue channel, respectively, and command (4) merges the channels into an image named merged.png. This script assumes a 16-bit input and forces a 16-bit output by using "-depth 16". Imagemagick will preserve the so-called quantum depth of an image by default, so these flags are not necessary. The contrast and brightness of the image can be adjusted within ArrayNinja. Contrast and brightness adjustments are for viewing only. The raw image data are quantified directly.

We do not generally find it easy to motivate use of 16-bit images over 8-bit images. Essentially, the experimental data would itself need to respect the change in precision from 8- to 16-bit for the change to make any impact on interpretation of results. Often in microarray experiments, uncertainty in mean spot intensity is far greater than the resolution of intensity in 16-bit data.

4.2 Quantification of Microarray Data

A unique feature of ArrayNinja is the ability to retrieve all the relevant information from the microarray planning stage (see Section 3) via the underlying SQL database, simplifying the process of array quantification. Below, we work step-by-step through a basic quantification example and discuss analysis details and some additional features of the analysis interface in Section 6. A working example is packaged with ArrayNinja, and the following instructions will reference that example. The example image must be saved from http://localhost:2080/abEX.png prior to working through the analysis example below.

Step-By-Step Quantification

1. Navigate the web browser to localhost:2080 and choose to quantify data. For the packaged example, choose the Aushon option.
2. Load plate: Enter the name that was used to store the source plate in the planning phase (Section 3) and press enter. The SQL data have been fetched and loaded. To work-up the packaged example, type abEX and press enter.
3. Use the "Choose File" button to select an image to load. Load abEX.png to work through the example.
4. Change the resolution to match your image. For the example, enter 20.
5. Adjust the margin side and margin top values until the circles are closely aligned to the spots. For the example, set the margin side value to 5350, and margin top to 6900.
6. Set the circle diameter slightly larger than the spot diameter. This sets the area that will be analyzed for each spot. For the example, enter 320.

7. If working with the packaged example, set Super Arrays to 1. Each of the three Super Arrays in this example was hybridized with a different antibody, so they should be analyzed individually. Default settings aggregate all active spots by ID. Otherwise, set Super Arrays according to your layout.
8. Click the "Toggle zoom" button to activate the zoom window. Mouseover a spot to show the spot in zoom. Toggle zoom applies a magnification to the spot and should only be activated after the resolution is set. Saturation in this zoom window does not imply saturation in the image.
9. Inspect the image for spots that have been compromised by handling of the slide or print artifacts. Printer drips, smudges, or debris can result in artifactual spots. ArrayNinja allows these spots to be manually excluded from analysis. To do this, click on the image anywhere (to be sure the cursor is not active in any input field) and mouse over a spot. Press the "a" key on the keyboard to in**a**ctivate a spot. Press "a" again to **a**ctivate the spot. Any in**a**ctive spot will be white and will not be quantified. In the example data, two of the H2AS1pCit3K5ac spots would be considered artifactual. Use the "Find" function to locate these features by typing the peptide name (case sensitive) in the Find field and pressing enter. The peptide spots will be highlighted with a red circle. Inactivate the two artifactual spots by mousing over and pressing "a" (Fig. 3A).
10. Click the "Spot Seek" button to automatically fine-tune the circle positions. Each circle will move toward maximum overlap with the spot on the image. This is achieved by calculating the "mass-weighted" average x and y positions of the pixels within each circle, where the "mass" of a pixel is taken to be the pixel (r,g,b) values. The center of each circle is updated to coincide with this average x and y, so that the center of the circle is iteratively driven to the "intensity-center" of the spot. This requires the earlier step of adjusting margins. Click the "Spot Seek" button three times for the example image.
11. The last step before quantification is the specification of a reference strategy. To set a **r**eference spot, mouse over any spot without a printed feature and press the "r" key. The spot will turn orange. Press "r" again to de-reference the spot. For the packaged example, set several (at least 3) **r**eferences across the slide where the spots are bright green with no red color, and have a "null" or "none" identity. Use the Find field to find all of the "none" spots.

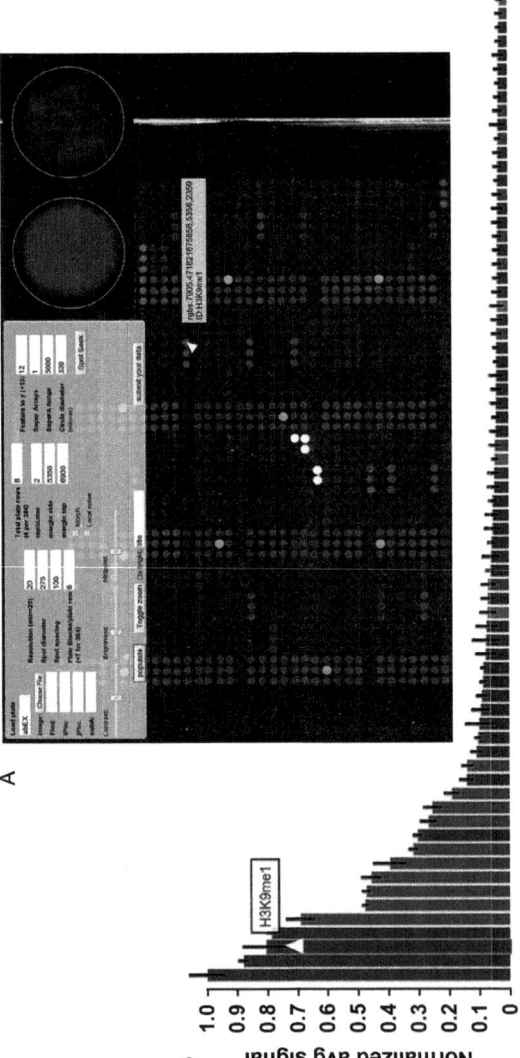

Fig. 3 Data analysis with ArrayNinja. (A) Screenshot of the ArrayNinja quantification interface. Interactive bar graphs (B) and tables (C) displaying quantified array results are automatically generated after clicking the "submit your data" button. (See the color plate.)

12. Click "populate." Once the pixels are counted, mousing over a circle will display the cumulative pixel counts for each color (Fig. 3A).
13. Enter a name for the experiment in the text field, and click "submit." An interactive bar graph will appear and a table of results will be shown (Fig. 3B and C). This data summary web page can be saved by right clicking on the page and choosing "view source." The resulting text can be pasted in a file named with the html extension. The file can then be viewed in a web browser at any time. The data table (Fig. 3C), containing averages and standard errors in the mean (s.e.m.), can be pasted into a spreadsheet by copying and pasting "as html." Additionally, a table of all spot intensities and identifiers is listed with the spots organized by k-means clustering. The clustering results provide all individual spot signals and help identify outliers.

5. BENCHMARKING ArrayNinja AGAINST ImageQuant TL

In this section, we validate ArrayNinja results by benchmarking against ImageQuant TL (GE Lifesciences) quantification for an image that has mostly homogeneous background and an image that has inhomogeneous background. These peptide microarray slides were printed on an OmniGrid 100 printer and imaged with a Typhoon Trio scanner. The raw images were processed as described earlier for ArrayNinja. For ImageQuant TL, images were directly loaded into the array analysis suite using the ds file generated by the Typhoon scanner. The ImageQuant dataset from Typhoon scans were deposited on www.histoneantibodies.com (Rothbart et al., 2015). For the ArrayNinja quantification, we did not inactivate any spots, so there is no quality control being asserted at the user level. It can be seen by the error bars (or lack-thereof) on some datapoints that the ImageQuant data have benefitted from user control. While the scans that produced this ImageQuant data are available, the user asserted controls are not. ArrayNinja makes recording these controls very simple. A screenshot of the user interface (as shown in Fig. 3A) before data submission records all details except how many times "spot seek" was used.

5.1 Homogeneous Noise

The ArrayNinja default mode requires the user to set references so that a nonlocal (ie, global) noise threshold can be computed. ImageQuant TL uses a local noise correction by default. In Section 5.3, we discuss why we chose a nonlocal noise threshold as the default for ArrayNinja.

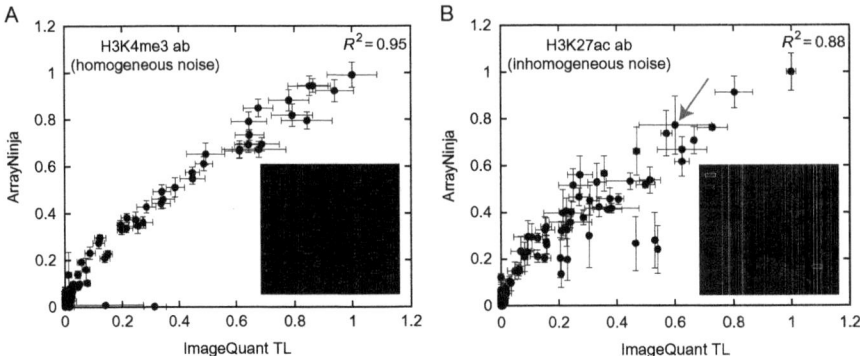

Fig. 4 Benchmarking ArrayNinja against ImageQuant TL. (A) Scatter plot comparing quantification methods of arrays with homogeneous noise between ImageQuant TL using local noise thresholding and ArrayNinja using nonlocal noise thresholding. Arrays were hybridized with an H3K4me3 antibody (Epicypher #13-0004). Correlation coefficient (R^2) was calculated by linear regression using gnuplot. (B) Scatter plot comparing quantification methods of arrays with inhomogeneous noise between ImageQuant TL using local noise thresholding and ArrayNinja using local noise thresholding. Arrays were hybridized with an H3K27ac antibody (Diagenode #C15410196). *Red arrow* points to a quantized data point from average intensity of spots marked by *white boxes*. R^2 was calculated as above. (See the color plate.)

Fig. 4A shows a comparison of ArrayNinja using a nonlocal noise correction to ImageQuant using a local noise correction. The local and nonlocal corrections should be identical when the noise is homogeneous over the image. Indeed, the results are strikingly similar, and all the same conclusions about the antibody behavior follow directly. In this case, with uniform background, the methods are nearly identical.

5.2 Inhomogeneous Noise

Fig. 4B shows quantification results for a noisy image where both ArrayNinja and ImageQuant use local corrections. The overall error estimates are larger than in Fig. 4A for both platforms, but both still lead to the same conclusions regarding the antibody behavior. We note that on the set of data points in Fig. 4B, where ArrayNinja estimates lower signals than ImageQuant, the ImageQuant results do not have error bars. In these data points, only one spot contributed to the ImageQuant results, reflecting some user-level data control (like spot inactivation) that is omitted from the ArrayNinja results.

As noted earlier, ArrayNinja outputs all individual nonzero spot intensities organized by the k-means clustering algorithm. This allows the experimentalist to check how spots of the same ID present signal. Inspecting the clustering results helps find replicate spots that present disparate signals, and going back to the image to find these spots can help identify the experimental conditions that cause the disparate results. Typical causes of this are printing errors or debris on the slide. The clustering results from the data in Fig. 4B suggest that local noise correction over-penalizes spots in noisier regions of the image, where we noted that replicate spots in high noise regions were clustered in low signal groups, and replicates in low noise regions were clustered in high signal groups. This suggests that the local correction may over-penalize spots in high noise areas. Using a nonlocal correction led to much less disparate clustering results for the same set of spots (eg, spots highlighted in Fig. 4B). In the next subsection, we present some formal details to explain these observations.

5.3 Implicit Assumption of Local Noise Correction

We can think of two reasons to support use of a local background correction: (1) the experiment was planned without incorporating reference spots, or (2) the belief that local noise corrections can "fix" inhomogeneous noise. Here, we provide mathematical reasoning to show that the local correction cannot "fix" inhomogeneous noise, in any sense, unless a strict assumption is met. This assumption was not met by the experiment in Fig. 4B.

The observed signal $\hat{S}(x, y)$ at any point (x, y) on the detector face can be written as $\hat{S}(x,y) = \int_X \int_Y [S(X, Y) + N(X, Y)] \mathrm{PSF}(x - X, y - Y) \mathrm{d}X \mathrm{d}Y$. The point spread function (PSF) of the detector is $\mathrm{PSF}(x, y)$, the signal emitted by the target fluorophore is $S(x, y)$, and the signal due to contaminant is $N(x, y)$. With a local correction, a given spot intensity is found by averaging this observed signal over the (x, y) points on the interior of a spot and subtracting the average observed signal found by averaging over the (x, y) points in the local exterior of the spot. For the interior of the spot, we have the average $\langle \hat{S} \rangle_{\mathrm{int}} = \langle S * \mathrm{PSF} \rangle_{\mathrm{int}} + \langle N * \mathrm{PSF} \rangle_{\mathrm{int}}$ and for the exterior region we have $\langle \hat{S} \rangle_{\mathrm{ext}} = \langle N * \mathrm{PSF} \rangle_{\mathrm{ext}}$, where there is no fluorophore on the exterior and all signal is due to noise. We use $*$ to indicate convolution with the PSF. If a local correction will "fix" inhomogeneous noise, then $\langle \hat{S} \rangle_{\mathrm{int}} - \langle \hat{S} \rangle_{\mathrm{ext}} = \langle S * \mathrm{PSF} \rangle_{\mathrm{int}}$. In order for this to work out correctly, the distribution $N(x, y)$ must be the same on the exterior boundary of the spot and on the interior of the spot in the mean sense, or else $\langle N * \mathrm{PSF} \rangle_{\mathrm{int}}$ and $\langle N * \mathrm{PSF} \rangle_{\mathrm{ext}}$ are not equal. Indeed, this assumption on $N(x, y)$ is implicit to the use of the local correction, owing to the idea that the correction should be stronger in regions of higher local noise. This basic assumption is not valid for this experiment, as can clearly be seen in Fig. 4B, where there are many dark spots where antibody (fluorophore) did not bind but are surrounded by high levels of background noise—the contaminant did not penetrate the spots, so $N(x, y)$ cannot be assumed similar inside and outside of the spots. This results in an unjustified coordinate-dependent bias in spot intensity.

The behavior of local corrections should be characterized case by case to assess the validity of the underlying assumptions. The experiment in Fig. 4B does not conform to those assumptions.

6. METHODOLOGICAL DETAILS

6.1 Default (Whole Spot) Quantification

Once the spot centers are found, the image can be quantified. Spots are quantified as follows: First, a spot-bounding domain is found by moving away from the spot center by a distance of c_r (plus a pixel) and then moving up a distance of c_r (plus a pixel), where c_r is the radius of the circle ($c_r = 320/2$ μm in the step-by-step example earlier). Second, every pixel inside the bounding domain is tested to find those pixels, no more than the circle radius c_r from the circle center. In other words, the interior of the circle is determined. Third, the color values of each interior pixel are summed into a red, blue, and green tally. Finally, the noise threshold value is subtracted from each spot value.

6.2 Nonlocal Noise Thresholding

The user must define reference spots, and each reference spot is quantified as described earlier. The noise threshold value is determined by averaging the cumulative tally from each reference spot over all user-defined references. The s.e.m. in this average is also computed. The noise threshold is defined by adding three times the s.e.m. to the mean intensity of the reference spots. The noise threshold is subtracted from all spots, and spots with negative signal are set to zero.

6.3 Local Noise Thresholding

ArrayNinja also supports local noise corrections. If the "Local noise" box is checked, all the pixels in the bounding domain, but outside the interior of the spot, are averaged and the s.e.m. is computed. The average plus three times the s.e.m. is taken as the noise threshold. The noise threshold is subtracted from all spots and spots with negative signal are set to zero. No references need to be set in this mode. When using this mode, keep in mind the implicit assumptions stated in Section 5.3. This is best for uniform or homogeneous noise.

6.4 Variegated Spot Morphology

ArrayNinja also includes a checkbox called "Morph." This checkbox enables a quantification method that corrects for variation in morphology, as shown in Fig. 5. This mode can only be used with nonlocal noise thresholding. The user defines several reference spots, and they are quantified with uniform circle size as described. Each reference spot's signal intensity is averaged over the number of pixels in that spot to define the average pixel intensity $\langle pi \rangle$. Each reference spot's average pixel intensity is then averaged over the set of references and the s.e.m. of this value is also computed. The noise threshold is set to $\langle pi \rangle + 3$ s.e.m. The more references, the smaller the s.e.m. will be. Likewise, fewer reference spots will lead to a larger background cutoff. This is the same noise threshold as in the default nonlocal method, except the default nonlocal method does not average over spot

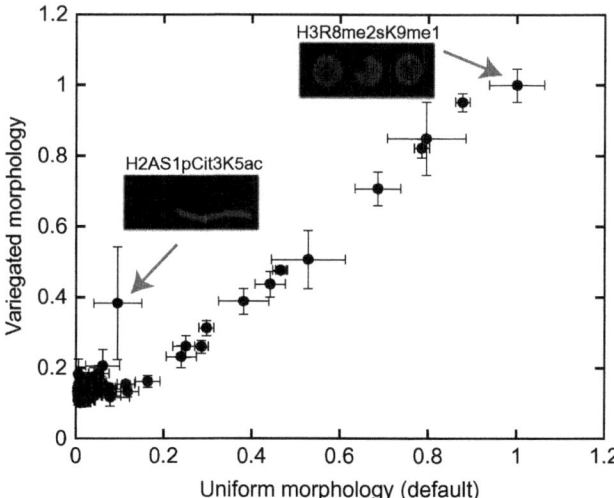

Fig. 5 Comparison of whole-spot and variegated quantification methods with ArrayNinja. Scatter plot comparison of whole-spot and variegated morphology quantification methods. Inset images illustrate how spot morphology impacts correlation between the two methods. (See the color plate.)

pixels. All spots have the same number of pixels in the default nonlocal method, so it amounts to division by a constant.

When the nonreference spots are quantified in the variegated method, the interior of each spot is identified as earlier, but only those pixels with a signal intensity greater than $\langle \text{pi} \rangle + 3$ s.e.m. will be included in this quantification. Each pixel with intensity less than this threshold is discarded, and the average pixel intensity of each nonreference spot minus the threshold is returned as the quantized spot value. Thus, only spot pixels brighter than the noise threshold are contributing to the spot intensity, and spots are no longer assumed to be circular.

The variegated morphology approach is attractive, because it implies that artifacts in spot morphology should be accounted for in the quantification. In Fig. 5, we show the results of the variegated morphology approach and the default approach for the packaged example image. In this quantification, no spots were discarded, and 11 references were set. Exactly the same references were used in each case. The "Morph" toggle allows for quantification with both methods without changing any other parameters.

Fig. 5 shows that for bright, well-resolved spots, the two methods agree. However, for spots with little to no structure that contain comet tails or artifacts (eg, H2AS1pCit3K5ac in the example), the variegated morphology method has the potential to produce different signals. This is because the average pixel intensity was calculated using only the brightest pixels. In the default method, all pixels on the interior of the circle are counted, resulting in a much lower effective average intensity. Thus, the variegated morphology method is not a standalone replacement for user-level quality control. The user should remain diligent in examining images for artifactual spots.

With modifications to the ArrayNinja core, the variegated methodology also opens avenues for quantifying cell and tissue microarrays, or other platforms presenting inherently irregular features.

7. LIMITATIONS, ASSUMPTIONS, OTHER FEATURES, AND FUTURE DEVELOPMENT

7.1 Limitations and Assumptions

Our microarray printer has a 4 × 4 pin configuration, so this is built into the code. Our tools also mandate that the source plates be read by the printer in a left-to-right, top-to-bottom pattern. We routinely use 384-well plates for our source plates. Accordingly, ArrayNinja assumes 24 columns by 16 rows for each source plate. ArrayNinja is built to talk to an SQL database, which allows integration of the planning and analysis stages, and forces certain formats on the tables where the feature identifiers, source plate layouts, and printing parameters are stored. The source code for all tools is freely available, so customization is permissible.

Lastly, we note that the OmniGrid 100 version of ArrayNinja may not count replicates in the same way as the OmniGrid printer. We have successfully tested the tools against a 48 × 48 spot layout that is commercially available through Epicypher and Millipore but not against any other formats. Robust stress testing of ArrayNinja for microarrayers other than the Aushon 2470 is recommended.

7.2 Other Features

The ArrayNinja quantification tool also includes a "Find" function and a set of inputs that activate spots that are specifically deposited by a given pin in a particular subarray. ArrayNinja also supports contrast and brightness adjustments to aid in spot finding. (Only raw image data are quantified.)

The "Find" function is straightforward. Enter a case-sensitive string and ArrayNinja will highlight all substrates that contain that string with a red circle. For the packaged example, try searching for K9me2. All peptides containing dimethylation at lysine 9 will be highlighted by a red boundary.

In the Aushon-specific quantization tool, the iPin, jPin, and subA fields allow pin isolation, which is useful when analyzing slides in which many subarrays were printed (eg, Fig. 2 in Cornett et al., 2016). Our printer uses a 4 × 4 pin configuration, which is assumed throughout ArrayNinja. Thus, the pins may be uniquely identified by specification of an (i,j) pair. Enter the i-th pin in the iPin field and the j-th pin in the jPin field and the requested subarray(s) in the subA field to activate only that pin. Lists of pins or subarrays can be selected by entering a space delimited series like "1 2 3 4" for both the iPin and jPin, to select the diagonal pins (1,1), (2,2), (3,3), (4,4).

7.3 Future Development

ArrayNinja can be extended to arbitrary printer architecture, and expansion to better cover existing commercial printers is a leading avenue of development. ArrayNinja could be extended to liquid handlers as well to allow the planning and use of custom source plates from master source plates that hold an entire substrate library. Where most other developments have come at the level of processing data, ArrayNinja represents development of the infrastructure that strings together all aspects of experiment planning and quantification. This unification and development of infrastructure is what makes ArrayNinja unique.

8. SUMMARY

ArrayNinja unifies the planning and analysis stages of microarray experimentation. Once the print layout is determined, and the source plate is populated, both the layout and plate are logged to a private database. After experimentation and slide imaging, the print parameters and source plate can be retrieved for automatic pairing of the scanned image with the print substrate identifiers. The quantification automatically aggregates spots with the same identifier and computes the average intensity along with s.e.m. (taking into account the number of spots for each identifier, which is not assumed constant). Spots that fall on regions of the slide that are damaged or confounded by obvious artifacts (see the included example) can be discarded easily by interacting with the imaged slide. The quantified data are displayed so that it can be inserted into the platform of choice (text files, spreadsheets, plotting softwares, etc.).

ArrayNinja is free to download (http://research.vai.org/Tools/arrayninja) and can also be obtained as a Docker app (bradleydickson/arrayninja). The tool is flexible, portable, and platform independent, and we have validated its quantification against commercially available software.

ACKNOWLEDGMENTS

We thank Dr Rochelle Tiedemann for piloting ArrayNinja and for providing useful comments on the chapter. This work was supported in part by grant CA181343 from the National Institutes of Health to S.B.R.

REFERENCES

Bock, I., Dhayalan, A., Kudithipudi, S., Brandt, O., Rathert, P., & Jeltsch, A. (2011). Detailed specificity analysis of antibodies binding to modified histone tails with peptide arrays. *Epigenetics*, 6(2), 256–263.

Bock, I., Kudithipudi, S., Tamas, R., Kungulovski, G., Dhayalan, A., & Jeltsch, A. (2011). Application of Celluspots peptide arrays for the analysis of the binding specificity of epigenetic reading domains to modified histone tails. *BMC Biochemistry*, 12, 48.

Bua, D. J., Kuo, A. J., Cheung, P., Liu, C. L., Migliori, V., Espejo, A., ... Gozani, O. (2009). Epigenome microarray platform for proteome-wide dissection of chromatin-signaling networks. *PLoS ONE*, 4(8), e6789.

Cornett, E. M., Dickson, B. M., Vaughan, R. M., Krishnan, S., Trievel, R. C., Strahl, B. D., & Rothbart, S. B. (2016). Substrate specificity profiling of histone-modifying enzymes by peptide microarray. *Methods in Enzymology*, 574, 31–52.

Dhayalan, A., Kudithipudi, S., Rathert, P., & Jeltsch, A. (2011). Specificity analysis-based identification of new methylation targets of the SET7/9 protein lysine methyltransferase. *Chemistry & Biology*, *18*(1), 111–120.
Fuchs, S. M., Krajewski, K., Baker, R. W., Miller, V. L., & Strahl, B. D. (2011). Influence of combinatorial histone modifications on antibody and effector protein recognition. *Current Biology*, *21*(1), 53–58.
Garske, A. L., Oliver, S. S., Wagner, E. K., Musselman, C. A., LeRoy, G., Garcia, B. A., …Denu, J. M. (2010). Combinatorial profiling of chromatin binding modules reveals multisite discrimination. *Nature Chemical Biology*, *6*(4), 283–290.
Hall, D. A., Zhu, H., Zhu, X., Royce, T., Gerstein, M., & Snyder, M. (2004). Regulation of gene expression by a metabolic enzyme. *Science*, *306*(5695), 482–484.
Hu, S., Xie, Z., Onishi, A., Yu, X., Jiang, L., Lin, J., …Zhu, H. (2009). Profiling the human protein-DNA interactome reveals ERK2 as a transcriptional repressor of interferon signaling. *Cell*, *139*(3), 610–622.
Kim, J., Daniel, J., Espejo, A., Lake, A., Krishna, M., Xia, L., …Bedford, M. T. (2006). Tudor, MBT and chromo domains gauge the degree of lysine methylation. *EMBO Reports*, *7*(4), 397–403.
Kudithipudi, S., Kusevic, D., Weirich, S., & Jeltsch, A. (2014). Specificity analysis of protein lysine methyltransferases using SPOT peptide arrays. *Journal of Visualized Experiments*, *93*, e52203.
Michaud, G. A., Salcius, M., Zhou, F., Bangham, R., Bonin, J., Guo, H., …Schweitzer, B. I. (2003). Analyzing antibody specificity with whole proteome microarrays. *Nature Biotechnology*, *21*(12), 1509–1512.
Moore, C. D., Ajala, O. Z., & Zhu, H. (2015). Applications in high-content functional protein microarrays. *Current Opinion in Chemical Biology*, *30*, 21–27.
Nady, N., Min, J., Kareta, M. S., Chedin, F., & Arrowsmith, C. H. (2008). A SPOT on the chromatin landscape? Histone peptide arrays as a tool for epigenetic research. *Trends in Biochemical Sciences*, *33*(7), 305–313.
Rothbart, S. B., Dickson, B. M., Raab, J. R., Grzybowski, A. T., Krajewski, K., Guo, A. H., …Strahl, B. D. (2015). An interactive database for the assessment of histone antibody specificity. *Molecular Cell*, *59*(3), 502–511.
Rothbart, S. B., Krajewski, K., Strahl, B. D., & Fuchs, S. M. (2012). Peptide microarrays to interrogate the "histone code". *Methods in Enzymology*, *512*, 107–135.
Rothbart, S. B., Lin, S., Britton, L. M., Krajewski, K., Keogh, M. C., Garcia, B. A., & Strahl, B. D. (2012). Poly-acetylated chromatin signatures are preferred epitopes for site-specific histone H4 acetyl antibodies. *Scientific Reports*, *2*, 489.
Smith, B. C., Settles, B., Hallows, W. C., Craven, M. W., & Denu, J. M. (2011). SIRT3 substrate specificity determined by peptide arrays and machine learning. *ACS Chemical Biology*, *6*(2), 146–157.

CHAPTER FOUR

Chemical Biology Approaches for Characterization of Epigenetic Regulators

D. Barsyte-Lovejoy*, M.M. Szewczyk*, P. Prinos*, E. Lima-Fernandes*, S. Ackloo*, C.H. Arrowsmith*,[†,1]
*Structural Genomics Consortium, University of Toronto, Toronto, ON, Canada
[†]Princess Margaret Cancer Centre, University of Toronto, Toronto, ON, Canada
[1]Corresponding author: e-mail address: carrow@uhnresearch.ca

Contents

1. Introduction 80
   1.1 Chemical Probes and Their Use in Biology 80
   1.2 Epigenetics and Protein Methyltransferases 81
2. Validation of Chemical Probes for Use in Cell-Based Experiments 81
   2.1 Requirements for Chemical Probes 81
   2.2 Biomarker Assays: How to Ensure Your Chemical Probe Is Active in Cells 82
   2.3 Assay Readout Choice 83
   2.4 Biomarker Assays for PMTs 84
   2.5 Timing and Other Considerations for Cell-Based Overexpression Assays 91
3. Inhibitor Enabled Discovery 92
   3.1 Inhibitor Handling and Inhibitor Libraries 92
   3.2 Phenotypic Assays Using Chemical Probes 93
   3.3 Demonstration of On-Target Phenotypic Effects 94
4. Conclusions 99
Acknowledgments 100
References 100

Abstract

Chemical biology approaches are a powerful means to functionally characterize epigenetic regulators such as histone modifying enzymes. We outline experimental protocols and best practices for the cellular characterization and use of "chemical probes" that selectively inhibit protein methyltransferases, many of which methylate histones to regulate heritable gene expression patterns. We describe biomarker assays to validate the probes in specific cellular systems, and provide guidelines for their use in functional characterization of methyltransferases including detailed protocols, examples, and controls. Together these techniques enable precision manipulation of cellular epigenomes and the exploration of the therapeutic potential of epigenetic targets in human disease.

ABBREVIATIONS

CETSA Cellular Thermal Stability Assays
EDTA ethylenediaminetetraacetic acid
GBM glioblastoma
HEK human embryonic kidney
Kme, Kme2, Kme3 lysine mono-, di-, and trimethylation, respectively
MES 2-(N-morpholino)ethanesulfonic acid
MOPS 3-(N-morpholino)propanesulfonic acid
PBS phosphate-buffered saline
PMTs protein methyltransferases
Rme arginine monomethylation
Rme2a asymmetric arginine dimethylation
Rme2s symmetric arginine dimethylation
RT room temperature
SDS sodium dodecyl sulfate
SGC Structural Genomics Consortium

1. INTRODUCTION
1.1 Chemical Probes and Their Use in Biology

Chemical biology methods using selective, cell-active chemical inhibitors (chemical probes) comprise a powerful approach for characterizing cellular signaling pathways and developing new therapeutics. For certain "druggable" protein classes, numerous inhibitors have been identified and characterized, perhaps best exemplified by protein kinases for which over 149 inhibitors have been or are being tested in clinical trials (Fedorov, Muller, & Knapp, 2010; Knapp et al., 2013). Chemical biology experiments are an essential first step for validating new therapeutic targets and require high quality inhibitors. "Chemical probes" are well-characterized drug-like small molecules that potently and selectively inhibit or antagonize the target protein with a defined mode of action in cellular studies (Bunnage, Chekler, & Jones, 2013; Frye, 2010). These probes enable the researcher to confidently link selective inhibition of a specific protein target with a biological or disease phenotype in cell-based assays. Chemical probes are primarily intended for initial target validation or phenotypic profiling studies in cell lines or primary patient samples cultured in vitro. As such, each probe needs to be confirmed to be cell-permeable and active in cells through direct inhibition or antagonism of the target. Chemical probes differ from drugs in that probes need not have the favorable pharmacokinetic or pharmacodynamic properties required for in vivo or clinical studies and therefore are much more economical to generate (Bunnage et al., 2013). In this chapter, we will focus on the cellular characterization and exploitation of chemical probes in the emerging therapeutic class of epigenetic regulatory enzymes, outlining aspects such as probe quality, selection of the most appropriate readout for cellar assays, and using the compounds in phenotypic discovery assays.

1.2 Epigenetics and Protein Methyltransferases

For the purpose of this chapter, we define epigenetics as heritable changes in phenotype or a gene expression program that are stably enforced by the chromatin state of a cell, and not due to changes in the DNA sequence itself. Distinct classes of proteins that regulate chromatin condensation or participate in DNA-templated processes have been described with the majority being classified as "writers," "erasers," and "readers" of chromatin marks. More specifically, these proteins acetylate/deacetylate histones, methylate/demethylate DNA or histones, or recognize acetyl or methyl "marks" (Arrowsmith, Bountra, Fish, Lee, & Schapira, 2012). Interestingly, many of what were initially called histone-specific proteins have since been shown to also modify or bind nonhistone substrates involved in transcription, splicing, and metabolism (Barsyte-Lovejoy et al., 2014; Mazur et al., 2014; Scholz et al., 2015; Zhu et al., 2015). In this chapter, we focus on, protein methyltransferases (PMTs), a group of enzymes that can mono-, di-, or trimethylate lysines, monomethylate arginine, or symmetrically/asymmetrically dimethylate arginine. Although the range of substrates for both lysine and arginine PMTs is rapidly growing, a complete identification of substrates and exploration of their activities is expected to be a longer term endeavor. High-quality chemical probes for PMTs enable the functional characterization of PMTs in chromatin-associated processes and facilitate the identification of new substrates. The sections below describe strategies and protocols for cell-based experiments to first validate and characterize the cellular biochemical activity of chemical probes, and then to exploit their activity to link the target with a phenotypic readout.

2. VALIDATION OF CHEMICAL PROBES FOR USE IN CELL-BASED EXPERIMENTS

2.1 Requirements for Chemical Probes

Although we stress the benefits of using chemical probes to explore target-related biology, our ability to reliably link inhibition of a target with specific biology or phenotype is critically dependent on the degree to which the probes have been characterized. The literature is littered with erroneous conclusions from the use of compounds that have been poorly characterized (that is, those that have poor selectivity or promiscuous chemical reactivity), and the reader is directed to a recent commentary (Arrowsmith et al., 2015) which addresses the caveats associated with using poor quality inhibitors. Just as with antibodies or shRNA reagents, it is essential that scientists validate small-molecule inhibitors for cellular use using appropriate controls as outlined below (Arrowsmith et al., 2015).

In this chapter, we will focus on the cellular characterization of high-quality inhibitors or chemical probes (Workman & Collins, 2010) which adhere to strict criteria for in vitro potency, selectivity, and on-target activity in cells (see for example, http://www.thesgc.org/chemical-probes/epigenetics). Further, we will present cell-based data which exemplify the biology that is achievable with chemical probes.

2.2 Biomarker Assays: How to Ensure Your Chemical Probe Is Active in Cells

When considering the use of a new chemical probe in a cellular assay, it is essential that scientists demonstrate "target engagement," showing that the inhibitor not only enters the cell system under study but also directly inhibits or antagonizes the target protein. Although target engagement is established as part of the criterion for the chemical probes referenced in this chapter (thesgc.org/chemical-probes/epigenetics), it is good practice to verify this for each cell system. Generic methods for confirming target engagement include Cellular Thermal Stability Assays (CETSA) in which binding of the inhibitor stabilizes the thermal stability of the target protein, pull down experiments using a "precipitatable," derivatized version of the inhibitor in the presence or absence of the free unmodified inhibitor as competitor, and detected by Western blot or mass spectrometry (MS) (Franken et al., 2015; Huber et al., 2015; Martinez Molina et al., 2013; Martinez Molina & Nordlund, 2016).

For enzymes, the most straightforward way to confirm direct cellular activity is to have a biomarker assay that measures, for example, lysine methylation of the substrate in cells, thus providing a direct readout of the protein lysine methyltransferase enzymatic activity in cells. This readout will reflect the cellular activity of the inhibitor, which, in turn, reflects the potency, cell permeability, and stability of the compound when measured over time. Two key parameters are important to determine prior to conducting such studies:

(1) the appropriate concentration range and (2) the time frame over which the cells are monitored. The former can be mapped by following the biomarker over a concentration range of several orders of magnitude (usually low nanomolar to mid-micromolar) and then choosing several concentrations spanning the response curve for further studies. The duration of the experiment is particularly important to consider for epigenetic targets as biomarker or phenotypic changes often take much longer (days or weeks) to manifest compared to typical signaling events such as phosphorylation-mediated signaling (minutes–hours).

Cellular inhibition of a PMT should result in reduction of substrate methylation much the same as knockdown of the PMT does (but not necessarily vice versa). Thus, reduction in the mark can serve as a biomarker for the cellular response to the chemical probe. Due to redundancy among enzymes or the opposing activity of demethylases, the knockdown or inhibition of a single PMT may not completely eliminate the mark on the histone substrate. In other cases, one may only observe loci-specific reductions in the mark that are difficult to observe globally in the cells. In the section below, we will outline how we measure the reduction in a specific substrate methyl mark using specific antibodies. In these experiments, it is important to compare the effect of the inhibitor to that of knocking down the target to ensure that the observed effect is specific.

2.3 Assay Readout Choice

Cellular substrates of epigenetic enzymes have typically been identified using antibody detection or MS approaches. Recently, due to vast improvements in MS technology especially in methylarginine detection, the network of known arginine methyltransferase cellular substrates has grown (Bremang et al., 2013; Guo et al., 2014; Soldi, Cuomo, Bremang, & Bonaldi, 2013; Sylvestersen & Nielsen, 2015). At the same time, the availability of antibodies highly specific for a given methyl mark and the ease of use of antibody-based fluorescence detection methods to quantitatively monitor methylation have also expanded our knowledge of PMT substrates. In some cases, these newer antibodies have helped to validate substrates identified by MS (Guo et al., 2014). The following sections will focus on outlining the strategies and tools for PMT cell-based assays employing methyl mark-specific antibodies.

A growing number of methylation-specific antibodies are commercially available; these antibodies can recognize either the methylation of a specific residue within a protein or methylation within specific arginine and glycine-rich motifs. In the case of arginine methylation, there are antibodies that can distinguish between mono- (Rme) and either symmetric (Rme2s) or asymmetric (Rme2a) dimethylation. For lysine methylation, mono- (Kme), di- (Kme2), and trimethyl- (Kme3) specific antibodies are available. The selectivity and specificity of histone-modification antibodies can be verified by peptide arrays, which consist of a library of various possible peptide modifications and sequences. Several publications and online databases (Imanishi et al., 2008; Liu et al., 2010) provide data on antibody validation in Western/dot blots and chromatin immunoprecipitation (ChIP) (Egelhofer et al., 2011). The user should consult these databases but should also keep in mind that polyclonal antibodies vary from batch to batch so rigorous in-house validation of your material is essential. If the specific mark for the target protein is known, the antibody validation can be done by knocking down the protein and examining the mark by Western blot, immunofluorescence, or ChIP. Sometimes posttranslational modifications such as methylation can affect total protein levels so it is important to have two antibodies: one to detect the methylated form and a second, which does not distinguish between modifications, to show the total levels of the protein (Fig. 1). If the two antibodies are from different host species (eg, rabbit and mouse), one can use secondary antibodies labeled with different fluorophores and perform the imaging on a fluorescence-based Li-COR® Imaging System. This will not only save time, since one can quantify methylated and total protein levels in one gel, but also minimize technical errors introduced by differences in loading or protein transfer. The dynamic range of the Li-COR system is much greater compared to the horse radish peroxidase-based X-ray film quantitation, thus making it the method of choice.

2.4 Biomarker Assays for PMTs

The simplest biomarker assays measure the modification of a cellular substrate using a modification-specific antibody. Several PMTs have unique substrates and their methylation can be detected in vitro and in cells so the modification of these substrates can provide a specific readout for the respective enzyme's activity in both settings. For example, PRMT5 methylates arginines in SmD and SmB proteins, and the G9a/GLP heterodimer is the sole enzyme that dimethylates H3K9 (Biggar & Li, 2015). Because these biomarkers are specific for their respective enzymes,

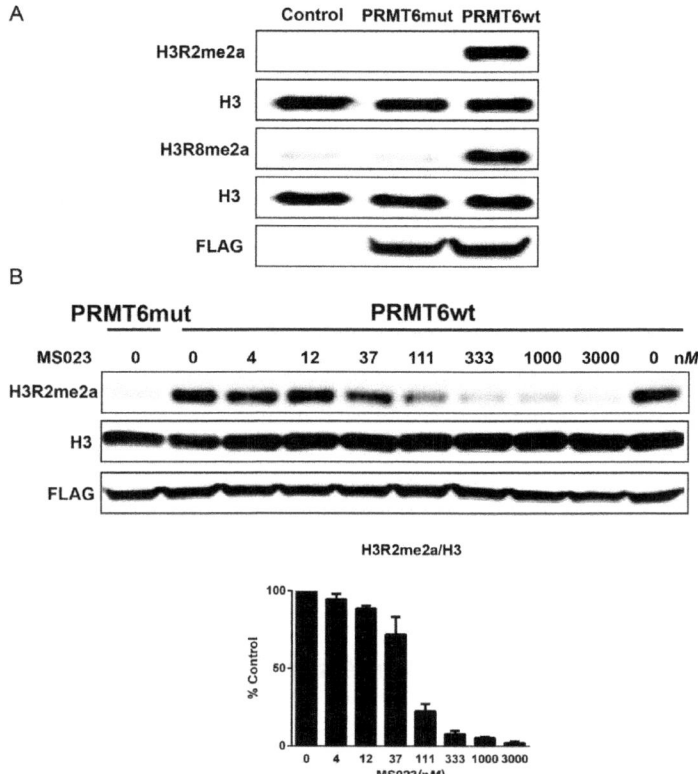

Fig. 1 Overexpression of a wild type but not catalytic PRMT6 mutant leads to increase in H3R2me2a and H3R8me2a levels. MS023 effectively inhibits the transfected PRMT6. (A) HEK293T cells were transfected with empty plasmid (control), FLAG-tagged PRMT6 or FLAG-tagged PRMT6 catalytically dead mutant (V86K/D88A) for 20 h and analyzed by Western blot with rabbit anti-H3R2me2a (Millipore #04-808), rabbit anti-H3R8me2a (Rockland, #600-401-I67), mouse anti-FLAG (Sigma, #F1804), mouse anti-H3 (Abcam #174628). (B) PRMT type I inhibitor, MS023, inhibits the increase in H3R2me2a levels caused by PRMT6 overexpression. The histogram represents quantitation of the band intensities normalized to control. H3R2me2a signal was normalized to total H3.

assays developed to detect them were used to confirm the cellular activities of chemical probes that inhibit PRMT5 or G9a/GLP (Chan-Penebre et al., 2015; Konze et al., 2013; Pappano et al., 2015; Sweis et al., 2014; Vedadi et al., 2011). Comparable biomarker assays can be developed for new PMTs. Importantly, these biomarker assays should be validated using target knockdown experiments. Below we provide a protocol for an in-cell Western assay for measuring a histone mark in a high-throughput manner.

Customarily, when substrates for PMTs are already known, chemical probe development starts within vitro biochemical assays, which are used to determine the "intrinsic" potency of chemical probes on the catalytic activity of the enzyme. Once activity and potency in vitro are established, scientists must ask—is the probe effective in cells? Maintaining continuity with the in vitro enzyme–substrate pair in cells is highly desirable. Ideally, genetic knockdown of the PMT would result in the reduction of the specific substrate mark; however, many PMTs contribute only a small amount of that cellular mark in specific areas of the genome, making these assays challenging to design. Below, we also outline a strategy to develop biomarker assays using ectopically expressed PMTs and substrates that mimic the in vitro biochemical activity of PMTs. The protocol and examples are provided. In addition to generating robust signals, this approach also has the advantage of specifically addressing the potential effectiveness of the inhibitor on the transfected cellular target. Once established, the ectopic system can also be used as a first step in identifying novel substrates in cells, as was recently demonstrated for PRMT3 (Kaniskan et al., 2015).

2.4.1 Protocol 1—In-Cell Western Assay for Histone/Protein Mark Detection

This is an immunofluorescence-based method that has been successfully used for characterizing G9a, EZH2, DOT1L, PRMT5, and SMYD2 chemical probes (Chan-Penebre et al., 2015; Konze et al., 2013; Liu et al., 2013; McCabe et al., 2012; Nguyen et al., 2015; Sweis et al., 2014; Vedadi et al., 2011; Verma et al., 2012; Yu et al., 2012)

1. Seed cells in clear bottom black wall 96-well plates, such as Thermo Scientific/NUNC (#165305). The cell density will depend on how long the experiment needs to be performed (for a 2-day experiment, seed 5–10 thousand cells/well, for a 7-day experiment, seed 1–2 thousand cells/well). For an even cell distribution that is desirable in imaging, we routinely place 50 µL of media first and then add 50 µL of media with cells using a multichannel pipette.
2. On the next day, add the chemical probes in a two- to fivefold dilution series. For example, if a well already contains 100 µL of cells with media, adding 50 µL of 3× concentration of the probes will ensure better accuracy when using multichannel pipettes.
3. Incubate the cells for the required time (see timing considerations below). Remove the media by shaking, wash with 200 µL phosphate-buffered saline (PBS), and add the fixation solution, 200 µL of 0.1%

Triton-X-100 in PBS. Some histone mark detection requires the presence of detergent in the fixation step (H3K79me2 staining fixation was performed with 0.3% Triton-X-100, 0.1% sodium dodecyl sulfate (SDS), 2% formaldehyde in PBS on MCF10A cells). Incubate for 10 min.
4. Remove the fixation solution by shaking and permeabilize/wash cells with 0.1% Triton-X-100 in PBS using 200 μL per well, four times.
5. Block by adding 200 μL of blocking buffer (5% BSA, 0.1% Tween-20 in PBS). It should be noted that blocking and incubation buffers may vary for different antibodies, for example, 1% BSA was used for H3K9me2 staining (Vedadi et al., 2011), 3% BSA and 5% goat serum for H3K27me3 (Konze et al., 2013), and 0.5% Triton-X-100, 3% BSA, 5% goat serum in PBS with 0.25 M glycine was used for H3K79me2 (Yu et al., 2012). These conditions should be established beforehand employing smaller scale microscopy experiments. Blocking can be done for 1 h at room temperature (RT) or overnight at 4°C.
6. Remove the blocking solution and add 60 μL per well of the antibody staining solution—1:800 for H3K9me2 (Abcam Ab8898), incubate 1–2 h RT; 1:4000 for H3K27me3 (Diagenode mAb-181-050), or 1:1000 for H3K79me2 (Abcam Ab3594)—and incubate in the refrigerator overnight.
7. Remove the antibody incubation solution and wash with PBS, 0.1% Tween-20 four times using 200 μL per well and a repeat multichannel pipette.
8. Incubate with 60 μL of secondary IR800 anti-rabbit (Li-COR® #926-32211) or anti-mouse (LI-COR®#926-32210) at 1:1000 and DRAQ5 (Cell Signaling Technologies) at 1:10,000 in LI-COR® blocking buffer (LI-COR®#927-40000) for 1–2 h at RT.
9. Scan using the Odyssey LI-COR® scanner using 3 mm offset for the above recommended plates, quantify using plate settings of the Image Studio software. Normalize the histone mark signal to the nucleic acid stain DRAQ5 signal to account for the effects on cell viability.

2.4.2 Protocol 2—Exogenous PMT Cell-Based Biomarker Assay

Depending on the cell line chosen to perform this assay, the density of cells seeded might need to be adjusted so they are 40–50% confluent for the transfection. This protocol below is an example for HEK293T cells.
1. Day 1: Seed 1×10^5 HEK293T cells per well of a 12-well plate.

2. Day 2: Transfect cells with 1 μg of Flag-tagged enzyme, its catalytically dead mutant or a Flag-tagged enzyme plus an ectopic substrate such as H3-GFP. For a co-transfection assay using a Flag-PMT and its exogenous substrate, it is important to use excess enzyme (0.6–0.9 μg) compared to substrate (0.1–0.3 μg) in order to maximize the effect of exogenous enzyme on the substrate methylation.

   Add the appropriate amount of DNA to a tube and mix with Jetprime buffer and 2 μL of Jetprime (Polyplus-transfection® SA). Vortex and incubate 10 min at RT before adding dropwise to the cells.
3. Four hours after transfection, remove the media, replace with fresh media containing the desired amount of inhibitor (or control such as DMSO alone) and incubate the cells for at least 20 h and up to 3–4 days, depending on the time it takes the mark to turnover in your cellular system (see discussion below).
4. Day 3: Remove the media and lyse the cells, as described below.

Western blot analysis
1. Total cell lysis. After removing the media and washing cells with PBS, lyse cells directly on a plate in 100 μL (per well of a 12-well plate) of total lysis buffer (20 mM Tris–HCl, pH 8, 150 mM NaCl, 1 mM ethylenediaminetetraacetic acid (EDTA), 10 mM MgCl$_2$, 0.5% Triton-X-100, 12.5 U/mL benzonase (Sigma), complete EDTA-free protease inhibitor cocktail (Roche)). Total lysis buffer is very important for solubilizing histone proteins and dissociating them from DNA. DNA increases sample viscosity; therefore, samples are incubated at RT for a few minutes in lysis buffer in order to allow benzonase to digest the DNA. After 3 min incubation, add SDS to a final 1% concentration.
2. SDS-PAGE. Total cell lysates can be resolved in 4–12% Bis–Tris Protein Gels (Invitrogen) with MOPS (3-(N-morpholino)propanesulfonic acid) or MES buffer (Invitrogen). When immunoblotting for histones, especially histone H4, with gradient gels, MOPS buffer compacts low-molecular-weight proteins, which yields easier analysis.
3. Immunoblotting. Transfer gels onto the PVDF membrane (0.2 μm, Millipore) for 1.5 h (80 V) in Tris–glycine transfer buffer containing 20% methanol and 0.05% SDS. Incubate membranes for 1 h in blocking buffer (5% milk in 0.1% Tween-20 PBS) and then incubate with primary antibodies in the blocking buffer overnight at 4°C. After five washes using 0.1% Tween-20 in PBS, incubate the blots with goat-anti rabbit (IR800 conjugated, LI-COR®#926-32211) and donkey anti-mouse (IR 680, LiCor #926-68072) antibodies (1:5000) in Odyssey blocking

buffer (LI-COR®) diluted 1:3 with 0.1% Tween-20 in PBS for 1 h at RT then wash five times for 5 min each with 0.1% Tween-20 in PBS. Visualize the signal on an Odyssey scanner (LI-COR®) at 800 and 700 nm.

2.4.3 Example 1: Enzyme Overexpression Assay

PRMT6 can methylate arginine in glycine and arginine-rich motifs (Frankel et al., 2002). Two of the well-characterized PRMT6 substrates are arginine 2 and 8 of histone 3, and it is believed that PRMT6 is responsible for regulating global H3R2me2a levels (Guccione et al., 2007). However, we observed that at least 4 days are needed in order to observe small quantitative changes in these marks. This extended time frame probably reflects a slow turnover of arginine-methylated H3 and H4. In an effort to increase the sensitivity of this assay, we transfected PRMT6 into HEK293 cells and used the Western blot protocol with antibodies specified in Fig. 1 legend to monitor H3R2me2a levels. The overexpression of PRMT6, but not a catalytically dead mutant of PRMT6, leads to an increase in H3R2me2a and H3R8me2a (Fig. 1A) making them both possible substrates for PRMT6 and candidate biomarkers for cell assay development. MS023, a pan-Type I PRMT inhibitor, decreased PRMT6-dependent H3R2me2a levels in a dose-dependent manner (Fig. 1B).

2.4.4 Example 2: Enzyme/Substrate Co-Overexpression Assay

The PRDM protein family is a poorly characterized family of proteins related to the SET domain methyltransferases. Only a few of the 17 PRDMs have been reported to methylate proteins such as histones (Fog, Galli, & Lund, 2012). One of them is PRDM9, whose expression is restricted to germ cells; PRDM9 has been reported to catalyze H3K4 trimethylation (Hayashi, Yoshida, & Matsui, 2005). In an effort to identify novel substrates for this protein family, we used purified proteins in vitro and a co-overexpression assay in HEK293T cells to show that PRDM9 also catalyzes H3K36 trimethylation (Eram et al., 2014).

Given its restricted, tissue-specific localization, a cell-based assay on endogenous PRDM9 was challenging. However, using Flag-tagged PRDM9 we performed the overexpression assay as described earlier and monitored the changes in H3K4me3 and H3K36me3 (Fig. 2A). Addition of exogenous PRDM9 increased H3K36me3 levels by 2.5-fold and H3K4me3 levels by 1.6-fold. This modest increase might be due to the presence, in the cells, of other PMTs known to methylate these two lysine

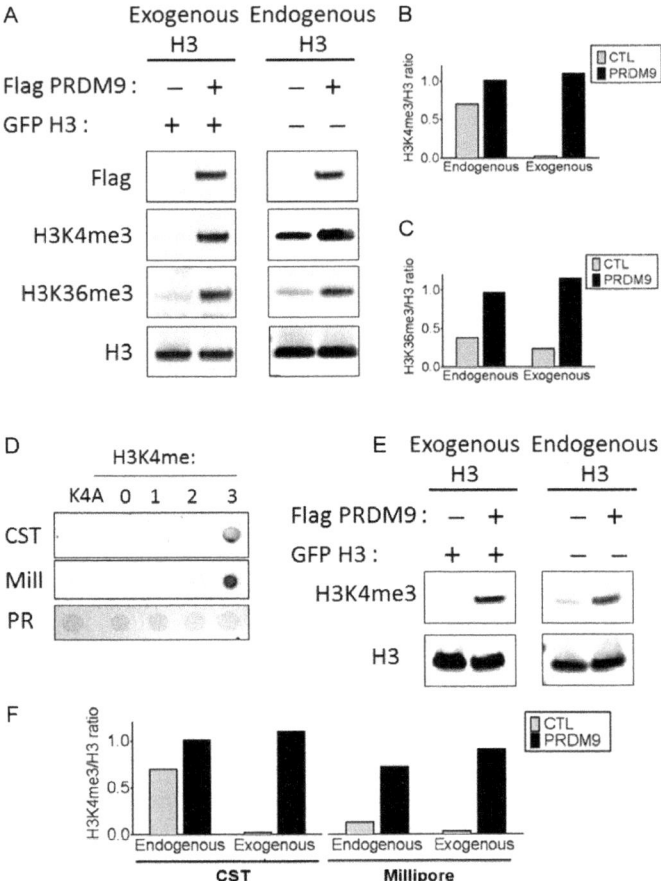

Fig. 2 Biomarker assay for PRDM9: measuring trimethylation of H3K4 and H3K36 on endogenous and exogenous H3 substrates. (A) Comparison between exogenous GFP-H3 and endogenous H3 as a substrate for Flag-tagged PRDM9. HEK293T cells were co-transfected with FLAG-PRDM9 and empty vector or with GFP-H3, respectively, and cells were processed for blotting 24 h after transfection. The membranes were probed with anti-H3K4me3 (Cell Signalling Technology #9723), anti-H3K36me3 (Cell Signalling Technology #9727S), mouse anti-FLAG (Sigma, #F1804), and mouse anti-H3 (Abcam #174628). The blots show an increase in H3K4me3 for both endogenous and exogenous substrates, with a more robust fold change for the exogenous substrate as quantified in (B) and (C). Quantifications represent the ratio between methylated lysine and total H3. (D) Dot blot performed with two antibodies against H3K4me3 to confirm the specificity and compare the intensity of the signal obtained (CST: Cell Signalling Technology #9727S; Mill: Millipore #04-745). Nitrocellulose membranes were dotted with 1 μL of 5 mM synthetic peptides (0: unmethylated, 1: mono-, 2: di-, 3: trimethylated, and H3K4A mutant). Loading of the peptides was confirmed by Ponceau red (PR) staining. The dot blot shows that both antibodies are specific, but that Millipore gives a stronger signal. (E) Exogenous and endogenous H3 were probed for H3K4me3 using the Millipore antibody tested in (D) and shows a stronger fold change for the endogenous assay when compared to the Cell Signalling antibody probed in (A) and quantified in (F).

residues, as reflected by the high basal level for these marks without PRDM9 overexpression (Fig. 2A). By co-transfecting PRDM9 with GFP-tagged H3, we increased the sensitivity of the assay and showed much more robust and significant changes in methylation levels; a 59-fold increase for H3K4me3 and a fivefold increase for H3K36me3 (Fig. 2A–C).

As described in Section 2.3, the choice of antibody is critical for the robustness of an assay, whether it is to test new inhibitors or to identify novel substrates. This defined transfection system allows the user to compare two antibodies for H3K4me3. We used a peptide dot blot using mono-, di-, tri-, or unmethylated H3K4 peptides to determine that both antibodies were equally specific but were differentially sensitive. The Millipore antibody was the most sensitive (Fig. 2D). In the ectopic overexpression assay (shown in Fig. 2A), the Millipore antibody proved superior in sensitivity (5.8-fold increase) and robustness (Fig. 2F). Thus, the transfection system is a more accurate test for antibody specificity than the peptide dot blot.

2.5 Timing and Other Considerations for Cell-Based Overexpression Assays

The section earlier provides protocols and illustrates some of the uses of overexpression systems in epigenetics cell-based assays. However, not all enzymes methylate their targets when they are overexpressed. For example, overexpressed PRMT4 is only weakly active on its known targets, possibly due to cytoplasmic localization of exogenous PRMT4 or other unknown factors. Thus, even in cases where exogenous enzyme or enzyme–substrate overexpression enhances the sensitivity of an assay, this method should be considered an initial step in further characterization of the endogenous epigenetic targets. The endogenous systems are much more complex; cellular enzyme levels are usually low, and substrates can be methylated by numerous enzymes. Thus, steady-state levels of methylation can range from low to high with turnover of the methyl mark passively diluted during the cell cycle (low) or actively removed by demethylases (high). This variation creates a challenge when trying to establish the appropriate time frame for the experiments and can confound data interpretation. For example, inhibiting G9a with UNC0638 for 48 h reduces the mark only 40%, whereas inhibition for 3–4 days reduces the mark to 20% relative to control, which matches outcomes with the stable knockdown of G9a and GLP (Vedadi et al., 2011) (Fig. 3). In another example, DOT1L must be inhibited for 7 days to achieve reduction of H3K79me2 in slow-cycling epithelial cells, while inhibition for only 4 days can achieve the same level of reduction in faster cycling leukemia

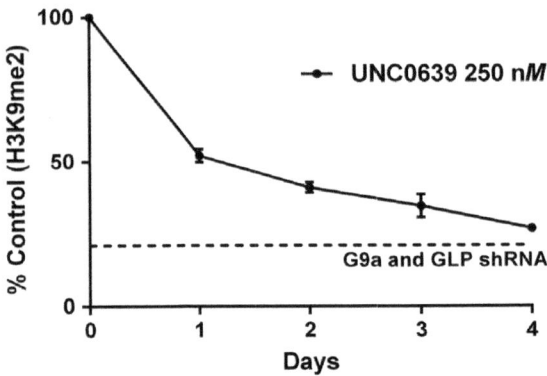

Fig. 3 Example of time dependence of G9a/GLP inhibitor effects. H3K9me2 levels were determined using the Protocol 1 earlier. Note that when breast cancer cell line MDA-MB231 was treated with 250 nM UNC0638 (representing IC_{90} of H3K9me2 response at 2 days), 4 days were required for the mark to decrease to the levels observed in G9a and GLP knockdowns.

cells (Yu et al., 2012). The repercussions of this slow kinetics on the downstream phenotypic effects will be discussed below.

3. INHIBITOR ENABLED DISCOVERY

The ultimate goal of generating chemical probes is to use them to discover new biological functions of the target protein and investigate the therapeutic relevance of the target. This implies that the target biomarker activity will need to be correlated with a functional readout. Many functional readouts are possible, from changes in cell viability to expression of specific apoptotic or differentiation markers and gene expression signatures. Here we focus on linking target inhibition with a phenotypic readout by making use of the biomarker assay to ensure quality functional discovery.

3.1 Inhibitor Handling and Inhibitor Libraries

In the previous section, we discussed the importance of using high-quality selective and potent inhibitors; likewise, it is essential to handle inhibitors properly. Information on how to solubilize and store compounds is usually provided by the supplier. Most frequently DMSO is used as a solvent and the dissolved stocks (10–20 mM) are kept frozen. Practical considerations that researchers then face are how many times the DMSO stock can be defrosted and how many stock tubes can be accommodated in the freezer (Liu, Richard, Kim, & Wojcik, 2014). In the absence of data on the stability of the compound, single aliquots of the stock solution should be prepared and stored in a −80°C freezer. Alternatively, the user can test the stability of the compound after freeze–thaw. One way to do this is to subject a set of stock solutions (from the same batch of compound) to freeze–thaw cycles, for example, one to five cycles then evaluate its activity in (i) a biochemical in vitro assay and (ii) the biomarker assay. Once this quality control is done, the user can determine appropriate conditions for stock storage. In case of larger screens involving multiple compounds where the 96-well plate format is used, it is convenient to array the compounds into that format. Ideally this process can be done using a liquid handler device such as an Echo 550 (Labcyte) and single-use plates stored in pressurized storage pods under inert nitrogen gas to ensure the stability of the compounds. However, most academic labs do not have access to this type of equipment. As an alternative, we recommend aliquoting a 1000× stock into sterile round bottom 96-well plates, using a multichannel pipette, sealing with an aluminum seal (for example, Axygen PCR-AS-200), and keeping the plates in the −80°C freezer until needed. These plates should be considered single-use dilution plates, as once the inhibitors are diluted with the cell media, stability may become an issue.

Another consideration for the user is how long the compound is stable in the cellular environment. This can also be experimentally tested using a biomarker; however, this assessment is complicated by compound potentially being used up by the cells and by the user not knowing what proportion of the target protein is bound by the compound. Some of these questions can be addressed using biotinylated compounds and target engagement assays. In general and for practical purposes in long-term assays, it is prudent to maintain the inhibitor at the same concentration with every change of cell media.

3.2 Phenotypic Assays Using Chemical Probes

Chemical probes are ideal reagents to link cellular phenotypes or signaling pathways with the chemical inhibition of a specific target. Phenotypic assays can be performed in a variety of ways including (i) testing the hypothesis that the inhibition of a given target will have an effect on a specific phenotype or pathway, (ii) "phenotype hunting" by testing a chemical probe against a variety of assays to understand the function of the target protein, and

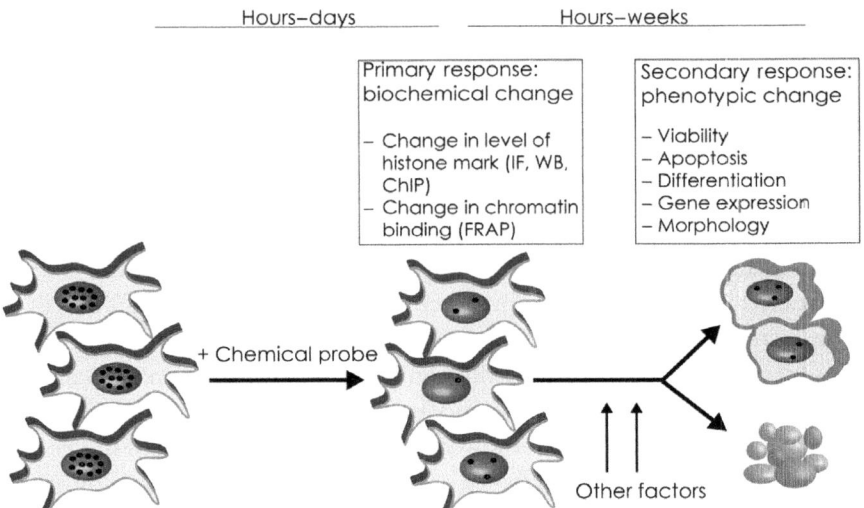

Fig. 4 Cellular responses to chemical probes.

(iii) "target hunting" assays, in which one phenotypic assays is used to assess a library of chemical probes to find a target that modifies the phenotype of interest. When planning such experiments, it is important to consider the nature of the responses being measured. Changes in the biomarker can be easily attributed to the function of the target, and interpretation is relatively straightforward as the biomarker is a direct product of the target activity (Fig. 4). Functional or phenotypic readouts, however, can be several biochemical steps removed from the target and multiple other proteins, pathways, and factors can influence the same functional outcome. It is therefore useful to classify responses to chemical probes as either primary or secondary. The primary response is measured by a biomarker (such as the cellular levels of a histone methyl mark deposited by an enzyme that is inhibited by the probe), while the secondary functional readouts, such as cell viability, often depend on many other factors including those that are activated in response to a general toxic response (Fig. 4). A general toxic response can be elicited by many chemotypes and may not necessarily be related to direct inhibition of the target.

3.3 Demonstration of On-Target Phenotypic Effects

To determine if the inhibition of the biomarker activity correlates with an observed functional effect, for example, cell death as is frequently studied in oncology, it is important to compare the dose–response curves.

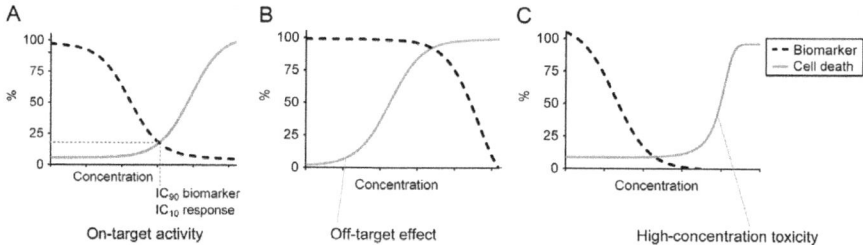

Fig. 5 Dose–response scenarios for the response biomarker and secondary events. (A) Correlated and likely causal relationship of biomarker and cell death responses. (B) An increase in cell death at concentrations lower than those that lead to target inhibition as measured by the biomarker indicates noncausal relationship. (C) A dose–response relationship where full biomarker inhibition does not elicit changes in cell death likely indicates a noncausal relationship between the target and cell death and indicates the compound has general off-target cytotoxicity at high concentrations.

Roughly three main scenarios of cellular dose responses are possible and are illustrated in Fig. 5.

If an enzyme is 90% inhibited at a particular concentration of the probe (as measured by a biomarker) and cell death starts to increase at this concentration (Fig. 5A), then this is likely to be a target-specific cell death. Such a conclusion can be further supported by genetic knockdown of the target. In another scenario, if the cell death occurs at probe concentrations lower than that required for target inhibition (measured by a biomarker), then the response is most likely not target related (Fig. 5B), and may be due to off-target or general cytotoxic effects of the probe. The use of chemically similar but target-inactive (or less active) "control" compounds is very helpful here. The shape of the cell viability curve can also reveal off-target effects, as excessively steep or shallow slopes have been associated with polypharmacology (Anighoro, Bajorath, & Rastelli, 2014; Merino, Bronowska, Jackson, & Cahill, 2010), population response heterogeneity or nonspecific toxicity (Fallahi-Sichani, Honarnejad, Heiser, Gray, & Sorger, 2013). The third scenario is one in which the cell death occurs at much higher concentrations than target inhibition (Fig. 5C), and most often this kind of response is caused by nonspecific compound toxicity at high concentrations. Again the slope of the toxicity curve can be revealing, and negative control compounds are important. A complementary strategy is to use two or more chemically unrelated probes for the same target; it is highly unlikely that two different compounds will show the same off-target activity.

Timing is another very important consideration in the phenotypic experiments. As mentioned earlier, epigenetic biomarker changes can take days to show the maximal response. Because the phenotype should "track" with the primary biomarker response, it should take just as long or longer to observe a change in phenotype, as illustrated in Figs. 3 and 4, where primary responses take hours–days, while phenotypic secondary responses may take weeks.

Many of the standard viability assays commonly used in oncology are typically performed for 48–72 h, the time frame that reliably captures the effects of kinase inhibition or interference of other signaling pathways as well as the cytotoxic effects of chemotherapy. However, epigenetic targets are unique in their functional aspect—epigenetic heritable changes are stable, meaning that they do not respond to transient signals but carry information from one generation of cells to the next, so the time to acquire or erase the epigenetic marks often follows cell division. The functional effects of changing epigenetic marks often trigger reprogramming events such as differentiation—phenotypic effects that typically take days to weeks. Some epigenetic chemical probes such as those for EZH2 (Konze et al., 2013; McCabe et al., 2012; Verma et al., 2012) and DOT1L (Daigle et al., 2011; Yu et al., 2012) take days for significant changes in the biomarker histone marks to occur and weeks for the subsequent effects on the MLL translocated leukemia cell viability to be apparent (Daigle et al., 2011; Yu et al., 2012). Others, such as MS023—a chemical probe for type I PRMTs—have faster kinetics. It is important to compare the chemical probe effect on a particular phenotype with those resulting from genetic knockdown of the target; however, this comparison may be complicated by the presence of multiple domains in the target protein or protein complex disruption. Given that it is challenging to determine the exact time frame of the phenotypic changes and early toxicity maybe an indicator of off-target toxicity, it is beneficial to investigate the kinetics of the phenotypic response. Below we discuss a protocol for a phenotypic continuous, live-cell assay measuring cell viability, and we provide some examples of specific and nonspecific responses to chemical probes.

3.3.1 Protocol 3: Measuring Early Toxicity/Apoptosis Response to Chemical Probes Using Incucyte

Protocols for the derivation, establishment, and propagation of adherent glioblastoma (GBM) stem cell cultures have been described previously (Danovi et al., 2010, Danovi, Folarin, Baranowski, & Pollard, 2012; Pollard et al., 2009). Furthermore, a detailed protocol for high-content screening of patient-derived GBM cells with chemical libraries has been described recently (Danovi et al., 2012). We use a slightly modified version of the Danovi protocol as outlined below:

1. We seed GBM511 cells in laminin-coated, clear 96-well plates at 2000 cells per well (Danovi et al., 2012; Gallo et al., 2015; Meyer et al., 2015; Pollard et al., 2009). We avoid seeding cells in the first and last columns and rows of each plate because we find these wells are subject to excessive evaporation during the long incubation periods required for some epigenetic inhibitors.
2. The media used for propagation and screening were Neurocult NS-A basal media (Stem Cell Technologies) supplemented with 10 ng/ml EGF and FGF plus N2 and B27 supplements (Life Technologies) (Danovi et al., 2012; Gallo et al., 2015; Pollard et al., 2009). Media was replenished every 5–6 days by topping-up with 100 μL per well.
3. For chemical probes, we typically prepare a master plate at a 10 mM stock solution in DMSO. Several daughter plates are prepared and kept at $-80°C$ for single use. We typically employ a randomized plate layout to minimize systematic or instrument errors.
4. Chemical probes were added at the indicated concentrations in media 12–24 h after seeding. In order to avoid edge effects, we recommend not screening probes in the first and last rows or columns of the plate.
5. For measuring apoptosis in real-time, we used the CellPlayer kinetic Caspase-3/7 reagent (Essen Bioscience) which couples the activated Caspase-3/7 DEVD recognition motif to NucView488, a DNA intercalating dye to label apoptotic cells. The reagent was added to tissue culture medium at a final concentration of 0.5 μM. Addition of this reagent is nonperturbing to cell growth and morphology and enables the kinetic detection of Caspase-3/7 activity.
6. A cell-impermeable stain like Yoyo-3 (Life Technologies) may be added at 1:10,000 dilution either initially or at the end of the assay to identify nonviable cells.
7. Allow for 20 min equilibration of the plates in the incubator before starting scanning in the Incucyte Zoom station (Essen Bioscience). Briefly, this is a microscope with a bright-field camera equipped with two fluorescent channels (green and red) housed within a standard tissue culture incubator that can capture images continuously at user defined time intervals. The image acquisition parameters were set up as recommended by the manufacturer (Essen Bioscience). We typically acquire four sets of images per well every 2–4 h using the bright field as well as the green and red fluorescence channels.

8. Cell growth, viability, and apoptosis were monitored continuously over a period of 10–12 days. Cell confluence was calculated using the Incucyte Zoom image analysis software following the manufacturer's recommendations. Similarly, red and green fluorescence images were processed using the Incucyte Zoom software and using the adaptive segmentation method. The image analysis pipeline typically consisted of optimization of different parameters using a representative image set (6–10 images) per experiment. Processed images were previewed in bright field, green and red channels, respectively, in order to evaluate accuracy of the processing definition. Subsequently, the image analysis was launched based on successful segmentation of this initial image set.

3.3.2 Example 3: Target Validation with Multiple Probes Per Target

Using this assay, we evaluated the effects of two different probes targeting the SMYD2 methyltransferase. The LLY-507 probe (Nguyen et al., 2015) resulted in rapid loss of cell viability and apoptosis of GBM primary cells, and these changes were evident just a few hours after drug treatment (see Fig. 6). In contrast, the second probe, BAY-598 (http://www.thesgc.org/chemical-probes/BAY-598), did not elicit any discernable effect on cell viability or apoptosis in the same cells (Fig. 6). Moreover in dose–response assays, the cytotoxic effect of LLY-507 was dose-independent and evident as low as 200 nM, whereas BAY-598 did not elicit any cytotoxicity in doses up to 10 μM. This pattern of cytotoxicity has been observed in all 10 GBM cell lines tested so far plus 2 normal fetal neural lines and is consistent with

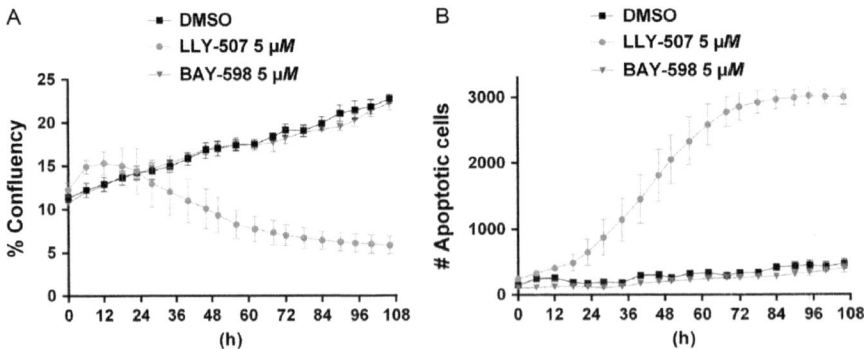

Fig. 6 LLY-507 impairs cell growth and induces rapid cell apoptosis in GBM primary cells. (A) Cell confluence and (B) Apoptosis measurements of GBM511 cells treated with the indicated probes at 5 μM over 10 days. Note the striking differences in the phenotypes observed with the two SMYD2 probes, LLY-507 and BAY-598 probes. (See the color plate.)

LLY-507 having widespread cytotoxicity in neuronal cells; in the original characterization of LLY-507, the probe was not tested in primary neuronal cells (Nguyen et al., 2015). By comparing the phenotypic (cell viability) response to LLY-507 at 200 nM with its potency (IC$_{90}$ = 1 µM) in the biomarker assay, based on the in-cell Western and Western blotting as described earlier (Nguyen et al., 2015), we showed that the cytotoxicity is likely an off-target effect of the compound. Thus, LLY-507 seems to promote off-target cytotoxicity in GBM and neural cells. This result underscores the importance of using multiple probes of different chemical structures, when available, as well as comparing the biomarker response to the phenotypic cell viability response.

Thus, kinetic responses can provide valuable information as to make decisions for primary hit selection and prioritization for subsequent target validation.

4. CONCLUSIONS

Applying chemical probes in an appropriate manner in complex biological systems such as collections of primary patient-derived cells (Edwards et al., 2015), which are more reflective of human disease than common cell lines, will enable future target discovery in many therapeutic areas. In this chapter, we outline strategies for designing and carrying out target-specific biomarker assays that ensure quality experiments and pave the way for exploring new phenotypes achieved with chemical probes for target validation and biological discovery. We illustrated these strategies with examples of epigenetic chemical probes and discussed the caveats and interpretation of the chemical probe phenotypic data in relation to the biomarker experiments.

We point the reader to the chemical probes portal (www.chemicalprobes.org) (Arrowsmith et al., 2015) which is a community-driven resource that provides information on the best available chemical probes for numerous protein families and how they should be used.

We also point the reader to Structural Genomics Consortium's (SGC's) Epigenetic Chemical Probe Library, a collection of epigenetic chemical probes developed and distributed by the SGC that currently comprises more than 35 well-characterized compounds that selectively and potently inhibit or antagonize specific chromatin regulatory proteins or domains. The collection includes probes that target PMTs, bromodomains, methyllysine binders, and demethylases. Furthermore, most chemical probes in the collection also have a "control compound" that is structurally similar to the active probe, but is inactive or much less active against the target protein. When used at appropriate concentrations and experimental time frames and with well-validated antibodies to monitor biomarkers, these probes can provide new and valuable insights in to epigenetic mechanisms and potential new therapeutic strategies.

ACKNOWLEDGMENTS

The Structural Genomics Consortium is funded by AbbVie, Bayer Pharma AG, Boehringer Ingelheim, Canada Foundation for Innovation, Eshelman Institute for Innovation, Genome Canada, Innovative Medicines Initiative (EU/EFPIA) [ULTRA-DD grant no. 115766], Janssen, Merck & Co., Novartis Pharma AG, Ontario Ministry of Economic Development and Innovation, Pfizer, São Paulo Research Foundation-FAPESP, Takeda, and the Wellcome Trust. E. L.-F. is the recipient of a Banting Postdoctoral Fellowship. We would like to thank Amy Donner, Marco Gallo, and Anthony Apostoli for critical reading of the chapter.

REFERENCES

Anighoro, A., Bajorath, J., & Rastelli, G. (2014). Polypharmacology: Challenges and opportunities in drug discovery. *Journal of Medicinal Chemistry*, 57, 7874–7887.
Arrowsmith, C. H., Audia, J. E., Austin, C., Baell, J., Bennett, J., Blagg, J., et al. (2015). The promise and peril of chemical probes. *Nature Chemical Biology*, 11, 536–541.
Arrowsmith, C. H., Bountra, C., Fish, P. V., Lee, K., & Schapira, M. (2012). Epigenetic protein families: A new frontier for drug discovery. *Nature Reviews. Drug Discovery*, 11, 384–400.
Barsyte-Lovejoy, D., Li, F. L., Oudhoff, M. J., Tatlock, J. H., Dong, A. P., Zeng, H., et al. (2014). (R)-PFI-2 is a potent and selective inhibitor of SETD7 methyltransferase activity in cells. *Proceedings of the National Academy of Sciences of the United States of America*, 111, 12853–12858.
Biggar, K. K., & Li, S. S. (2015). Non-histone protein methylation as a regulator of cellular signalling and function. *Nature Reviews Molecular Cell Biology*, 16, 5–17.
Bremang, M., Cuomo, A., Agresta, A. M., Stugiewicz, M., Spadotto, V., & Bonaldi, T. (2013). Mass spectrometry-based identification and characterisation of lysine and arginine methylation in the human proteome. *Molecular bioSystems*, 9, 2231–2247.
Bunnage, M. E., Chekler, E. L., & Jones, L. H. (2013). Target validation using chemical probes. *Nature Chemical Biology*, 9, 195–199.
Chan-Penebre, E., Kuplast, K. G., Majer, C. R., Boriack-Sjodin, P. A., Wigle, T. J., Johnston, L. D., et al. (2015). A selective inhibitor of PRMT5 with in vivo and in vitro potency in MCL models. *Nature Chemical Biology*, 11, 432–437.
Daigle, S. R., Olhava, E. J., Therkelsen, C. A., Majer, C. R., Sneeringer, C. J., Song, J., et al. (2011). Selective killing of mixed lineage leukemia cells by a potent small-molecule DOT1L inhibitor. *Cancer Cell*, 20, 53–65.
Danovi, D., Falk, A., Humphreys, P., Vickers, R., Tinsley, J., Smith, A. G., et al. (2010). Imaging-based chemical screens using normal and glioma-derived neural stem cells. *Biochemical Society Transactions*, 38, 1067–1071.
Danovi, D., Folarin, A. A., Baranowski, B., & Pollard, S. M. (2012). High content screening of defined chemical libraries using normal and glioma-derived neural stem cell lines. *Methods in Enzymology*, 506, 311–329.

Edwards, A. M., Arrowsmith, C. H., Bountra, C., Bunnage, M. E., Feldmann, M., Knight, J. C., et al. (2015). Preclinical target validation using patient-derived cells. *Nature Reviews. Drug Discovery, 14*, 149–150.
Egelhofer, T. A., Minoda, A., Klugman, S., Lee, K., Kolasinska-Zwierz, P., Alekseyenko, A. A., et al. (2011). An assessment of histone-modification antibody quality. *Nature Structural & Molecular Biology, 18*, 91–93.
Eram, M. S., Bustos, S. P., Lima-Fernandes, E., Siarheyeva, A., Senisterra, G., Hajian, T., et al. (2014). Trimethylation of histone H3 lysine 36 by human methyltransferase PRDM9 protein. *The Journal of Biological Chemistry, 289*, 12177–12188.
Fallahi-Sichani, M., Honarnejad, S., Heiser, L. M., Gray, J. W., & Sorger, P. K. (2013). Metrics other than potency reveal systematic variation in responses to cancer drugs. *Nature Chemical Biology, 9*, 708–714.
Fedorov, O., Muller, S., & Knapp, S. (2010). The (un)targeted cancer kinome. *Nature Chemical Biology, 6*, 166–169.
Fog, C. K., Galli, G. G., & Lund, A. H. (2012). PRDM proteins: Important players in differentiation and disease. *Bioessays, 34*, 50–60.
Frankel, A., Yadav, N., Lee, J., Branscombe, T. L., Clarke, S., & Bedford, M. T. (2002). The novel human protein arginine N-methyltransferase PRMT6 is a nuclear enzyme displaying unique substrate specificity. *The Journal of Biological Chemistry, 277*, 3537–3543.
Franken, H., Mathieson, T., Childs, D., Sweetman, G. M., Werner, T., Tögel, I., et al. (2015). Thermal proteome profiling for unbiased identification of direct and indirect drug targets using multiplexed quantitative mass spectrometry. *Nature Protocols, 10*, 1567–1593.
Frye, S. V. (2010). The art of the chemical probe. *Nature Chemical Biology, 6*, 159–161.
Gallo, M., Coutinho, F. J., Vanner, R. J., Gayden, T., Mack, S. C., Murison, A., et al. (2015). MLL5 orchestrates a cancer self-renewal state by repressing the histone variant H3.3 and globally reorganizing chromatin. *Cancer Cell, 28*, 715–729.
Guccione, E., Bassi, C., Casadio, F., Martinato, F., Cesaroni, M., Schuchlautz, H., et al. (2007). Methylation of histone H3R2 by PRMT6 and H3K4 by an MLL complex are mutually exclusive. *Nature, 449*, 933–937.
Guo, A. L., Gu, H. B., Zhou, J., Mulhern, D., Wang, Y., Lee, K. A., et al. (2014). Immunoaffinity enrichment and mass spectrometry analysis of protein methylation. *Molecular & Cellular Proteomics, 13*, 372–387.
Hayashi, K., Yoshida, K., & Matsui, Y. (2005). A histone H3 methyltransferase controls epigenetic events required for meiotic prophase. *Nature, 438*, 374–378.
Huber, K. V., Olek, K. M., Müller, A. C., Tan, C. S., Bennett, K. L., Colinge, J., et al. (2015). Proteome-wide drug and metabolite interaction mapping by thermal-stability profiling. *Nature Methods, 12*, 1055–1057.
Imanishi, M., Tomishima, Y., Itou, S., Hamashima, H., Nakajima, Y., Washizuka, K., et al. (2008). Discovery of a novel series of biphenyl benzoic acid derivatives as potent and selective human beta3-adrenergic receptor agonists with good oral bioavailability. Part 1. *Journal of Medicinal Chemistry, 51*, 1925–1944.
Kaniskan, H. Ü., Zhengtian, Y., Eram, M. S., Luo, X., Yang, X., Schmidt, K., et al. (2015). A potent, selective and cell-active allosteric inhibitor of protein arginine methyltransferase 3 (PRMT3). *Angewandte Chemie, 54*, 5166–5170.
Knapp, S., Arruda, P., Blagg, J., Burley, S., Drewry, D. H., Edwards, A., et al. (2013). A public-private partnership to unlock the untargeted kinome. *Nature Chemical Biology, 9*, 3–6.
Konze, K. D., Ma, A., Li, F. L., Barsyte-Lovejoy, D., Parton, T., MacNevin, C. J., et al. (2013). An orally bioavailable chemical probe of the lysine methyltransferases EZH2 and EZH1. *ACS Chemical Biology, 8*, 1324–1334.

Liu, F., Barsyte-Lovejoy, D., Li, F., Xiong, Y., Korboukh, V., Huang, X. P., et al. (2013). Discovery of an in vivo chemical probe of the lysine methyltransferase G9a and GLP. *Journal of Medicinal Chemistry, 56*, 8931–8942.

Liu, L., Richard, J., Kim, S., & Wojcik, E. J. (2014). Small molecule screen for candidate antimalarials targeting Plasmodium Kinesin-5. *The Journal of Biological Chemistry, 289*, 16601–16614.

Liu, G., Zhang, J., Larsen, B., Stark, C., Breitkreutz, A., Lin, Z. Y., et al. (2010). ProHits: Integrated software for mass spectrometry-based interaction proteomics. *Nature Biotechnology, 28*, 1015–1017.

Martinez Molina, D., Jafari, R., Ignatushchenko, M., Seki, T., Larsson, E. A., Dan, C., et al. (2013). Monitoring drug target engagement in cells and tissues using the cellular thermal shift assay. *Science, 341*, 84–87.

Martinez Molina, D., & Nordlund, P. (2016). The cellular thermal shift assay: A novel biophysical assay for in situ drug target engagement and mechanistic biomarker studies. *Annual Review of Pharmacology and Toxicology, 56*, 141–161.

Mazur, P. K., Reynoird, N., Khatri, P., Jansen, P. W. T. C., Wilkinson, A. W., Liu, S. C., et al. (2014). SMYD3 links lysine methylation of MAP3K2 to Ras-driven cancer. *Nature, 510*, 283–287.

McCabe, M. T., Ott, H. M., Ganji, G., Korenchuk, S., Thompson, C., Van Aller, G. S., et al. (2012). EZH2 inhibition as a therapeutic strategy for lymphoma with EZH2-activating mutations. *Nature, 492*, 108–112.

Merino, A., Bronowska, A. K., Jackson, D. B., & Cahill, D. J. (2010). Drug profiling: Knowing where it hits. *Drug Discovery Today, 15*, 749–756.

Meyer, M., Reimand, J., Lan, X., Head, R., Zhu, X., Kushida, M., et al. (2015). Single cell-derived clonal analysis of human glioblastoma links functional and genomic heterogeneity. *Proceedings of the National Academy of Sciences of the United States of America, 112*, 851–856.

Nguyen, H., Allali-Hassani, A., Antonysamy, S., Chang, S., Chen, L. H., Curtis, C., et al. (2015). LLY-507, a cell-active, potent, and selective inhibitor of protein-lysine methyltransferase SMYD2. *The Journal of Biological Chemistry, 290*, 13641–13653.

Pappano, W. N., Guo, J., He, Y., Ferguson, D., Jagadeeswaran, S., Osterling, D. J., et al. (2015). The histone methyltransferase inhibitor A-366 uncovers a role for G9a/GLP in the epigenetics of leukemia. *PLoS One, 10*. e0131716.

Pollard, S. M., Yoshikawa, K., Clarke, I. D., Danovi, D., Stricker, S., Russell, R., et al. (2009). Glioma stem cell lines expanded in adherent culture have tumor-specific phenotypes and are suitable for chemical and genetic screens. *Cell Stem Cell, 4*, 568–580.

Scholz, C., Weinert, B. T., Wagner, S. A., Beli, P., Miyake, Y., Qi, J., et al. (2015). Acetylation site specificities of lysine deacetylase inhibitors in human cells. *Nature Biotechnology, 33*, 415–423.

Soldi, M., Cuomo, A., Bremang, M., & Bonaldi, T. (2013). Mass spectrometry-based proteomics for the analysis of chromatin structure and dynamics. *International Journal of Molecular Sciences, 14*, 5402–5431.

Sweis, R. F., Pliushchev, M., Brown, P. J., Guo, J., Li, F. L., Maag, D., et al. (2014). Discovery and development of potent and selective inhibitors of histone methyltransferase G9a. *ACS Medicinal Chemistry Letters, 5*, 205–209.

Sylvestersen, K. B., & Nielsen, M. L. (2015). Large-scale identification of the arginine methylome by mass spectrometry. *Current Protocols in Protein Science, 82*, 24.7.1–24.7.17.

Vedadi, M., Barsyte-Lovejoy, D., Liu, F., Rival-Gervier, S., Allali-Hassani, A., Labrie, V., et al. (2011). A chemical probe selectively inhibits G9a and GLP methyltransferase activity in cells. *Nature Chemical Biology, 7*, 566–574.

Verma, S. K., Tian, X. R., LaFrance, L. V., Duquenne, C., Suarez, D. P., Newlander, K. A., et al. (2012). Identification of potent, selective, cell-active inhibitors of the histone lysine methyltransferase EZH2. *ACS Medicinal Chemistry Letters*, *3*, 1091–1096.

Workman, P., & Collins, I. (2010). Probing the probes: Fitness factors for small molecule tools. *Chemistry & Biology*, *17*, 561–577.

Yu, W., Chory, E. J., Wernimont, A. K., Tempel, W., Scopton, A., Federation, A., et al. (2012). Catalytic site remodelling of the DOT1L methyltransferase by selective inhibitors. *Nature Communications*, *3*, 1288–1298.

Zhu, J., Sammons, M. A., Donahue, G., Dou, Z., Vedadi, M., Getlik, M., et al. (2015). Gain-of-function p53 mutants co-opt chromatin pathways to drive cancer growth. *Nature*, *525*, 206–211.

CHAPTER FIVE

Mapping Lysine Acetyltransferase–Ligand Interactions by Activity-Based Capture

D.C. Montgomery, J.L. Meier[1]

Chemical Biology Laboratory, National Cancer Institute, Frederick, MD, United States
[1]Corresponding author: e-mail address: jordan.meier@nih.gov

Contents

1. Introduction	106
2. Technical Aspects	110
2.1 Synthesis of KAT Capture Probes	110
2.2 Preparation of Proteomes	113
2.3 Activity-Based Capture of KATs and Competitive Profiling of KAT–Ligand Interactions	114
2.4 Analysis of KAT–Ligand Interactions by Immunoblot	117
2.5 Analysis of KAT Capture by LC-MS/MS	118
3. Discussion	118
3.1 Critical Parameters and Troubleshooting	118
3.2 Future Applications and Directions	120
References	121

Abstract

Changes in reversible protein acetylation mediate many key aspects of genomic regulation and enzyme function. The catalysts for this posttranslational modification, lysine acetyltransferases (KATs), have been difficult targets for characterization due to their complex architecture and challenging reconstitution. To address this challenge, here we describe methods to profile endogenous KAT activities using activity-based probes. This method facilitates the targeted analysis of several cellular KATs and can be used to study their interactions with many different types of ligands, including acyl-CoA metabolites. This competitive activity-based capture approach provides a method to assess the selectivity of ligands for different KAT families in complex proteomic settings, and thus has the potential to offer substantial insights into the regulation of cellular KAT function.

1. INTRODUCTION

Lysine acetyltransferase (KAT) enzymes catalyze protein acetylation, a central and abundant posttranslational modification (PTM) in all living organisms. Acetylation is best known for its role in transcription, where it can reduce electrostatic histone–DNA interactions and thus relax the chromatin fiber (Allfrey, Faulkner, & Mirsky, 1964), and also serve as a platform for signal transduction via the recruitment of bromodomain-containing proteins (Dhalluin et al., 1999). More recently, lysine acetylation has been shown to play an important role in the regulation of many nonnuclear functions, including metabolism. Consistent with the central role of acetylation in eukaryotic cell biology, disruption or dysregulation of KAT activity can profoundly impact disease phenotypes. For example, loss-of-function mutations in the KATs EP300/CREBBP are a known cause of the developmental disorder Rubinstein–Taybi syndrome (Petrij et al., 1995) and have also been implicated as drivers of oncogenesis in several types of cancer (Gayther et al., 2000; Pasqualucci et al., 2011; Peifer et al., 2012). Aberrant gene fusions of the KAT enzyme MOZ are observed in acute myeloid leukemia and can trigger dedifferentiation of white blood cells to a degree exceeding that of the well-known oncogenic gene fusion BCR-ABL (Carapeti, Aguiar, Goldman, & Cross, 1998; Huntly et al., 2004). These findings, as well as others, implicate regulation of KAT activity as a key driver of disease pathology.

In addition to mutations, cellular KAT activity can also be regulated by more elusive factors. These include regulatory PTMs, acetyl-CoA levels, integration into multiprotein complexes, and inhibition by endogenous metabolites or synthetic ligands (Fig. 1A; Lee et al., 2014; Lee & Workman, 2007; Thompson et al., 2004). Many of these features are challenging to recapitulate in vitro, where KAT enzymes are often studied as recombinant domains excised from their full-length protein (Fig. 1B). Methods to profile the activity of native cellular KAT enzymes thus have the potential to complement in vitro studies and provide new insights into the regulation of KAT function. Activity-based protein profiling (ABPP) represents one powerful method to study endogenous enzyme families (Cravatt, Wright, & Kozarich, 2008). In this approach, synthetic active site probes directed toward an enzyme class of interest are modified with chemical handles enabling their detection or affinity enrichment from

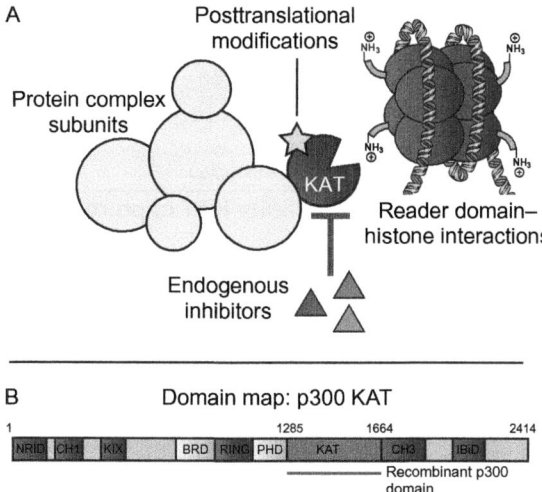

Fig. 1 Differences in the regulation of lysine acetyltransferase (KAT) activity in biochemical and cellular contexts. (A) Cellular acetylation is regulated by a multitude of different KAT activities, which are present as members of multiprotein complexes and regulated by expression level, substrate availability, endogenous metabolic inhibitors, and reversible posttranslational modifications. (B) KATs are large, multidomain enzymes whose native activity is challenging to reconstitute in vitro. Shown is the domain architecture of the prototypical KAT p300. Underlined in *red* is the commonly studied, commercially available p300 KAT domain. (See the color plate.)

complex proteomes. Since enzymes must be properly folded and have a free active site to bind the activity-based probe, their enrichment (as assessed by western blot or LC-MS/MS) can be used as a proxy for the enzyme's activity in the cellular sample of interest. ABPP has been applied to a variety of enzyme families, including hydrolases, proteases, kinases, and histone deacetylases (Bantscheff et al., 2011; Greenbaum, Medzihradszky, Burlingame, & Bogyo, 2000; Liu, Patricelli, & Cravatt, 1999). The first exploration of an ABPP method to study KAT enzymes utilized a novel sulfoxycarbamate analogue of the CoA cofactor (Fig. 2A). Upon incubation with the KAT enzyme p300, this probe was shown to transfer a desthiobiotin affinity handle to an active site cysteine, allowing p300 detection via western blot (Hwang et al., 2007). However, applications of this probe were limited by its dependence on the presence of a nucleophilic active site cysteine, which many KATs lack (Berndsen & Denu, 2008), as well as demonstrated cross-reactivity with non-KAT enzymes (Hwang et al., 2007).

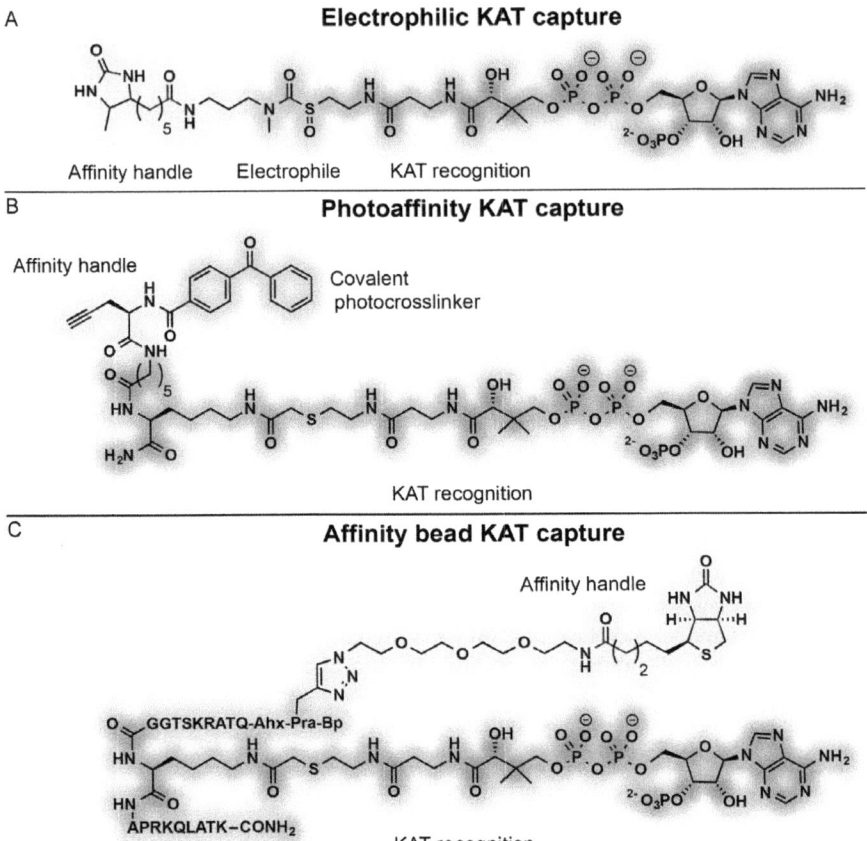

Fig. 2 Overview of activity-based probes for studying cellular KAT activity. (A) KATs can be labeled covalently with sulfoxide thiocarbamates, which transfer a reactive acyl group to the KAT active sites. This was the first method shown to be capable of detecting KAT active sites in cell lysates. (B) KATs can be labeled covalently using high-affinity bisubstrate inhibitors containing clickable photoaffinity groups. This approach allows stringent washing, but is limited by the low yield of the photocrosslinking step. (C) KATs can be enriched via a noncovalent capture method using bead-based capture. This approach enables low-abundance nuclear KATs to be studied and is the primary method described herein. (See the color plate.)

To address these challenges, our group recently developed an alternative activity-based approach to profile KAT activity in complex proteomes utilizing bisubstrate inhibitors (Fig. 2B and C). Pioneered by Cole and coworkers, KAT bisubstrate inhibitors are synthetic probes consisting of CoA linked to the ε-amino group of a lysine-containing peptide through a nonhydrolyzable thioether linkage (Lau et al., 2000). The resulting bidentate ligand can interact with both the substrate and cofactor-binding pocket of KAT active sites, driving high binding affinity and inhibitory potency (Liu et al., 2008). In order to adapt this classic inhibitor scaffold for activity-based profiling, our laboratory performed biochemical studies and found that modification of KAT bisubstrate inhibitors at the peptide N-terminus was well tolerated (Montgomery, Sorum, & Meier, 2014). This led us to develop a first-generation KAT activity-based profiling approach, where this site was used to incorporate clickable photoaffinity tags that could facilitate KAT detection (Fig. 2B). These probes enabled facile, activity-dependent profiling of KATs in vitro and in complex proteomes and also provided enrichment of endogenous KAT enzymes. However, a limitation was their dependence on a low-yielding photocrosslinking step, which impeded the detection of low-abundance nuclear KATs. Inspired by alternative chemoproteomic strategies, we recently developed a second-generation KAT profiling approach that provides the sensitivity required to enrich low-abundance nuclear KATs from reasonable quantities of native cell lysates ($\sim$5–10 $\times$ 10^6 cell equivalents) (Montgomery, Sorum, Guasch, Nicklaus, & Meier, 2015). This method utilizes biotin-labeled KAT bisubstrate inhibitors, immobilized on streptavidin-modified agarose beads (Fig. 2C). Incubation with KAT-containing proteomes, followed by a mild wash and elution, allows activity-based capture of low-abundance nuclear KATs including Gcn5, pCAF, and Mof. In addition, this approach also captures members of KAT complexes, as well as additional CoA-utilizing enzymes. This latter class includes N-terminal protein acetyltransferases, some of which have been posited to harbor orphan KAT activities (Evjenth et al., 2009; Montgomery, Sorum, & Meier, 2015). In our first application of this technology, we applied it in a competitive format to study KAT–ligand interactions (Fig. 3). This led to the unanticipated finding that palmitoyl-CoA binds to members of the GCN5 KAT family with high affinity, an observation that may have implications for their metabolic regulation (Montgomery, Sorum, Guasch, et al., 2015). In addition to metabolites, this approach also allows the study of KAT–small molecule interactions and thus provides a useful tool for studying the selectivity of synthetic inhibitors with KATs and specific KAT complexes. Here, we describe detailed procedures for activity-based capture of KAT enzymes from complex proteomes, highlighting key steps as well as future opportunities for the application of this powerful technique in the study of acetylation biology.

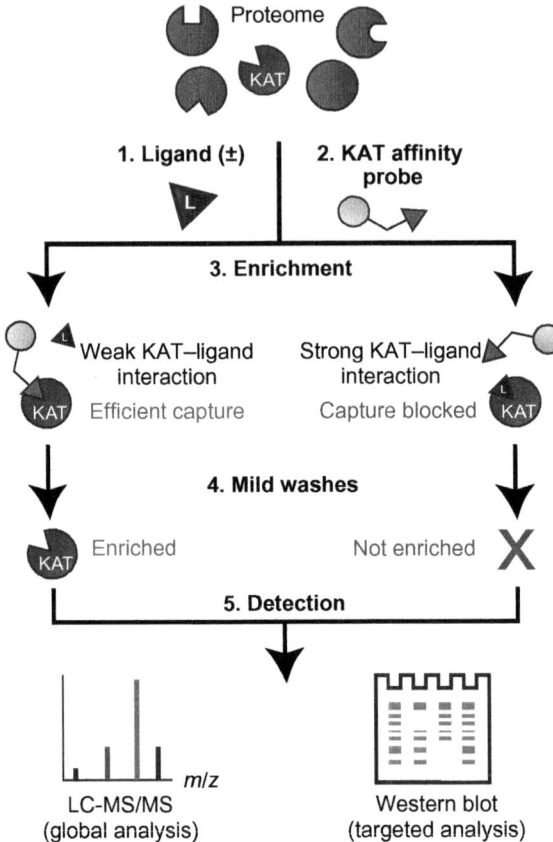

Fig. 3 Analyzing KAT–ligand interactions by competitive chemical proteomic profiling. In a two-step procedure, KAT-containing proteomes are incubated with a ligand of interest, and then these proteomes are mixed with a streptavidin-immobilized KAT capture probe. Ligands that interact strongly with KATs block affinity capture. Analysis of treated and untreated samples by LC-MS/MS or western blot enables quantitative rank-ordering of how KATs interact with ligands of interest. (See the color plate.)

2. TECHNICAL ASPECTS
2.1 Synthesis of KAT Capture Probes

Biotinylated KAT bisubstrate inhibitors for use in the described assay can be synthesized by a variety of routes. Peptide portions of bisubstrate inhibitors are readily synthesized using conventional Fmoc solid-phase peptide synthesis. From a design perspective, it is important that the lysine where CoA is to be covalently attached to the peptide bear an orthogonal protecting group (ie, Dde or ivDde) that is stable to standard deprotection conditions on the ε-amino lysine side chain. This facilitates selective on-resin deprotection of the Dde group, followed by coupling of the ε-amine to a haloacetic acid, in order to produce an electrophilic peptide amenable to coupling with the CoA thiol. Detailed procedures for the synthesis of KAT bisubstrate inhibitors have been excellently described elsewhere (Zheng et al., 2004). For incorporation of the biotin affinity handle, we utilize a propargyl glycine located at the N-terminal region of the bisubstrate peptide sequence. This allows straightforward incorporation of affinity handles using a preparative Cu-catalyzed [3 + 2] cycloaddition (click chemistry) reaction. Overall, the procedure consists of preparing the haloacetamide-containing peptide on resin, followed by two sequential solution-phase couplings, and provides ready access to high-purity KAT capture probes.

2.1.1 Materials
Peptide synthesis vessel
Standard Fmoc-amino acids (EMD Millipore)
N,N-Dimethylformamide (DMF)
Dichloromethane (DCM)
Rink Amide resin (100–200 mesh, EMD Millipore)
Fmoc-L-propargylglycine (Anaspec, catalog number AS-26264-F1)
Fmoc-Lys(ivDde)-OH (EMD Millipore, catalog number 852082)
Bromoacetic acid (Sigma-Aldrich catalog number 17000)
N,N'-Diisopropylcarbodiimide (Sigma-Aldrich catalog number D125407)
Coenzyme A lithium salt (US Biological, catalog number C7505-51)
Preparative HPLC
Lyophilizer
Phosphate-buffered saline (PBS) (Thermo Fisher, catalog number 10010)
Thermo Fisher Nanodrop 2000 spectrophotometer
Azide-PEG3-biotin conjugate (Sigma-Aldrich, catalog number 762024)
Tris(3-hydroxypropyltriazolylmethyl)amine (THPTA; Click Chemistry Tools, catalog number 1010-100)

2.1.2 Synthesis of Alkyne-Modified KAT Bisubstrate Inhibitors
1. Prepare the peptide precursor of the bisubstrate inhibitor via solid-phase peptide synthesis on Rink Amide resin according to the method of Cole and coworkers (Zheng et al., 2004). This protocol gives general guidelines for performing syntheses on a 50–500 μmol scale. Take care to incorporate the Fmoc-Lys(Dde)-OH amino acid at the position to be modified with CoA and the Fmoc-L-propargylglycine at the N-terminus. Cap with acetic anhydride using a mixture of 4:2:1 DMF:DIEA:Ac$_2$O.

2. To deprotect the Dde group, add a solution of 2% hydrazine in DMF to peptide resin, shake 2 h, and wash with DMF and DCM. Calculating based on resin loading, add five equivalents each of bromoacetic acid and N,N'-diisopropylcarbodiimide in DMF. Shake overnight, monitor by the Kaiser test until the reaction is complete, and wash resin well with DMF and DCM.
3. To cleave the peptide from resin, add a 5 mL solution of 95% trifluoroacetic acid (TFA), 2.5% triisopropylsilane, and 2.5% water. Shake for 1.5 h, filter the cleavage solution, and wash the resin with an additional 1 mL TFA. Collect and pool the filtrate in a 50-mL conical tube.
4. Remove excess TFA with a gentle stream of air to reduce the volume to 1–2 mL (caution: TFA is corrosive and hazardous).
5. Precipitate the haloacetamide-modified peptide by addition of diethyl ether (chilled on dry ice). Centrifuge at 4300 rcf for 15 min, discard supernatant, and air-dry pellet. Redissolve in aqueous solution and purify by preparative HPLC. Pool and lyophilize product-containing fractions.
6. To conjugate the haloacetamide peptide to CoA, dissolve the lyophilized peptide in 0.5 mL 100 mM NaHCO$_3$ [pH 8]. Add 1 equivalent coenzyme A trilithium salt and stir for 30 min. The progress of the reaction can be monitored by analytical HPLC.
7. Purify the reaction mixture by preparative HPLC. Pool and lyophilize product-containing fractions.
8. Dissolve in PBS and quantify bisubstrate inhibitor via Nanodrop 2000 (or equivalent UV–vis spectrophotometer) using the molar extinction coefficient (ε) for CoA of 15,000 M^{-1} cm^{-1} at λ_{max} of 259 nm (Killenberg & Dukes, 1976). This solution can be stored at −80°C or carried forward immediately for biotin conjugation via click chemistry.

2.1.3 Synthesis of Biotinylated KAT Bisubstrate Inhibitors

1. To a solution of the alkyne-modified bisubstrate inhibitor in PBS (2 mM, 250 µL), sequentially add sodium phosphate buffer (2 mL 100 mM [pH 7]), biotin-PEG-azide (40 µL of 25 mM in DMSO), and a premixed solution of CuSO$_4$/THPTA (5 µL of 50 mM CuSO$_4$, 25 µL of 50 mM THPTA) (Hong, Presolski, Ma, & Finn, 2009).
2. Initiate the cycloaddition reaction by adding sodium ascorbate (125 µL of 100 mM sodium ascorbate).
3. Rotate the mixture for 1 h at room temperature. The progress of the reaction can be monitored by analytical HPLC.

4. Purify reaction via reverse-phase preparative HPLC. Pool and lyophilize product-containing fractions.
5. Dissolve in PBS and quantify biotinylated bisubstrate inhibitor via Nanodrop 2000 (or equivalent UV–vis spectrophotometer) using the molar extinction coefficient (ε) for CoA of 15,000 $M^{-1} cm^{-1}$ at λ_{max} of 259 nm. This solution can be stored at $-80°C$.

2.2 Preparation of Proteomes

Proteomes are prepared from cells grown via standard cell culture methods. For all activity-based profiling methods, it is essential to minimize protein precipitation and denaturation. This necessitates that lysis be performed in cold, nondenaturing buffers such as PBS. Cells are lysed using sonication on ice, and the resulting solution is cleared via centrifugation at 4°C. Soluble proteomes may then be stored at $-80°C$ in a solution of PBS/5% glycerol and thawed on ice immediately prior to KAT capture experiments. Protease inhibitors may be added if critical to the specific experiment (eg, when studying the interaction of KATs with protease-sensitive ligands), but in our experience can generally be omitted with minimal effects.

2.2.1 Materials
Harvested cell pellet
PBS (Thermo Fisher, catalog number 10010)
Ultrasonicator (Qsonica Q700)
Refrigerated benchtop centrifuge
Qubit Protein Assay Kit (catalog number Q33212)
Qubit Fluorometer (catalog number Q32866)
Glycerol (Sigma-Aldrich, catalog number G5516)

2.2.2 Cell Lysate Preparation
1. Wash harvested cell pellet twice with chilled PBS, pelleting by centrifugation for 3 min at 1140 rcf, and discarding supernatant.
2. Add cold PBS to cell pellet (~10 mL PBS per 2.5 mL packed cells).
3. Lyse cells via ultrasonication on ice, using 15 pulses (1 s each at 5% amplitude) with 1 min between each pulse
4. Centrifuge for 30 min at 20,800 rcf at 4°C.
5. Transfer supernatant to fresh tube.
6. Quantify by Qubit Protein Assay according to the manufacturer's instructions.

7. Dilute to desired final concentration (~1–2 mg/mL) and add glycerol to a final concentration of 5%.
8. Aliquot proteomes to avoid freeze–thaw cycles (~0.5–2 mg depending on application) and store at −80°C.

2.3 Activity-Based Capture of KATs and Competitive Profiling of KAT–Ligand Interactions

The following protocol represents a standard procedure for activity-based capture, enrichment, and elution of KAT enzymes. The presence of KATs in these enriched samples can be determined in either a targeted manner by western blotting (Section 2.4) or an unbiased manner via LC-MS/MS (Section 2.5). Performing multiple parallel capture experiments in the presence of a small molecule that competes with the capture probe for the KAT active site (competitive profiling) enables KAT-ligand interactions to be quantitatively assessed. For these competitive profiling experiments, it is important that all enrichments (with and without competitor) be performed in parallel from the same lysate stock to minimize variability. In our experience, the parallel analysis of 6–12 conditions is relatively manageable; expanding beyond this number increases the lag time between steps (ie, capture, wash) for each sample and can introduce artifacts and error into KAT–ligand interaction experiments. An additional technical consideration is that in order to minimize background, we have found that it is useful to clarify lysates through a syringe filter immediately prior to KAT capture experiments. The following protocol is written to perform 10 conditions in parallel (ie, 1 KAT capture "no competitor" condition + 9 competitive profiling conditions). This scale enables samples to be analyzed on a conventional 12-lane gel, leaving room for an input and ladder.

2.3.1 Materials

Cellular proteomes
0.22 μm PVDF syringe filter, ultra-low protein binding (EMD Millipore, catalog number SLGV033RS)
Competitor ligand
Streptavidin-agarose resin (Thermo Fisher Scientific, catalog number 20353)
PBS (Thermo Fisher Scientific, catalog number 10010)
Refrigerated benchtop centrifuge

Wash buffer: 50 mM Tris–HCl [pH 7.5], 5% glycerol (note: omit glycerol for LC–MS/MS samples), 1.5 mM MgCl$_2$, and 150 mM NaCl (Bantscheff et al., 2007)

Centrifugal filters (VWR, catalog number 82031-356)

2.3.2 Enrichment of KATs in the Presence or Absence of Active Site Competitive Ligands

1. Thaw 500 μL of 1.5 mg/mL proteome per experimental condition (ie, 5 mL for 10 conditions) on ice.
2. Separately, add ~110 μL of streptavidin-agarose resin (50% slurry) per condition to a 1.7-mL centrifuge tube (ie, 1.1 mL of 50% resin slurry for 10 conditions). Centrifuge for 3 min at 1400 rcf to pellet resin.
3. Wash resin three times with 1 mL PBS. Between washes, centrifuge for 1 min at 1400 rcf and discard supernatant.
4. Add PBS to regenerate a 50% streptavidin-agarose slurry (ie, add 550 μL PBS for 10 samples for a total volume of 1.1 mL).
5. Aliquot 100 μL of 50% streptavidin-agarose slurry to individual 1.7-mL centrifuge tubes. To ensure equal volumes of resin are used in each condition, the slurry must be well mixed prior to every aliquot, and pipette tips should be cut with clean scissors or razor blade to facilitate transfer. Note: All further steps will be performed in parallel for each condition, so care must be taken that all samples are handled in a consistent fashion.
6. Centrifuge samples for 3 min at 1400 rcf.
7. Remove and discard the supernatants with a pipette tip, taking care not to disturb the resin.
8. Add 500 μL of biotinylated KAT bisubstrate inhibitor (10 μM in PBS) to each sample (total volume ~550 μL per condition, including resin).
9. Rotate samples for 1 h at room temperature.
10. While KAT capture resins are equilibrating, separately filter proteome using syringe filter. Aliquot 500 μL of filtered lysate (~1.5 mg/mL) per condition into individual, well-labeled 1.7-mL centrifuge tubes.
11. For competitive profiling experiments, add KAT-interacting ligands (typically 1–1000 μM in PBS) to the desired final concentration and allow to equilibrate with proteomes. Note: Care should be taken to ensure ligands are fully soluble at all tested concentrations and do not cause precipitation upon addition. For example, acidic ligands (such as HPLC-purified acyl-CoA analogues) can alter sample pH

and should be appropriately buffered. To minimize differences in KAT capture resulting from variations in DMSO/buffer conditions, an equivalent vehicle solution should be added to "no competitor" samples.
12. Allow lysates to equilibrate with competitor ligands for 30 min on ice.
13. Returning to the streptavidin-agarose resin: after 1 h incubation with probe is complete, centrifuge for 3 min at 1400 rcf. Discard the supernatants. Note: If using a concentrated solution of biotinylated KAT capture probe (ie, 100–1000 μM), the streptavidin-agarose resin can be subjected to an additional PBS wash in order to avoid potential competition by any remaining excess, unbound probe.
14. Combine each individual lysate/competitor mixture with a single KAT capture resin sample, bringing the volume of each sample to ~550 µL including resin.
15. Rotate for 1 h at room temperature.
16. Centrifuge for 3 min at 1400 rcf.
17. Remove supernatant. Note: These supernatants can be saved and analyzed for depletion of KATs from whole cell lysates by immunoblot, which can provide a useful secondary measure of KAT capture efficiency.
18. Add 500 µL ice-cold wash buffer. Note: All wash steps should be performed quickly and care should be taken to avoid sample warming in order to preserve noncovalent KAT capture. For this purpose, we recommend either performing washes in a cold room or working quickly using a cooled centrifuge and keeping individual samples on ice between wash steps.
19. Rotate for 3 min.
20. Centrifuge for 1 min at 1400 rcf at 4°C.
21. Discard supernatants.
22. Repeat steps 18–21.
23. Repeat steps 18–20. Do not discard supernatants from final wash.
24. Using cut pipette tips, transfer resin and a small portion of the final supernatant to labeled centrifugal filters. Use remaining supernatant solution to wash down sides of tube and transfer to centrifugal filter.
25. Centrifuge for 3 min at 1400 rcf at 4°C.
26. Discard filtrate, collecting enriched resin in centrifugal filters. Note: If necessary, this can serve as a pause point where enriched resins can be stored overnight at 4°C without significant protein degradation.

2.4 Analysis of KAT–Ligand Interactions by Immunoblot

This procedure details the elution of the enriched KAT samples from resin for quantitative analysis of competitive KAT capture experiments by western blot. Treatment of KAT capture resins with a detergent-containing buffer and heat is used to denature bound proteins, disrupting the interaction of KATs with immobilized bisubstrate inhibitors and providing enriched samples amenable to standard immunoblotting methods.

2.4.1 Materials
NuPAGE 4× LDS sample buffer (Thermo Fisher Scientific, catalog number NP0007)
Dithiothreitol (Sigma-Aldrich, catalog number D0632)
Eppendorf thermomixer
Benchtop centrifuge
4–12% Bis–Tris SDS-PAGE gel (Thermo Fisher Scientific, catalog number NP0322)
XCell II Blot Module (Thermo Fisher Scientific, catalog number EI9051)

2.4.2 Elution and Western Blot Sample Preparation
1. To enrich resin collected in centrifugal filters, add 40 μL of 1× LDS sample buffer (diluted from 4× stock and supplemented with 100 mM DTT).
2. Heat with shaking at 95°C for 10 min.
3. Centrifuge for 3 min at 1400 rcf to collect enriched KATs eluted from resin in filtrates.
4. Transfer filtrates to fresh, well-labeled 1.7-mL centrifugal tubes.
5. Repeat steps 1–3, collecting a second filtrate from KAT capture resin (using resin still contained in original centrifugal filter tubes).
6. For each condition, combine filtrates from both elution steps, giving a total volume of 60–80 μL per condition.
7. Load 10–20 μL per lane on a 4–12% Bis–Tris SDS-PAGE gel (200 V constant, 35 min).
8. Transfer gel to western blot. Block, probe with anti-KAT antibodies, wash, and develop immunoblot according to the manufacturer's protocol.

2.5 Analysis of KAT Capture by LC-MS/MS

Analyzing KAT capture experiments by LC-MS/MS provides orthogonal information to immunoblotting, enabling analysis of a wider range of CoA-dependent proteins and their interactions with KAT-competitive ligands. In these experiments, care must be taken to avoid any sources of sample contamination by abundant protein sources (such as keratin or BSA) in order to maximize sensitivity. Similarly, all reagents and buffers should be mass spectrometry grade.

2.5.1 Materials
Trypsin buffer: 50 mM Tris–HCl [pH 8.0], 1 M urea
1 M CaCl$_2$
Sequencing grade modified trypsin (Promega, catalog number V5111)
Formic acid, mass spectrometry grade (Sigma-Aldrich, catalog number 94318)

2.5.2 On-Bead Trypsin Digest and LC-MS/MS Sample Preparation
1. To resin in centrifugal filters, add 200 μL trypsin buffer.
2. Using a cut pipette tip, transfer solution and resin to fresh 1.7-mL centrifugal tubes.
3. Wash any remaining resin from filter using an additional 200 μL trypsin buffer for a total volume of ~450 μL including resin.
4. Add 0.4 μL of 1 M CaCl$_2$ to each sample.
5. Add 4 μL of a 0.25 mg/mL solution of trypsin to each sample.
6. Incubate with shaking at 37°C overnight.
7. Add formic acid to a final concentration of 5% to quench the trypsin.
8. Centrifuge for 3 min at 1400 rcf.
9. Transfer supernatants to fresh 1.7-mL centrifugal tubes.
10. Store samples at −80°C prior to proteomic analysis. Note: At this point samples may be desalted and analyzed by mass spectrometry according to the end user's preferred protocols.

3. DISCUSSION
3.1 Critical Parameters and Troubleshooting

One aspect critical to the success of activity-based capture is the quality of the cellular proteomes. For large-scale competitive activity-based capture experiments, we typically use lysates prepared from HeLa cells harvested at mid-log phase ($\sim$0.5–0.65 $\times$ 10^6 cells/mL, >98% viability) cells grown in 8-L spinner flasks. Cells are collected by centrifugation, washed twice with PBS, and resuspended in a Tris–glycerol buffer (50 mM Tris pH 7.5, 1 mM EDTA, 30% glycerol, fresh protease inhibitor) at a ratio of 2 mL for each mL of cell mass. This last step has been found to improve the activity of nuclear extracts commonly used for in vitro transcription assays (Theisen, Gucwa, Yusufzai, Khuong, & Kadonaga, 2013), and we have likewise found that it facilitates increased activity-based capture of KATs and KAT-containing nuclear protein complexes. These glycerol-stabilized cell pellets can be stored for months at $-80°C$ and freshly lysed in PBS immediately prior to individual experiments. An additional quality-control step that can be used to verify integrity of cellular proteome activities is to perform gel-based labeling experiments using commercially available activity-based probes such as TAMRA-FP (Thermo Fisher), a fluorophosphonate probe that will readily label active serine hydrolases (Kidd, Liu, & Cravatt, 2001). Competitive chemoproteomic analyses of KAT–ligand interactions will also be highly dependent on the stability of the ligand in a proteomic setting. For analysis of KAT/acyl-CoA interactions, we have used coupled enzyme methods that detect CoA to assess the hydrolysis of acyl-CoAs in cell lysates (Montgomery, Sorum, Guasch, et al., 2015). Another important metric to assess in any activity-based capture experiment is background, which can hinder analyses of KAT–ligand interactions. For example, we have observed that even slight protein precipitation at any point in the experiment can cause substantial background. This can be reduced by filtering freshly prepared lysates prior to initiating the pull-down experiment. Background due to precipitation (or other factors) can also be quantitatively assessed by immunoblotting-enriched samples against proteins such as cyclooxygenase IV (CoxIV), which do not interact with our KAT activity-based capture probes but which are present at high levels in precipitated protein fractions. A final consideration is the choice of internal controls to ensure any differences in enrichment are due to authentic KAT–ligand interactions, and not simply a result of technical issues such as sample loss or unequal loading. Here again, CoxIV immunoblot can serve as a valuable internal control in KAT capture experiments since it nonspecifically interacts with streptavidin beads and is not competed by KAT-interacting ligands. Quantitative gel densitometry can be used to normalize KAT capture in the presence or absence of a ligand relative to CoxIV capture, thus providing a more reliable portrait of KAT–ligand interactions in the cellular sample of interest.

3.2 Future Applications and Directions

While this chapter has focused on the applications and technical aspects of current activity-based KAT capture methods, it is also important to highlight some limitations. A central limitation is the fact that reported ABPP methods have been shown to target only a relatively small number of validated and putative KAT family members (Fig. 4). In the future, a number of improvements may help address this constraint. For example, in our studies of clickable photoaffinity capture reagents, we found that molecular recognition driven by the peptide moiety of KAT bisubstrate inhibitors can direct the efficiency of activity-based labeling toward discrete KAT enzyme families (Montgomery et al., 2014). In future studies, it will be important to assess whether expanding the range of peptidyl-bisubstrate inhibitors used

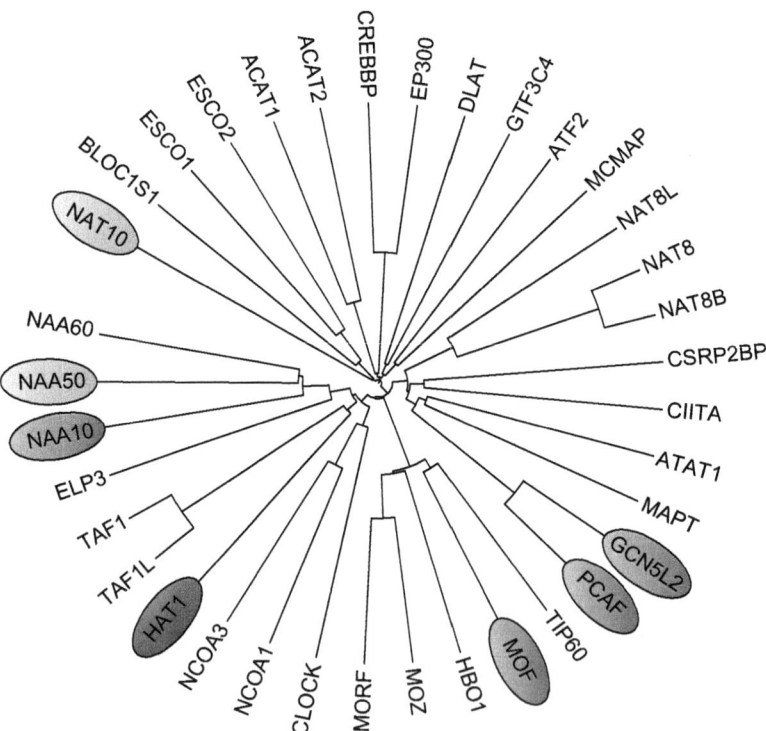

Fig. 4 Phylogenetic tree of putative and validated KAT enzymes (http://apps.thesgc.org/resources/phylogenetic_trees/). *Circled* enzymes represent KATs whose activity has been monitored in an endogenous context using chemical proteomics. *Orange circle*, KAT sulfoxide thiocarbamate probe; *yellow circles*, KAT-clickable photoaffinity bisubstrate probe; *green circles*, KAT-biotin bisubstrate probe. (See the color plate.)

in KAT enrichment can similarly increase the scope of the KAT-ome subject to activity-based capture. Choice of capture platform may also play a major role in profiling low-abundance KATs. In addition to the biotin–streptavidin approach used here, recently groups have reported activity-based capture methods that utilize Sepharose (Bantscheff et al., 2007, 2011) or HaloTag-derivatized beads (Friedman Ohana et al., 2015) in which the ligand is covalently immobilized to the bead surface. How differences in the density of bead surface functionalization alter the sensitivity of KAT pull-down will also need to be systematically explored, and the protocols and technical considerations outlined here should aid such studies. Finally, integrating activity-based KAT capture with quantitative proteomic technologies such as SILAC (Martin, Wang, Adibekian, Tully, & Cravatt, 2012; Ong et al., 2002) should significantly aid the ability to quantify changes in KAT activity, by providing a sensitive, digital readout which requires fewer peptides than other quantification methods (eg, spectral counting). Continued applications of activity-based KAT capture methods have the potential to expand our understanding of acetylation by revealing new candidate acetylation catalysts, illuminating global principles of metabolic regulation, and facilitating efforts to identify and characterize small molecule KAT inhibitors.

REFERENCES

Allfrey, V. G., Faulkner, R., & Mirsky, A. E. (1964). Acetylation and methylation of histones and their possible role in the regulation of RNA synthesis. *Proceedings of the National Academy of Sciences of the United States of America, 51*, 786–794.

Bantscheff, M., Eberhard, D., Abraham, Y., Bastuck, S., Boesche, M., Hobson, S., et al. (2007). Quantitative chemical proteomics reveals mechanisms of action of clinical ABL kinase inhibitors. *Nature Biotechnology, 25*, 1035–1044.

Bantscheff, M., Hopf, C., Savitski, M. M., Dittmann, A., Grandi, P., Michon, A. M., et al. (2011). Chemoproteomics profiling of HDAC inhibitors reveals selective targeting of HDAC complexes. *Nature Biotechnology, 29*, 255–265.

Berndsen, C. E., & Denu, J. M. (2008). Catalysis and substrate selection by histone/protein lysine acetyltransferases. *Current Opinion in Structural Biology, 18*, 682–689.

Carapeti, M., Aguiar, R. C., Goldman, J. M., & Cross, N. C. (1998). A novel fusion between MOZ and the nuclear receptor coactivator TIF2 in acute myeloid leukemia. *Blood, 91*, 3127–3133.

Cravatt, B. F., Wright, A. T., & Kozarich, J. W. (2008). Activity-based protein profiling: From enzyme chemistry to proteomic chemistry. *Annual Review of Biochemistry, 77*, 383–414.

Dhalluin, C., Carlson, J. E., Zeng, L., He, C., Aggarwal, A. K., & Zhou, M. M. (1999). Structure and ligand of a histone acetyltransferase bromodomain. *Nature, 399*, 491–496.

Evjenth, R., Hole, K., Karlsen, O. A., Ziegler, M., Arnesen, T., & Lillehaug, J. R. (2009). Human Naa50p (Nat5/San) displays both protein N alpha- and N epsilon-acetyltransferase activity. *The Journal of Biological Chemistry, 284*, 31122–31129.

Friedman Ohana, R., Kirkland, T. A., Woodroofe, C. C., Levin, S., Uyeda, H. T., Otto, P., et al. (2015). Deciphering the cellular targets of bioactive compounds using a chloroalkane capture tag. *ACS Chemical Biology, 10*, 2316–2324.

Gayther, S. A., Batley, S. J., Linger, L., Bannister, A., Thorpe, K., Chin, S. F., et al. (2000). Mutations truncating the EP300 acetylase in human cancers. *Nature Genetics, 24*, 300–303.

Greenbaum, D., Medzihradszky, K. F., Burlingame, A., & Bogyo, M. (2000). Epoxide electrophiles as activity-dependent cysteine protease profiling and discovery tools. *Chemistry & Biology, 7*, 569–581.

Hong, V., Presolski, S. I., Ma, C., & Finn, M. G. (2009). Analysis and optimization of copper-catalyzed azide-alkyne cycloaddition for bioconjugation. *Angewandte Chemie (International ed. in English), 48*, 9879–9883.

Huntly, B. J., Shigematsu, H., Deguchi, K., Lee, B. H., Mizuno, S., Duclos, N., et al. (2004). MOZ-TIF2, but not BCR-ABL, confers properties of leukemic stem cells to committed murine hematopoietic progenitors. *Cancer Cell, 6*, 587–596.

Hwang, Y., Thompson, P. R., Wang, L., Jiang, L., Kelleher, N. L., & Cole, P. A. (2007). A selective chemical probe for coenzyme A-requiring enzymes. *Angewandte Chemie (International ed. in English), 46*, 7621–7624.

Kidd, D., Liu, Y., & Cravatt, B. F. (2001). Profiling serine hydrolase activities in complex proteomes. *Biochemistry, 40*, 4005–4015.

Killenberg, P. G., & Dukes, D. F. (1976). Coenzyme A derivatives of bile acids-chemical synthesis, purification, and utilization in enzymic preparation of taurine conjugates. *Journal of Lipid Research, 17*, 451–455.

Lau, O. D., Kundu, T. K., Soccio, R. E., Ait-Si-Ali, S., Khalil, E. M., Vassilev, A., et al. (2000). HATs off: Selective synthetic inhibitors of the histone acetyltransferases p300 and PCAF. *Molecular Cell, 5*, 589–595.

Lee, J. V., Carrer, A., Shah, S., Snyder, N. W., Wei, S., Venneti, S., et al. (2014). Akt-dependent metabolic reprogramming regulates tumor cell histone acetylation. *Cell Metabolism, 20*, 306–319.

Lee, K. K., & Workman, J. L. (2007). Histone acetyltransferase complexes: One size doesn't fit all. *Nature Reviews. Molecular Cell Biology, 8*, 284–295.

Liu, Y., Patricelli, M. P., & Cravatt, B. F. (1999). Activity-based protein profiling: The serine hydrolases. *Proceedings of the National Academy of Sciences of the United States of America, 96*, 14694–14699.

Liu, X., Wang, L., Zhao, K., Thompson, P. R., Hwang, Y., Marmorstein, R., et al. (2008). The structural basis of protein acetylation by the p300/CBP transcriptional coactivator. *Nature, 451*, 846–850.

Martin, B. R., Wang, C., Adibekian, A., Tully, S. E., & Cravatt, B. F. (2012). Global profiling of dynamic protein palmitoylation. *Nature Methods, 9*, 84–89.

Montgomery, D. C., Sorum, A. W., Guasch, L., Nicklaus, M. C., & Meier, J. L. (2015). Metabolic regulation of histone acetyltransferases by endogenous acyl-CoA cofactors. *Chemistry & Biology, 22*, 1030–1039.

Montgomery, D. C., Sorum, A. W., & Meier, J. L. (2014). Chemoproteomic profiling of lysine acetyltransferases highlights an expanded landscape of catalytic acetylation. *Journal of the American Chemical Society, 136*, 8669–8676.

Montgomery, D. C., Sorum, A. W., & Meier, J. L. (2015). Defining the orphan functions of lysine acetyltransferases. *ACS Chemical Biology, 10*, 85–94.

Ong, S. E., Blagoev, B., Kratchmarova, I., Kristensen, D. B., Steen, H., Pandey, A., et al. (2002). Stable isotope labeling by amino acids in cell culture, SILAC, as a simple and accurate approach to expression proteomics. *Molecular & Cellular Proteomics: MCP, 1*, 376–386.

Pasqualucci, L., Dominguez-Sola, D., Chiarenza, A., Fabbri, G., Grunn, A., Trifonov, V., et al. (2011). Inactivating mutations of acetyltransferase genes in B-cell lymphoma. *Nature, 471*, 189–195.

Peifer, M., Fernandez-Cuesta, L., Sos, M. L., George, J., Seidel, D., Kasper, L. H., et al. (2012). Integrative genome analyses identify key somatic driver mutations of small-cell lung cancer. *Nature Genetics, 44*, 1104–1110.

Petrij, F., Giles, R. H., Dauwerse, H. G., Saris, J. J., Hennekam, R. C., Masuno, M., et al. (1995). Rubinstein-Taybi syndrome caused by mutations in the transcriptional co-activator CBP. *Nature, 376*, 348–351.

Theisen, J. W., Gucwa, J. S., Yusufzai, T., Khuong, M. T., & Kadonaga, J. T. (2013). Biochemical analysis of histone deacetylase-independent transcriptional repression by MeCP2. *The Journal of Biological Chemistry, 288*, 7096–7104.

Thompson, P. R., Wang, D., Wang, L., Fulco, M., Pediconi, N., Zhang, D., et al. (2004). Regulation of the p300 HAT domain via a novel activation loop. *Nature Structural & Molecular Biology, 11*, 308–315.

Zheng, Y., Thompson, P. R., Cebrat, M., Wang, L., Devlin, M. K., Alani, R. M., et al. (2004). Selective HAT inhibitors as mechanistic tools for protein acetylation. *Methods in Enzymology, 376*, 188–199.

CHAPTER SIX

Investigating Histone Acetylation Stoichiometry and Turnover Rate

J. Fan*, J. Baeza*, J.M. Denu*,†,1

*School of Medicine and Public Health, Wisconsin Institute for Discovery, University of Wisconsin, Madison, WI, United States
†Morgridge Institute for Research, Madison, WI, United States
1Corresponding author: e-mail address: jmdenu@wisc.edu

Contents

1. Introduction 126
2. Labeling and Methods for Sample Preparation 128
    2.1 General Experimental Design 128
    2.2 Incubating Cells with Isotopically Labeled Precursors 129
    2.3 Extracting Metabolites 131
    2.4 Extracting Histones 131
3. Sample Analysis 132
    3.1 Analyzing Metabolite Labeling 132
    3.2 Analyzing Overall Histone Acetylation 133
    3.3 Analyzing Site-Specific Acetylation Stoichiometry 134
    3.4 Analyzing Site-Specific Acetylation-Labeling Kinetics 139
4. Data Analysis and Kinetic Modeling 140
    4.1 Analyzing Small-Molecule Data 140
    4.2 Analyzing Site-Specific Histone Acetylation 141
    4.3 Quantifying Histone Acetylation Turnover Rate 143
5. Discussion and Perspective 145
Acknowledgments 146
References 146

Abstract

Histone acetylation is a dynamic epigenetic modification that functions in the regulation of DNA-templated reactions, such as transcription. This lysine modification is reversibly controlled by histone (lysine) acetyltransferases and deacetylases. Here, we present methods employing isotopic labeling and mass spectrometry (MS) to comprehensively investigate histone acetylation dynamics. Turnover rates of histone acetylation are determined by measuring the kinetics of labeling from ^{13}C-labeled precursors of acetyl-CoA, which incorporates ^{13}C-carbon onto histones via the acetyltransferase reaction. Overall histone acetylation states are assessed from complete protease digestion to single amino acids, which is followed by MS analysis. Determination of site-specific acetylation stoichiometry is achieved by chemically acetylating endogenous histones with isotopic acetic anhydride, followed by trypsin digestion and LC–MS analysis. Combining metabolic labeling with stoichiometric analysis permits determination of both acetylation level and acetylation dynamics. When comparing genetic, diet, or environmental perturbations, these methods permit both a global and site-specific evaluation of how histone acetylation is dynamically regulated.

1. INTRODUCTION

Histones are subject to a myriad of covalent posttranslational modifications (PTMs), especially on the lysine-rich N-terminal tails. The PTMs of histones, together with the differential deposition of histone variants, affect chromatin structure and play an essential role in regulating transcription. Particularly, acetylation of lysine residues on histones neutralizes the positive charge, resulting in a more open and accessible chromatin conformation. Additionally, lysine acetylation can serve as a recognition "mark" for the recruitment of acetyl-lysine-binding domains, thus recruiting large multi-subunit complexes that regulate transcription and other DNA-templated processes. Dysregulation of histone PTMs is associated with many human diseases, such as cancer and diabetes (Berdasco & Esteller, 2010; Leroy et al., 2013; Mellor, Brimble, & Delbridge, 2015; Miao et al., 2014).

Mass spectrometry (MS)-based methods have provided powerful tools to analyze histone PTM states and to identify novel histone modifications (Arnaudo & Garcia, 2013; Dai et al., 2014; Tan et al., 2011). Bottom-up approaches for quantifying histone modifications utilize chemical labeling of the unmodified lysine on histones, allowing trypsin-digested histone peptides to be resolved by reverse-phase chromatography and analysis by MS. With this approach, it has been possible to quantify different histone PTM states, and the fraction of a lysine that is modified with a particular mark, i.e., stoichiometry. Measuring the stoichiometry of a given modification is essential to interpret the biological significance of that modification. Recently we reported a method for quantifying site-specific acetylation stoichiometry (Baeza et al., 2014).

Histone PTMs are dynamically regulated by many enzyme complexes that "read," "write," or "erase" histone marks. The addition of histone PTMs requires small metabolites as substrates. For the case of acetylation, the acetyl donor is acetyl-CoA, a central metabolite produced by multiple metabolic pathways including glucose metabolism, fatty acid β-oxidation, and amino acid degradation. The acetyl group is directly transferred onto lysine residues by various histone acetyltransferases (HATs). Deacetylation occurs through both NAD^+-dependent sirtuins and NAD^+-independent histone deacetylases (HDACs). The direct transfer of the acetyl moiety from acetyl-CoA to lysine residues allows for the quantification of histone acetylation turnover rate by isotopic labeling with a metabolic precursor of acetyl-CoA and following the kinetic incorporation of the labeled acetyl group onto histone. Quantifying the turnover rate of histone PTMs provides a dynamic view of epigenetic regulation and is critical to understanding the mechanisms by which histone PTMs are controlled under different biological conditions. For example, analyzing the dynamic change of histone PTMs on newly replicated DNA compared to parental histones during mitotic cell division revealed the different modes that histone marks propagate (Alabert et al., 2015), quantifying H3K9 methylation turnover in the presence or absence of H3K14 acetylation revealed the influence of neighboring PTM on the modification of a specific site (Zee et al., 2010), and measuring acetylation turnover rate with HDAC inhibitors demonstrated the contribution of these HDACs in controlling histone acetylation (Evertts et al., 2013; Zheng, Thomas, & Kelleher, 2013).

A challenge in quantifying histone PTM turnover with metabolic labeling is that many precursors, which generate metabolites used as substrates for histone modifications, also result in the labeling of amino acids that can subsequently be incorporated into newly synthesized histone proteins and generate multiple isotopic permutations of the same protein. In the case of acetylation, glucose is the precursor that contributes to the majority of acetyl-CoA production in most mammalian cell lines, while uniformly labeled ^{13}C-glucose ([U-^{13}C] glucose) also labels alanine, aspartate, asparagine, serine, and glutamate within a relatively short-time period. The labeling of amino acids in addition to acetylation complicates the MS analysis and can make it difficult to quantify the overall histone acetylation rate that includes newly synthesized histone.

In addition to investigating site-specific acetylation stoichiometry and turnover rate under various conditions, other important questions have been raised in recent reports concerning histone PTMs as a whole, and its connection with metabolism: How much of the acetyl group is in the covalently bonded form on histones compared to free acetyl-CoA? Does this covalently modified form serve as a reserve pool for acetyl metabolism or a substrate for acetate production (McBrian et al., 2013)? And does the acetylation turnover on histones pose a significant demand for acetyl-CoA supply, or does the flux through protein PTMs impact cellular metabolism by consuming or producing key metabolites (Martinez-Pastor, Cosentino, & Mostoslavsky, 2013)? To answer these questions, as well as to understand the influence of different metabolic states on protein PTMs in general, a method to quantify overall histone PTM stoichiometry and turnover rate is needed.

In this chapter, we describe methods to quantify both overall, as well as site-specific, histone acetylation turnover with metabolic labeling. This method specifically traces the labeling kinetics of acetylation, regardless of the labeling of amino acids on histone peptides. We also describe methods to quantify overall and site-specific stoichiometry of histone acetylation. The stoichiometry together with the turnover rate provides a comprehensive characterization of histone acetylation under the investigated condition.

2. LABELING AND METHODS FOR SAMPLE PREPARATION

2.1 General Experimental Design

The histone acetylation rate is analyzed by feeding cells an isotopic-labeled precursor of acetyl-CoA and following the dynamic incorporation of the labeled acetyl group onto histones. The observed labeling of histone acetylation is dependent upon the actual rate of histone acetylation as well as the labeling kinetics of acetyl-CoA from its isotopic precursors, which is usually not negligible compared to histone-labeling rate. Thus, to accurately quantify the histone acetylation rate, it is necessary to measure the labeling time courses of both histone and small metabolites. Similarly, histone turnover rate can be quantified by comparing the kinetic labeling of histone peptides and its direct metabolic precursors, free amino acids. To acquire these data, we take an approach as shown in Fig. 1. Cells of interest are incubated with media containing isotopic-labeled substrate, and metabolite and histone samples are extracted at various time points in parallel, based on methods previously established (Krautkramer, Reiter, Denu, & Dowell, 2015; Lin & Garcia, 2012; Lu et al., 2010). To analyze both overall and site-specific histone acetylation, histone samples are divided into two parts after acid extraction: For quantification of overall histone acetylation stoichiometry and turnover rates, histone samples are completely digested to single amino acids with a combination of proteases (Walker, 1996). To quantify the site-specific acetylation rate, histone samples are chemically derivatized and digested to peptides. Additionally, to quantify the stoichiometry of histone acetylation at different sites, unlabeled histone samples are chemically acetylated using ^{13}C-labeled acetic anhydride for the comparison of light and

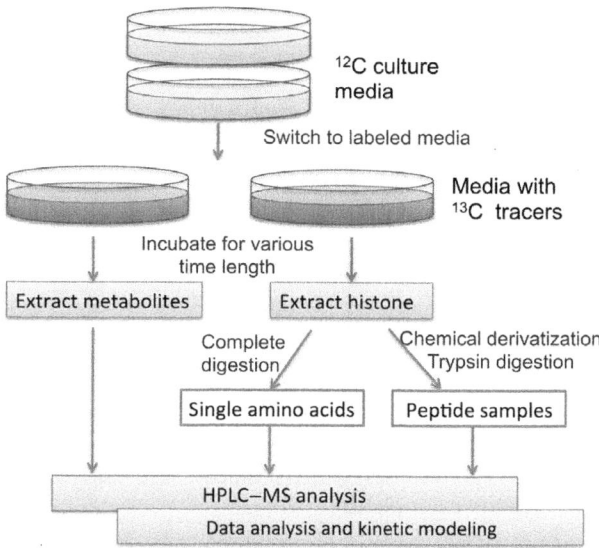

Fig. 1 General experimental workflow.

heavy acetylpeptides (Baeza et al., 2014). Abundance and isotopic labeling patterns of key metabolites (acetyl-CoA, CoA, and free amino acids), amino acids from histone digestion (acetyl-lysine, lysine, and other amino acids), and histone peptides are analyzed by different HPLC–MS-based methods designed for each case. Finally, histone acetylation and turnover rates are quantified by fitting the labeling kinetics of histones and the corresponding metabolic precursors with a differential equation-based model.

2.2 Incubating Cells with Isotopically Labeled Precursors

In order to isotopically label acetyl groups in cells, several labeled nutrients can be used, including glucose, glutamine, and acetate. Most culture media does not contain acetate, therefore to avoid changing the chemical composition of standard culture media, it is preferable to label cells with [U-^{13}C] glucose or glutamine replacing the regular glucose or glutamine in media. Both tracers can be metabolized to various intracellular amino acids in addition to acetyl-CoA. [U-^{13}C] glucose can label alanine, serine, aspartate, and asparagine, while glutamine mainly labels intracellular glutamine, glutamate, proline, aspartate, and asparagine. These labeled amino acids can be incorporated into histones by new histone synthesis and deposition, which allows for the quantification of histone turnover rate in the same labeling experiment. In most cell lines, the majority of

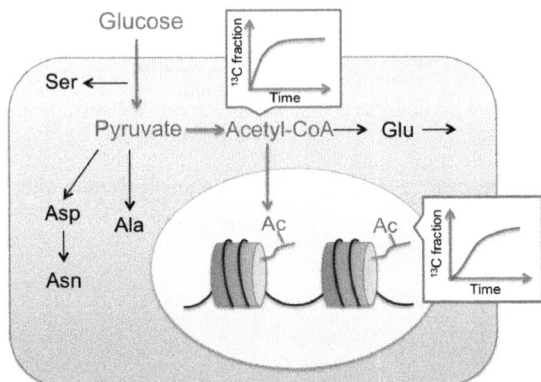

Fig. 2 Quantifying histone acetylation rate by metabolic labeling from ^{13}C-glucose. Labeling from glucose is first incorporated into acetyl-CoA by glucose metabolism and then incorporated onto histones by HATs. Glucose also labels various amino acids at the same time. (See the color plate.)

acetyl-CoA is produced from glucose, therefore we chose [U-^{13}C] glucose as the metabolic tracer (Fig. 2).

To acquire high-quality labeling time courses for both histone and small metabolites, we aim at choosing the time points that best define the kinetics, i.e., time points where the compound of interest labels at various degrees until isotopic steady state is reached (Fig. 2). Metabolites label faster than histones, thus for metabolite samples the incubation time in labeled media is relatively shorter (e.g., 0, 5, 15, 30, 70 min), while histones generally have a half-maximal labeling time of 0.5–2 h, thus the incubation time is accordingly longer (e.g., 0, 20, 40, 80, 160 min). In most cases, it is sufficient to obtain enough material for metabolite analysis from cells in an 80% confluent 60 mm tissue culture plate, while it is optimal to harvest more cells for histone extraction (an 80% confluent 10 cm or 15 cm plate). The choice of time points and sample sizes may need to be adapted to the cell type and condition under investigation.

1. Maintain cell culture (e.g., we have used MCF7, HepG2, and HEK293) in Dulbecco's Modified Eagle's Medium supplied with 10% dialyzed fetal bovine serum at 37°C.
2. Before labeling, seed cells at the same density (i.e., same cell number per area) into 16 60 mm tissue culture plates (Thermo Fisher Scientific) for metabolite samples and 16 10 cm or 15 cm tissue culture plates for histone samples. Culture cells until they reach 80% confluency. Under this exponential growth condition, cells are in steady state, metabolites levels and histone PTM states remain stable.

3. 2 h prior to labeling, change to fresh unlabeled media to replenish nutrients and to remove cellular waste products in the old media.
4. At time 0, completely remove unlabeled media and replace with prewarmed media containing [U-^{13}C] glucose instead of regular glucose at the same concentration of 4.5 g/L.
5. Incubate cells in labeled media at 37°C for various time duration before extracting metabolite or histone.

2.3 Extracting Metabolites

1. Prepare extraction solvent by mixing LC–MS grade methanol and water at 80:20 (v:v). Prechill the solvent in −80°C freezer before metabolite extraction.
2. After incubation in labeled media for designated time, quickly remove all media from tissue culture plate with an aspirator.
3. [Optional] Rinse plate with 1.5 mL PBS 3× and remove PBS completely. This step is only needed to quantify the labeling of amino acids contained in culture media, so that the amino acids from the residual media will not contaminate the signal for intracellular amino acids.
4. Add 1.5 mL extraction solvent and set the plate on dry ice right away. Incubate the plates in −80°C freezer for 15 min.
5. Scrape down the cells and transfer everything into a clean 2 mL Eppendorf tube®.
6. Spin at maximum speed using a chilled microcentrifuge at 4°C for 5 min. Transfer the supernatant into a clean tube.
7. Add 0.5 mL extraction solvent to pellet for reextraction, then vortex the pellet and mix well.
8. Spin the reextraction at maximum speed in 4°C centrifuge for 5 min, take the supernatant, and combine the reextraction with the first extraction.
9. [Optional] Quantify cell number or packed cell volume from a parallel 60 mm plate to normalize metabolite abundance.
10. Dry down cell extraction under N_2 flow using a Techne sample concentrator at ∼5 psi. Resuspend sample in LC–MS grade water (about 50 μL water per 1 μL packed cell volume or one million cells).

2.4 Extracting Histones

1. Prepare the hypotonic lysis buffer with 10 mM Tris–HCl (pH 7.5), 10 mM KCl, and 3 mM MgCl$_2$. Right before experiment, add protease inhibitors: 10 μg/mL leupeptin, 10 μg/mL aprotinin, and 100 μM

phenylmethanesulfonyl fluoride; deacetylase inhibitors: 1 mM sodium butyrate, 4 μM trichostatin A, 10 mM nicotinamide; and 1 mM DTT.
2. After labeling cells for designated times, quickly remove all media. Rinse cells in plate with PBS 3×.
3. Harvest cells by scraping with a cell lifter. Transfer cells suspension in PBS to a 15-mL tube. Spin at 1000 × g for 1 min and remove PBS.
4. Resuspend cell pellet in hypotonic lysis buffer (use 300 μL for a 10-cm plate, 750 μL for a 15-cm plate) and incubate for 20 min on ice.
5. Prewash 21 G needle with hypotonic buffer and pass the cell suspension through 5× using slow strokes to break cells.
6. Spin the cell lysate at 700 × g, 4°C, for 10 min to pellet nuclei.
7. [Optional] Transfer the supernatant to a clean tube and proceed to further fractionation, if cytosolic or mitochondrial fraction is desired.
8. Wash the nuclear pellet with ice cold PBS supplemented with protease and deacetylase inhibitors. Repellet nuclei by spinning at 1000 × g for 10 min at 4°C. Discard supernatant.
9. Slowly add about five volumes of 0.4 N H_2SO_4 to the nuclear pellet. Resuspend the pellet by gentle pipetting.
10. Incubate the sample with constant rotation or gentle shaking for 4 h at 4°C.
11. Spin the suspension at 3400 × g, 4°C, for 5 min. Transfer supernatant to a new 1.5-mL microcentrifuge tube.
12. Add 1/4 volume of 100% trichloroacetic acid to supernatant and mix. Incubate the mixture at 4°C overnight to precipitate histones.
13. Spin the mixture at 3400 × g, 4°C, for 5 min. Discard supernatant.
14. Rinse the pellet twice with acetone. Air dry the pellet.
15. Dissolve the histones with water. Spin the mixture at maximum speed for 2 min at 4°C. Transfer supernatant, which contains histones, to a new tube.
16. Measure protein concentration using a Bradford assay kit (Bio-Rad).

3. SAMPLE ANALYSIS
3.1 Analyzing Metabolite Labeling

To analyze the labeling time course of acetyl-CoA and free amino acids from metabolite extraction, many HPLC–MS methods are available (Lu, Bennett, & Rabinowitz, 2008; Lu et al., 2010). Our current setup is a Thermo Q-Exactive Orbitrap mass spectrometer coupled to a UPLC (Dionex 3000). Metabolites are separated with an 1.7 μm particle

2.1 × 100 mm ACQUITY UPLC® BEH C18 column, with a gradient of solvent A (95% H_2O, 5% methanol, 10 mM tributanolamine, 9 mM acetate, pH 8.2) and solvent B (100% methanol) at 0.2 mL/min flow rate. The gradient is: 0 min, 5% B; 2.5 min, 5% B; 5 min, 20% B; 7.5 min, 20% B; 13 min, 55% B; 15.5 min, 95% B; 18.5 min, 95% B; 19 min, 5% B; and 25 min, 5% B. Data were collected on full scan negative-ion mode at a resolution of 70 K with a maximum injection time of 40 ms and automatic gain control (AGC) of 1E6.

3.2 Analyzing Overall Histone Acetylation

3.2.1 Digesting Histones into Single Amino Acids

1. Dilute 20 μg histone sample in 50 μL digestion buffer (50 mM NH_4HCO_3, pH 7.5, 5 mM DTT, using LC–MS grade water). Include a procedure blank with 50 μL digestion buffer and no histone.
2. Add 0.4 μg Pronase to each sample and incubate for 24 h at 37°C.
3. Heat samples to 95°C for 5 min. Cool down to room temperature.
4. Add 0.8 μg aminopeptidase and incubate at 37°C for 18 h.
5. Heat samples to 95°C for 5 min. Cool down to room temperature.
6. Add 0.4 μg prolidase and incubate at 37°C for 3 h.
7. Add 200 μL LC–MS grade acetonitrile (ACN) to quench reaction, then vortex for 5 s. Spin at maximal speed for 5 min and transfer supernatant to a clean vial.

3.2.2 Analyzing Histone Digest by HPLC–MS

The enzymatic digestion converts histones into single amino acids. Acetylation sites are converted to N-ε-acetyl-lysine, which is labeled on the acetyl moiety when cells are fed [U-^{13}C] glucose. (Note that in human cell lines, lysine is not labeled from glucose, as lysine is an essential amino acid.) To measure the labeling of acetyl-lysine and other amino acids, digested histone samples (including procedure blank) are analyzed by LC–MS. To quantify the total amount of acetyl-lysine and lysine, a series of external standards are run in the same sequence with the samples, with lysine standard range from 10 to 200 μM, and acetyl-lysine range from 0.5 to 10 μM. Acetyl-lysine is isobaric with the dipeptides Gly-Leu, Gly-Ile, and Ala-Glu. Thus, it is important to make sure that the histone digestion is complete, and any possible contamination of these dipeptides is separated from acetyl-lysine in the LC–MS method.

Amino acids in complete histone digest are analyzed using the same LC–MS unit as small metabolites (Thermo Q-Exactive Orbitrap mass spectrometer coupled to a Dionex 3000 UPLC). The compounds are separated with a 5-μm polymer 150 × 2.1 mm SeQuant® ZIC®-pHILIC column with a gradient of two solvents. Solvent A is ACN and solvent B is 10 mM ammonium acetate in water, pH 5.5. The gradient profile for chromatography is as follows: 10% solvent B for 2 min, linear increase in solvent B to 90% over 12 min, isocratic 90% solvent B for 3 min, and then equilibration with 10% solvent B for 2 min at a flow rate of 0.3 mL/min. Compounds separated by HPLC are detected by heated electrospray ionization high-resolution mass spectroscopy. Analysis is performed under positive ionization mode. Settings for the ion source are: 10 aux gas flow rate, 35 sheath gas flow rate, 1 sweep gas flow rate, 3.5 kV spray voltage, 320°C capillary temperature, and 300°C heater temperature. Data-dependent acquisition mode with a dynamic exclusion of 10 s is enabled. In every cycle, one full MS scan is collected with a scan range of 88–500 m/z, resolution of 70 K, maximum injection time of 40 ms, and AGC of 1E6. Then three MS2 scans are followed on parent ions from the most intense peaks, with NCE 25. Resolution is set at 17,500, AGC target 2E4, maximum injection time 40 ms, and isolation width 1 m/z.

3.3 Analyzing Site-Specific Acetylation Stoichiometry

Two general MS methods for quantifying acetylation are available: relative quantitation- and stoichiometry-based methods. In relative quantitation-based method, a sample is first digested into peptides followed by immunoenrichment of acetylated peptides and subsequently analyzed by MS (Choudhary, Weinert, Nishida, Verdin, & Mann, 2014; Olsen & Mann, 2013). Coupling this method with either stable isotope labeling by amino acids in cell culture (SILAC) or isobaric tags (TMT or iTRAQ) can provide relative quantitation between two or more biological conditions. An alternative method measures the stoichiometry of acetylation at individual lysine residues. This approach utilizes a chemical-labeling step with isotopic acetic anhydride at the protein level, which labels all unmodified lysine residues. Using this strategy, all lysines will bear an acetyl group: a "light" acetyl group from endogenous acetylation or a "heavy" derived from in vitro chemical acetylation. Trypsin digestion generates chemically identical peptides that can be measured using MS. The general scheme for acetylation stoichiometry is displayed in Fig. 3.

In our lab, we employ different isotopes of acetic anhydride for chemical acetylation, which is dictated by the particular experimental design. When using an MS2-based acetyl stoichiometry analysis, we use $^{13}C_2$-acetic

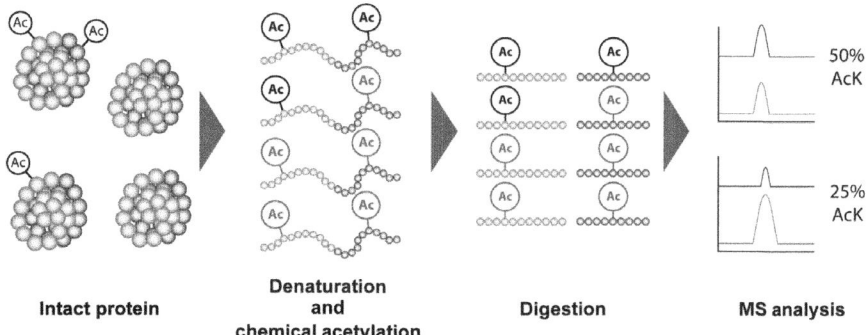

Fig. 3 General scheme for determining acetylation stoichiometry. A protein population is denatured and chemically acetylated using isotopic acetic anhydride followed by trypsin digestion. The heavy and light acetyl peptides can be measured by high-resolution mass spectrometry. (See the color plate.)

anhydride. This is useful for large-scale acetyl proteomic analysis as well as site-specific acetylation when a tryptic peptide has more than one lysine residue. During higher-energy collisional dissociation fragmentation, the acetyl-lysine generates an immonium ion observed in the low-mass region of the spectrum. The population of an acetylated peptide produces heavy and light immonium ions, which are coisolated and cofragmented in the collision cell, giving rise to a combined MS/MS spectrum. The reporter immonium ion intensity is measured and stoichiometry is determined (Fig. 4). An MS1-based method of determining stoichiometry is accomplished by chemically acetylating protein samples with D_6-acetic anhydride or $^{13}C_2,D_6$-acetic anhydride and measuring the area under the curve for the precursor peak of the light and heavy peptides. Table 1 outlines the various isotopes used and their utility.

3.3.1 Acetylation Stoichiometry Sample Preparation for LC–MS/MS

1. Begin by diluting 10 μg of unlabeled histone sample into 25 μL of urea buffer (8 M urea, 100 mM ammonium bicarbonate, pH ~8.5, 5 mM DTT, prepared fresh).
2. Heat denature proteins by incubating at 60°C on a Thermomixer® C while mixing at 1000 RPM for 30 min.
   (*Note*: We utilize a Thermomixer® C because this setup allows us to use a heated lid to prevent condensation along with constant agitation of the sample. If a Thermomixer® C is not available, we have also utilized a thermocycler for denaturation with similar results.)

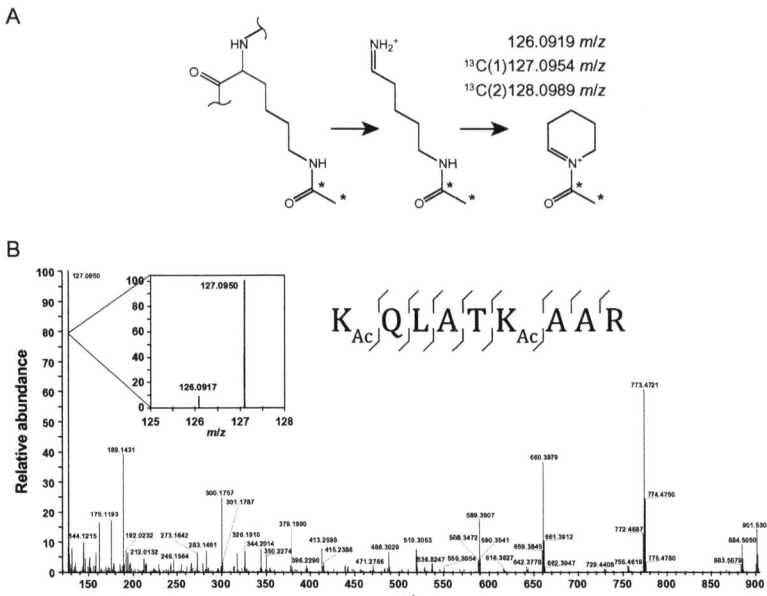

Fig. 4 MS2-based method for determining acetylation stoichiometry. (A) Generation of the acetyl-lysine immonium reporter ion used for quantitation. A free acetyl-lysine generates its corresponding immonium ion, which cyclizes to form the stable acetyl-lysine reporter ion. Displayed on the *top right* are monoisotopic masses of isotopic reporter ions. (B) MS/MS spectrum of the histone H3 acetylated tryptic peptide, $K_{18}QLATK_{23}AAR$. Measuring the immonium reporter ions from peptide fragmentation, the acetylation stoichiometry of the histone H3 K18, K23 peptide is 7.8%.

Table 1 Chemical Acetylation-Labeling Reagents

Chemical-Labeling Reagent	Acetyl-Lysine Mass (Monoisotopic)	Acetyl-Lysine (Composition)	MS Quant	Reporter Ion
Acetic anhydride	42.010565	C2 H(2) O		126.0919
$^{13}C_2$-acetic anhydride	43.013920	^{13}C C H(2) O	MS2	127.0954
$^{13}C_4$-acetic anhydride	44.017274	$^{13}C(2)$ H(2) O	MS2	128.0989
D_6-acetic anhydride	45.029395	C(2) $^2H(3)$ H(−1) O	MS1	
$^{13}C_4,D_6$-acetic anhydride	47.036105	$^{13}C(2)$ $^2H(3)$ H(−1) O	MS1	

3. [Optional] If measuring acetylation stoichiometry of nonhistone proteins, it is necessary to alkylate cysteine residues. Add 40 mM of Iodoacetamide and incubate at 60°C, 1000 RPM, for 30 min on Thermomixer® C.
4. For the chemical acetylation of lysine residues, add 2 µL of $^{13}C_2$-acetic anhydride to the sample and incubate on the Thermomixer® C at 60°C, 1000 RPM, 30 min.
5. Add ~5–10 µL of ammonium hydroxide to raise the pH ~8.5. Checking the pH can be accomplished by spotting 0.5 µL of procedural blank without protein onto litmus paper and checking for the appropriate color change.
6. To ensure complete lysine acetylation, perform another round of chemical acetylation using 2 µL of $^{13}C_2$-acetic anhydride. Perform the incubation and raise the pH as before.
7. Once the pH of the sample has been raised to ~8.5, incubate the sample at 60°C, for 20 min. This step hydrolyzes any acetyl-esters that occurred during the chemical acetylation steps.
8. Add up to 100 µL using 50 mM ammonium bicarbonate buffer, pH ~8.5.
9. Add sequencing grade trypsin at a 1:40–1:100 ratio (trypsin:histone). Incubate 4 h to overnight at 37°C.
10. Quench the reaction by adding 10 µL of 1% formic acid.

3.3.2 Peptide Cleanup Prior to MS Analysis

Sample preparation for MS introduces many hydrophilic molecules, which interfere with MS analysis. It is necessary to remove any salts prior to LC–MS analysis. In our lab, we routinely generate our own desalting tips for peptide cleanup based on the protocol described previously (Rappsilber, Mann, & Ishihama, 2007). The benefits of this system include minimizing sample loss as well as being cost-effective.

Materials
    3 M Empore™ C-18 47 mm extraction disk (Model 2215)
    Hamilton™ 16 G blunt point leur lock needles (Model 90516)
    19 G steel wire
    P200 pipette tips
    Glass petri dish
    16 G needle

Solvents
    Activation solvent: 100% methanol
    Wash solvent: 5% acetic acid in LC–MS grade H_2O

Elution solvent: 5% acetic acid in 80% ACN, in LC–MS grade H_2O
Sample diluent: 2% ACN, 0.1% formic acid in LC–MS grade H_2O

3.3.2.1 Generating Desalting Tips
1. Place the Empore™ C-18 extraction disk into the glass petri dish (C-18 disk can be stored in petri dish inside a dessicator).
2. Using the Hamilton™ 16 G blunt needle, punch a plug of the C-18 disk. Punch enough plugs for the initial amount of starting material. It is estimated that each plug has a binding capacity for ∼8 µg of peptide (Rappsilber et al., 2007).
3. Place the Hamilton™ needle that contains the C-18 plugs into a clean p200 pipette tip to the point where the needle hits the pipette tip.
4. Using the 19 G steel wire, push the C-18 plugs into the pipette tip to the point where the plugs are slightly snug.
    (*Note*: Do not push too forcefully on the C-18 plugs, as this will create a higher backpressure during the wash and elution steps.)
5. Take a 2-mL microcentrifuge tube and make a hole on the lid.
    (*Note*: The size of the hole must be such that the pipette tip fits snug.)
6. Place the desalting tip into the hole and proceed with the desalting procedure in Section 3.3.2.2.

3.3.2.2 Desalting Procedure
1. *Activation*: Add 50 µL methanol to the desalting tip and centrifuge $1000 \times g$ for ∼1 min.
2. *Conditioning*: Add 50 µL of elution solvent to the desalting tip and centrifuge $1000 \times g$ for ∼1 min.
3. *Preequilibration*: Add 50 µL of wash solvent to the desalting tip and centrifuge $1000 \times g$ for ∼1 min.
4. *Sample loading*: Load the sample volume to the desalting tip.
    (*Note*: Sample must be at low pH prior to loading.)
5. Load the sample by passing it through the desalting tip using centrifugation at $1000 \times g$ for ∼1 min.
    (*Note*: Centrifugation times may vary depending on sample volume. To ensure desalting tip does not dry out, set aside those tips in which the sample has passed completely through.)
    (*Note*: Centrifugation speed has been increased to $2000 \times g$.)
6. *Wash*: Wash the sample by adding 50 µL wash solvent and centrifuge for ∼1 min at $1000 \times g$.

7. *Elution*: Add 50 μL of elution solvent. Confirm that the solvent is completely at the bottom of the desalting tip. Do so by lightly snapping the tip so as to force the solvent down. Place the desalting tip at the end of the 20 mL syringe with leur lock attachment. Force the elution solvent through the tip by pushing on the syringe plunger. Collect the eluent into a new 1.5-mL tube.
8. Dry down sample using speedvac and prepare the samples for LC–MS/MS analysis.

3.3.3 LC–MS/MS Data Analysis

Our lab utilizes a Thermo Q-Exactive Orbitrap mass spectrometer connected to a Dionex Ultimate 3000 RSLC nano HPLC using a Waters Atlantis reverse-phase column (100 μm × 150 mm). Mobile phase consists of (A) 0.1% formic acid in HPLC grade H_2O and (B) 0.1% formic acid in HPLC grade ACN. Peptides are eluted in a linear gradient of 2–40% B at 700 nL/min over 60 min with a column temperature of 60°C and introduced into the Thermo Q-Exactive by nanoelectrospray ionization. Survey scan is performed in positive ion mode with a 70,000 resolution, AGC of 1E6, max fill time 250 ms, and a scan range of 350–2000 m/z. Data-dependent MS/MS is performed with a resolution of 17,500, AGC of 1E5, max fill time 100 ms, isolation window of 2.0 m/z, and a loop count of 10. For great coverage of b and y ions series as well as the acetyl-lysine immonium reporter ion (see later), we use a stepped NCE of 26, 32, and 40. The source voltage is set at 2.3 kV and capillary temperature at 250°C.

3.4 Analyzing Site-Specific Acetylation-Labeling Kinetics
3.4.1 Sample Preparation for LC–MS/MS

In order to measure site-specific acetylation-labeling kinetics, we chose to propionylate the lysine residues. This will prevent trypsin digestion on lysine residues, which can result in peptides that are too small for mass spectrometer detection in lysine-enriched histone proteins (Lin & Garcia, 2012). Following trypsin digestion, N-termini are further derivatized with phenylisocyanate (PIC) (Maile et al., 2015).

1. Dissolve 5 μg histones in a final volume of 9 μL ddH_2O.
2. Add 1 μL of 1 *M* triethylammonium bicarbonate buffer and mix by pipetting.
3. Prepare fresh propionic anhydride solution by mixing propionic anhydride 1:100 with ddH_2O.

4. Add 1 μL of this propionic anhydride solution to the histone sample and vortex.
5. Incubate at room temperature (20–25°C) for 2 min.
6. Quench with 1 μL 80 mM hydroxylamine. Incubate for 20 min at room temperature.
7. Trypsinize for 4 h to overnight with 0.1 μg trypsin.
8. Bring pH up to ~9–10 with ~1–3 μL 0.02 M sodium hydroxide.

(*Caution*: From this point onward, phenylisocyanate is used. Phenylisocyanate is extremely and acutely toxic. All steps must be performed in the fume hood, and proper personal protection equipment must be worn.)

9. Prepare a 2% (v/v) PIC in ACN solution.
10. Add 3 μL of this PIC solution to trypsinized histone peptides.
11. Incubate at 37°C for 1 h.
12. Acidify samples with 8 μL 1% TFA and perform peptide cleanup, as described earlier (in the fume hood).

3.4.2 LC–MS/MS Analysis for Histone Acetylation-Labeling Kinetics

Propionylated histone peptide samples are analyzed by an LC–MS/MS method similar to what is used for the chemically acetylated samples, as described in Section 3.3.3. We only changed the following few parameters: (1) HPLC gradient is 2% B–35% B over 50 min and 35% B–40% B over 10 min at 700 nL/min flow rate to ensure sufficient separation of different histone peptides. (2) For MS/MS, we use an isolation window of 8.0 m/z. In most cases, this will include all the permutations of isotopic peptides across all experimental conditions. These isotopic peptides may result from metabolic labeling on the acetyl moiety, amino acids (eg, alanine), or a combination of both. The specific labeling fraction on acetyl moiety regardless of amino acid labeling is analyzed from the immonium reporter ion in MS2. Note that for a single peptide in our experimental design, the chance of isotopic incorporation by acetylation and amino acids that results in a mass shift greater than 8 m/z is minimized. On a peptide-by-peptide basis, we also check if the isolation window covers all the significant isotopes and whether it includes unique coeluting peptides during data analysis.

4. DATA ANALYSIS AND KINETIC MODELING

4.1 Analyzing Small-Molecule Data

LC–MS data from metabolite analysis and completed histone digest are analyzed using metabolomics analysis and visualization engine (Melamud,

Vastag, & Rabinowitz, 2010). For histone digests, signals in the procedure blank indicate the level of amino acids that result from enzyme self-digestion. This signal should be a very small fraction compared to the corresponding signal in histone samples. The procedure blank signals are deducted in the data analysis. For all labeling data, we adjust for natural ^{13}C abundance.

To calculate average acetylation stoichiometry on histone, calibration curves of acetyl-lysine and lysine are acquired from the standard series. And total amount of acetyl-lysine and lysine in histone digest samples are analyzed according to the calibration curve.

Average histone acetylation stoichiometry

$$= \frac{\text{Acetyl-Lys(nmole)}}{\text{Acetyl-Lys(nmole)} + \text{Lys(nmole)}}. \quad (1)$$

Note that this average includes both lysine sites with the histone tails and core domains.

4.2 Analyzing Site-Specific Histone Acetylation

To analyze acetylation stoichiometry, we typically use the Thermo Scientific Proteome Discoverer™ software; however, other software packages may be available that select and report the immonium ion reporter intensities. For instances where these software packages are unavailable, we present a general method for manually selecting the reporter ion intensities. Using this method, we can quantitate the acetylation-labeling kinetics in histone peptides shown in Table 2.

1. Perform a database search to identify peptides. Designate acetyl, isotopic acetyl, monomethyl, dimethyl, trimethyl, propionyl, amino acid labeling that can be derived from the tracer (e.g., $^{13}C_3$-alanine label, and $^{13}C_2$-glycine label) as variable modifications. Designate N-terminal PIC as a fixed modification.
2. Using the Xcalibur™ software, find the MS/MS scan corresponding to the peptide of interest. For example, as shown in Fig. 5, we used the monoacetylated, monopropionylated H3 peptide, and KQLATKAAR, corresponding to either K18 or K23 is acetylation.
3. In the "Spectrum" view, zoom in to the region of the reporter ions and in the toolbar, change the view to "Spectrum List" in order to get a list of fragment ions with their corresponding absolute intensities. The intensities of the 126.0919, 127.0954, and 128.0989 (± 10 ppm) peaks from all the MS runs indicate the abundance of ^{12}C, $^{13}C_1$, and $^{13}C_2$ isotopes of lysine acetylation in the peptide.

Table 2 List of Detected Acetylated Histone Peptides

Histone	Sequence	Modification
H2A	AK(13)AK(15)TR	Acetyl
H2A	GK(5)QGGK(9)AR	Acetyl
H3.1	K(27)SAPATGGVK(36)K(37)PHR	Acetyl
H3.1	K(27)SAPATGGVK(36)K(37)PHR	Acetyl, trimethyl
H3.1	K(27)SAPATGGVK(36)K(37)PHR	Acetyl, dimethyl
H3.1 H3.3	K(18)QLATK(23)AAR	Acetyl
H3.1 H3.3	K(9)STGGK(14)APR	Acetyl
H3.1 H3.3	K(9)STGGK(14)APR	Acetyl, trimethyl
H3.1 H3.3	K(9)STGGK(14)APR	Acetyl, dimethyl
H3.1 H3.3	K(9)STGGK(14)APR	2 Acetyl
H3.1 H3.3	K(18)QLATK(23)AAR	Acetyl
H3.1 H3.3	K(18)QLATK(23)AAR	2 Acetyl
H4	GK(5)GGK(8)GLGK(12)GGAK(16)R	Acetyl
H4	GK(5)GGK(8)GLGK(12)GGAK(16)R	2 Acetyl

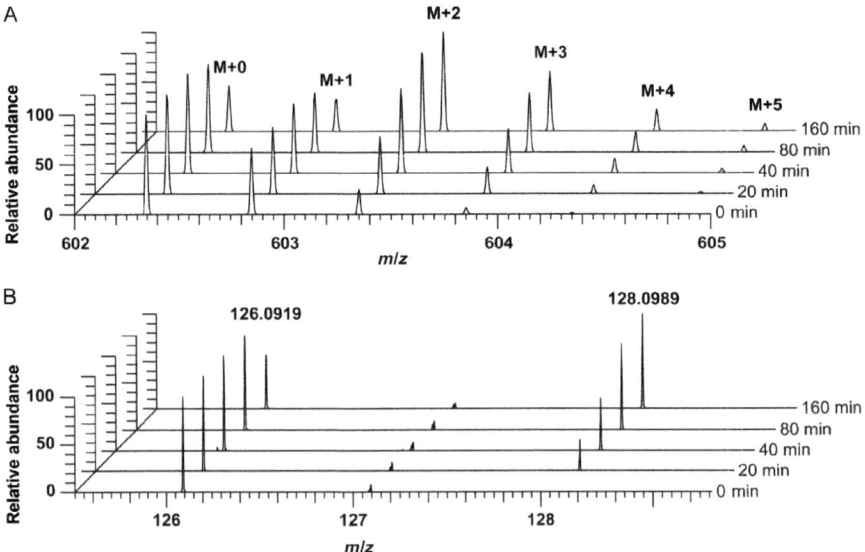

Fig. 5 Metabolic labeling of H3 peptide $K_{18}QLATK_{23}AAR$ in which either K18 or K23 is acetylated. (A) Relative abundance of isotopes in the $[M+2H]^{+2}$ spectra change over time during $[U-^{13}C]$ glucose incubation. The shift toward the heavy-labeled form is a result of an isotopic acetyl group or amino acid incorporation into the peptide. (B) The relative abundance of isotopes in acetyl-lysine immonium ion in MS2 spectra. The increase of the 128.0989 m/z species indicates the incorporation of $^{13}C_2$ acetyl group onto lysine residues by metabolic labeling.

4. The absolute intensities of the reporter immonium ions are used in the subsequent steps for analysis of acetylation turnover rates.

4.3 Quantifying Histone Acetylation Turnover Rate

Histone lysine acetylation is dynamically controlled by three processes, as shown in Fig. 6: (1) acetylation by HATs, with flux of F_{HAT}; (2) deacetylation by various HDACs, with deacetylation flux of F_{HDAC}; and (3) decay of acetylated histones during histone turnover, i.e., newly synthesized histones are deposited onto chromatin, as old histone is diluted out during cell growth or degraded. Histone turnover removes old acetylated histone with flux of $F_{histone}$.

In cells maintained in steady state, histone acetylation level remains constant, thus

$$F_{HAT} = F_{HDAC} + F_{histone}. \tag{2}$$

When cells are labeled with [U-^{13}C] glucose, acetyl-CoA becomes labeled, and this isotopic acetyl group is incorporated onto histones by HAT activity. How closely the labeling of histone acetyl group follows the labeling of acetyl-CoA reflects the rate of histone acetylation turnover, which can be quantified by kinetic flux profiling (Yuan, Bennett, & Rabinowitz, 2008).

Mathematically, the concentration change of acetylpeptide labeled on the acetyl moiety is a function of the new acetylpeptide formation using

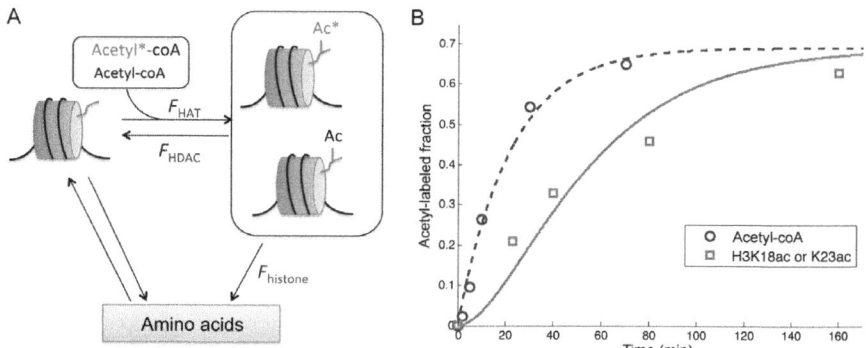

Fig. 6 Quantifying histone acetylation rate with metabolic labeling, using model presented in (A). The isotopic acetyl group (*red with* *) is incorporated onto histone lysine sites by HATs and removed by HDACs or turnover of old acetylated histones. Example of acetyl-CoA and monoacetylated H3 peptide K$_{18}$QLATK$_{23}$AAR labeling kinetics is shown in (B). *Circles and squares* are experimental data and *lines* are fitting result. (See the color plate.)

labeled acetyl-CoA as substrate, and the decay of old labeled acetylpeptide by deacetylation or histone turnover, as following:

$$\frac{d[\text{Ac-P}]^L}{dt} = F_{\text{HAT}} \frac{[\text{Acetyl-CoA}]^L_{(t)}}{[\text{Acetyl-CoA}]^T} - (F_{\text{HDAC}} + F_{\text{histone}}) \frac{[\text{Ac-P}]^L_{(t)}}{[\text{Ac-P}]^T} \quad (3)$$

$[\text{Ac-P}]^L$ is the concentration of labeled acetylpeptide, which is a function of time, and $[\text{Ac-P}]^T$ is the total concentration of the acetylpeptide, which is constant.

Given the balance between histone acetylation and histone turnover and deacetylation, plugging Eq. (1) into Eq. (3) yields:

$$\frac{d[\text{Ac-P}]^L}{dt} = F_{\text{HAT}} \frac{[\text{Acetyl-CoA}]^L_{(t)}}{[\text{Acetyl-CoA}]^T} - F_{\text{HAT}} \frac{[\text{Ac-P}]^L_{(t)}}{[\text{Ac-P}]^T}. \quad (4)$$

We directly measure the fraction of acetyl-CoA and acetylpepetide labeled on the acetyl moiety. To directly reflect the changes of these labeling fractions over time, the equation above is rearranged as:

$$\frac{d\left([\text{Ac-P}]^L/[\text{Ac-P}]^T\right)}{dt} = k_{\text{acetylation}} \left(\frac{[\text{Acetyl-CoA}]^L_{(t)}}{[\text{Acetyl-CoA}]^T}\right) - k_{\text{acetylation}} \left(\frac{[\text{Ac-P}]^L_{(t)}}{[\text{Ac-P}]^T}\right), \quad (5)$$

where $k_{\text{acetylation}}$ is acetylation turnover rate, $k_{\text{acetylation}} = F_{\text{HAT}}/[\text{Ac-P}]^T$.

Thus, acetylation rate $k_{\text{acetylation}}$ can be quantified by fitting the labeling kinetics of acetyl-CoA and acetylpeptide. It is worth noting that when labeled with glucose, acetyl-CoA may be labeled in many positions in addition to the acetyl moiety. This is because coenzyme A is synthesized from ATP, which can be labeled from glucose. In most cases, acetyl moiety in acetyl-CoA turns over much faster than coenzyme A synthesis. As a result, at the end of the acetyl-CoA-labeling time course (e.g., 70 min), the labeling on coenzyme A is still negligible. However, it is necessary to confirm this assumption by measuring the labeling of coenzyme A. When the acetyl group is labeled by ^{13}C glutamine, coenzyme A is not labeled.

Similarly, we can analyze the overall histone acetylation rate by measuring the labeling on the acetyl moiety of acetyl-lysine from a complete histone digest. When considering all histone acetylated sites as one pool, acetyl-lysine-labeling kinetics can be expressed as a function of overall acetylation rate and acetyl-CoA-labeling kinetics, as follows:

$$\frac{d\left([\text{Acetyl-lysine}]^L/[\text{Acetyl-lysine}]^T\right)}{dt} = k\left(\frac{[\text{Acetyl-CoA}]^L_{(t)}}{[\text{Acetyl-CoA}]^T}\right) - k\left(\frac{[\text{Acetyl-lysine}]^L_{(t)}}{[\text{Acetyl-lysine}]^T}\right). \quad (6)$$

Here, k is the overall acetylation turnover rate, which provides a general estimation of histone acetylation. Note that the acetyl-lysine labeling measured from a histone digest is a weighted average of labeling in all histone acetylation sites, with more abundant acetylation sites carrying more weight.

Additionally, the turnover rate of histone can be quantified from the labeling of free amino acids (e.g., alanine) and the labeling of amino acids in histone, which can be measured from the histone digest as well, using similar kinetics:

$$\frac{dY_{\text{histone-}^{13}\text{C-Ala}}}{dt} = k_{\text{histone}} X_{\text{free-}^{13}\text{C-Ala}}(t) - k_{\text{histone}} Y_{\text{histone-}^{13}\text{C-Ala}}(t) \quad (7)$$

Here k_{histone} is histone turnover rate, $X_{\text{free-}^{13}\text{C-Ala}}$ is the fraction of labeled free alanine, and $Y_{\text{histone-}^{13}\text{C-Ala}}$ is the fraction of labeled alanine in histone. The difference of acetylation turnover and histone turnover indicates the contribution of histone deacetylation reactions.

5. DISCUSSION AND PERSPECTIVE

We have discussed methods to quantify the overall and site-specific dynamics and stoichiometry of histone acetylation. These approaches give a comprehensive characterization of histone acetylation under a given condition. And because the abundance and kinetics of acetyl-CoA are measured at the same time, these methods allow a researcher to relate histone acetylation to acetyl-CoA metabolism. This approach provides a useful tool to investigate the impact of different factors on histone acetylation, for instance, to characterize the changes in histone acetylation when a nuclear sirtuin is knocked down.

Building on the previously established methods for quantifying histone acetylation rates using isotopic labeling, the method herein takes the labeling kinetics of acetyl-CoA into consideration. In the example shown in Fig. 6, acetyl group on histone peptide has half-maximum labeling time of less than

1 h, while the half-maximum labeling time of acetyl-CoA is around 20 min, which significantly delays the incorporation of ^{13}C from glucose onto histone. Thus, it is important to adjust for acetyl-CoA-labeling kinetics for the accurate quantification of histone acetylation rate, particularly for the acetylation sites with fast turnover rates. Additionally, this method directly follows the labeling on the acetyl moiety regardless of labeling in amino acids. Thus, instead of specifically quantifying the acetylation in old histone peptide, this approach measures the effective average of acetylation rate on both old and newly synthesized histones.

Similar to the acetylation methods described here, these approaches can be applied to investigate other histone PTMs. For example, another histone modification that plays an important role in transcriptional regulation is methylation. To investigate methylation, cells would be fed using 5-^{13}C-methionine, which labels S-adenosylmethionine (SAM), the substrate used for histone methylation. Very similar histone sample preparation and analyses can be applied to measure histone methylation at the peptide level. The abundance and labeling of SAM, as well as methyl-lysine, dimethyl-lysine, and trimethyl-lysine, which are the products of complete digestion of methylated histone, can also be quantified by LC–MS-based methods. Thus, the overall and site-specific stoichiometry and dynamic turnover of histone methylation can be investigated by employing similar analytical methods and kinetic models as outlined here.

ACKNOWLEDGMENTS

We would like to thank Kimberly Krautkramer and Dr. James Dowell for helpful discussions and generous help with the PIC labeling. We would also like to thank the UW-Madison Biotechnology Center Mass Spectrometry/Proteomics Facility for use of the Mascot server. This work was supported, in whole or in part, by National Institutes of Health (NIH) Grant GM065386-14 (J.M.D.), GM059785-15 (J.M.D.), NIH National Research Service Award T32 GM007215 (J.B.), and National Science Foundation (NSF) GRFP DGE-1256259 (J.B.).

Discloser: J.M.D. consults for Bio-Techne and FORGE Life Science.

REFERENCES

Alabert, C., Barth, T. K., Reveron-Gomez, N., Sidoli, S., Schmidt, A., Jensen, O. N., et al. (2015). Two distinct modes for propagation of histone PTMs across the cell cycle. *Genes & Development, 29*, 585–590.

Arnaudo, A. M., & Garcia, B. A. (2013). Proteomic characterization of novel histone post-translational modifications. *Epigenetics & Chromatin, 6*, 24.

Baeza, J., Dowell, J. A., Smallegan, M. J., Fan, J., Amador-Noguez, D., Khan, Z., et al. (2014). Stoichiometry of site-specific lysine acetylation in an entire proteome. *The Journal of Biological Chemistry, 289*, 21326–21338.

Berdasco, M., & Esteller, M. (2010). Aberrant epigenetic landscape in cancer: How cellular identity goes awry. *Developmental Cell*, *19*, 698–711.
Choudhary, C., Weinert, B. T., Nishida, Y., Verdin, E., & Mann, M. (2014). The growing landscape of lysine acetylation links metabolism and cell signalling. *Nature Reviews. Molecular Cell Biology*, *15*, 536–550.
Dai, L., Peng, C., Montellier, E., Lu, Z., Chen, Y., Ishii, H., et al. (2014). Lysine 2-hydroxyisobutyrylation is a widely distributed active histone mark. *Nature Chemical Biology*, *10*, 365–370.
Evertts, A. G., Zee, B. M., Dimaggio, P. A., Gonzales-Cope, M., Coller, H. A., & Garcia, B. A. (2013). Quantitative dynamics of the link between cellular metabolism and histone acetylation. *Journal of Biological Chemistry*, *288*, 12142–12151.
Krautkramer, K. A., Reiter, L., Denu, J. M., & Dowell, J. A. (2015). Quantification of SAHA-dependent changes in histone modifications using data-independent acquisition mass spectrometry. *Journal of Proteome Research*, *14*, 3252–3262.
Leroy, G., Dimaggio, P. A., Chan, E. Y., Zee, B. M., Blanco, M. A., Bryant, B., et al. (2013). A quantitative atlas of histone modification signatures from human cancer cells. *Epigenetics & Chromatin*, *6*, 20.
Lin, S., & Garcia, B. A. (2012). Examining histone posttranslational modification patterns by high-resolution mass spectrometry. *Methods in Enzymology*, *512*, 3–28.
Lu, W., Bennett, B. D., & Rabinowitz, J. D. (2008). Analytical strategies for LC-MS-based targeted metabolomics. *Journal of Chromatography. B, Analytical Technologies in the Biomedical and Life Sciences*, *871*, 236–242.
Lu, W., Clasquin, M. F., Melamud, E., Amador-Noguez, D., Caudy, A. A., & Rabinowitz, J. D. (2010). Metabolomic analysis via reversed-phase ion-pairing liquid chromatography coupled to a stand alone orbitrap mass spectrometer. *Analytical Chemistry*, *82*, 3212–3221.
Maile, T. M., Izrael-Tomasevic, A., Cheung, T., Guler, G. D., Tindell, C., Masselot, A., et al. (2015). Mass spectrometric quantification of histone post-translational modifications by a hybrid chemical labeling method. *Molecular & Cellular Proteomics*, *14*, 1148–1158.
Martinez-Pastor, B., Cosentino, C., & Mostoslavsky, R. (2013). A tale of metabolites: The cross-talk between chromatin and energy metabolism. *Cancer Discovery*, *3*, 497–501.
McBrian, M. A., Behbahan, I. S., Ferrari, R., Su, T., Huang, T. W., Li, K., et al. (2013). Histone acetylation regulates intracellular pH. *Molecular Cell*, *49*, 310–321.
Melamud, E., Vastag, L., & Rabinowitz, J. D. (2010). Metabolomic analysis and visualization engine for LC-MS data. *Analytical Chemistry*, *82*, 9818–9826.
Mellor, K. M., Brimble, M. A., & Delbridge, L. M. (2015). Glucose as an agent of post-translational modification in diabetes—New cardiac epigenetic insights. *Life Sciences*, *129*, 48–53.
Miao, F., Chen, Z., Genuth, S., Paterson, A., Zhang, L., Wu, X., et al. (2014). Evaluating the role of epigenetic histone modifications in the metabolic memory of type 1 diabetes. *Diabetes*, *63*, 1748–1762.
Olsen, J. V., & Mann, M. (2013). Status of large-scale analysis of post-translational modifications by mass spectrometry. *Molecular & Cellular Proteomics*, *12*, 3444–3452.
Rappsilber, J., Mann, M., & Ishihama, Y. (2007). Protocol for micro-purification, enrichment, pre-fractionation and storage of peptides for proteomics using StageTips. *Nature Protocols*, *2*, 1896–1906.
Tan, M., Luo, H., Lee, S., Jin, F., Yang, J. S., Montellier, E., et al. (2011). Identification of 67 histone marks and histone lysine crotonylation as a new type of histone modification. *Cell*, *146*, 1016–1028.
Walker, J. M. (1996). *The protein protocols handbook*. Totowa, NJ: Humana Press.
Yuan, J., Bennett, B. D., & Rabinowitz, J. D. (2008). Kinetic flux profiling for quantitation of cellular metabolic fluxes. *Nature Protocols*, *3*, 1328–1340.

Zee, B. M., Levin, R. S., Xu, B., LeRoy, G., Wingreen, N. S., & Garcia, B. A. (2010). In vivo residue-specific histone methylation dynamics. *The Journal of Biological Chemistry, 285*, 3341–3350.

Zheng, Y., Thomas, P. M., & Kelleher, N. L. (2013). Measurement of acetylation turnover at distinct lysines in human histones identifies long-lived acetylation sites. *Nature Communications, 4*, 2203.

CHAPTER SEVEN

Rapid Semisynthesis of Acetylated and Sumoylated Histone Analogs

A. Dhall, C.E. Weller, C. Chatterjee[1]

University of Washington, Seattle, WA, United States
[1]Corresponding author: e-mail address: chatterjee@chem.washington.edu

Contents

1. Introduction — 150
2. Materials and Methods — 152
   2.1 General Materials and Methods — 152
3. Semisynthesis of Sumoylated Histone H4 — 153
   3.1 Overall Design of the Semisynthesis — 153
   3.2 Preparation of Recombinant Histone H4 K12C — 154
   3.3 Preparation of Recombinant SUMO-3-Aminoethanethiol — 155
   3.4 Generation of Sumoylated Histone H4 — 156
4. Preparation of Acetylated Histone H3 Analogs — 156
   4.1 Overall Design of the Semisynthesis — 156
   4.2 Preparation of Recombinant Histone H3 — 157
   4.3 Generation of Thialysine Analogs of Acetylated Histone H3 — 159
5. Generation of Designer MNs — 159
   5.1 Octamer Formation Using Modified Histones — 159
   5.2 Generation of 147 bp *601* DNA — 160
   5.3 Reconstitution of MNs — 160
   5.4 Preparation of 177 bp Repeat *601* DNA — 161
   5.5 Generation of 12-mer Nucleosome Arrays — 161
6. Summary and Conclusions — 163
Acknowledgments — 164
References — 164

Abstract

The density and diversity of posttranslational modifications (PTMs) observed in histone proteins typically limit their purification to homogeneity from biological sources. Access to quantities of uniformly modified histones is, however, critical for investigating the downstream effects of histone PTMs on chromatin-templated processes. Therefore, a number of semisynthetic methodologies have been developed to generate histones bearing precisely defined PTMs or close analogs thereof. In this chapter, we present two optimized and rapid strategies for generating functional analogs of site-specifically acetylated and sumoylated histones. First, we describe a convergent strategy to site-specifically attach the small ubiquitin-like modifier-3 (SUMO-3) protein to the site of Lys12 in histone H4 by means of a disulfide linkage. We then describe the generation of thialysine analogs of histone H3 acetylated at Lys14 or Lys56, using thiol-ene coupling chemistry. Both strategies afford multimilligram quantities of uniformly modified histones that are easily incorporated into mononucleosomes and nucleosome arrays for biophysical and biochemical investigations. These methods are readily extendable to any desired sites in the four core nucleosomal histones and their variant forms.

1. INTRODUCTION

Posttranslational modifications (PTMs) of amino acid side chains in eukaryotic histones add epigenetic diversity to the chromatin landscape (Allis, Jenuwein, Reinberg, & Caparros, 2007; Kouzarides, 2007). Histone PTMs can directly influence chromatin structure and/or mediate protein–protein interactions that drive the transcription, replication, and repair of genetic material (Badeaux & Shi, 2013). Histone PTMs range from small chemical moieties such as methyl, acetyl, phosphoryl, and glycosyl groups to entire proteins such as ubiquitin and the small ubiquitin-like modifier protein (SUMO) (Dhall & Chatterjee, 2011). Given their central role in regulating gene function and repair, the misregulation or misinterpretation of histone PTMs is closely associated with human diseases such as cancers, ataxias, and muscular dystrophy (Portela & Esteller, 2010).

Early genomic approaches with modification-specific antibodies identified distinct sets of histone PTMs at specific chromatin loci and correlated these with transcriptional activity of the associated genes. These observations inspired the histone code hypothesis for gene regulation, which postulates that different histone PTMs may act individually or in combination to regulate chromatin-templated functions (Strahl & Allis, 2000). Advances in mass spectrometric techniques have led to the identification of about 15 chemically unique PTMs in histones, and combinations of these modifications lead to hundreds of uniquely identifiable histone *states* (Garcia, Pesavento, Mizzen, & Kelleher, 2007). Therefore, synthetic and semisynthetic techniques that yield rapid access to various modified histones are crucial when testing elements of the histone code hypothesis.

Peterson and coworkers first demonstrated the power of protein semisynthesis in interrogating the effects of acetylation at lysine 16 in H4 (H4 K16ac) on chromatin structure (Shogren-Knaak et al., 2006). In their protocol, a 22-amino acid long histone H4 N-terminal peptide containing acetylated lysine 16 and a C-terminal α-thioester were produced by solid-phase peptide synthesis. Full-length acetylated H4 was then generated by native chemical ligation (NCL) (Dawson, Muir, Clark-Lewis, & Kent, 1994) of the acetylated α-thioester peptide with a heterologously expressed C-terminal fragment of histone H4 (residues 23–102) containing an arginine-to-cysteine mutation at the N-terminus. The authors successfully incorporated acetylated H4 protein into nucleosome arrays and observed that H4 K16ac significantly impairs chromatin condensation and higher-order fiber formation. Despite its utility, the synthesis of long peptide thioesters required for NCL continues to be technically challenging and the semisynthesis of acetylated histones by NCL has remained limited to specialized laboratories (Kent, 2009).

In contrast with acetylation, the modification of histone lysines by conjugation with the proteins ubiquitin and SUMO leads to dramatic changes in the overall physical and chemical properties of nucleosomes. However, the fact that ubiquitylation and sumoylation (modification by Ub and SUMO, respectively) are transient modifications that mark a small fraction of core histones in cells (typically <5%, with ubiquitylated H2A being the exception at ~10%) has limited the isolation of quantities of ubiquitylated and sumoylated histones for in vitro mechanistic studies (Davies & Lindsey, 1994; Nathan et al., 2006). Recent progress toward understanding the roles for these modifications has been made through the efforts of several research groups in developing novel chemical handles and ligation auxiliaries that permit NCL at lysine side chains (Long et al., 2014; McGinty, Kim, Chatterjee, Roeder, & Muir, 2008). However, the technical challenges in generating wild-type ubiquitylated and sumoylated histones go beyond those encountered in generating semisynthetic acetylated histones.

A key turning point in facile semisynthetic access to posttranslationally modified histones was the discovery that thialysine analogs of methylated lysine side chains in histones are reasonable substrates for chromatin-modifying enzymes (Simon et al., 2007). Thialysine analogs are readily generated by alkylating genetically encoded cysteine residues in full-length histones with N-methylated forms of 2-aminoethyl halides, thereby obviating the need for preparing peptide thioesters for NCL. Several thiol-directed strategies provide rapid access to close functional analogs of wild-type PTMs of lysines and arginines (Dhall & Chatterjee, 2015). In this chapter, we provide detailed methods for chemical strategies that yield analogs of acetylated and sumoylated histones (Dhall et al., 2014). A high-yielding disulfide forming reaction (Rabanal, DeGrado, & Dutton, 1996) is used to generate H4 site-specifically sumoylated at Lys12 (suH4$_{ss}$) and radical-mediated thiol-ene coupling chemistry is used to introduce acetylated thialysine in histone H3 at positions 14 and 56 (Li et al., 2011). Both strategies provide versatile and straightforward approaches to generate multimilligram quantities of modified histones suitable for a wide range of biophysical and biochemical studies. Furthermore, we describe the incorporation of these modified histones into octamers, which are applied to reconstitute both mononucleosomes (MNs) and chromatin-like nucleosome arrays.

2. MATERIALS AND METHODS
2.1 General Materials and Methods

All commonly used chemical reagents and solvents were purchased from either Sigma-Aldrich Chemical Company (Milwaukee, WI) or Fischer Scientific (Pittsburgh, PA). 2xYT medium was reconstituted by mixing 16 g Bacto Tryptone, 10 g Bacto Yeast Extract, and 5 g NaCl per liter of water. A Superdex S-200 10/300 GL size-exclusion column was purchased from GE Healthcare (Waukesha, WI). Chemically competent DH5α and BL21 (DE3) cells were purchased from Novagen (Madison, WI). T4 DNA ligase and restriction enzymes were purchased from New England BioLabs (Ipswich, MA). DNA primer synthesis and gene sequencing were performed by Integrated DNA Technologies (Coralville, IA) and Genewiz (South Plainfield, NJ), respectively. Site-directed mutagenesis was performed with a QuikChange Site-Directed mutagenesis kit from Agilent Technologies (Santa Clara, CA). Criterion 5% TBE gels were purchased from Bio-Rad (Hercules, CA). Centrifugal filtration units were from Sartorius (Göttingen, Germany) and Slide-A-Lyzer dialysis cassettes and MINI dialysis units were from Pierce (Rockford, IL). PCR purification and gel extraction kits were purchased from Qiagen (Valencia, CA). Size-exclusion chromatography was performed on an AKTA FPLC system (GE Healthcare) equipped with a P-920 pump and UPC-900 monitor. Analytical reversed-phase HPLC (RP-HPLC) was performed on a Varian ProStar instrument with Vydac C18 or C4 columns (5 μm, 4 × 150 mm), employing 0.1% trifluoroacetic acid (TFA) in water (HPLC buffer A), and 90% acetonitrile and 0.1% TFA in water (HPLC buffer B) as the mobile phases. Typical analytical gradients were 30–70% buffer B over 30 min at a flow rate of 1 mL/min. Preparative scale purifications were conducted on Vydac C18 or C4 preparative columns (15–20 μm, 50 × 250 mm) over 60 min at a flow rate of 9 mL/min. Electrospray ionization mass spectrometry (ESI-MS) analysis was performed on a Bruker Esquire LC-ion trap spectrometer (Bruker Daltonics, Billerica, Massachusetts). All protein starting materials and ligation products were analyzed by both C18 or C4 analytical RP-HPLC and ESI-MS.

3. SEMISYNTHESIS OF SUMOYLATED HISTONE H4
3.1 Overall Design of the Semisynthesis

Histones that are heterologously produced in *Escherichia coli* are ideal candidates for installing cysteine-derived PTM analogs. *E. coli* does not have any known histone-modifying enzymes, which ensures that histones expressed in this organism are devoid of PTMs. Furthermore, of the four core histones, H3 is the only histone in higher eukaryotes with a cysteine at position 110 (Sullivan et al., 2002). This residue can be mutated to an alanine without discernible effects on nucleosome structure or function. Working in this mutant background ensures that cysteine residues introduced by mutagenesis at desired positions in the core histones are the only thiols present, thereby allowing site-selective chemical manipulations.

Histone sumoylation is a well-documented but poorly understood modification (Dhall et al., 2014; Nathan et al., 2006; Shilo & Eisenman, 2003). Similar to ubiquitylation, sumoylation has been observed on all four of the core histones as well as the linker histone H1. Although proteomic studies have identified Lys12 in the H4 tail as a site of modification (Galisson et al., 2011; Hendriks et al., 2014), the consequences on chromatin structure and function remain unknown. The first reported semisynthesis of a sumoylated histone H2B C-terminal peptide employed a photocleavable NCL auxiliary group appended to the side chain of a lysine in the target peptide (Chatterjee, McGinty, Pellois, & Muir, 2007). Ligation between the peptide and a C-terminally truncated yeast SUMO(1–97)-α-thioester followed by UV irradiation resulted in a site-specifically sumoylated peptide. While this method affords a native isopeptide linkage, the complex synthetic route poses challenges for obtaining bulk quantities of sumoylated proteins.

In order to address the gap in our understanding of the roles for histone sumoylation, we employed a disulfide-directed approach to generate milligram quantities of histone H4 uniformly sumoylated at Lys12 (suH4$_{ss}$). The semisynthesis is achieved in two parts (Fig. 1). First, the target Lys12 in H4 is replaced with a cysteine by site-directed mutagenesis. After affinity-tag enrichment and purification of H4 K12C, the sulfhydryl group of the lone cysteine is activated using a small molecule disulfide. This generates a

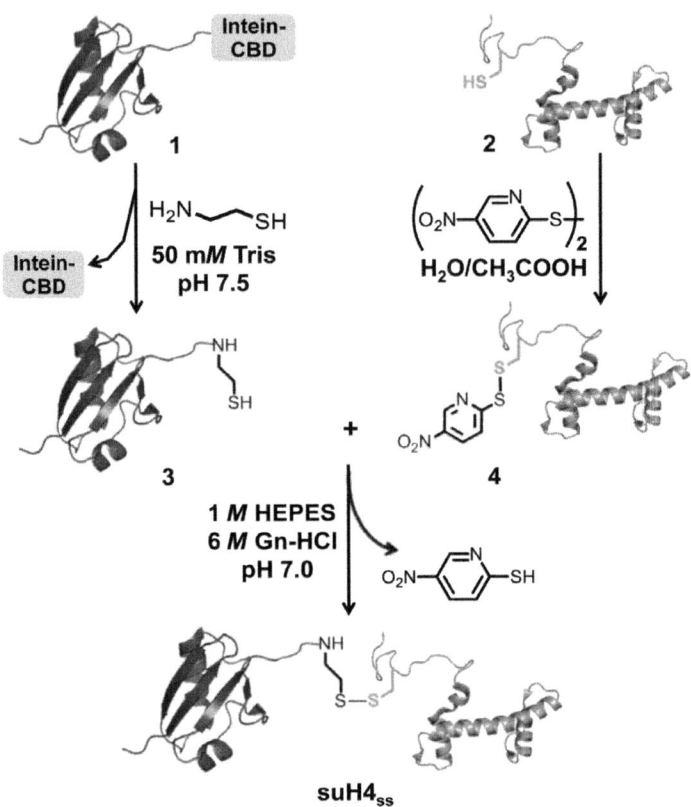

Fig. 1 Semisynthesis of disulfide-linked sumoylated histone H4 (suH4$_{ss}$). Synthetic scheme for the generation of SUMO-3 C47S-aminoethanethiol, **3**, using intein-mediated thiolysis and its subsequent reaction with the activated H4 K12C mutant, **4**, to generate suH4$_{ss}$.

reactive asymmetric disulfide on H4, poised for nucleophilic attack by an external thiol. Additionally, a free nucleophilic thiol is introduced at the C-terminus of SUMO-3 C47S by intein-mediated chemistry (Muir, Sondhi, & Cole, 1998). In the second step, mixing the SUMO-3 C47S with activated H4 K12C leads to the desired product in near-quantitative conversion over the course of 2–3 h.

3.2 Preparation of Recombinant Histone H4 K12C

Site-directed mutagenesis is employed to generate the mutant H4 K12C from human histone H4 cloned between the *Nde*I and *Bam*HI restriction sites in a pET15b vector. Surprisingly, simply overexpressing the H4 K12C mutant in BL21(DE3) cells leads to very poor yields. Hence, we employ a His$_6$-tag strategy to purify the mutant. A tobacco etch virus (TEV) protease cleavage sequence is introduced between the His$_6$-tag and the N-terminus of H4. The following primers are used with the QuikChange Site-Directed mutagenesis kit to introduce the K12C mutations: 5′-GGT AAA GGT GGT AAA GGT CTG GGT TGC GGT GGT GCT AAA CGT CAC CGT AAA-3′ and 5′-TTT ACG GTG ACG TTT AGC ACC ACC GCA ACC CAG ACC TTT ACC ACC TTT ACC-3′. The resulting plasmid is verified by gene sequencing. *E. coli* BL21(DE3) cells are transformed with the mutant plasmids and grown at 37°C in 6 L of 2xYT medium until OD$_{600}$ ~ 0.7. Protein expression is induced by the addition of 0.3 mM IPTG for 3 h at 37°C. Cells are harvested by centrifugation at 7000 × g, resuspended in 150 mM NaCl, 50 mM Tris at pH 7.5, and lysed by sonication. The lysate is centrifuged at 20,000 × g for 20 min and insoluble histones are recovered from inclusion bodies with an extraction buffer consisting of 6 M Gn–HCl, 100 mM NaCl, 50 mM Tris at pH 7.5. The resolubilized histones are purified by Ni^{2+}-affinity chromatography. Briefly, proteins are bound to the Ni^{2+}-column overnight at 4°C with gentle shaking and the column subsequently washed with 3 volumes of extraction buffer containing 25 mM imidazole. Immobilized proteins are eluted using 500 mM imidazole in the extraction buffer. Histone-containing fractions are identified by 15% SDS-PAGE, combined and dialyzed into 1 mM DTT overnight at 4°C. Purified His$_6$-TEV protease is then added in a 1:10 M ratio (enzyme:substrate) and the His$_6$-tag is cleaved overnight in a buffer containing 1 mM EDTA, 10 mM DTT, 10 mM Cys, 50 mM Tris, pH 6.9. The cleavage products are dialyzed into extraction buffer and reapplied to a Ni^{2+}-column. This leads to retention of the cleaved His$_6$-tag, His$_6$-TEV, and any uncleaved histone on the column. The eluted His$_6$-tag-free H4 K12C is further purified to homogeneity by preparative RP-HPLC and characterized by ESI-MS. Purified H4 K12C (Fig. 1, protein **2**) is then dissolved in a 1:3 mixture of water:acetic acid (v/v) and reacted with an excess of 2,2′-dithiobis(5-nitropyridine) for 2 h at 25°C with vigorous shaking. The desired product H4 K12C-3-nitro-2-pyridinesulfenyl (Npys) disulfide (Fig. 1, protein **4**) is purified away from unreacted starting materials by RP-HPLC and characterized by ESI-MS.

3.3 Preparation of Recombinant SUMO-3-Aminoethanethiol

Site-directed mutagenesis is employed to generate the SUMO-3 C47S mutant from human SUMO-3 cloned in the pTXB1 vector. The following primers are used with the QuikChange Site-Directed mutagenesis kit: 5′-AGC AAG CTG ATG AAG GCC TAC TCT GAG AGG CAG GGC TTG TCA ATG-3′ and 5′-CAT TGA CAA GCC CTG CCT CTC AGA GTA GGC CTT CAT CAG CTT GCT-3′. The resulting plasmid is verified by gene sequencing. *E. coli* BL21(DE3) cells are transformed with SUMO-3 C47S plasmids and grown to an OD_{600} of ~0.5 at 37°C followed by the addition of 0.3 m*M* IPTG. The cells are grown for another 4 h at 25°C, harvested by centrifugation at 7000 × *g*, resuspended in lysis buffer containing 200 m*M* NaCl, 20 m*M* Tris, pH 7.5, and lysed by sonication. The lysate is centrifuged at 20,000 × *g* for 20 min and the soluble supernatant containing SUMO-3 C47S-intein–chitin binding domain fusion protein (Fig. 1, protein **1**) is bound to chitin beads at 4°C overnight. The column is subsequently washed with 3 column volumes of lysis buffer and intein-mediated production of SUMO-3 C47S-aminoethanethiol (Fig. 1, protein **3**) is initiated by incubating the column with lysis buffer containing 100 m*M* cysteamine, pH 7.5, for 48 h at 4°C. After this period, the desired adducts are eluted from the column with lysis buffer, further purified by preparative RP-HPLC and characterized by ESI-MS.

3.4 Generation of Sumoylated Histone H4

One equivalent of H4 K12C-Npys (Fig. 1, protein **4**) and two equivalents of SUMO-3 C47S-aminoethanethiol (Fig. 1, protein **3**) are dissolved in a reaction buffer consisting of 6 *M* Gn–HCl, 1 *M* HEPES, pH 6.9 and allowed to react for 1 h at 25°C with continuous shaking. The desired product, $suH4_{ss}$, is purified away from starting materials by RP-HPLC and characterized by ESI-MS.

4. PREPARATION OF ACETYLATED HISTONE H3 ANALOGS

4.1 Overall Design of the Semisynthesis

The incorporation of acetyllysine into histones can be achieved by protein semisynthesis (Shogren-Knaak et al., 2006) or amber suppression (Neumann, Peak-Chew, & Chin, 2008), but these approaches can be low yielding and complex. In contrast, a method reported by Liu and coworkers in 2011 is uniquely well suited for generating functional analogs of acetyllysine in a single step on recombinant histone proteins, which can be purified in large quantities (Li et al., 2011). The acetyllysine analog is installed via a thiol-ene reaction between the sulfhydryl group of Cys and

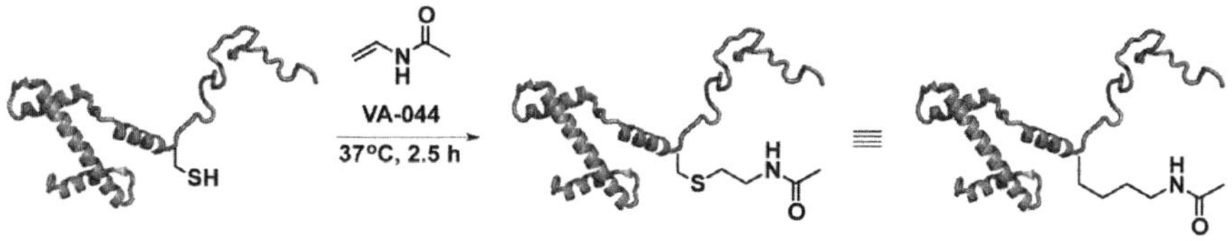

Fig. 2 Thiol-ene coupling chemistry at Cys with N-vinylacetamide to generate acetylated thialysine.

N-vinylacetamide; thus the target protein must contain a Cys residue only at the desired site of Lys acetylation (Fig. 2). The radical thiol-ene click reaction is fast, site-specific, and gives high product yields. Further, we observe that initiation with heat rather than UV light reduces the possibility of undesired side reactions (Dhall et al., 2014). In addition, multiple sites may be simultaneously and efficiently modified. The resulting N-acetyl thialysine, an isostere for acetyllysine at the side-chain γ position, is functionally similar to the natural modification, as evidenced by its identical effect on nucleosome array compaction, α-acetyllysine antibody reactivity, and susceptibility to enzymatic deacetylation (Li et al., 2011). As an example of site-specific incorporation of acetyllysine analogs, we describe the generation of H3 C110A bearing this modification at position 14 or 56, each a site of functionally important H3 acetylation (Shahbazian & Grunstein, 2007).

4.2 Preparation of Recombinant Histone H3

We have reported a pET3a vector containing the gene for human histone variant H3.2, *HIST2H3C*, bearing a C110A point mutation (Dhall et al., 2014). This plasmid is subjected to QuikChange Mutagenesis (Qiagen) to incorporate the K14C or K56C point mutation utilizing the primers: For K14C, 5′-CAG ACG GCT CGG AAA TCC ACC GGC GGT TGC GCG CCA CGC AAG CAG CTG GCT ACC AAG-3′ and 5′-CTT GGT AGC CAG CTG CTT GCG TGG CGC GCA ACC GCC GGT GGA TTT CCG AGC CGT CTG-3′; and for K56C, 5′-GCT CTG CGC GAG ATC CGC CGC TAC CAA TGC TCG ACC GAG TTG CTG ATT CGG AAG CTG-3′ and 5′-CAG CTT CCG AAT CAG CAA CTC GGT CGA GCA TTG GTA GCG GCG GAT CTC GCG CAG AGC-3′. The resulting plasmids are verified by gene sequencing. *E. coli* BL21(DE3) cells transformed with the pET3a-H3 C110A, K14C, or pET3a-H3 C110A, K56C vector are grown in 2 L of 2xYT broth supplemented with 100 μg/mL ampicillin at 37°C and with shaking at 250 rpm until the OD_{600} reaches 0.6–0.8. Overexpression is then induced by the addition of IPTG from a stock solution (1 M in H_2O) to a concentration of 0.3 mM, and growth is continued for no longer than 2 h at 37 °C. Protein overexpression is confirmed by SDS-PAGE analysis of the lysed cells at the end of growth (Fig. 3A). The cells are harvested by centrifugation at 7000 × g for 15 min and then resuspended in 4 M Gn–HCl, 20 mM Tris, 1 mM DTT, pH 7.5 (pH adjusted for use at 4°C). The cells are kept on ice and lysed by sonication. The presence of Gn–HCl causes significant frothing of the buffer, and longer sonication times may be required. If the cell suspension becomes warm, it can be allowed to rest on ice before further sonication to prevent overheating. After lysis, the cells are centrifuged at 20,000 × g for 15 min. The lysate supernatant is passed through a 0.45-μm filter and then applied to a 500-mL Superdex S-200 size-exclusion column equilibrated with lysis buffer eluting at 1 mL/min at 4°C. Fractions are collected and evaluated for the presence of the desired protein by SDS-PAGE. Note that Gn–HCl in the buffer precipitates in the presence of SDS and causes gels to run irregularly. To avoid this, samples can be centrifuged at 20,000 × g for 5 min after boiling with loading dye, and the supernatant loaded on the gel. Fractions containing a disulfide of the product may be present. This can be evaluated by adding 10 mM DTT to the loading dye. Product fractions are pooled and dialyzed against 4 L of H_2O containing

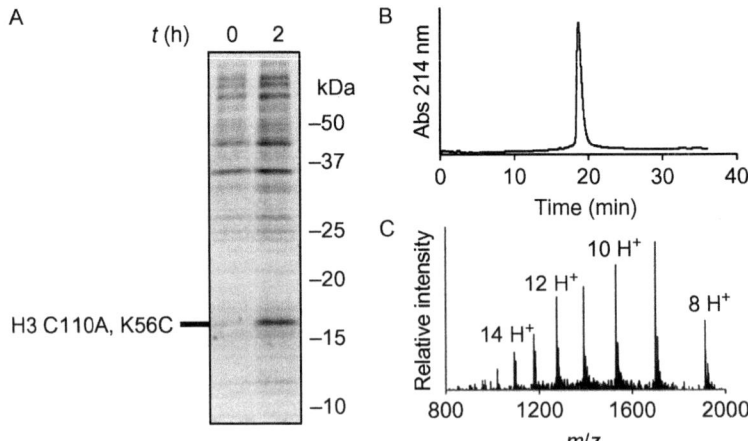

Fig. 3 Generation of H3 (C110A)Kc56ac analog. (A) SDS-PAGE of *E. coli* cultures containing pET3a-H3 C110A, K56C at 0 and 2 h after the induction of overexpression with IPTG. 15% SDS-PAGE gel run at 200 V for 45 min and stained with Coomassie. (B) RP-HPLC chromatogram of pure H3 (C110A)KC56ac analog. (C) ESI-MS of pure H3 (C110A)KC56ac analog.

1 mM DTT, pH 7.2, at 4°C for 6 h. The sample is then lyophilized and redissolved in 6 M Gn–HCl, 100 mM Na$_2$HPO$_4$, 100 mM TCEP, pH 7.0–7.5 and incubated at 4°C for 30 min. The reduced product is then purified by RP-HPLC and characterized by ESI-MS. ESI-MS for H3 C110A, K14C, and H3 C110A, K56C. Calculated m/z [M+H]$^+$ 15,200.7 Da, observed m/z [M+H]$^+$ 15,204.3 ± 4.4 and 15,203.8 ± 2.9 Da, respectively.

4.3 Generation of Thialysine Analogs of Acetylated Histone H3

Purified H3 C110A, K14C, or H3 C110A, K56C is dissolved at 0.1 mM in 6 M Gn–HCl, 200 mM sodium acetate, pH 5.0. An inert atmosphere is crucial for this reaction, so the buffer is freeze-thaw degassed thrice under N$_2$, and the reaction kept under an atmosphere of N$_2$. To this solution is added 15 mM L-glutathione, 50 mM N-vinylacetamide, 100 mM dimethyl sulfide, and 50 mM of the azo radical initiator VA-044 (2,2′-azobis[2-(2-imidazolin-2-yl)propane]dihydrochloride). The thiol-ene click reaction is initiated by incubation at 37°C for 2.5 h. Longer incubation times should be avoided as they may lead to alkylation at undesired sites, such as the protein N-terminus. The product is then purified by RP-HPLC and characterized by ESI-MS (Fig. 3B and C). ESI-MS for H3 (C110A)K$_c$14Ac and H3 (C110A)K$_c$56Ac, calculated m/z [M+H]$^+$ 15,284.8 Da, observed m/z [M+H]$^+$ 15,287.3 ± 2.6 and 15,288.1 ± 3.6 Da, respectively.

5. GENERATION OF DESIGNER MNS

5.1 Octamer Formation Using Modified Histones

Histone octamers are assembled with minor modification of previous reports with the strict exclusion of reducing agents (Dyer et al., 2004). Briefly, each core histone is dissolved at ~4 mg/mL in an unfolding buffer containing 7 M Gn–HCl, 20 mM Tris, pH 7.5. It is important to use Ultrapure Gn–HCl at this step to allow accurate protein quantitation by Abs$_{280}$. The web-based ExPASy Protein Parameters Tool at http://web.expasy.org/protparam/ is used to compute the extinction coefficient for each protein. The histones are mixed in equimolar amounts and the resulting mixture is dialyzed into a refolding buffer (3 × 1 L) consisting of 2 M NaCl, 1 mM EDTA, 10 mM Tris, pH 7.5 at 4°C. Crude octamers are concentrated by Vivaspin 500 concentrators and purified by size exclusion on a Superdex S-200 column equilibrated in the octamer refolding buffer. Fractions

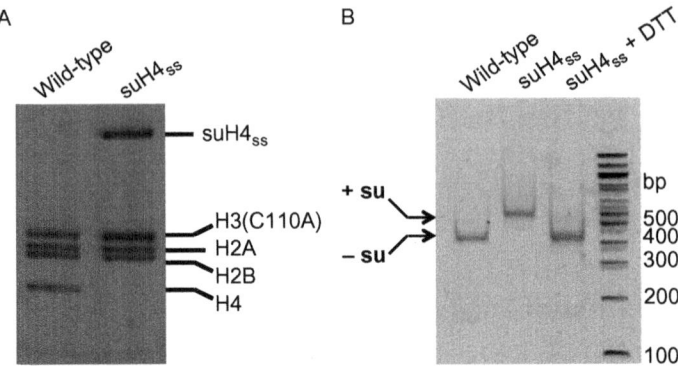

Fig. 4 Generation of designer mononucleosomes. (A) Wild-type and symmetrically sumoylated octamers visualized by 15% SDS-PAGE gel run at 200 V for 45 min and stained with Coomassie. (B) Wild-type and symmetrically sumoylated octamers visualized by 5% TBE gel run at 180 V for 50 min and stained with ethidium bromide. *su*, SUMO-3 C47S.

containing pure histone octamers are identified by 15% SDS-PAGE stained with Coomassie, combined and concentrated prior to nucleosome assembly (Fig. 4A).

5.2 Generation of 147 bp *601* DNA

The 147 bp *601* DNA is amplified from a plasmid containing the 1_147_601 fragment (Lowary & Widom, 1998) by PCR in 5 mL net volume using the forward primer 5′-CTG GAG AAT CCC GGT GCC GAG G-3′ and reverse primer 5′-ACA GGA TGT ATA TAT CTG ACA CG-3′. The PCR product is purified using a QIAquick PCR purification kit and eluted in sterile water (pH 7.5). The eluate is lyophilized to concentrate the DNA and then resuspended in sterile water to a final concentration of 20–25 μM. The DNA is visualized for purity by running on a 2% Agarose gel at 150 V for 30 min in TBE buffer and staining with ethidium bromide.

5.3 Reconstitution of MNs

Following a modified stepwise dilution protocol reported by Workman and coworkers (Utley et al., 1996), pure histone octamers and 147 bp *601* DNA are combined in 10 μL of a high-salt refolding buffer consisting of 2 M NaCl, 1 mM EDTA, 10 mM Tris, pH 7.5, to a final concentration of 2 μM in octamers and DNA. After incubation at 37°C for 15 min, 3.3 μL of dilution buffer 1 containing 1 mM EDTA, 0.5 mM PMSF, 10 mM HEPES, pH 7.9, is added and the temperature dropped to 30°C. Further dilutions of 6.7, 5, 3.6, 4.7, 6.7, 10, 30, and 20 μL, respectively, are performed every 15 min. A final dilution is undertaken with 100 μL of dilution buffer 2 containing 1 mM EDTA, 0.1% (v/v) NP-40, 0.5 mM PMSF, 20% (v/v) glycerol, 10 mM Tris, pH 7.5. After an additional 15 min at 30°C, the MNs are concentrated and analyzed by separation on a Criterion 5% TBE gel run in 0.5× TBE, followed by staining with ethidium bromide. For sumoylated MNs, reduction with 1 mM DTT for 30 min on ice leads to the removal of SUMO and comigration of the reduced MNs with unmodified wild-type MNs. This is used to confirm the core histone composition as being identical to that of wild-type MNs (Fig. 4B).

5.4 Preparation of 177 bp Repeat *601* DNA

A plasmid containing 12 copies of a 177 bp repeat of the *601* nucleosome positioning sequence (12_177_601) (Dorigo, Schalch, Bystricky, & Richmond, 2003) flanked by *EcoRV* sites is purified from a 6-L culture of DH5α cells using a Qiafilter Plasmid Giga kit. The 12_177_601 sequence is obtained by preparative-scale digestion of the plasmid with *EcoRV*. This is followed by selective precipitation of the 12_177_601 fragment with 6% polyethylene glycol (PEG)-6000 on ice and centrifugation at 26,000 × g for 30 min at 4°C. After phenol extraction and ethanol precipitation, the DNA is redissolved in TE buffer (1 mM EDTA, 10 mM Tris, pH 7.5) and quantified by Abs_{260} prior to array formation.

5.5 Generation of 12-mer Nucleosome Arrays

Pure histone octamers (2 μM) and 12_177_601 DNA (0.17 μM in DNA, 2 μM in octamer binding sites) are combined in 75 μL of reconstitution buffer consisting of 2 M KCl, 0.1 mM EDTA, 10 mM Tris, pH 7.8. Additionally, 0.7 μM of a weaker-binding 147 bp fragment of the MMTV DNA is added to prevent overloading of the array with octamers. Stepwise dialysis is performed at 4°C against reconstitution buffer containing 1.4 M NaCl, 1.2 M NaCl, 1 M NaCl, 0.8 M NaCl, 0.5 M NaCl, and 10 mM NaCl for 90 min each, followed by a final dialysis step against reconstitution buffer containing 10 mM NaCl. DNA fragments are removed by selective precipitation of the arrays with $MgCl_2$. The crude array mixture is incubated with 4 mM $MgCl_2$ for 10 min on ice to facilitate aggregation and precipitation of the arrays. The mixture is then spun at 15,000 × g at 4°C for 10 min and the supernatant is immediately removed. The MMTV DNA does not aggregate under these conditions and remains in the supernatant. It is important to remove the buffer from the centrifuged pellet immediately because chromatin aggregation is a reversible and dynamic process, which leads to the resolubilization of some pelleted material upon standing in buffer. The final concentration of $MgCl_2$ used to precipitate arrays often varies with the nature of chemical modification in the histone tails. For example, arrays containing acetylation at Lys16 of H4 should be precipitated at 6 mM $MgCl_2$, due to the known inhibition of chromatin aggregation by acetylation. Therefore, as far as possible, the minimal salt concentration to be used for each array should be determined empirically with 4 mM $MgCl_2$ as a reasonable starting point. After removal of the supernatant, the arrays are resuspended in TEN buffer (10 mM Tris, 1 mM EDTA, 10 mM NaCl, pH 8.0), dialyzed against fresh TEN buffer, and quantified by Abs_{260}. Wild-type and modified arrays are visualized on 1% *agarose*–2% *polyacrylamide gel* electrophoresis (APAGE) by running gels at 180 V for 20 min at 4°C followed by staining with ethidium bromide (Fig. 5A).

The saturation of octamer binding sites in 12-mer arrays is confirmed by the digestion of 0.17 pmol of arrays with 10 U of *Sca*I restriction enzyme in NEB buffer 3 (50 mM Tris, 100 mM NaCl, 10 mM $MgCl_2$, 1 mM DTT, pH 7.9) at 25°C for 12 h, followed by separation on a Criterion 5% TBE gel run in 0.5 × TBE buffer and staining with ethidium bromide. The

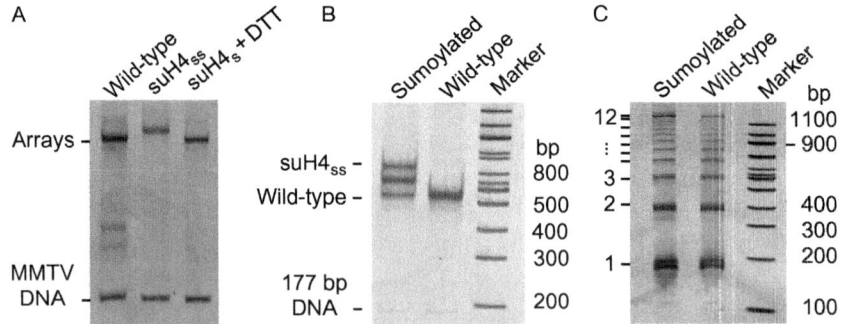

Fig. 5 Generation and characterization of designer nucleosome arrays. (A) 1% APAGE gel stained with ethidium bromide showing wild-type and uniformly sumoylated 12-mer arrays and the desumoylation product upon addition of a reducing agent. (B) 5% TBE gel stained with ethidium bromide showing *Sca*I digestion products of wild-type and sumoylated 12-mer arrays. The presence of three bands in the sumoylated sample corresponds to di-, mono-, and nonsumoylated mononucleosomes arising from partial reduction in the digestion buffer. (C) 5% TBE gel stained with ethidium bromide showing digested DNA arising from limited *Micrococcal nuclease* digestion of wild-type and sumoylated 12-mer arrays. The presence of 12 DNA bands of increasing size is indicated.

presence of an MN band as well as the absence of free DNA and higher molecular weight species indicates full array occupancy. DTT in the digestion and restriction enzyme buffer leads to partial loss of the SUMO moiety in sumoylated arrays resulting in three distinct MN bands corresponding to the doubly, singly, and nonsumoylated species (Fig. 5B). The presence of 12 MNs per array is also confirmed by partial digestion with the enzyme *Micrococcal nuclease*. In a typical assay, 0.17 pmol of 12-mer arrays are digested with 0.2 U of *Micrococcal nuclease* for 60 s on ice. The reaction was stopped by the addition of 0.2% (w/v) SDS and 20 mM EDTA followed by DNA purification using a QIAquick PCR purification kit (Qiagen). The DNA fragments were separated on a Criterion 5% TBE gel run in 0.5 × TBE buffer and visualized after ethidium bromide staining (Fig. 5C). The presence of 12 distinct DNA bands indicates the uniform distribution of nucleosomes on the 12_177_601 DNA.

6. SUMMARY AND CONCLUSIONS

Synthetic access to uniformly and site-specifically modified histones is key when investigating the molecular details of chromatin-mediated mechanisms that drive gene regulation, replication, and repair. Entirely synthetic (Shimko, North, Bruns, Poirier, & Ottesen, 2011) and combinations of synthetic and recombinant technologies, embodied by the technique of expressed protein ligation (Muir et al., 1998), allow exquisite chemical control of both modification site and type in full-length histones. However, these methodologies remain challenging for laboratories not equipped for advanced chemical synthesis. The discovery of multiple functional analogs of modified histones provides convenient alternatives that are accessible by one- or two-step transformations on recombinant proteins with commercially available reagents (Dhall & Chatterjee, 2015). The two methods presented to generate analogs of acetylated and sumoylated histones are readily accomplished in a 2-week period, starting from heterologous expression, and provide access to milligram quantities of the modified histones. Although disulfide-linked sumoylated histones cannot be used in the presence of reducing agents, these can easily be excluded from in vitro assays and may even be tolerated based on the solvent accessibility of the modified residue. Importantly, these methods can be easily adapted toward histone ubiquitylation as well as for generating multiply acetylated histones (Chatterjee, McGinty, Fierz, & Muir, 2010).

ACKNOWLEDGMENTS

We would like to thank the Department of Chemistry and the Royalty Research Fund at the University of Washington for generous support. C.C. is supported by NIGMS Grant 1R01M110430. C.E.W. gratefully acknowledges support from the NSF GRFP (Grant No. DGH-1256082) and an ARCS foundation fellowship.

REFERENCES

Allis, C. D., Jenuwein, T., Reinberg, D., & Caparros, M.-L. (2007). *Epigenetics* (1st ed.). New York: Cold Spring Harbor Laboratory Press. 502 pp.

Badeaux, A. I., & Shi, Y. (2013). Emerging roles for chromatin as a signal integration and storage platform. *Nature Reviews Molecular Cell Biology, 14*, 211.

Chatterjee, C., McGinty, R. K., Fierz, B., & Muir, T. W. (2010). Disulfide-directed histone ubiquitylation reveals plasticity in hDot1L activation. *Nature Chemical Biology, 6*, 267.

Chatterjee, C., McGinty, R. K., Pellois, J.-P., & Muir, T. W. (2007). Auxiliary-mediated site-specific peptide ubiquitylation. *Angewandte Chemie (International Ed in English), 46*, 2814.

Davies, N., & Lindsey, G. G. (1994). Histone H2B (and H2A) ubiquitination allows normal histone octamer and core particle reconstitution. *Biochimica et Biophysica Acta, 1218*, 187.

Dawson, P. E., Muir, T. W., Clark-Lewis, I., & Kent, S. B. H. (1994). Synthesis of proteins by native chemical ligation. *Science, 266*, 776.

Dhall, A., & Chatterjee, C. (2011). Chemical approaches to understand the language of histone modifications. *ACS Chemical Biology, 6*, 987.

Dhall, A., & Chatterjee, C. (2015). Chapter 8: Chemical and genetic approaches to study histone modifications. In Y. G. Zheng (Ed.), *Epigenetic technological applications* (p. 149). New York: Academic Press.

Dhall, A., Wei, S., Fierz, B., Woodcock, C. L., Lee, T. H., & Chatterjee, C. (2014). Sumoylated human histone H4 prevents chromatin compaction by inhibiting long-range internucleosomal interactions. *Journal of Biological Chemistry, 289*, 33827.

Dorigo, B., Schalch, T., Bystricky, K., & Richmond, T. J. (2003). Chromatin fiber folding: Requirement for the histone H4 N-terminal tail. *Journal of Molecular Biology, 327*, 85.

Dyer, P. N., Edayathumangalam, R. S., White, C. L., Bao, Y., Chakravarthy, S., Muthurajan, U. M., et al. (2004). Reconstitution of nucleosome core particles from recombinant histones and DNA. *Methods in Enzymology, 375*, 23.

Galisson, F., Mahrouche, L., Courcelles, M., Bonneil, E., Meloche, S., Chelbi-Alix, M. K., et al. (2011). A novel proteomics approach to identify SUMOylated proteins and their modification sites in human cells. *Molecular and Cellular Proteomics, 10*. M110.004796.

Garcia, B. A., Pesavento, J. J., Mizzen, C. A., & Kelleher, N. L. (2007). Pervasive combinatorial modification of histone H3 in human cells. *Nature Methods, 4*, 487.

Hendriks, I. A., D'Souza, R. C., Yang, B., Verlaan-de Vries, M., Mann, M., & Vertegaal, A. C. (2014). Uncovering global SUMOylation signaling networks in a site-specific manner. *Nature Structural and Molecular Biology, 21*, 927.

Kent, S. B. H. (2009). Total chemical synthesis of proteins. *Chemical Society Reviews, 38*, 338.

Kouzarides, T. (2007). Chromatin modifications and their function. *Cell, 128*, 693.

Li, F., Allahverdi, A., Yang, R., Lua, G. B. J., Zhang, X., Cao, Y., et al. (2011). A direct method for site-specific protein acetylation. *Angewandte Chemie (International Ed. in English), 50*, 9611.

Long, L., Thelen, J. P., Furgason, M., Haj-Yahya, M., Brik, A., Cheng, D., et al. (2014). The U4/U6 recycling factor SART3 has histone chaperone activity and associates with USP15 to regulate H2B deubiquitination. *Journal of Biological Chemistry, 289*, 8916.

Lowary, P. T., & Widom, J. (1998). New DNA sequence rules for high affinity binding to histone octamer and sequence-directed nucleosome positioning. *Journal of Molecular Biology, 276,* 19.

McGinty, R. K., Kim, J., Chatterjee, C., Roeder, R. G., & Muir, T. W. (2008). Chemically ubiquitylated histone H2B stimulates hDot1L-mediated intranucleosomal methylation. *Nature, 453,* 812.

Muir, T. W., Sondhi, D., & Cole, P. A. (1998). Expressed protein ligation: A general method for protein engineering. *Proceedings of the National academy of Sciences of the United States of America, 95,* 6705.

Nathan, D., Ingvarsdottir, K., Sterner, D. E., Bylebyl, G. R., Dokmanovic, M., Dorsey, J. A., et al. (2006). Histone sumoylation is a negative regulator in Saccharomyces cerevisiae and shows dynamic interplay with positive-acting histone modifications. *Genes and Development, 20,* 966.

Neumann, H., Peak-Chew, S. Y., & Chin, J. W. (2008). Genetically encoding N(epsilon)-acetyllysine in recombinant proteins. *Nature Chemical Biology, 4,* 232.

Portela, A., & Esteller, M. (2010). Epigenetic modifications and human disease. *Nature Biotechnology, 28,* 1057.

Rabanal, F., DeGrado, W., & Dutton, P. (1996). Use of 2,2'-dithiobis(5-nitropyridine) for the heterodimerization of cysteine containing peptides. Introduction of the 5-nitro-2-pyridinesulfenyl group. *Tetrahedron Letters, 37,* 1347.

Shahbazian, M. D., & Grunstein, M. (2007). Functions of site-specific histone acetylation and deacetylation. *Annual Review of Biochemistry, 76,* 75.

Shilo, Y., & Eisenman, R. (2003). Histone sumoylation is associated with transcriptional repression. *Proceedings of the National academy of Sciences of the United States of America, 100,* 13225.

Shimko, J. C., North, J. A., Bruns, A. N., Poirier, M. G., & Ottesen, J. J. (2011). Preparation of fully synthetic histone H3 reveals that acetyl-lysine 56 facilitates protein binding within nucleosomes. *Journal of Molecular Biology, 408,* 187.

Shogren-Knaak, M., Ishii, H., Sun, J.-M., Pazin, M. J., Davie, J. R., & Peterson, C. L. (2006). Histone H4-K16 acetylation controls chromatin structure and protein interactions. *Science, 311,* 844.

Simon, M., Chu, F., Racki, L., Delacruz, C., Burlingame, A., Panning, B., et al. (2007). The site-specific installation of methyl-lysine analogs into recombinant histones. *Cell, 128,* 1003.

Strahl, B. D., & Allis, C. D. (2000). The language of covalent histone modifications. *Nature, 403,* 41.

Sullivan, S., Sink, W. D., Trout, K. L., Makalowska, I., Taylor, P. M., Baxevanis, A. D., et al. (2002). The histone database. *Nucleic Acids Research, 30,* 341–342.

Utley, R. T., Owen-Hughes, T. A., Juan, L. J., Cote, J., Adams, C. C., & Workman, J. L. (1996). In vitro analysis of transcription factor binding to nucleosomes and nucleosome disruption/displacement. *Methods in Enzymology, 274,* 276.

CHAPTER EIGHT

An IF–FISH Approach for Covisualization of Gene Loci and Nuclear Architecture in Fission Yeast

K.-D. Kim, O. Iwasaki, K. Noma[1]

The Wistar Institute, Philadelphia, PA, United States
[1]Corresponding author: e-mail address: noma@wistar.org

Contents

1. Introduction   168
2. Case Studies in the Application of the IF–FISH Approach   168
       2.1 Scoring Associations Between Two Gene Loci   168
       2.2 Visualizing Clusters of Centromeres and Telomeres   170
       2.3 Coordinating Gene Positioning to the Nuclear Architecture   170
3. Supplies   171
       3.1 Equipment   171
       3.2 Materials   172
       3.3 Buffers   173
4. Protocol   174
       4.1 Preparation of the FISH Probe Templates   174
       4.2 Fluorescent Labeling of the Templates   174
       4.3 Fixation of the Fission Yeast Cells   175
       4.4 Permeabilization of the Cells   175
       4.5 Antigen–Antibody Reactions (IF) and Fixation   176
       4.6 Hybridization (FISH)   177
       4.7 Preparation of the Cells for Microscopy   177
5. Notes   178

Acknowledgments   179
References   179

Abstract

Recent genomic studies have revealed that chromosomal structures are formed by a hierarchy of organizing processes ranging from gene associations, including interactions among enhancers and promoters, to topologically associating domain formations. Gene associations identified by these studies can be characterized by microscopic analyses. Fission yeast is a model organism, in which gene associations have been broadly mapped across the genome, although many of those associations have not been further examined by cell biological approaches. To address the technically challenging process of the visualization of associating gene loci in the fission yeast nuclei, we provide, in detail, an IF–FISH procedure that allows for covisualizing both gene loci and nuclear structural markers such as the nuclear membrane and nucleolus.

1. INTRODUCTION

Next-generation DNA sequencing combined with the molecular biology procedure called chromosome conformation capture (3C), referred to as Hi-C, allows one to investigate a series of genome-organizing events and to map gene associations throughout the genomes of various organisms (Lieberman-Aiden et al., 2009; Tanizawa & Noma, 2012). Hi-C-related approaches have been successfully applied to fission yeast cells to map the yeast global gene association network (Grand et al., 2014; Mizuguchi et al., 2014; Tanizawa et al., 2010). An alternative approach, chromatin interaction analysis by paired-end tag sequencing (ChIA-PET), has also been employed to identify genome-wide associations mediated by particular proteins (Fullwood et al., 2009).

It is becoming clear that the organization of the genome is tightly connected to nuclear activities such as transcriptional regulation, repair, and DNA replication (Misteli, 2007). The nucleus consists of various well-defined nuclear domains including nucleoli, Cajal bodies, and PML bodies, which are also linked to nuclear activities (Spector, 2001). How gene associations and global genome architecture, as determined by genomic approaches, are coupled to nuclear activities and various subnuclear domains remain largely unknown. The application of cell biological approaches to the study of these processes can help fill this significant gap. In this chapter, we describe a practical protocol combining IF (immunofluorescence) and FISH (fluorescence in situ hybridization), which allows investigators to covisualize gene loci and nuclear structural components such as the nuclear membrane and nucleoli. The application of this method will contribute to the understanding of how and where gene associations occur in the fission yeast nucleus.

2. CASE STUDIES IN THE APPLICATION OF THE IF–FISH APPROACH

2.1 Scoring Associations Between Two Gene Loci

An IF–FISH approach can be used to detect and confirm gene associations predicted by a Hi-C analysis. The IF–FISH images shown in this chapter were captured by a Zeiss Axioimager Z1 fluorescence microscope with an oil immersion objective lens (Plan Apochromat, 100×, NA 1.4, Zeiss). Multicolor FISH can be used to covisualize two different gene loci in a nucleus (Fig. 1A). IF can be used to visualize α-tubulin in the same cell using 1:10-diluted antitubulin TAT1 monoclonal antibody (Woods et al., 1989). Based on tubulin staining patterns, cell cycle phases can be estimated (Funabiki, Hagan, Uzawa, & Yanagida, 1993). Since gene associations are highly transient, these associations are scored as frequencies of nearby localization of two gene loci (Iwasaki et al., 2015; Kim et al., 2013). Centers of FISH foci are defined as positions of the loci, and the distance between two foci present on the same focal layer must be measured in order to minimize errors derived from low resolution along the z-axis (Fig. 1B). The distance between two loci should be measured in more than 100 cells for reliability. It is also important to have a negative control, where two loci separated by the same genomic distance are predicted not to associate with each other (Fig. 1A). In order to examine whether a specific mutation affects a particular gene association in a statistically significant manner, distributions of distances between two loci in wild-type and mutant cells can be subjected to the Mann–Whitney U test (Fig. 1C).

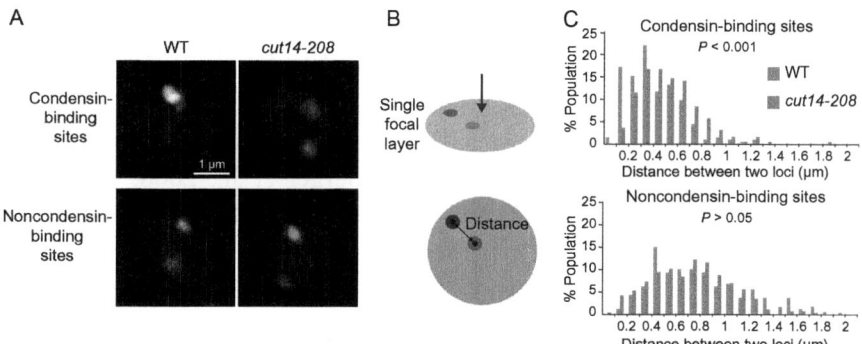

Fig. 1 Analysis on FISH foci reflecting positions of two gene loci. (A) The two paired gene loci were visualized in wild-type and *cut14-208* condensin mutant cells. Because the *cut14-208* mutation is temperature sensitive, wild-type and mutant cells were cultured at the restrictive temperature (36°C) for 1 h. The FISH foci (*green* and *red*) representing the two loci bound by condensin were positioned nearby in wild-type cells but not in the mutant (*top*), whereas the two control loci (noncondensin-binding sites) were consistently separated (*bottom*). DAPI signals are shown in *blue*. (B) Two FISH foci are present on a single focal layer (*top*), and the distance between centers of the two FISH foci is defined as a physical distance between the two loci (*bottom*). (C) Distributions of distances between the two gene loci in wild-type and *cut14-208* mutant cells. (See the color plate.)

2.2 Visualizing Clusters of Centromeres and Telomeres

Fission yeast centromeres and telomeres are known to form respective clusters at the nuclear periphery, and this clustering is linked to chromosome dynamics during mitosis (Funabiki et al., 1993). The plasmid pRS140 is used for preparing FISH probes specific to centromeres (Chikashige et al., 1989). FISH foci represent centromeres of all the three chromosomes (Fig. 2A). The cosmid cos212 is used to prepare a telomere-specific FISH probe, which visualizes the left and right subtelomeric regions of the chromosomes 1 and 2 (Sadaie, Naito, & Ishikawa, 2003). To visualize every FISH focus reflecting centromeres or telomeres, images are acquired at 0.2 μm intervals in the z-axis, and z-stack images should be used to count the total number of FISH foci in a nucleus as demonstrated (Fig. 2A).

To demonstrate the disruption of the centromeric clustering we used the *sad1* gene mutant. Sad1 is a nuclear membrane protein bearing the Sad1-UNC-84 (SUN) domain, and the temperature-sensitive *sad1-1* mutation is known to disrupt the clustering of centromeres in fission yeast (Hagan & Yanagida, 1995; Hou et al., 2012). As an example, we visualized centromeres in wild-type and *sad1-1* mutant cells (Fig. 2B). The difference in numbers of centromeric FISH foci in wild-type and mutant cell populations ($n > 100$) was evaluated by the chi-square test (Fig. 2C).

2.3 Coordinating Gene Positioning to the Nuclear Architecture

In yeast, gene loci tend to be positioned within subnuclear domains, and gene territories are often coordinated with two nuclear structural markers, such as the nuclear membrane and nucleolus (Berger et al., 2008; Kim et al., 2013). A particular gene locus and nuclear structural components can be

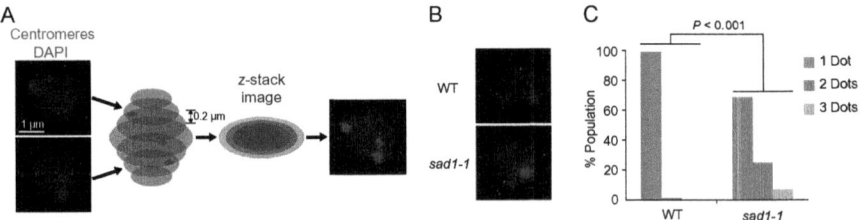

Fig. 2 Analysis on centromeric clusters. (A) FISH visualization of centromeric foci (*red*). Centromeric clusters are captured by a z-stack projection. DAPI signals are shown in *blue*. (B) Centromeric clusters in the wild-type and *sad1-1* mutant. Since the *sad1-1* is a temperature-dependent mutation, wild-type and mutant cells were cultured at 36° C for 1 h, and subjected to FISH analysis. (C) Numbers of centromeric clusters in wild-type and mutant cells. (See the color plate.)

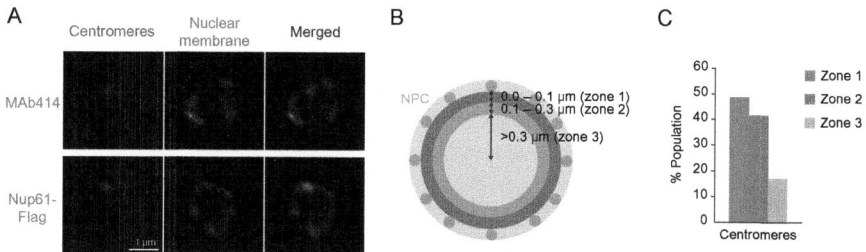

Fig. 3 Gene positioning relative to the nuclear architecture. (A) Anti-NPC protein antibody (MAb414) was used to visualize the nuclear membrane (*top*). Alternatively, a NPC subunit (Nup61) fused to a Flag epitope was expressed from the endogenous locus with its own promoter, and Nup61-Flag proteins were visualized by an IF approach (*bottom*). In the same cells, centromeres were covisualized by FISH (*red*). (B) NPC signals are used to trace the nuclear membrane, and the nucleus is divided into the three zones based on distance from the membrane. (C) The distance between the centromeric signal and the nuclear membrane was measured in more than 100 cells, and assigned to one of the nuclear zones. (See the color plate.)

covisualized by an IF–FISH approach (Fig. 3A). To visualize the nuclear membrane, a nuclear pore complex (NPC) component (Nup61) tagged with Flag can be visualized using 1:500-diluted mouse monoclonal anti-Flag antibody (F1804, Sigma–Aldrich). Alternatively, 1:100-diluted mouse monoclonal anti-NPC proteins (MAb414, Covance) can be used. The nucleolus can also be visualized using 1:20-diluted mouse monoclonal anti-Nop1 (Encor Biotechnology). Nop1 is known to function in the processing of ribosomal RNA and localizes in the nucleolus (Henriquez, Blobel, & Aris, 1990).

The nucleus is divided into three zones according to distance from the nuclear periphery visualized by IF (Fig. 3B). The distance between the nuclear membrane and a gene locus should be measured in more than 100 cells and binned into one of the assigned zones. As an example, the distance between the centromeric FISH foci and the nuclear periphery was measured and summarized in a graph (Fig. 3C). IF–FISH images carrying both a centromeric focus and clear NPC signals should be employed for this analysis.

3. SUPPLIES

3.1 Equipment

1. Water bath (37°C and 65°C)
2. Heat block (75°C and 95°C)
3. Air incubator (37°C and 40°C)

4. 18°C Water bath shaker in a cold room
5. 37°C Water bath shaker
6. Rotator
7. Thermomixer (Eppendorf)
8. Vacufuge (Eppendorf)
9. Centrifuges, micropipettes, and other items associated with standard molecular biology experiments.

3.2 Materials

1. Restriction enzymes (*Alu*I, *Rsa*I, *Hae*III, *Bfu*CI, and *Dde*I; New England BioLabs)
2. Phenol chloroform isoamyl alcohol (P3803, Sigma-Aldrich)
3. Amicon Ultra-30K and 10K centrifugal filters (UFC503024 and UFC501024, Millipore)
4. Random primer DNA labeling kit (TAK6045, Takara)
5. Cy3-dCTP or Cy5-dCTP (PA53021 and PA55021, GE Healthcare)
6. 30% Paraformaldehyde in YEA medium
7. 2.4 M Sorbitol in YEA medium (Alfa, Fantes, Hyams, McLeod, & Warbrick, 1993)
   YEA: 5 g yeast extract, 30 g glucose, and 75 mg adenine for 1 L (pH 5.5)
8. 2.5 M Glycine (filtered)
9. 1 mg/mL Zymolyase 100 T (07665-55, Seikagaku) in PEMS (stored at −80°C)
10. Primary antibodies described earlier
11. Secondary antibodies that successfully worked for our experiments
    1:500-Diluted Alexa Flour 488-conjugated anti-mouse IgG (Life Technologies)
    1:400-Diluted Alexa Flour 594-conjugated anti-mouse IgG (Life Technologies)
    1:400-Diluted Alexa Flour 488-conjugated anti-rabbit IgG (Life Technologies)
    1:4000-Diluted Cy3-conjugated anti-rabbit IgG (Jackson ImmunoResearch)
12. 30 μg/μL RNase A solution (R4642, Sigma-Aldrich)
13. 1 μg/mL DAPI in PBS buffer (1 mg/mL DAPI-PBS stock solution at −20°C)
14. Antifade solution (P6001, Sigma-Aldrich)

Prepare 10 mg/mL p-phenylenediamine in 1 M Tris–HCl (pH 8.0) and store at $-20°C$. Mix 5 µL solution with 45 µL glycerol before use
15. Poly-L-lysine (P8920, Sigma-Aldrich)
16. Glass slides and coverslips
17. $1 \times$ PBS and $1 \times$ PBS + 0.1% NaN_3
18. Nail polish.

3.3 Buffers

1. PEM

Component	Final Conc. (mM)	Stock (mM)	Amount
PIPES (pH 6.9)	100		30.233 g
EGTA	1	500	2 mL
MgSO$_4$	1	500	2 mL

Adjust to pH 6.9 with 10 N NaOH. Add H_2O up to 1 L.

2. PEMS
   1 M Sorbitol in PEM buffer.
3. PEMBAL

Component	Final Conc.	Stock	Amount
BSA	1%		1 g
L-Lysine	0.1 M		1.46 g
NaN$_3$	0.1%	10%	1 mL

Add PEM buffer up to 100 mL.

4. Hybridization buffer

Component	Final Conc.	Stock	Amount (µL)
Formamide			500
SSC	$2\times$	$20\times$	100
Denhardt's solution	$5\times$	$50\times$	100
Dextran sulfate	10%	50%	200
DW			100

4. PROTOCOL

4.1 Preparation of the FISH Probe Templates

1. Digest 1 µg of DNA fragments (see *Note 1*) with 4-base cutters (*Alu*I, *Rsa*I, *Hae*III, *Bfu*CI, and *Dde*I) (see *Note 2*) in 100 µL total volume at 37°C for 3 h.
2. Add 130 µL H_2O and extract DNA by phenol/chloroform.
3. Transfer aqueous phase (200 µL) to Amicon Ultra-30K centrifugal filter.
4. Spin at 13,000 rpm for 5 min and discard the flow-through (avoid completely drying the filter).
5. Add 200 µL H_2O to the filter column.
6. Spin at 13,000 rpm for 5 min and discard the flow-through.
7. Repeat four more times (steps 5 and 6).
8. Invert the filter column and spin at 3000 rpm for 3 min.
9. Reduce the sample volume to 20 µL using Vacufuge for approximately 12 min.
10. Store the FISH probe template at 4°C.

4.2 Fluorescent Labeling of the Templates

1. Prepare 16.5 µL of the FISH probe template in 1.5-mL microcentrifuge tube.
2. Add 2 µL of random primer mix and mix well.
3. Denature DNA at 95°C for 3 min.
4. Immediately transfer to ice and leave for 5 min to prevent renaturation of single-stranded DNA. Flash spin at 13,000 rpm.
5. Add random labeling mix (total volume 6.5 µL) and mix by pipetting. 2.5 µL 10× buffer, 2.5 µL dNTP mix, 0.5 µL Cy3- or Cy5-dCTP, 1.0 µL Klenow enzyme.
6. Incubate at 37°C for 1 h with shaking at 500 rpm using Thermomixer. Protect sample from light using aluminum foil from here on.
7. Inactivate Klenow enzyme at 65°C for 10 min and incubate on ice for 3 min.
8. Add 150 µL TE buffer and apply to Amicon Ultra-10K centrifugal filter.
9. Spin at 13,000 rpm at 4°C for 10 min and discard the flow-through.
10. Apply 200 µL TE buffer to the filter column.

11. Spin at 13,000 rpm at 4°C for 10 min and discard the flow-through.
12. Repeat three more times (steps 10 and 11).
13. Invert the column and spin at 3000 rpm for 3 min to recover the FISH probe.
14. Adjust to 100 µL with TE buffer.
15. Store the FISH probe at 4°C in dark (stable for a few years).

4.3 Fixation of the Fission Yeast Cells

1. Prepare logarithmically growing cells in 20 mL YEA medium at 30°C ($OD_{595} = 0.5$).
2. Prepare 30% paraformaldehyde in YEA medium (see *Note 3*).
3. Add 1 volume (20 mL) of 2.4 M sorbitol in YEA medium to the culture.
4. Agitate the culture at 18°C for 5 min using water bath shaker (cold room).
5. Add 6 mL of 30% paraformaldehyde and shake at 18°C for 10 min.
6. Add 2.4 mL of 2.5 M glycine and shake at 18°C for 5 min.
7. Transfer to 50-mL centrifuge tube.
8. Spin at 3000 rpm at 4°C for 5 min and remove the supernatant.

4.4 Permeabilization of the Cells

1. Resuspend the cells in 1 mL PEM buffer and transfer the cells to a 1.5-mL tube.
2. Spin down the cells at 7000 rpm for 1 min and remove the supernatant (see *Note 4*).
3. Resuspend the cells in 1 mL PEMS buffer and let stand at room temperature for 5 min.
4. Spin down the cells at 7000 rpm for 1 min and remove the supernatant.
5. Resuspend the cells in 1 mL of 1 mg/mL Zymolyase 100 T in PEMS. Cells become fragile after this step. Therefore, cut 2–3 mm from the end of pipette tips or mix by tapping.
6. Incubate the cells at 37°C (water bath) for 30 min with occasional tapping. Check the cell wall digestion with 2 µL sample mixed with 2 µL of 10% SDS (see *Note 5*).
7. Spin at 7000 rpm for 1 min and remove the supernatant.
8. Resuspend the cells in 1 mL PEMS buffer.
9. Spin down the cells at 7000 rpm for 1 min and remove the supernatant.
10. Resuspend the cells in 500 µL PEMS containing 1% Triton X-100 and leave at room temperature for 30 s.

11. Spin at 7000 rpm for 1 min and remove the supernatant.
12. Resuspend the cells with 1 mL PEMS.
13. Spin at 7000 rpm for 1 min and remove the supernatant.
14. Repeat steps 12 and 13 (namely, wash the cells with 1 mL PEMS twice).
15. Wash the cells with 1 mL PEM twice.

4.5 Antigen–Antibody Reactions (IF) and Fixation

1. Resuspend the cells in 500 µL PEMBAL and rotate at room temperature for at least 1 h (see *Note 6*).
2. Spin at 7000 rpm for 1 min and remove the supernatant.
3. Resuspend the cells in 200 µL PEMBAL containing primary antibodies as detailed earlier.
4. Rotate the tube at room temperature for 14–16 h.
5. Spin at 7000 rpm for 1 min and remove the supernatant.
6. Wash the cells with 1 mL PEMBAL three times.
7. Add 200 µL PEMBAL containing secondary antibodies as detailed earlier.
8. Wrap the tube in foil and rotate at room temperature for 4 h (see *Note 7*).
9. Spin at 7000 rpm for 1 min and remove the supernatant.
10. Resuspend the cells in 500 µL PEMBAL and rotate at room temperature for 15 min.
11. Spin at 7000 rpm for 1 min and remove the supernatant.
12. Repeat steps 10 and 11 for a total of three times.
13. Wash the cells with 500 µL PEM twice.
14. Resuspend the cells in 900 µL PEM and add 100 µL of 30% paraformaldehyde in PEM (see *Note 8*).
15. Rotate the tube for 20 min at room temperature in the dark.
16. Add 40 µL 2.5 M glycine.
17. Wash with 1 mL PEM twice.
18. Resuspend the cells in 500 µL of 0.1 N HCl and rotate at room temperature for 5 min (see *Note 9*).
19. Wash the cells with 1 mL PEM twice.
20. Resuspend the cells with 500 µL PEMBAL and add 1.6 µL of RNase A solution (30 µg/µL).
21. Cover the sample with aluminum foil and rotate at 37°C (air incubator) for 1.5 h (see *Note 10*).

4.6 Hybridization (FISH)

1. Prepare the hybridization buffer as described earlier.
2. Mix 10 μL (100–150 ng) FISH probe with 100 μL hybridization buffer.
3. Incubate at 75°C (heat block) for 15 min.
4. Spin the RNase-treated cells at 7000 rpm for 1 min and remove the supernatant.
5. Wash the cells with 500 μL PEM twice.
6. Resuspend the cells in 110 μL hybridization buffer containing the FISH probe.
7. Incubate the cells at 75°C (heat block) for 5 min (see *Note 11*).
8. Wrap the tube in aluminum foil and rotate at 40°C (air incubator) for 12–14 h.
9. Add 100 μL 2× SSC and mix well by pipetting.
10. Spin at 7000 rpm for 2 min and decant. Spin again at 7000 rpm for 2 min and remove the remaining supernatant (see *Note 12*).
11. Resuspend the cells with 100 μL 2× SSC and rotate at room temperature for 30 min in the dark.
12. Spin at 7000 rpm for 30 s and decant. Spin again at 7000 rpm for 30 s and remove the remaining supernatant.
13. Repeat steps 11 and 12 twice (three times total).

4.7 Preparation of the Cells for Microscopy

1. Resuspend the cells with 200 μL of 1 μg/mL DAPI in PBS buffer.
2. Incubate at room temperature for 5 min with rotation.
3. Wash the cells with 100 μL PBS twice.
4. Resuspend the cells in 10–100 μL of PBS containing 0.1% NaN_3 solution.
5. Spread 2 μL of the cells onto a poly-L-lysine-coated coverslip (see *Note 13*).
6. Leave for 5 min in the dark.
7. Apply 5 μL of antifade solution to the glass slide.
8. Place the coverslip on the glass slide as the cells touch the antifade solution.
9. Leave for 5 min in the dark.
10. Seal the coverslip with nail polish.
11. Leave for 5–10 min and investigate the cells under a fluorescent microscope.

5. NOTES

1. To generate FISH probes, cosmids, plasmids, and PCR products (~15 kb) can be used. Fission yeast cosmid information is found at the Pombase website (http://www.pombase.org/tools/clone-and-mapping-resources), and clones can be obtained from the Sanger Institute. Three 5 kb DNA fragments adjoined but not overlapping each other are amplified by PCR and can be used for generating locus-specific FISH probes.
2. DNA fragments should be 50–200 bp. For generating a centromeric probe using pRS140, the combination of *Alu*I, *Rsa*I, *Hae*III, *Bfu*CI, *Dde*I, and *Sau*3AI are optimal for the digestion.
3. Prepare a fresh solution for each experiment. Add 3 g of paraformaldehyde and 40 µL of 10 N NaOH to about 7 mL of YEA medium and mix well. Incubate at 65°C and occasionally vortex until the solution becomes clear. It takes approximately 20 min. Cool to room temperature and filter with 0.22 µm Steriflip (SE1M179M6, Millipore). Be cautious, as paraformaldehyde powder and its fume are toxic.
4. Spin at 7000 rpm for 30 s and decant the supernatant. To remove the buffer completely, spin again at 7000 rpm for 30 s and remove the residual supernatant by pipetting.
5. Keep the sample on ice and check digestion efficiency under the microscope. If cell walls are not digested completely, incubate again at 37°C for 5 min and check the cells. Repeat this step until digestion is complete.
6. Blocking time is dependent upon antibody specificity. If a background signal is high, increase the blocking time to 2–4 h.
7. Protect samples against light from this step on because the secondary antibodies are fluorescence conjugated. Wrap tubes in foil.
8. 30% Paraformaldehyde in PEM is prepared by the same procedure as described in *Note 3*, except for using PEM instead of YEA.
9. Acid treatment is recommended, because this step helps deproteinize and facilitates penetration of FISH probes (Syrjanen, 1992).
10. Excess RNase A or prolonged incubation may affect the quality of FISH images, potentially resulting from a DNase activity contaminated in the RNase A solution. Prepare the hybridization buffer and preheat at 75°C during the RNase A treatment.

11. Resuspend the cells with the hybridization buffer by pipetting. Cut the end of tips to handle the sticky hybridization buffer. By handling multiple samples at 30 s intervals, you can make sure that every sample is treated at 75°C for exactly 5 min.
12. Mix well with 100 µL 2 × SSC (step 7) and do not remove the supernatant completely until the second spin.
13. Apply 300 µL of poly-L-lysine to the coverslip and wait for 20 min. Remove poly-L-lysine by vacuum aspiration and protect from dust.

ACKNOWLEDGMENTS

We would like to thank the Wistar Imaging Facility for microscopic analysis. We also thank Louise Showe for critically reading the manuscript and Sylvie Shaffer for editorial assistance. Research reported in this publication was supported by the G. Harold and Leila Y. Mathers Charitable Foundation and the NIH Director's New Innovator Award Program of the National Institutes of Health under award number (DP2-OD004348 to K.N.). Support for Shared Resources utilized in this study was provided by Cancer Center Support Grant (CCSG) P30CA010815 to The Wistar Institute.

REFERENCES

Alfa, C., Fantes, P., Hyams, J., McLeod, M., & Warbrick, E. (1993). *Experiments with fission yeast*. New York: Cold Spring Harbor Laboratory Press.

Berger, A. B., Cabal, G. G., Fabre, E., Duong, T., Buc, H., Nehrbass, U., et al. (2008). High-resolution statistical mapping reveals gene territories in live yeast. *Nature Methods, 5*(12), 1031–1037.

Chikashige, Y., Kinoshita, N., Nakaseko, Y., Matsumoto, T., Murakami, S., Niwa, O., et al. (1989). Composite motifs and repeat symmetry in S. pombe centromeres: Direct analysis by integration of NotI restriction sites. *Cell, 57*(5), 739–751.

Fullwood, M. J., Liu, M. H., Pan, Y. F., Liu, J., Xu, H., Mohamed, Y. B., et al. (2009). An oestrogen-receptor-alpha-bound human chromatin interactome. *Nature, 462*(7269), 58–64.

Funabiki, H., Hagan, I., Uzawa, S., & Yanagida, M. (1993). Cell cycle-dependent specific positioning and clustering of centromeres and telomeres in fission yeast. *The Journal of Cell Biology, 121*(5), 961–976.

Grand, R. S., Pichugina, T., Gehlen, L. R., Jones, M. B., Tsai, P., Allison, J. R., et al. (2014). Chromosome conformation maps in fission yeast reveal cell cycle dependent sub nuclear structure. *Nucleic Acids Research, 42*(20), 12585–12599.

Hagan, I., & Yanagida, M. (1995). The product of the spindle formation gene sad1 + associates with the fission yeast spindle pole body and is essential for viability. *The Journal of Cell Biology, 129*(4), 1033–1047.

Henriquez, R., Blobel, G., & Aris, J. P. (1990). Isolation and sequencing of NOP1. A yeast gene encoding a nucleolar protein homologous to a human autoimmune antigen. *The Journal of Biological Chemistry, 265*(4), 2209–2215.

Hou, H., Zhou, Z., Wang, Y., Wang, J., Kallgren, S. P., Kurchuk, T., et al. (2012). Csi1 links centromeres to the nuclear envelope for centromere clustering. *The Journal of Cell Biology, 199*(5), 735–744.

Iwasaki, O., Tanizawa, H., Kim, K. D., Yokoyama, Y., Corcoran, C. J., Tanaka, A., et al. (2015). Interaction between TBP and condensin drives the organization and faithful segregation of mitotic chromosomes. *Molecular Cell, 59*(5), 755–767.

Kim, K. D., Tanizawa, H., Iwasaki, O., Corcoran, C. J., Capizzi, J. R., Hayden, J. E., et al. (2013). Centromeric motion facilitates the mobility of interphase genomic regions in fission yeast. *Journal of Cell Science, 126*(Pt. 22), 5271–5283.

Lieberman-Aiden, E., van Berkum, N. L., Williams, L., Imakaev, M., Ragoczy, T., Telling, A., et al. (2009). Comprehensive mapping of long-range interactions reveals folding principles of the human genome. *Science, 326*(5950), 289–293.

Misteli, T. (2007). Beyond the sequence: Cellular organization of genome function. *Cell, 128*(4), 787–800.

Mizuguchi, T., Fudenberg, G., Mehta, S., Belton, J. M., Taneja, N., Folco, H. D., et al. (2014). Cohesin-dependent globules and heterochromatin shape 3D genome architecture in S. pombe. *Nature, 516*(7531), 432–435.

Sadaie, M., Naito, T., & Ishikawa, F. (2003). Stable inheritance of telomere chromatin structure and function in the absence of telomeric repeats. *Genes & Development, 17*(18), 2271–2282.

Spector, D. L. (2001). Nuclear domains. *Journal of Cell Science, 114*(Pt. 16), 2891–2893.

Syrjanen, S. (1992). *Viral gene detection by in situ hybridization* (pp. 103–137). Oxford: Oxford University Press.

Tanizawa, H., Iwasaki, O., Tanaka, A., Capizzi, J. R., Wickramasinghe, P., Lee, M., et al. (2010). Mapping of long-range associations throughout the fission yeast genome reveals global genome organization linked to transcriptional regulation. *Nucleic Acids Research, 38*(22), 8164–8177.

Tanizawa, H., & Noma, K. (2012). Unravelling global genome organization by 3C-seq. *Seminars in Cell & Developmental Biology, 23*(2), 213–221.

Woods, A., Sherwin, T., Sasse, R., MacRae, T. H., Baines, A. J., & Gull, K. (1989). Definition of individual components within the cytoskeleton of Trypanosoma brucei by a library of monoclonal antibodies. *Journal of Cell Science, 93*(Pt. 3), 491–500.

PART II

Small Molecule Epigenetic Regulators

CHAPTER NINE

Biology, Chemistry, and Pharmacology of Sirtuins

A. Bedalov, S. Chowdhury, J.A. Simon[1]

Fred Hutchinson Cancer Research Center, Seattle, WA, United States
[1]Corresponding author: e-mail address: jsimon@fhcrc.org

Contents

1. Introduction — 184
2. Sirtuins and Metabolism — 187
3. Sirtuins and Regulation of Cellular NAD$^+$ Levels — 189
4. Sirtuin Functions — 189
5. Sirtuins and Metabolic Disorders — 190
6. Sirtuins and Cancer — 191
7. Sirtuin Activity Assays — 194
8. Identification of First-Generation Sirtuin Inhibitors — 197
9. Second-Generation Splitomicin Inhibitors — 200
10. Other Sirtuin Inhibitors — 201
11. Concluding Remarks — 205
References — 205

Abstract

Sirtuins are a family of protein deacylases related by amino acid sequence and cellular function to the yeast *Saccharomyces cerevisiae* protein Sir2 (Silent *I*nformation *R*egulator-2), the first of this class of enzymes to be identified and studied in detail. Based on its initially discovered activity, Sir2 was classified as a histone deacetylase that removes acetyl groups from histones H3 and H4. The acetylation/deacetylation of these particular substrates leads to changes in transcriptional silencing at specific loci in the yeast genome, hence its name. Sirtuins, however, have been shown to regulate a wide variety of cellular processes beyond transcriptional repression in varied subcellular compartments and in different cell types. Mechanistically distinct from Zn^{2+}-dependent deacylases, sirtuins use nicotinamide adenine dinucleotide as a cofactor in the removal of acetyl and other acyl groups linking metabolic status and posttranslational modification. Sirtuins' unique position has made them attractive targets for small-molecule drug development. In this chapter, we describe the biological roles, therapeutic areas in which sirtuins may play a role and development of small-molecule inhibitors of sirtuins employing phenotypic screening technologies ranging from assays in yeast, as well as biochemical screens to yield lead drug development candidates targeting a broad spectrum of human diseases.

1. INTRODUCTION

Protein acylation is a reversible posttranslational modification that alters protein surface charge and, like phosphorylation, regulates protein conformational states, or protein/protein interactions. Additionally, acylation with long-chain fatty acids may regulate subcellular localization. Acyl transferases use acyl-coenzyme A (acyl–CoA) as acyl group donors to modify the lysine ε-amine of surface residues to convert positively charged lysine amino acid side chains into neutral ε-acyl-amine-lysine side chains. Acetyl transferases that use acetyl-CoA, including p300 and CREBP among others, are largely regulated through protein/protein interactions. Opposing the acyl transferases are the deacylases, enzymes that remove acyl groups from Nε-lysine residues. Overall acetylation status of a given target protein is regulated by the balance of enzymatic activities of the acetyl transferases and deacetylases as well as the levels of acyl-CoA. Acyl groups other than acetyl such as succinyl, crotonyl, myristoyl, and lipoyl can also be removed by members of the sirtuin family; however, the exact functional significance and regulation of these modifications is still an active area of investigation.

The larger family of protein deacylases can be grouped phylogenetically based on sequence similarity. There are four main branches of deacylases with class I, II, and IV enzymes (classical HDACs) sharing a metallohydrolase mechanism that uses a divalent zinc metal in the active site to activate the amide functionality for hydrolysis (Aka, Kim, & Yang, 2011). Class III deacylases, the sirtuins, also contain an obligatory zinc metal; however, unlike class I, II, and IV enzymes, the sirtuins' zinc serves a structural rather than catalytic function (Min, Landry, Sternglanz, & Xu, 2001). The sirtuins use a radically different mechanism from that of the classic HDACs linking the removal of the Nε-acyl group to hydrolysis of nicotinamide adenine dinucleotide (NAD^+) (Landry et al., 2000). The final products of this reaction are nicotinamide, adenine diphosphate ribose (ADPr), acetate, and the deacylated protein. The chemical steps in the transfer of the acyl group from lysine to the ultimate nucleophile water have been worked out through the elegant studies of Denu and others (Borra, Langer, Slama, & Denu, 2004; Sauve & Schramm, 2004; Tanner, Landry, Sternglanz, & Denu, 2000; Zhao, Chai, & Marmorstein, 2003) (Fig. 1). Briefly, in the sirtuin active site, the carbonyl oxygen of the acyl group is positioned to act as a nucleophile and displace the glycosidic nicotinamide by attack at the $1'$ carbon of the ribose ring. This step has generated a significant amount of interest. Theoretically,

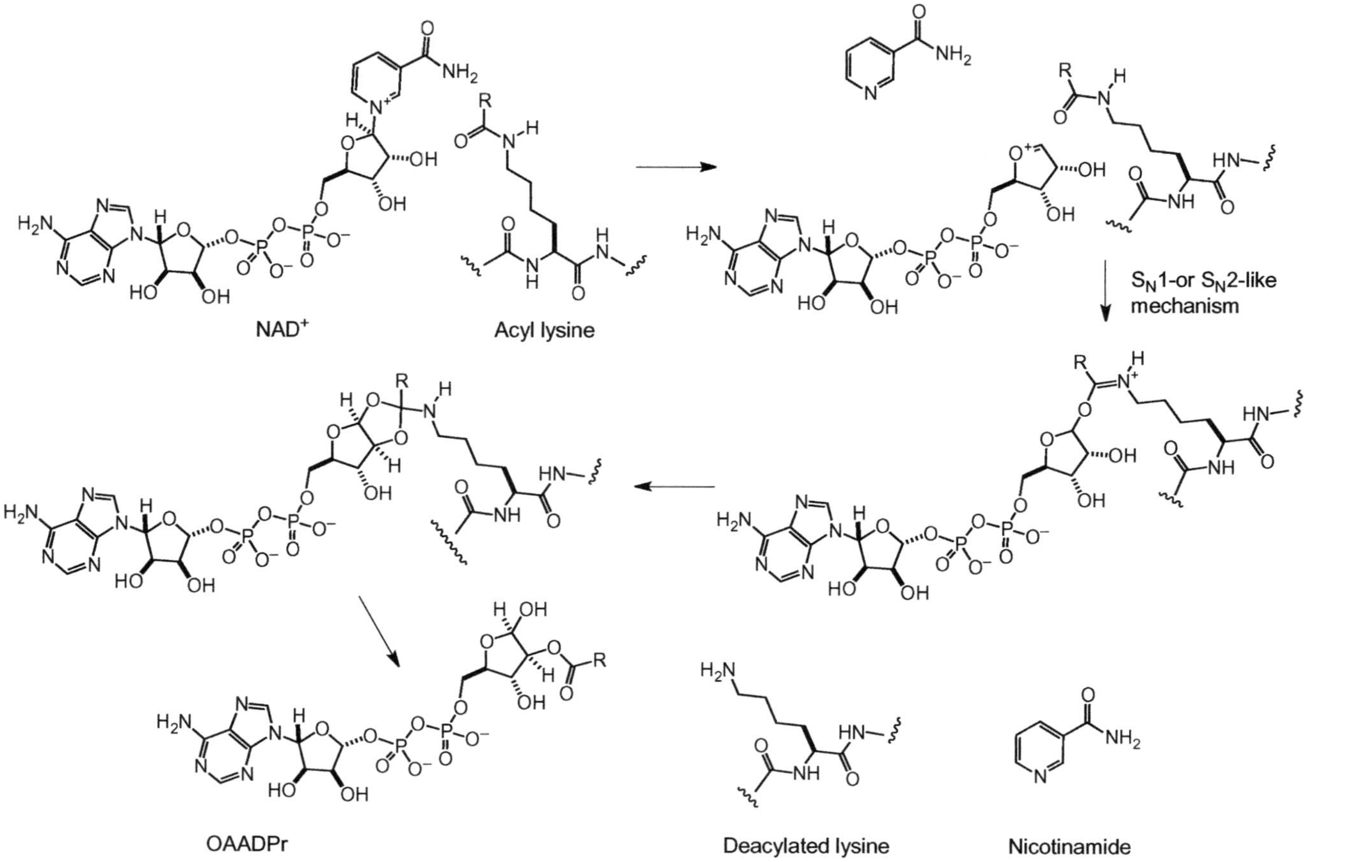

Fig. 1 Mechanism of sirtuin-catalyzed, NAD$^+$-dependent deacylation reaction.

two possible routes lead to the next intermediate. An S_N1 mechanism proceeds through an ionized oxocarbenium ion that is subsequently trapped by the nucleophilic oxygen of the amide. Alternately, an S_N2 mechanism in which the formation of the amide oxygen–ribose carbon bond is concurrent with the displacement of the nicotinamide group and proceeds through a pentavalent transition state. The majority of the evidence points to an intermediate answer where partial ionization (S_N1-like) of nicotinamide is followed closely by amide attack (S_N2-like) on the anomeric carbon in a single step. The O-alkylamidate intermediate that is formed is in turn attacked by the $2'$ hydroxyl on the amidate carbon to form a bicyclic intermediate that collapses to yield $2'$-O-acetyl adenine diphosphate ribose (OAADPr) and the deacylated lysine protein. The mechanistic details of this step are still being investigated. The final step of the deacylase mechanism of action is the release of the $2'$-acyl group from OAADPr, presumably by uncatalyzed hydrolysis by water. Detailed biochemical, crystallographic, and analogue studies by Denu, Wolberger, Pavletich, Marmorstein, Schramm, and others have illuminated this complex and mechanistically unique activity (Denu, 2005; Finnin, Donigian, & Pavletich, 2001; Hoff, Avalos, Sens, & Wolberger, 2006; Min et al., 2001; Sauve et al., 2001; Sauve & Schramm, 2004; Zhao, Chai, Clements, & Marmorstein, 2003; Zhao, Chai, & Marmorstein, 2003).

The gene encoding the founding member of the sirtuin family, Sir2, was discovered in the 1970s as one of the genes involved in mating competence and mating type switching in *S. cerevisiae* (Klar, Fogel, & Macleod, 1979; Rine & Herskowitz, 1987). Subsequent work showed that Sir2 participates in transcriptional repression of Pol2 at the silent mating loci (Klar, Strathern, Broach, & Hicks, 1981), telomeres (Aparicio, Billington, & Gottschling, 1991), and at the ribosomal DNA (Smith, Brachmann, Pillus, & Boeke, 1998), which is linked to hypoacetylation of histones at these loci (Braunstein, Rose, Holmes, Allis, & Broach, 1993; Moazed, 2001). These cellular roles of Sir2 in yeast were established before the breakthrough discovery of enzymatic function of Sir2 as an NAD^+-dependent protein deacetylase by the Guarente, Sternglanz, and Boeke groups in 2000 (Imai, Armstrong, Kaeberlein, & Guarente, 2000; Landry et al., 2000; Smith et al., 2000). This key discovery propelled the pace of research regarding the role of sirtuins in other species, but the direction of the research efforts, which heavily focused on aging-related phenomena, was largely a result the initial discoveries made in the yeast model of replicative aging.

Yeast replicative aging and Sir2 are linked by the functions of Sir2 at the rDNA locus where, besides transcriptional repression, Sir2 also suppresses unequal sister chromatid exchange among the rDNA repeats and generation of extrachromosomal rDNA circles (ERCs). During asymmetric mitotic division, which generates a larger mother and smaller daughter cells, ERCs preferentially accumulate in mother cells. Excessive accumulation of ERCs in mother cells is one of the factors that limits the number of daughter cells a mother cell can produce (Sinclair & Guarente, 1997), a measure of its replicative lifespan. An additional copy of Sir2 reduces rDNA accumulation in mother cells and extends replicative lifespan, whereas deletion of *SIR2* exerts the opposite effect (Kaeberlein, McVey, & Guarente, 1999). Moreover, calorie restriction-induced delay in aging and the extension of lifespan were proposed to be mediated by activation of *SIR2* (Lin, Defossez, & Guarente, 2000), leading to the idea that Sir2 activation is a key molecular event that mediates the beneficial effects of calorie restriction across the species. Initial studies in nematodes (Tissenbaum & Guarente, 2001) and flies (Rogina & Helfand, 2004) showed increased lifespan upon overexpression of Sir2 orthologues in these organisms. As accumulation of rDNA does not occur in flies and nematodes and yet overexpression of sirtuins extends their lifespan, a hypothesis was put forth that the role of sirtuins in delaying aging has been preserved during evolution, regardless of the differences in specific aging-related degenerative processes between the species. Although several key findings that supported this line of reasoning, both in yeast (Kaeberlein, Kirkland, Fields, & Kennedy, 2004) and in metazoans (Burnett et al., 2011) were later disputed, this idea generated a high level of enthusiasm for evaluating the role of sirtuins in many aging-related processes in higher vertebrates including humans, from diabetes and metabolic syndrome to neurodegeneration. Over the past 15 years, a large body of data has accumulated regarding a role of sirtuins in metabolism and many disease states including diabetes and insulin resistance, neurodegeneration, and cancer. This body of knowledge suggests that pharmacologic interventions with either sirtuin activators or inhibitors, depending on the condition, can have beneficial effects in many diseases.

2. SIRTUINS AND METABOLISM

The use of NAD^+ as a cofactor by sirtuins raises the intriguing possibility that fluctuations in NAD^+ levels related to different metabolic states or cellular stresses contribute to variation in sirtuin activity thus making protein acylation part of the cellular metabolic sensor network. One particularly attractive mechanism that was considered as a link between metabolism and sirtuin activity were changes in cellular redox state and resulting alterations in the $NAD^+/NADH$ ratio. However, close examination of cellular NAD^+ concentrations and biochemical characterization of sirtuins refute this hypothesis. NAD^+ is a key coenzyme in oxidoreductive metabolism where it serves an oxidizing agent in electron transfer reactions. While the ratio of NAD^+ to NADH fluctuates according to the cellular redox state, almost all of the free cellular NAD^+ pool is in the oxidized (NAD^+) state, with the ratio of free cellular $NAD^+/NADH$ measuring up to 700-fold under normal growth conditions (Williamson, Lund, & Krebs, 1967). Changes in redox state cannot, therefore, appreciably alter the pool of free NAD^+, and thereby change sirtuin activity. The changes in the level of free NADH, on the other hand, are directly proportional to changes in the cellular redox state, and NADH can inhibit sirtuin activity. However, at concentrations found in cells, which are estimated to be <10 μM (Patterson, Knobel, Arkhammar, Thastrup, & Piston, 2000; Zhang, Piston, & Goodman, 2002), free NADH is not expected to inhibit sirtuin activity. Systematic characterization of coenzyme specificity of yeast sirtuin homologue Hst2 by Schmidt, Smith, Jackson, and Denu (2004), revealed that the IC_{50} for NADH is in the 28 mM range compared to >1000-fold lower K_M for NAD^+. Based on these considerations, neither the alterations in the NAD^+ level nor the inhibition by NADH in response to changes in cellular redox state is expected to directly change cellular sirtuin activity.

Cellular NAD^+ levels are maintained by reutilization of nicotinamide via the salvage pathway or by de novo NAD^+ synthesis from tryptophan via the kynurenine pathway. Increased availability of cellular NAD^+ requires either increased de novo synthesis or salvage (Bogan & Brenner, 2008). NAD^+ consuming reactions, which similarly to NAD^+-dependent protein deacylation, cleave the glycosidic bond in NAD^+ and release nicotinamide, include protein mono- and poly-ADP ribosylation which is carried out by >20 different enzymes, and the generation of signaling molecules such as cyclic ADP-ribose, which participates in intracellular Ca^{2+} mobilization (Opitz & Heiland, 2015). NAD^+-consuming enzymes, such as PARPs, can reduce cellular sirtuin activity by reducing NAD^+ levels and by increasing the local concentration of nicotinamide, a sirtuin inhibitor. An acute depletion of NAD^+ can occur during genotoxic stress as a result of activation of PARPs. Reutilization of nicotinamide via the salvage pathway serves to restore cellular NAD^+ levels and to remove nicotinamide, which both stimulate sirtuin activity. Maintaining adequate NAD^+ supply is therefore critically important for assuring optimal sirtuin activity and for cellular and organismal health (Verdin, 2015).

3. SIRTUINS AND REGULATION OF CELLULAR NAD⁺ LEVELS

Given NAD^+ dependence and direct involvement in transcriptional regulation, sirtuins are well positioned to serve as both NAD^+ sensors and regulators of NAD^+ biosynthesis genes. Consistent with this idea, feedback loops involving sirtuins and NAD^+ biosynthesis genes have been described both in yeast (Bedalov, Hirao, Posakony, Nelson, & Simon, 2003) and in mammals (Nakahata et al., 2008; Ramsey et al., 2009), with different pathways targeted in the two organisms. In yeast the target of sirtuin-mediated transcriptional control is the battery of the de novo NAD^+ biosynthesis genes, while in mammals, the target is the rate-limiting enzyme in the salvage pathway.

Salvage pathway genes in yeast are constitutively active regardless of the cellular NAD^+ status. In contrast, in the NAD^+-replete state, genes encoding for de novo biosynthesis genes are directly repressed by the Sir2 homologue, Hst1, recruited to their promoters by the transcriptional repressor Sum1 (Bedalov et al., 2003). Reduction in NAD^+ levels abrogates Hst1-mediated repression leading to upregulation of genes in the de novo biosynthesis pathway and to restoration of NAD^+ levels. Consistent with its role as an NAD^+ sensor, Hst1 has low affinity for NAD^+ with an NAD^+ K_M of 94 μM, which is three- and sixfold lower than the NAD^+ K_M for Sir2 and Hst2, respectively.

NAD^+ levels in mammalian cells exhibit circadian oscillations, as a result of changes in the activity of the NAD^+ salvage pathway. Transcription of nicotinamide nucleotide adenylyl-transferase (NMNAT), that encodes for the rate-limiting enzyme in the NAD^+ salvage pathway, is activated by the core circadian regulator CLOCK and is repressed by SIRT1. As the magnitude of SIRT1-mediated repression of NMNAT is responsive to NAD^+ levels, SIRT1 contributes to circadian fluctuations of NAD^+ levels (Nakahata et al., 2008; Ramsey et al., 2009).

4. SIRTUIN FUNCTIONS

The seven human sirtuins localize to distinct subcellular compartments and share partially overlapping acyl-protein substrate specificities (Frye, 2000; Milne & Denu, 2008). SIRT1, the closest orthologue of yeast Sir2, shuttles between the cytoplasm and nucleus and displays a preference for acetyl-lysine substrates. Its deacetylase activity has been linked to a variety of processes ranging from gene expression by acting directly on a variety of transcription factors (eg, p53, HIF-1a, and FOXO-4) and histone proteins (eg, histones H1, H3, and H4), to cellular responses to metabolic stress and DNA damage to yet more global processes like aging. SIRT2, like SIRT1, shuttles between the cytoplasm and nucleus but is primarily cytoplasmic. In addition to deacetylase activity, SIRT2 has been shown to remove myristoyl acyl groups. The mitochondrial isoform SIRT3 also displays a preference for removal of lysine acetyl groups as well as several long-chain fatty acid modifications. SIRT4 and SIRT5 are also localized to the mitochondria and like SIRT3 display low deacylase activity in vitro compared to that of SIRT1–3. Interestingly, SIRT5 displays a marked preference for succinyl-modified lysine. Of all of the acyl modifications identified to date, the succinylation of lysine is unique in that it converts a positively charged lysine amino acid side chain to one that is negatively charged. SIRT6 and SIRT7 reside in the nucleus and nucleolus, respectively, and also display low enzymatic activity (Feldman et al., 2015). The modification of lysine residues with long-chain fatty acids is intriguing since attachment of a single myristoyl group is sufficient to induce relocalization of a protein from the cytoplasm to the plasma membrane. Reversible myristoylation may play a role in recruitment of target proteins to sites of activity within the cell.

5. SIRTUINS AND METABOLIC DISORDERS

Metabolic syndrome is a set of related conditions including central (abdominal) obesity, insulin resistance, hypertension, proinflammatory, and prothrombotic states, which markedly increases the risk of cardiovascular disease. Type 2 diabetes mellitus is a more serious, advanced form of metabolic syndrome in which tissues lose the ability to respond to insulin, and is the most common form of diabetes. Lack of physical activity combined with excessive calorie intake is major culprit for the high prevalence of these conditions, which affects up to a quarter of the population in industrial societies. The health detriment due to excessive calorie intake and salutary effects of calorie restriction were proposed to be mediated through the same molecular circuitry that includes sirtuins and the fuel sensing enzyme AMP-activated protein kinase (Canto et al., 2009; Guarente, 2006; Ruderman, Carling, Prentki, & Cacicedo, 2013). SIRT1 has been implicated in several aspects of metabolic regulation in different tissues, including liver, where it promotes gluconeogenesis; pancreas, where it promotes insulin secretion;

muscle, where SIRT1 promotes fatty acid oxidation and mitochondrial respiration; and in adipose tissues, where SIRT1 promotes fatty acid mobilization. These metabolic activities of SIRT1 raised the possibility that pharmacologic manipulation of SIRT1 can be used to mimic the calorie-restricted state and ameliorate obesity-associated type 2 diabetes mellitus. However, while activation of SIRT1 in muscle and adipose tissue would exert outcomes that would be desirable in patients with type 2 diabetes mellitus (eg, fatty acid mobilization and oxidation and mitochondrial respiration) (Gerhart-Hines et al., 2007; Picard et al., 2004); activation of SIRT1 in the liver increases gluconeogenesis and would therefore exacerbate the problem of already elevated glucose production in the liver observed in diabetes (Rodgers et al., 2005). In liver, inhibition, rather than activation of SIRT1 would be desirable for ameliorating hyperglycemia in type 2 diabetes mellitus. In support of this view, in two separate studies, liver-specific SIRT1 inactivation improved glucose tolerance and fasting glucose levels in mouse model of type 2 diabetes and protected animals from fat diet-induced weight gain and hepatic steatosis (Chen et al., 2008; Rodgers & Puigserver, 2007).

In contrast to SIRT1, which promotes hepatic gluconeogenesis, a different NAD^+-dependent deacetylase, SIRT6, inhibits hepatic gluconeogenesis (Dominy et al., 2012). Both enzymes participate in gluconeogenesis by regulating the acetylation state of PGC-1alpha. PGC-1alpha is kept in the hypoacetylated, inactive state by the HAT GCN5. SIRT1 directly deacetylates and activates PGC1-alpha, whereas SIRT6 deacetylates and activates GCN5, which renders PGC-1alpha in the hypoacetylated, inactive state.

6. SIRTUINS AND CANCER

Sirtuins have been implicated in cancer through deacetylation and regulation of tumor suppressor and oncoproteins and more recently through regulation of metabolism. Sirtuins' role in either preventing or promoting tumorigenesis has been established in a variety of mouse tumor models reviewed in Chalkiadaki and Guarente (2015); however, loss-of-function mutations or amplification of sirtuin genes are not very frequently observed in human cancers as judged by the results of large cancer sequencing projects. The most likely reason for this is that sirtuins modulate the function of driver genes in context-specific modes and loss or gain of function of sirtuins alone is not sufficient to impact cancer development and progression directly.

Furthermore, sirtuin protein level and activity is frequently altered in cancer without mutations in the corresponding genes.

Among the seven sirtuins, the role of SIRT1 in tumorigenesis has been most extensively evaluated, with studies pointing both to its tumor suppressive and oncogenic roles (Chalkiadaki & Guarente, 2015). SIRT1 was first linked to tumorigenesis by the discovery of p53 as its deacetylation substrate (Luo et al., 2001; Vaziri et al., 2001). Acetylation, carried out by several cellular histone acetyl transferases (HATs) (eg, CBP p300), is a key posttranslational modification that regulates p53 stability and that promotes its function as a transcriptional activator (Tang, Zhao, Chen, Zhao, & Gu, 2008). SIRT1 deacetylates p53 and suppresses its activity. Consequently, inhibition of SIRT1 promotes p53 function and sensitizes cells to genotoxic agents. Interestingly, the sensitization to genotoxic stress is only partially dependent of p53, as many p53-negative cancer cell lines are also sensitized suggesting involvement of other acetylation-dependent regulatory pathways.

While the genes encoding sirtuin isoforms are rarely mutated, the expression levels of SIRT1 are elevated relative to normal tissues in numerous type of human cancer (Li et al., 2015; Stenzinger et al., 2013; Wauters et al., 2013). Elevation of SIRT1 protein level leads to dysregulation of acetylation homeostasis. The critical role of dysregulated acetylation in tumorigenesis is most evident in germinal center derived B-cell lymphoma. Loss-of-function mutations in genes encoding HATs, another path to an acetylation imbalance, occur in up to 40% of diffuse large B-cell lymphoma and follicular lymphoma, presenting one of the most common genetic alterations in these tumor types (Pasqualucci et al., 2011). HAT loss subverts normal B-cell maturation pathways by deregulating critical proteins involved in lymphomagenesis. For example, hypoacetylation activates the transcriptional repressor BCL6 and inactivates p53, which are both associated with lymphomagenesis. The SIRT1/SIRT2 inhibitor cambinol exerts potent antilymphoma activity possibly through the restoration of the acetylation imbalance in lymphoma cells, leading to activation of p53 and inactivation of BCL6 (Heltweg et al., 2006).

SIRT1 plays important oncogenic roles in myeloid malignancies, including chronic myeloid leukemia (CML), the Abl kinase driven cancer (Li et al., 2012), and acute myeloid leukemia carrying activating mutations in the FLT3 kinase gene (Li et al., 2014). While Abl-kinase inhibitors such as imatinib have been effective treatments in CML, these drugs require chronic use, presumably because they fail to eliminate quiescent leukemia stem cells. Both in CML and FLT3 mutation-driven AML, combined inhibition of

SIRT1 and the respective kinase, eradicate leukemia stem cells through activation of p53. Elimination of leukemia stem cells using a combination of imatinib and a SIRT1 inhibitor suggests a therapeutic strategy that might enable CML patients to discontinue therapy without a risk of disease recurrence (Li et al., 2012).

SIRT1 was also implicated as an oncogene in several mouse cancer models including colon cancer (Leko, Park, Lao, Simon, & Bedalov, 2013), prostate (Byles et al., 2012), and thyroid cancer (Herranz et al., 2013).

However, several studies document tumor suppressive roles of SIRT1 in mouse models and consistent with a tumor suppressor role, downregulation of SIRT1 is observed in several human cancer types reviewed in Chalkiadaki and Guarente (2015). A possible explanation for tumor suppressive roles of SIRT1 is the known involvement of SIRT1 in genome stability and DNA repair including nucleotide excision repair, homologous recombination, and nonhomologous end joining. One way to reconcile the opposing effect of SIRT1 in different tumors is that depending on the tumor type, cellular context or different stage of tumor formation SIRT1 might have either tumor suppressing or oncogenic roles.

SIRT2, already mentioned in the context of germinal center derived B-cell lymphoma (Chalkiadaki & Guarente, 2015; Heltweg et al., 2006), shares the tumor suppressor/oncogene duality observed with SIRT1. SIRT2 null mice develop tumors in various tissues that exhibit signs of chromosomal instability, consistent with SIRT2's role in regulating the fidelity of chromosome segregation. However, SIRT2 has been shown to control acetylation and promote oncogenic activity of mutant K-RAS (Yang et al., 2013), suggesting that inhibition of SIRT2 might be used in treatment of K-RAS-driven cancers (Liu et al., 2013).

SIRT3, a mitochondrial sirtuin, regulates energy metabolism through control over fatty acid oxidation, energy balance, and mitochondrial biogenesis. Loss of SIRT3 function leads to an increase in reactive oxygen species (ROS) a known driver of tumorigenesis. Through its control over ROS, SIRT3 appears to play a tumor suppressor function. SIRT4 is another mitochondrial sirtuin that acts as a tumor suppressor. Its tumor suppressive activity has been linked to its ability to inhibit glutamine utilization (Jeong et al., 2013), which is important for cancer cell growth.

As with other sirtuin isoforms, SIRT6 demonstrates a pattern of activity that is difficult to reduce to simple terms (Lerrer, Gertler, & Cohen, 2015). This nuclear isoform participates in a broad range of processes including metabolism, transcriptional regulation, DNA integrity, and aging. SIRT6 null mice display profound aging phenotypes and die at an early age. Mice-overexpressing SIRT6 conversely have a longer than wild-type lifespan. A tumor suppressive role of SIRT6 may depend on its ability to inhibit the shift of hypoxic cells to anaerobic glycolysis. Additional functions of SIRT6 in promoting apoptosis may also be relevant.

SIRT7 is the most enigmatic of the sirtuin isoforms. It localizes to the nucleolus and appears to play a role in regulating ribosomal gene expression through RNA polymerase I. It is overexpressed in breast and liver cancer and its downregulation induces apoptosis in cancer cell lines and reverses metastatic phenotype in both epithelial and mesenchymal tumors (Malik et al., 2015). As noted below, SIRT7 failed to demonstrate appreciable enzymatic activity against a panel of acyl-lysine-containing peptides in vitro making identification of bone fide targets challenging.

7. SIRTUIN ACTIVITY ASSAYS

Cloning and bacterial expression of intact proteins or isolated catalytic domains of sirtuins allowed for the in vitro determination of deacylase enzymatic activity. The most direct method for measuring the deacetylation of a peptide substrate is the determination of starting material and product by high-performance liquid chromatography (HPLC). The reaction of an acetylated peptide is carried out in the presence of sirtuin and NAD^+. The reaction is quenched by the addition of a denaturant like trifluoroacetic acid and the entire mixture is subjected to separation on a C18 reversed-phase HPLC column. Because the only peptide components of the mixture are the acetylated starting material and deacetylated product, the quantification of the degree of deacetylation is straightforward. Initial studies used low-throughput methods such as release of ^3H-acetate from synthetic peptides from validated sirtuin substrate proteins such as histone H3 or p53. For detailed protocols, see Emiliani, Fischle, Van Lint, Al-Abed, and Verdin (1998) and Tanner et al. (2000). In these assays, radiolabeled peptide, sirtuin enzyme, and NAD^+ are incubated for a specified amount of time. The reaction is acidified and ^3H acetate is extracted with organic solvent, typically ethyl acetate. The immiscible organic phase is separated from the aqueous layer and the amount of acetate extracted is determined scintillation counting. Alternately, following acidification, reaction products including deacetylated peptide, nicotinamide, and 2′-O-acetyl-ADP-ribose (OAADPr) are separated by HPLC and UV active fractions are collected and identified by mass spectrometry. ^3H acetate is quantified by liquid scintillation counting. These are fairly labor-intensive process that involve several liquid transfers and are not suitable for high-throughput applications. Although OAADPr is labile to aqueous hydrolysis, some concerns have been raised about how incomplete hydrolysis may affect the quantification of the deacetylation reaction since some of the acetate is present as OAADPr and some as free acetate. A variant of the radiolabeled substrate assay is the use of radiolabeled ^{32}P-NAD$^+$ in the reaction. NAD$^+$ is converted to ADP-ribose (ADPr) in the deacetylation reaction and several methods, including thin layer chromatography, exist for the quantification of the reaction products (Tanny & Moazed, 2001).

Another variant of the ^{3}H-acetate extraction method is the charcoal-binding assay (Borra & Denu, 2004). Activated charcoal exhibits a strong affinity for proteins, peptides, and small molecules, especially if they are charged (eg, NAD$^+$ and ADPr), but not acetate. Following the reaction, the entire mixture is incubated with an aqueous slurry of charcoal at a moderately high pH to facilitate the hydrolysis of intermediates like OAADPr. The charcoal slurry is pelleted in a benchtop centrifuge and the supernatant containing released ^{3}H-acetate is analyzed by scintillation counting.

Several criteria need to be met for a sirtuin activity assay to be useful for high-throughput applications. Among these, a minimum number of manipulations, low reagent cost, and reproducibility are foremost. Additional considerations such as low background and high signal-to-noise ratio are also critical. A number of assays designed to function in a high-throughput format have been devised. While an exhaustive cataloging of these is beyond the scope of the current review, several deserve mention. The scintillation proximity assay (SPA) relies on the energy transfer, typically radioactive decay from a source that is attached to a bead containing a fluor, which converts this energy into a more easily measured form such as light. For example, a synthetic ^{3}H-acetylated peptide containing a biotin moiety removed from the critical acetyl lysine by several amino acids or a synthetic linker is loaded onto SPA beads decorated with streptavidin. The proximity of the ^{3}H-acetate label to the bead leads to a steady-state background fluorescence. Addition of the sirtuin and NAD$^+$ leads to cleavage of the labeled acetate and although still in solution with the SPA bead, the energy transfer of freely dissolved ^{3}H-acetate to the bead is infinitesimal compared to the proximity facilitated energy transfer. The result is a decrease in fluorescence intensity from that well. The fluorescence is directly proportional to the amount of ^{3}H-acetate bound to the bead and consequently, the determination of deacetylase activity is straightforward (Nare et al., 1999).

A number of enzyme-linked deacetylase assays have been devised to facilitate quantification of the deacetylation reaction. The first of these was developed by Hoffaman, Jung, and colleagues (Heltweg, Dequiedt, Verdin, & Jung, 2003). It relies on the ability of trypsin to cleave lysine-containing peptides but not acetyl-lysine peptides. A minimal deacetylase substrate is composed of N-α-BOC-N-ε-acetyl-lysine amino-coumarin amide. Incubation of this substrate with sirtuin in the presence of NAD^+ leads to unmasking of the lysine. Subsequent treatment with trypsin cleaves the amino-coumarin from the peptide but only in the deacetylated product. The fluorescent-free amino-coumarin is quantified spectrophotometrically (Heltweg et al., 2003). This assay has been commercialized and is widely used in HTS applications. However, its use by Sinclair and colleagues in identifying compounds that activate sirtuins has generated controversy. Several groups have attempted to validate SIRT1 activation by the naturally occurring polyphenol resveratrol that was identified using this assay, and were unable to do so. A structural resolution to the controversy was put forth by Xu et al. (Cao et al., 2015). The crystal structure shows that several molecules of resveratrol bind to SIRT1 and two of these stabilize binding of the substrate, N-α-BOC-N-ε-acetyl-lysine amino-coumarin amide, to the amino-terminal domain of SIRT1 resulting in tighter binding and increased enzymatic activity. In spite of this and other reports challenging the validity of small-molecule activators of sirtuins, widespread confusion persists in the literature. A similar enzyme-linked assay that has gained popularity recently is the SIRT-Glo assay that couples deacetylation of a substrate to proteolytic cleavage that releases amino-luciferin, a substrate for luciferase (Promega Corp., Fitchburg, WI).

A variant on the fluorophore-release strategy exploits fluorescence quenching. A p53-derived peptide contains a carboxy-terminal fluorophore (6-TAMRA), a central acetylated lysine residue and an amino-terminal nonfluorescent quencher derived from 9-phenyl-xanthene. In the unreacted peptide, fluorophore fluorescence is effectively quenched. Deacetylation followed by tryptic cleavage separates the dyes, liberating TAMRA fluorescence (Marcotte et al., 2004). A theoretical advantage of the fluorescence quenching assay is that the acetylated lysine does not need to be located immediately proximal to the fluorophore, as it does in the amino-coumarin-based assay described earlier. The physical separation between acetylated lysine and fluorophore makes the substrate more similar to the native deacetylation substrate and therefore less prone to artifacts. However, the screening artifacts similar to those observed with the amino-coumarin-based assays can be found even when acetyl-lysine is separated from the fluorophore by several amino acids. In the fluorescence quenching-based screen for SIRT1 activators that employed acetyl-lysine at position six relative to the TAMRA fluorophore (Milne et al., 2007), none of the hits that promoted deacetylation of the fluorophore-containing peptide were later found to have any activity against the native peptide (Pacholec et al., 2010). Counter screens with the native peptide substrates are therefore essential.

A continuous assay, requiring no manipulation and direct quantification, was developed by Denu and colleagues. In addition to generating free acetate, deacetylated peptide, and ADPr, the deacetylation reaction generates free nicotinamide. Nicotinamide is a component of several biosynthetic pathways and a substrate for enzymes. In the deacetylase assay, Denu et al. exploit conversion of nicotinamide to nicotinic acid by the action of nicotinamidase generating ammonia, which, in turn is a substrate for the conversion of α-ketoglutarate to glutamate by glutamate dehydrogenase with the concomitant oxidation of NADP(H) to $NADP^+$. The latter is quantified spectrophotometrically by absorbance at 340 nm (Smith, Hallows, & Denu, 2009).

The majority of sirtuin activity assays have focused on the removal of an acetyl group from a native or pseudosubstrate. Several human sirtuins, including SIRT4, 5, 6, and 7, however, lack robust deacetylase activity. Denu et al. recently went back to the distinctly low throughput ^{32}P-NAD^+ to ^{32}P-O-acyl-ADPr conversion assay relying on thin layer chromatography to investigate the substrate specificity of sirtuins 1–7 against a panel of histone H3 lysine 9 (H3K9) peptides with 13 different short, medium, and long-chain fatty acid N-ε-lysine modifications. The results clearly indicate that only SIRT1, 2, and 3 are robust deacetylases, although even these are quite active against longer chain modifications while SIRT4, 5, and 6 show a distinct preference for long-chain acyl-lysine substrates (Feldman, Baeza, & Denu, 2013). Surprisingly, SIRT7 demonstrated no activity under these assay conditions against this panel of substrates and its native substrates remain as yet unidentified.

8. IDENTIFICATION OF FIRST-GENERATION SIRTUIN INHIBITORS

The first sirtuin inhibitors, splitomicin and sirtinol (**1** and **2** in Fig. 2), were both discovered and characterized using cell-based screens in yeast *S. cerevisiae* (Bedalov, Gatbonton, Irvine, Gottschling, & Simon, 2001;

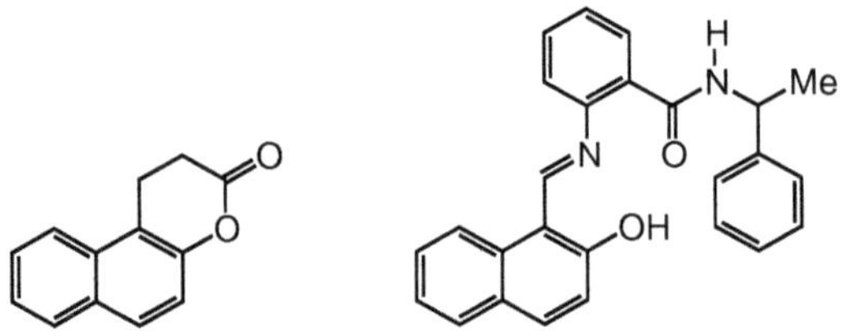

Fig. 2 Structures of certain inhibitors splitomicin (1) and sirtinol (2).

Grozinger, Chao, Blackwell, Moazed, & Schreiber, 2001). A full panel of strains and the assay conditions that can be used for screening and characterizing sirtuin inhibitors is described by Newcomb and Bedalov (2009). Yeast Sir2 functions in transcriptional silencing at specific loci in the yeast genome, namely the silent mating loci (Klar et al., 1979), ribosomal DNA (Smith et al., 1998), and subtelomeric regions (Aparicio et al., 1991). Elegant genetic studies by Rine and others showed that this effect was positional rather than sequence specific, in other words any gene, regardless of promoter or function, would be transcriptionally repressed if located within one of these regions. This property allowed us to exploit auxotrophy—the inability of an organism to synthesize an essential nutrient and requirement for dietary supplementation to enable growth—as the basis for the small-molecule screen that identified the first inhibitors of sirtuins. Yeast mutant strains deficient for selected amino acid (eg, leucine, methionine, or tryptophan) or nucleobase (eg, adenine or uracil) biosynthesis can be grown if the final product of the disabled pathway is supplied in the medium. Conversely, the same mutant yeast cannot grow in defined media if the nutrient is omitted. This property can be used as the basis for positive (ie, growth) selection. If the mutant strain were to express a functional copy of the disabled gene, supplied on an exogenous plasmid or inserted into the genome, it would be able to grow on media lacking the essential nutrient. We constructed yeast strains with genes for uracil (*URA3*), tryptophan (*TRP1*) biosynthesis at telomeres, rDNA, and silent mating type locus. The yeast grew well on media containing these nutrients but failed to grow if uracil or tryptophan was lacking. Genetic inactivation of *SIR2* led to reexpression of the reporter genes and enabled yeast growth on defined media lacking uracil and tryptophan. It was our expectation that pharmacological inhibition of the catalytic activity of the Sir2 protein would have the same effect.

A library of 6000 compounds from the National Cancer Institute's Developmental Therapeutics Program's repository was tested in a phenotypic screen (Bedalov et al., 2001). A compound, which was named splitomicin, enabled yeast with subtelomeric *URA3* to grow on uracil deficient media. Follow-up studies showed that splitomicin allowed yeast harboring the *TRP2* gene at the silent mating type locus to grow on tryptophan-deficient media. This proved two important points; first, splitomicin was not substituting for uracil deficiency directly as addition of uracil would since it is highly unlikely that a single compound could supplement for both uracil, which is needed for RNA and DNA synthesis and tryptophan, which is needed for protein synthesis; and second, since *SIR2* in a nonessential gene, it was expected that a specific Sir2 inhibitor to be nontoxic and allow cell growth in general. Next, the rate of DNA recombination at the highly repetitive rDNA locus was determined since, in addition to suppressing transcription, Sir2 also suppresses recombination at this locus.

Silencing at the mating type locus and subtelomeric regions depends on the function of Sir2 in complex with Sir3 and Sir4, while silencing within the rDNA gene array depends on Sir2 and Net1. This relationship reinforces the conclusion that splitomicin acts on Sir2 rather than Sir3, Sir4, or Net1 as the transcriptional derepression at all three loci (ie, telomeres, silent mating type loci, and rDNA) depends singularly on Sir2. In addition to the three genomic locations already mentioned, Sir2 regulates transcription of individual genes throughout the genome. To determine how closely splitomicin treatment of wild-type cells phenocopies the loss of Sir2 function, global transcriptional changes under the two conditions were compared. First, messenger RNA (mRNA) was isolated from *sir2Δ* and wild-type cells. Using competitive hybridization of fluorescently labeled cDNAs derived from the mRNAs on whole-genome microarray chips, genes were identified whose expression level is altered in *sir2Δ* relative to wild-type cells. As expected, many genes are upregulated, or derepressed in *sir2Δ* cells relative to wild type. However, a significant number of genes are also downregulated, or repressed in *sir2Δ* cells, presumably through indirect or secondary effects. Next, expression levels in wild-type cells treated with splitomicin were compared to those of untreated wild-type cells. As in the genetic inactivation of *SIR2*, its pharmacological inactivation with splitomicin resulted in transcriptional changes relative to wild type that were nearly identical. Of the 160 genes that were down- or upregulated by more than twofold in the *sir2Δ* strain, 95 were also down- or upregulated in the splitomicin-treated wild-type cells. Even though the yeast genome contains four additional genes encoding proteins with high sequences homology to Sir2, Hst (Homologue of Sir Two) 1–4, significantly, only two genes were similarly regulated between strains in which the closely related SIR2 (sir2Δ) and the HST1 (hst1Δ) were deleted.

9. SECOND-GENERATION SPLITOMICIN INHIBITORS

Splitomicin served as a useful tool to demonstrate that pharmacological inhibition of sirtuins was feasible and could shed light on sirtuin biology. However, the δ-lactone functionality was found to be labile to aqueous hydrolysis under neutral conditions, giving the compound a short half-life in mammalian tissue culture media. To identify suitable analogues for use in mammalian tissue culture and in vivo studies, a related compound from the NCI collection, which we named cambinol (**3** in Fig. 3) was tested (Heltweg et al., 2006). Like splitomicin, cambinol has a β-naphthol core but lacks the lactone moiety. Instead, a thio-pyrimidione functionality connects the β-naphthol to a phenyl substituent. Cambinol is very stable under physiological conditions, has reasonable aqueous solubility and is nontoxic to mice when administered in acute or chronic dosing regimens. Cambinol was tested against SIRT1, SIRT2, SIRT3, and SIRT5 in a biochemical assay using the commercially available Fluor-de-Lys assay and showed that it is a moderately potent but nonselective inhibitor with IC_{50} values of 56 and 59 μM, for SIRT1 and SIRT2, respectively, has weak activity against SIRT5 and no inhibitory activity against SIRT3. Double reciprocal plots of activity with varying substrate (acetyl histone H-4 peptide) and NAD^+ concentration revealed that cambinol is competitive with peptide and noncompetitive with NAD^+. Deacetylase inhibitory activity of cambinol in human NCI H460 lung cancer cells was evidenced by increased acetylation

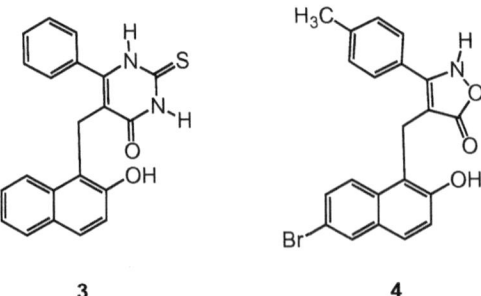

Fig. 3 Structures of cambinol (**3**) and its SIRT2-selective analogue (**4**).

of p53 (a SIRT1 target) and α-tubulin (a SIRT2 target) following drug treatment. Additionally, cambinol treatment led to hyperacetylation of BCL6, a transcriptional repressor and key driver of B-cell lymphoma. Hyperacetylation of BCL6 correlated with decreased repressor activity in a reporter assay as well as increased toxicity of cambinol in lymphoma cells. A survey of human normal and cancer cell lines revealed that cambinol was selectively toxic to B-cell lymphoma cell lines. To validate pharmacological inhibition of sirtuin activity in vivo, cambinol was tested in a Burkitt lymphoma xenograft model. Over the course of 2-week treatment, cambinol (i.v. daily) reduced the growth of established tumors relative to untreated control animals.

The demonstration of cambinol's in vivo antilymphoma activity left unanswered whether the antitumor effect required inhibition of SIRT1, SIRT2, or both. In an effort to expand the chemical space of our sirtuin inhibitors, a series of five-membered ring pyrazolone and isoxazol-5-one cambinol analogues (eg, **4** in Fig. 3) were prepared (Mahajan et al., 2014). This series proved fruitful, yielding compounds with >8-fold selectivity for SIRT1 (ie, compound 17) and >15-fold selectivity for SIRT2 (ie, 24) as well as more potent but relatively nonselective sirtuin inhibitors (ie, 8). Next, in vitro inhibition of specific isoforms was compared with cytotoxicity in lymphoma cells. The scatter plot of in vitro IC_{50} vs percentage growth inhibition yielded a clear correlation between SIRT2 inhibition and cytotoxicity. In spite of these promising results, the clinical utility of SIRT2 inhibitors as antilymphoma agents is yet to be demonstrated.

10. OTHER SIRTUIN INHIBITORS

Initial reports of identification of yeast and mammalian sirtuin inhibitors coupled with illumination of sirtuins' roles in several pathological conditions led to an increased interest in identifying and developing small-molecule sirtuin-modulating compounds as mechanistic tools and leads for drug development. The first known small-molecule inhibitor is the byproduct of the deacetylation reaction, nicotinamide. Nicotinamide (**5** in Fig. 4), the amide of nicotinic acid (niacin or vitamin B3) is a weak sirtuin inhibitor (SIRT1 IC_{50} 1.2 mM). Although it lacks potency and specificity, nicotinamide has been used extensively in human clinical trials. Nicotinamide has an extensive and largely side effect-free history of human use as a dietary supplement. Recent clinical studies in Australia demonstrated that oral nicotinamide can reduce the occurrence of actinic

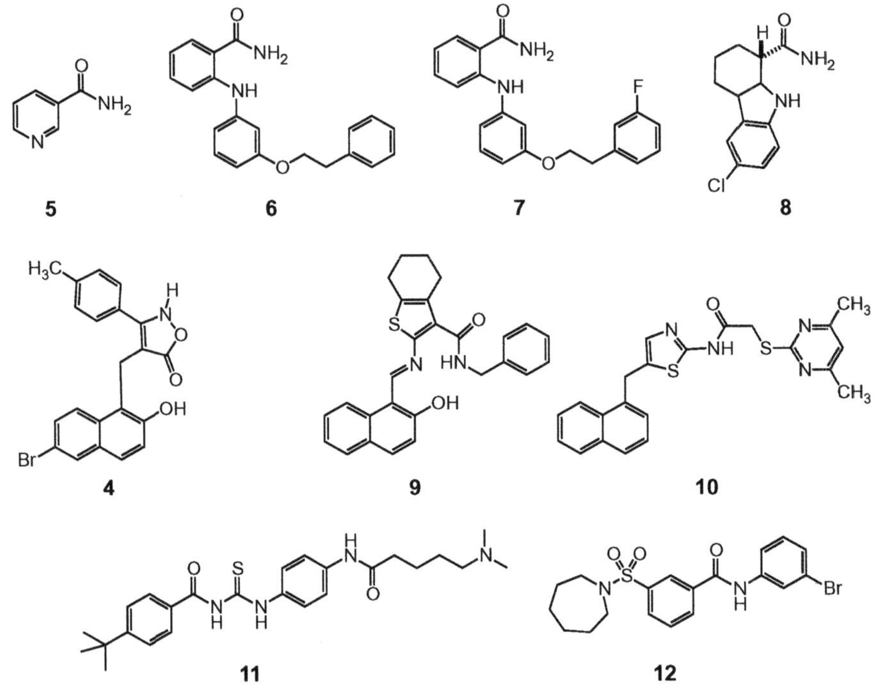

Fig. 4 Structures of certain inhibitors: nicotinamide (**5**), compounds 33a (**6**) and 33i (**7**), EX-527 (**8**), cambinol analogue (**4**), JGB-1741 (**9**), SirReal-2 (**10**), tennovin-6 (**11**) and AK-7 (**12**).

keratoses, potentially premalignant skin lesions. Additionally, nicotinamide used in combination with the classic HDAC inhibitor vorinostat produced a 24% response rate and 57% disease stabilization rate in a trial in germinal center derived diffuse large B-cell lymphoma (Amengual et al., 2013). However, the lack of potency and selectivity of nicotinamide will likely limit its clinical utility.

Several groups have used the pyridyl carboxamide or phenyl carboxamide (benzamide) functionality found in nicotinamide as the starting point for development of sirtuin inhibitors. Among the most successful of these, compounds 33a and 33i (**6** and **7** in Fig. 4) reported by Suzuki et al. (2012) show high potency with IC_{50} values against SIRT2 of 1.0 and 0.57 μM, respectively and high selectivity. Indole-containing compounds, including the carboxamide EX-527 (Selisistat) (**8** in Fig. 4), show potent in vitro SIRT1 inhibitory activity (IC_{50} 60–100 nM) (Napper et al., 2005) and are mixed-type inhibitors against both substrate and NAD^+. The compounds are orally bioavailable and have good metabolic properties making them well suited to explore the therapeutic targeting of SIRT1 in vivo. Selisistat was recently shown to be well tolerated in a Phase I trial in Huntington's disease (Sussmuth et al., 2015).

A series of mechanism-based deacetylase inhibitors have been prepared based on the observation that thioamides can act as semireversible, inhibitors of proteases, and consequently, N-thioacetyl lysine-containing substrate mimetics have been developed as sirtuin inhibitors. The thioacetyl moiety in these compounds acts as the nucleophile in the displacement of nicotinamide from NAD^+, much in the same manner as the normal acetyl group in the first step of the deacetylation reaction. The immediate intermediate that is formed, $1'$-S-alkylimidate-ADPr, is far more stable than it's oxygen analogue and does not undergo the subsequent steps that lead to acetate release and ADPr dissociation form the active site. Simple mono-, di-, or oligopeptide pseudosubstrates incorporating the thioacetyl group have shown low- to mid-micromolar inhibitory activity and modest specificity for SIRT1.

Another class of mechanism-based sirtuin inhibitors is N-ε-trifluoroacetyl-lysine pseudosubstrate compounds (Chen et al., 2015). Among the more innovative methods for the identification of SIRT isoform-specific inhibitors is the application of flexible in vitro translation, or FIT, published by Suga and colleagues (Morimoto, Hayashi, & Suga, 2012). Briefly, an in vitro translation/ribosome display system capable of incorporating unnatural amino acids as well as degenerate codons was used to create a large library of N-ε-trifluoroacetyl-lysine-containing cyclic peptides. In this approach, the peptide remains covalently attached to the RNA that carries the peptide-encoding sequence, in a methodology called random, nonstandard peptide integrated discovery system. Peptides are selected based on their ability to bind affinity-tagged SIRT proteins, the sequence is determined by cDNA synthesis and sequencing. This approach yielded a 14 amino acid, cyclic peptide with impressive IC_{50} values for SIRT1 (47 nM), SIRT2 (3.2 nM), and SIRT3 (480 nM). These are by far the most potent inhibitors of SIRT2 reported to date. However, the high-molecular-weight and proteolytic liability of peptides make it unlikely that these agents will advance to animal or human trials.

The first synthetic small-molecule sirtuin inhibitors were reported by Bedalov et al. (2001) and Grozinger et al. (2001). Both groups exploited the transcriptional repression activity of yeast Sir2 to screen libraries of small, drug like, nonbiased compounds. Interestingly, both groups identified compounds based on a β-naphthol scaffold. The two inhibitors, splitomicin (described in detail earlier) and sirtinol, showed modest potency and specificity but were useful to demonstrate the utility of small-molecule sirtuin inhibitors to unravel the complex biology of this class of enzymes. Several other small-molecule inhibitors of sirtuins share the naphthyl group as part of the pharmacophore found in splitomicin (1), sirtinol (2), cambinol (3), and its analogues (eg, 4) also include JGB-1741 (**9** in Fig. 4) (Kalle et al., 2010) and saleramide (Lara et al., 2009), a reversed amide analogue of sirtinol. Among these, the recently reported SirReal2 (**10**) and analogues by Jung and colleagues (Rumpf et al., 2015) provide a detailed structural analysis of the compounds' interaction with SIRT2 (IC_{50} 0.14 μM). Unlike typical active site targeting agents, SirReal compounds appear to induce conformational changes that inhibit structural rearrangements required for the deacetylation reaction to proceed. Interestingly, SirReal2, the most potent and selective SIRT2 inhibitor is competitive with the acetyl-lysine substrate, like splitomicin and cambinol; it is also a partial noncompetitive inhibitor with regard to NAD^+. In contrast, the closely related SirReal1, in which the naphthyl group is replaced by phenyl, is competitive with both substrate and NAD^+. These subtle mechanistic differences may reflect different modes of binding.

Several sirtuin inhibitors were discovered in phenotypic screens based on activation or stabilization of the human tumor suppressor p53. Westwood et al. carried out a cell-based screen to identify agents that increase expression of p53-regulated genes. One of the hits was termed tennovin-1 and its more water-soluble analogue tennovin-6 (**11** in Fig. 4) was used to query a collection of heterozygous diploid yeast strains lacking one copy of each gene. Among the strains that were sensitized by the deletion of one gene copy was the *SIR2* sirtuin. Additional in vitro assays confirmed that tennovin-6 inhibited sirtuins including human SIRT1 (IC_{50} 37.5 μM) and SIRT2 (IC_{50} 10.4 μM) (Lain et al., 2008). AK-7, a benzamide-containing compound (**12** in Fig. 4), has also shown modest SIRT2 activity (IC_{50} 15.5 μM) and neuroprotective activity in vivo (Chopra et al., 2012; Taylor et al., 2011).

A number of compounds that often appear on lists of screening hits have been reported to inhibit sirtuins. Among these are suramin, bis-indoylmaleamide (Ro31-8220), the arylidene-indanones (GW-5074), and cyanine dye derivatives (AC-93253). It is difficult to assess whether these compounds are specific sirtuin inhibitors or nonspecific screening artifacts.

11. CONCLUDING REMARKS

The seven human sirtuins comprise an intriguing and challenging panel of potential drug targets. The developing understanding of their roles in a variety of human diseases makes a strong case for the development of sirtuin-targeting agents. However, the intimidating complexity of their regulation, cellular localization, and substrate specificity make it exceptionally difficult to assign specific roles in healthy and diseased tissues. Yet the assignment of isoform-specific roles is a critical requirement for rationalizing drug development efforts. Substrate specificity provides an excellent example of the challenge. High-throughput proteomic technologies have enabled identification of posttranslational modifications on an unprecedented scale and identification of proteins whose activity is regulated by acetylation has grown exponentially. Initially, sirtuins were classified as histone deacetylases. And while they do indeed remove acetyl groups from several histone isoforms, the number of nonhistone substrates far outnumbers histones. A recent review listed more than 40 SIRT1 targets, almost 20 SIRT2 and SIRT3 substrates and 10 or so substrate proteins each for the other sirtuins. The diversity of substrates is complicated by the fact that many sirtuin isoforms can deacylate the same target. This diversity is multiplied by the variety of acyl–lysine substrates that can be processed by each isoform. One potential solution to this conundrum is the development of probe reagents, molecules that target each isoform with high potency and specificity. Perhaps once such reagents are available, the true therapeutic potential of sirtuin modulators can be determined.

REFERENCES

Aka, J. A., Kim, G. W., & Yang, X. J. (2011). K-acetylation and its enzymes: Overview and new developments. *Handbook of Experimental Pharmacology*, *206*, 1–12. http://dx.doi.org/10.1007/978-3-642-21631-2_1.

Amengual, J. E., Clark-Garvey, S., Kalac, M., Scotto, L., Marchi, E., Neylon, E., ... O'Connor, O. A. (2013). Sirtuin and pan-class I/II deacetylase (DAC) inhibition is synergistic in preclinical models and clinical studies of lymphoma. *Blood*, *122*(12), 2104–2113. http://dx.doi.org/10.1182/blood-2013-02-485441.

Aparicio, O. M., Billington, B. L., & Gottschling, D. E. (1991). Modifiers of position effect are shared between telomeric and silent mating-type loci in S. cerevisiae. *Cell*, *66*(6), 1279–1287.

Bedalov, A., Gatbonton, T., Irvine, W. P., Gottschling, D. E., & Simon, J. A. (2001). Identification of a small molecule inhibitor of Sir2p. *Proceedings of the National Academy of Sciences of the United States of America*, *98*(26), 15113–15118. http://dx.doi.org/10.1073/pnas.261574398.

Bedalov, A., Hirao, M., Posakony, J., Nelson, M., & Simon, J. A. (2003). NAD$^+$-dependent deacetylase Hst1p controls biosynthesis and cellular NAD+ levels in Saccharomyces cerevisiae. *Molecular and Cellular Biology*, *23*(19), 7044–7054.

Bogan, K. L., & Brenner, C. (2008). Nicotinic acid, nicotinamide, and nicotinamide riboside: A molecular evaluation of NAD+ precursor vitamins in human nutrition. *Annual Review of Nutrition*, *28*, 115–130. http://dx.doi.org/10.1146/annurev.nutr.28.061807.155443.

Borra, M. T., & Denu, J. M. (2004). Quantitative assays for characterization of the Sir2 family of NAD(+)-dependent deacetylases. *Methods in Enzymology*, *376*, 171–187. http://dx.doi.org/10.1016/S0076-6879(03)76011-X.

Borra, M. T., Langer, M. R., Slama, J. T., & Denu, J. M. (2004). Substrate specificity and kinetic mechanism of the Sir2 family of NAD$^+$-dependent histone/protein deacetylases. *Biochemistry*, *43*(30), 9877–9887. http://dx.doi.org/10.1021/bi049592e.

Braunstein, M., Rose, A. B., Holmes, S. G., Allis, C. D., & Broach, J. R. (1993). Transcriptional silencing in yeast is associated with reduced nucleosome acetylation. *Genes & Development*, *7*(4), 592–604.

Burnett, C., Valentini, S., Cabreiro, F., Goss, M., Somogyvari, M., Piper, M. D., ... Gems, D. (2011). Absence of effects of Sir2 overexpression on lifespan in C. elegans and Drosophila. *Nature*, *477*(7365), 482–485. http://dx.doi.org/10.1038/nature10296.

Byles, V., Zhu, L., Lovaas, J. D., Chmilewski, L. K., Wang, J., Faller, D. V., & Dai, Y. (2012). SIRT1 induces EMT by cooperating with EMT transcription factors and enhances prostate cancer cell migration and metastasis. *Oncogene*, *31*(43), 4619–4629. http://dx.doi.org/10.1038/onc.2011.612.

Canto, C., Gerhart-Hines, Z., Feige, J. N., Lagouge, M., Noriega, L., Milne, J. C., ... Auwerx, J. (2009). AMPK regulates energy expenditure by modulating NAD+ metabolism and SIRT1 activity. *Nature*, *458*(7241), 1056–1060. http://dx.doi.org/10.1038/nature07813.

Cao, D., Wang, M., Qiu, X., Liu, D., Jiang, H., Yang, N., & Xu, R. M. (2015). Structural basis for allosteric, substrate-dependent stimulation of SIRT1 activity by resveratrol. *Genes & Development*, *29*(12), 1316–1325. http://dx.doi.org/10.1101/gad.265462.115.

Chalkiadaki, A., & Guarente, L. (2015). The multifaceted functions of sirtuins in cancer. *Nature Reviews. Cancer*, *15*(10), 608–624. http://dx.doi.org/10.1038/nrc3985.

Chen, D., Bruno, J., Easlon, E., Lin, S. J., Cheng, H. L., Alt, F. W., & Guarente, L. (2008). Tissue-specific regulation of SIRT1 by calorie restriction. *Genes & Development*, *22*(13), 1753–1757. http://dx.doi.org/10.1101/gad.1650608.

Chen, B., Zang, W., Wang, J., Huang, Y., He, Y., Yan, L., ... Zheng, W. (2015). The chemical biology of sirtuins. *Chemical Society Reviews*, *44*(15), 5246–5264. http://dx.doi.org/10.1039/c4cs00373j.

Chopra, V., Quinti, L., Kim, J., Vollor, L., Narayanan, K. L., Edgerly, C., ... Kazantsev, A. G. (2012). The sirtuin 2 inhibitor AK-7 is neuroprotective in Huntington's disease mouse models. *Cell Reports*, *2*(6), 1492–1497. http://dx.doi.org/10.1016/j.celrep.2012.11.001.

Denu, J. M. (2005). The Sir 2 family of protein deacetylases. *Current Opinion in Chemical Biology*, *9*(5), 431–440. http://dx.doi.org/10.1016/j.cbpa.2005.08.010.

Dominy, J. E., Jr., Lee, Y., Jedrychowski, M. P., Chim, H., Jurczak, M. J., Camporez, J. P., ... Puigserver, P. (2012). The deacetylase Sirt6 activates the acetyltransferase GCN5 and suppresses hepatic gluconeogenesis. *Molecular Cell*, *48*(6), 900–913. http://dx.doi.org/10.1016/j.molcel.2012.09.030.

Emiliani, S., Fischle, W., Van Lint, C., Al-Abed, Y., & Verdin, E. (1998). Characterization of a human RPD3 ortholog, HDAC3. *Proceedings of the National Academy of Sciences of the United States of America*, *95*(6), 2795–2800.

Feldman, J. L., Baeza, J., & Denu, J. M. (2013). Activation of the protein deacetylase SIRT6 by long-chain fatty acids and widespread deacylation by mammalian sirtuins. *The Journal of Biological Chemistry*, *288*(43), 31350–31356. http://dx.doi.org/10.1074/jbc.C113.511261.

Feldman, J. L., Dittenhafer-Reed, K. E., Kudo, N., Thelen, J. N., Ito, A., Yoshida, M., & Denu, J. M. (2015). Kinetic and structural basis for acyl-group selectivity and NAD(+) dependence in sirtuin-catalyzed deacylation. *Biochemistry*, *54*(19), 3037–3050. http://dx.doi.org/10.1021/acs.biochem.5b00150.

Finnin, M. S., Donigian, J. R., & Pavletich, N. P. (2001). Structure of the histone deacetylase SIRT2. *Nature Structural Biology*, *8*(7), 621–625. http://dx.doi.org/10.1038/89668.

Frye, R. A. (2000). Phylogenetic classification of prokaryotic and eukaryotic Sir2-like proteins. *Biochemical and Biophysical Research Communications*, *273*(2), 793–798. http://dx.doi.org/10.1006/bbrc.2000.3000.

Gerhart-Hines, Z., Rodgers, J. T., Bare, O., Lerin, C., Kim, S. H., Mostoslavsky, R., ... Puigserver, P. (2007). Metabolic control of muscle mitochondrial function and fatty acid oxidation through SIRT1/PGC-1alpha. *The EMBO Journal*, *26*(7), 1913–1923. http://dx.doi.org/10.1038/sj.emboj.7601633.

Grozinger, C. M., Chao, E. D., Blackwell, H. E., Moazed, D., & Schreiber, S. L. (2001). Identification of a class of small molecule inhibitors of the sirtuin family of NAD-dependent deacetylases by phenotypic screening. *The Journal of Biological Chemistry*, *276*(42), 38837–38843. http://dx.doi.org/10.1074/jbc.M106779200.

Guarente, L. (2006). Sirtuins as potential targets for metabolic syndrome. *Nature*, *444*(7121), 868–874. http://dx.doi.org/10.1038/nature05486.

Heltweg, B., Dequiedt, F., Verdin, E., & Jung, M. (2003). Nonisotopic substrate for assaying both human zinc and NAD + − dependent histone deacetylases. *Analytical Biochemistry*, *319*(1), 42–48.

Heltweg, B., Gatbonton, T., Schuler, A. D., Posakony, J., Li, H., Goehle, S., ... Bedalov, A. (2006). Antitumor activity of a small-molecule inhibitor of human silent information regulator 2 enzymes. *Cancer Research*, *66*(8), 4368–4377. http://dx.doi.org/10.1158/0008-5472.CAN-05-3617.

Herranz, D., Maraver, A., Canamero, M., Gomez-Lopez, G., Inglada-Perez, L., Robledo, M., ... Serrano, M. (2013). SIRT1 promotes thyroid carcinogenesis driven by PTEN deficiency. *Oncogene*, *32*(34), 4052–4056. http://dx.doi.org/10.1038/onc.2012.407.

Hoff, K. G., Avalos, J. L., Sens, K., & Wolberger, C. (2006). Insights into the sirtuin mechanism from ternary complexes containing NAD+ and acetylated peptide. *Structure*, *14*(8), 1231–1240. http://dx.doi.org/10.1016/j.str.2006.06.006.

Imai, S., Armstrong, C. M., Kaeberlein, M., & Guarente, L. (2000). Transcriptional silencing and longevity protein Sir2 is an NAD-dependent histone deacetylase. *Nature*, *403*(6771), 795–800. http://dx.doi.org/10.1038/35001622.

Jeong, S. M., Xiao, C., Finley, L. W., Lahusen, T., Souza, A. L., Pierce, K., ... Haigis, M. C. (2013). SIRT4 has tumor-suppressive activity and regulates the cellular metabolic response to DNA damage by inhibiting mitochondrial glutamine metabolism. *Cancer Cell*, *23*(4), 450–463. http://dx.doi.org/10.1016/j.ccr.2013.02.024.

Kaeberlein, M., Kirkland, K. T., Fields, S., & Kennedy, B. K. (2004). Sir2-independent life span extension by calorie restriction in yeast. *PLoS Biology*, *2*(9), E296. http://dx.doi.org/10.1371/journal.pbio.0020296.

Kaeberlein, M., McVey, M., & Guarente, L. (1999). The SIR2/3/4 complex and SIR2 alone promote longevity in Saccharomyces cerevisiae by two different mechanisms. *Genes & Development*, *13*(19), 2570–2580.

Kalle, A. M., Mallika, A., Badiger, J., Alinakhi, Talukdar, P., & Sachchidanand (2010). Inhibition of SIRT1 by a small molecule induces apoptosis in breast cancer cells. *Biochemical and Biophysical Research Communications, 401*(1), 13–19. http://dx.doi.org/10.1016/j.bbrc.2010.08.118.

Klar, A. J., Fogel, S., & Macleod, K. (1979). MAR1—A regulator of the HMa and HMalpha loci in Saccharomyces cerevisiae. *Genetics, 93*(1), 37–50.

Klar, A. J., Strathern, J. N., Broach, J. R., & Hicks, J. B. (1981). Regulation of transcription in expressed and unexpressed mating type cassettes of yeast. *Nature, 289*(5795), 239–244.

Lain, S., Hollick, J. J., Campbell, J., Staples, O. D., Higgins, M., Aoubala, M., ... Westwood, N. J. (2008). Discovery, in vivo activity, and mechanism of action of a small-molecule p53 activator. *Cancer Cell, 13*(5), 454–463. http://dx.doi.org/10.1016/j.ccr.2008.03.004.

Landry, J., Sutton, A., Tafrov, S. T., Heller, R. C., Stebbins, J., Pillus, L., & Sternglanz, R. (2000). The silencing protein SIR2 and its homologs are NAD-dependent protein deacetylases. *Proceedings of the National Academy of Sciences of the United States of America, 97*(11), 5807–5811. http://dx.doi.org/10.1073/pnas.110148297.

Lara, E., Mai, A., Calvanese, V., Altucci, L., Lopez-Nieva, P., Martinez-Chantar, M. L., ... Fraga, M. F. (2009). Salermide, a sirtuin inhibitor with a strong cancer-specific proapoptotic effect. *Oncogene, 28*(6), 781–791. http://dx.doi.org/10.1038/onc.2008.436.

Leko, V., Park, G. J., Lao, U., Simon, J. A., & Bedalov, A. (2013). Enterocyte-specific inactivation of SIRT1 reduces tumor load in the APC(+/min) mouse model. *PLoS One, 8*(6), e66283. http://dx.doi.org/10.1371/journal.pone.0066283.

Lerrer, B., Gertler, A. A., & Cohen, H. Y. (2015). The complex role of SIRT6 in carcinogenesis. *Carcinogenesis, 37*, 108–118. http://dx.doi.org/10.1093/carcin/bgv167.

Li, L., Osdal, T., Ho, Y., Chun, S., McDonald, T., Agarwal, P., ... Bhatia, R. (2014). SIRT1 activation by a c-MYC oncogenic network promotes the maintenance and drug resistance of human FLT3-ITD acute myeloid leukemia stem cells. *Cell Stem Cell, 15*(4), 431–446. http://dx.doi.org/10.1016/j.stem.2014.08.001.

Li, L., Wang, L., Li, L., Wang, Z., Ho, Y., McDonald, T., ... Bhatia, R. (2012). Activation of p53 by SIRT1 inhibition enhances elimination of CML leukemia stem cells in combination with imatinib. *Cancer Cell, 21*(2), 266–281. http://dx.doi.org/10.1016/j.ccr.2011.12.020.

Li, C., Wang, L., Zheng, L., Zhan, X., Xu, B., Jiang, J., & Wu, C. (2015). SIRT1 expression is associated with poor prognosis of lung adenocarcinoma. *Onco Targets and Therapy, 8*, 977–984. http://dx.doi.org/10.2147/OTT.S82378.

Lin, S. J., Defossez, P. A., & Guarente, L. (2000). Requirement of NAD and SIR2 for lifespan extension by calorie restriction in Saccharomyces cerevisiae. *Science, 289*(5487), 2126–2128.

Liu, P. Y., Xu, N., Malyukova, A., Scarlett, C. J., Sun, Y. T., Zhang, X. D., ... Liu, T. (2013). The histone deacetylase SIRT2 stabilizes Myc oncoproteins. *Cell Death and Differentiation, 20*(3), 503–514. http://dx.doi.org/10.1038/cdd.2012.147.

Luo, J., Nikolaev, A. Y., Imai, S., Chen, D., Su, F., Shiloh, A., ... Gu, W. (2001). Negative control of p53 by Sir2alpha promotes cell survival under stress. *Cell, 107*(2), 137–148.

Mahajan, S. S., Scian, M., Sripathy, S., Posakony, J., Lao, U., Loe, T. K., ... Simon, J. A. (2014). Development of pyrazolone and isoxazol-5-one cambinol analogues as sirtuin inhibitors. *Journal of Medicinal Chemistry, 57*(8), 3283–3294. http://dx.doi.org/10.1021/jm4018064.

Malik, S., Villanova, L., Tanaka, S., Aonuma, M., Roy, N., Berber, E., ... Chua, K. F. (2015). SIRT7 inactivation reverses metastatic phenotypes in epithelial and mesenchymal tumors. *Scientific Reports, 5*, 9841. http://dx.doi.org/10.1038/srep09841.

Marcotte, P. A., Richardson, P. L., Guo, J., Barrett, L. W., Xu, N., Gunasekera, A., & Glaser, K. B. (2004). Fluorescence assay of SIRT protein deacetylases using an acetylated peptide substrate and a secondary trypsin reaction. *Analytical Biochemistry*, *332*(1), 90–99. http://dx.doi.org/10.1016/j.ab.2004.05.039.

Milne, J. C., & Denu, J. M. (2008). The sirtuin family: Therapeutic targets to treat diseases of aging. *Current Opinion in Chemical Biology*, *12*(1), 11–17. http://dx.doi.org/10.1016/j.cbpa.2008.01.019.

Milne, J. C., Lambert, P. D., Schenk, S., Carney, D. P., Smith, J. J., Gagne, D. J., ... Westphal, C. H. (2007). Small molecule activators of SIRT1 as therapeutics for the treatment of type 2 diabetes. *Nature*, *450*(7170), 712–716. http://dx.doi.org/10.1038/nature06261.

Min, J., Landry, J., Sternglanz, R., & Xu, R. M. (2001). Crystal structure of a SIR2 homolog-NAD complex. *Cell*, *105*(2), 269–279.

Moazed, D. (2001). Enzymatic activities of Sir2 and chromatin silencing. *Current Opinion in Cell Biology*, *13*(2), 232–238.

Morimoto, J., Hayashi, Y., & Suga, H. (2012). Discovery of macrocyclic peptides armed with a mechanism-based warhead: Isoform-selective inhibition of human deacetylase SIRT2. *Angewandte Chemie (International Ed. in English)*, *51*(14), 3423–3427. http://dx.doi.org/10.1002/anie.201108118.

Nakahata, Y., Kaluzova, M., Grimaldi, B., Sahar, S., Hirayama, J., Chen, D., ... Sassone-Corsi, P. (2008). The NAD+-dependent deacetylase SIRT1 modulates CLOCK-mediated chromatin remodeling and circadian control. *Cell*, *134*(2), 329–340. http://dx.doi.org/10.1016/j.cell.2008.07.002.

Napper, A. D., Hixon, J., McDonagh, T., Keavey, K., Pons, J. F., Barker, J., ... Curtis, R. (2005). Discovery of indoles as potent and selective inhibitors of the deacetylase SIRT1. *Journal of Medicinal Chemistry*, *48*(25), 8045–8054. http://dx.doi.org/10.1021/jm050522v.

Nare, B., Allocco, J. J., Kuningas, R., Galuska, S., Myers, R. W., Bednarek, M. A., & Schmatz, D. M. (1999). Development of a scintillation proximity assay for histone deacetylase using a biotinylated peptide derived from histone-H4. *Analytical Biochemistry*, *267*(2), 390–396. http://dx.doi.org/10.1006/abio.1998.3038.

Newcomb, B., & Bedalov, A. (2009). Identification of inhibitors of chromatin modifying enzymes using the yeast phenotypic screens. *Methods in Molecular Biology*, *548*, 145–160. http://dx.doi.org/10.1007/978-1-59745-540-4_8.

Opitz, C. A., & Heiland, I. (2015). Dynamics of NAD-metabolism: Everything but constant. *Biochemical Society Transactions*, *43*(6), 1127–1132. http://dx.doi.org/10.1042/BST20150133.

Pacholec, M., Bleasdale, J. E., Chrunyk, B., Cunningham, D., Flynn, D., Garofalo, R. S., ... Ahn, K. (2010). SRT1720, SRT2183, SRT1460, and resveratrol are not direct activators of SIRT1. *The Journal of Biological Chemistry*, *285*(11), 8340–8351. http://dx.doi.org/10.1074/jbc.M109.088682.

Pasqualucci, L., Dominguez-Sola, D., Chiarenza, A., Fabbri, G., Grunn, A., Trifonov, V., ... Dalla-Favera, R. (2011). Inactivating mutations of acetyltransferase genes in B-cell lymphoma. *Nature*, *471*(7337), 189–195. http://dx.doi.org/10.1038/nature09730.

Patterson, G. H., Knobel, S. M., Arkhammar, P., Thastrup, O., & Piston, D. W. (2000). Separation of the glucose-stimulated cytoplasmic and mitochondrial NAD(P)H responses in pancreatic islet beta cells. *Proceedings of the National Academy of Sciences of the United States of America*, *97*(10), 5203–5207. http://dx.doi.org/10.1073/pnas.090098797.

Picard, F., Kurtev, M., Chung, N., Topark-Ngarm, A., Senawong, T., Machado De Oliveira, R., ... Guarente, L. (2004). Sirt1 promotes fat mobilization in white adipocytes by repressing PPAR-gamma. *Nature*, *429*(6993), 771–776. http://dx.doi.org/10.1038/nature02583.

Ramsey, K. M., Yoshino, J., Brace, C. S., Abrassart, D., Kobayashi, Y., Marcheva, B., ... Bass, J. (2009). Circadian clock feedback cycle through NAMPT-mediated NAD+ biosynthesis. *Science, 324*(5927), 651–654. http://dx.doi.org/10.1126/science.1171641.

Rine, J., & Herskowitz, I. (1987). Four genes responsible for a position effect on expression from HML and HMR in Saccharomyces cerevisiae. *Genetics, 116*(1), 9–22.

Rodgers, J. T., Lerin, C., Haas, W., Gygi, S. P., Spiegelman, B. M., & Puigserver, P. (2005). Nutrient control of glucose homeostasis through a complex of PGC-1alpha and SIRT1. *Nature, 434*(7029), 113–118. http://dx.doi.org/10.1038/nature03354.

Rodgers, J. T., & Puigserver, P. (2007). Fasting-dependent glucose and lipid metabolic response through hepatic sirtuin 1. *Proceedings of the National Academy of Sciences of the United States of America, 104*(31), 12861–12866. http://dx.doi.org/10.1073/pnas.0702509104.

Rogina, B., & Helfand, S. L. (2004). Sir2 mediates longevity in the fly through a pathway related to calorie restriction. *Proceedings of the National Academy of Sciences of the United States of America, 101*(45), 15998–16003. http://dx.doi.org/10.1073/pnas.0404184101.

Ruderman, N. B., Carling, D., Prentki, M., & Cacicedo, J. M. (2013). AMPK, insulin resistance, and the metabolic syndrome. *The Journal of Clinical Investigation, 123*(7), 2764–2772. http://dx.doi.org/10.1172/JCI67227.

Rumpf, T., Schiedel, M., Karaman, B., Roessler, C., North, B. J., Lehotzky, A., ... Jung, M. (2015). Selective Sirt2 inhibition by ligand-induced rearrangement of the active site. *Nature Communications, 6*, 6263. http://dx.doi.org/10.1038/ncomms7263.

Sauve, A. A., Celic, I., Avalos, J., Deng, H., Boeke, J. D., & Schramm, V. L. (2001). Chemistry of gene silencing: The mechanism of NAD^+-dependent deacetylation reactions. *Biochemistry, 40*(51), 15456–15463.

Sauve, A. A., & Schramm, V. L. (2004). SIR2: The biochemical mechanism of NAD(+)-dependent protein deacetylation and ADP-ribosyl enzyme intermediates. *Current Medicinal Chemistry, 11*(7), 807–826.

Schmidt, M. T., Smith, B. C., Jackson, M. D., & Denu, J. M. (2004). Coenzyme specificity of Sir2 protein deacetylases: Implications for physiological regulation. *The Journal of Biological Chemistry, 279*(38), 40122–40129. http://dx.doi.org/10.1074/jbc.M407484200.

Sinclair, D. A., & Guarente, L. (1997). Extrachromosomal rDNA circles—A cause of aging in yeast. *Cell, 91*(7), 1033–1042.

Smith, J. S., Brachmann, C. B., Celic, I., Kenna, M. A., Muhammad, S., Starai, V. J., ... Boeke, J. D. (2000). A phylogenetically conserved NAD^+-dependent protein deacetylase activity in the Sir2 protein family. *Proceedings of the National Academy of Sciences of the United States of America, 97*(12), 6658–6663.

Smith, J. S., Brachmann, C. B., Pillus, L., & Boeke, J. D. (1998). Distribution of a limited Sir2 protein pool regulates the strength of yeast rDNA silencing and is modulated by Sir4p. *Genetics, 149*(3), 1205–1219.

Smith, B. C., Hallows, W. C., & Denu, J. M. (2009). A continuous microplate assay for sirtuins and nicotinamide-producing enzymes. *Analytical Biochemistry, 394*(1), 101–109. http://dx.doi.org/10.1016/j.ab.2009.07.019.

Stenzinger, A., Endris, V., Klauschen, F., Sinn, B., Lorenz, K., Warth, A., ... Weichert, W. (2013). High SIRT1 expression is a negative prognosticator in pancreatic ductal adenocarcinoma. *BMC Cancer, 13*, 450. http://dx.doi.org/10.1186/1471-2407-13-450.

Sussmuth, S. D., Haider, S., Landwehrmeyer, G. B., Farmer, R., Frost, C., Tripepi, G., ... Consortium, Paddington. (2015). An exploratory double-blind, randomized clinical trial with selisistat, a SirT1 inhibitor, in patients with Huntington's disease. *British Journal of Clinical Pharmacology, 79*(3), 465–476. http://dx.doi.org/10.1111/bcp.12512.

Suzuki, T., Khan, M. N., Sawada, H., Imai, E., Itoh, Y., Yamatsuta, K., ... Miyata, N. (2012). Design, synthesis, and biological activity of a novel series of human sirtuin-2-selective inhibitors. *Journal of Medicinal Chemistry, 55*(12), 5760–5773. http://dx.doi.org/10.1021/jm3002108.

Tang, Y., Zhao, W., Chen, Y., Zhao, Y., & Gu, W. (2008). Acetylation is indispensable for p53 activation. *Cell, 133*(4), 612–626. http://dx.doi.org/10.1016/j.cell.2008.03.025.

Tanner, K. G., Landry, J., Sternglanz, R., & Denu, J. M. (2000). Silent information regulator 2 family of NAD- dependent histone/protein deacetylases generates a unique product, 1-O-acetyl-ADP-ribose. *Proceedings of the National Academy of Sciences of the United States of America, 97*(26), 14178–14182. http://dx.doi.org/10.1073/pnas.250422697.

Tanny, J. C., & Moazed, D. (2001). Coupling of histone deacetylation to NAD breakdown by the yeast silencing protein Sir2: Evidence for acetyl transfer from substrate to an NAD breakdown product. *Proceedings of the National Academy of Sciences of the United States of America, 98*(2), 415–420. http://dx.doi.org/10.1073/pnas.031563798.

Taylor, D. M., Balabadra, U., Xiang, Z., Woodman, B., Meade, S., Amore, A., ... Kazantsev, A. G. (2011). A brain-permeable small molecule reduces neuronal cholesterol by inhibiting activity of sirtuin 2 deacetylase. *ACS Chemical Biology, 6*(6), 540–546. http://dx.doi.org/10.1021/cb100376q.

Tissenbaum, H. A., & Guarente, L. (2001). Increased dosage of a sir-2 gene extends lifespan in Caenorhabditis elegans. *Nature, 410*(6825), 227–230. http://dx.doi.org/10.1038/35065638.

Vaziri, H., Dessain, S. K., Ng Eaton, E., Imai, S. I., Frye, R. A., Pandita, T. K., ... Weinberg, R. A. (2001). hSIR2(SIRT1) functions as an NAD-dependent p53 deacetylase. *Cell, 107*(2), 149–159.

Verdin, E. (2015). NAD(+) in aging, metabolism, and neurodegeneration. *Science, 350*(6265), 1208–1213. http://dx.doi.org/10.1126/science.aac4854.

Wauters, E., Sanchez-Arevalo Lobo, V. J., Pinho, A. V., Mawson, A., Herranz, D., Wu, J., ... Rooman, I. (2013). Sirtuin-1 regulates acinar-to-ductal metaplasia and supports cancer cell viability in pancreatic cancer. *Cancer Research, 73*(7), 2357–2367. http://dx.doi.org/10.1158/0008-5472.CAN-12-3359.

Williamson, D. H., Lund, P., & Krebs, H. A. (1967). The redox state of free nicotinamide-adenine dinucleotide in the cytoplasm and mitochondria of rat liver. *The Biochemical Journal, 103*(2), 514–527.

Yang, M. H., Laurent, G., Bause, A. S., Spang, R., German, N., Haigis, M. C., & Haigis, K. M. (2013). HDAC6 and SIRT2 regulate the acetylation state and oncogenic activity of mutant K-RAS. *Molecular Cancer Research, 11*(9), 1072–1077. http://dx.doi.org/10.1158/1541-7786.MCR-13-0040-T.

Zhang, Q., Piston, D. W., & Goodman, R. H. (2002). Regulation of corepressor function by nuclear NADH. *Science, 295*(5561), 1895–1897. http://dx.doi.org/10.1126/science.1069300.

Zhao, K., Chai, X., Clements, A., & Marmorstein, R. (2003a). Structure and autoregulation of the yeast Hst2 homolog of Sir2. *Nature Structural Biology, 10*(10), 864–871. http://dx.doi.org/10.1038/nsb978.

Zhao, K., Chai, X., & Marmorstein, R. (2003b). Structure of the yeast Hst2 protein deacetylase in ternary complex with 2'-O-acetyl ADP ribose and histone peptide. *Structure, 11*(11), 1403–1411.

CHAPTER TEN

Synthesis and Assay of SIRT1-Activating Compounds

H. Dai*, J.L. Ellis*, D.A. Sinclair[†,‡], B.P. Hubbard[§,1]

*Sirtuin DPU, GlaxoSmithKline (GSK), Collegeville, PA, United States
[†]Glenn Labs for the Biological Mechanisms of Aging, Harvard Medical School, Boston, MA, United States
[‡]The University of New South Wales, Sydney, NSW, Australia
[§]University of Alberta, Edmonton, AB, Canada
[1]Corresponding author: e-mail address: bphubbard@ualberta.ca

Contents

1. Introduction	214
2. Materials	215
2.1 Synthesis of SIRT1-Activating Compounds	215
2.2 Expression and Purification of Recombinant His-Tagged SIRT1	216
2.3 Assay of SIRT1 Activators Using the PNC1-OPT Assay	216
2.4 Assay of SIRT1 Activators Using the RapidFire Mass Spectrometry Assay	217
3. Methods	217
3.1 Synthesis of SIRT1-Activating Compounds	217
3.2 Expression and Purification of Recombinant His-Tagged SIRT1	234
3.3 Assay of SIRT1 Activators Using the PNC1-OPT Assay	236
3.4 Assay of SIRT1 Activators Using the RapidFire O-Ac-ADPR Detection Assay	239
4. Notes	240
Acknowledgments	242
References	242

Abstract

The NAD^+-dependent deacetylase SIRT1 plays key roles in numerous cellular processes including DNA repair, gene transcription, cell differentiation, and metabolism. Overexpression of SIRT1 protects against a number of age-related diseases including diabetes, cancer, and Alzheimer's disease. Moreover, overexpression of SIRT1 in the murine brain extends lifespan. A number of small-molecule sirtuin-activating compounds (STACs) that increase SIRT1 activity in vitro and in cells have been developed. While the mechanism for how these compounds act on SIRT1 was once controversial, it is becoming increasingly clear that they directly interact with SIRT1 and enhance its activity through an allosteric mechanism. Here, we present detailed chemical syntheses for four STACs, each from a distinct structural class. Also, we provide a general protocol for purifying active SIRT1 enzyme and outline two complementary enzymatic assays for characterizing the effects of STACs and similar compounds on SIRT1 activity.

1. INTRODUCTION

Overexpression of the NAD^+-dependent histone deacetylase Sir2 extends the lifespan of diverse model organisms including yeast, worms, and flies (Hubbard & Sinclair, 2013, 2014; Morris, 2013). In mammals, seven Sir2 homologs have been identified (SIRT1-7), with SIRT1 bearing the closest phylogenetic relationship to yeast Sir2 (Hubbard & Sinclair, 2014). SIRT1 is involved in mediating numerous critical cellular processes such as DNA repair and apoptosis, muscle and fat differentiation, neurogenesis, mitochondrial biogenesis, and various aspects of cell metabolism (Morris, 2013). Interestingly, deregulation of SIRT1 activity has been implicated in a number of age-related diseases including heart disease, diabetes, Alzheimer's disease, and cancer (Hubbard & Sinclair, 2014). Moreover, overexpression of SIRT1 has been shown to be protective in models of colon cancer (Firestein et al., 2008), neurodegeneration (Kim et al., 2007), and age-related metabolic decline (Feige et al., 2008; Gomes et al., 2013; Price et al., 2012), and to extend the lifespan of mice when overexpressed in neurons (Satoh et al., 2013). In light of the wide array of health benefits it confers, SIRT1 has emerged as an attractive drug target for treating a variety of diseases in humans (Hubbard & Sinclair, 2014).

While a number of different strategies to increase SIRT1 enzymatic activity in cells have been proposed (Hubbard & Sinclair, 2014), including pan-sirtuin activators that block binding of the sirtuin inhibitor nicotinamide (Sauve, Moir, Schramm, & Willis, 2005), and molecules that displace the protein inhibitor of SIRT1, DBC1 (Hubbard, Loh, et al., 2013), the majority of research has focused on the development of allosteric SIRT1 activators (STACs) (Hubbard & Sinclair, 2014). The first STACs were discovered in 2003 using a high-throughput screen employing a fluorophore-conjugated peptide (Howitz et al., 2003). Several classes of polyphenols, including flavones, stilbenes, and anthocyanidins, were shown to increase SIRT1 activity through a mechanism involving a lowering of peptide substrate K_m (Bhullar & Hubbard, 2015; Howitz et al., 2003). Resveratrol, the most efficacious SIRT1 activator identified in this screen, activated SIRT1 by up to 10-fold (Howitz et al., 2003). Later, chemically distinct compounds based on an imidazothiazole scaffold (eg, SRT1460, SRT1720) that are more potent were developed by Sirtris Pharmaceuticals, and these were also shown to elicit changes in cells consistent with SIRT1 activation (Milne et al., 2007). However, the validity of these early compounds was challenged by a series of reports showing that while STACs activated SIRT1 deacetylation on the fluorophore-tagged peptides used in the initial screen, no activity enhancement was observed when the corresponding untagged peptides were used (Borra, Smith, & Denu, 2005; Kaeberlein et al., 2005; Pacholec et al., 2010). Furthermore, these studies led to speculation that the effects of STACs in vivo might be due to off-target effects (Chung, 2012).

Recent work has supported the original assertion that STACs do indeed act as direct allosteric SIRT1 activators (Dai et al., 2015, 2010; Gertz et al., 2012; Hubbard, Gomes, et al., 2013). For example, STACs can activate SIRT1 deacetylation of natural amino acid peptides (Dai et al., 2010; Lakshminarasimhan, Rauh, Schutkowski, & Steegborn, 2013) and can enhance SIRT1 deacetylation of native peptides bearing hydrophobic amino acids adjacent to the acetyl-lysine (Hubbard, Gomes, et al., 2013). A model of "assisted allosteric activation" has been proposed to account for how STACs activate SIRT1 (Hubbard, Gomes, et al., 2013). In this model, binding of specific peptide substrate to SIRT1 induces the formation of an exosite that enhances drug binding. Once bound to SIRT1, the activator is thought to stabilize the substrate binding, resulting in a lowering of apparent peptide K_m (Hubbard, Gomes, et al., 2013). This model has now been supported by additional crystallographic data (Cao et al., 2015; Dai et al., 2015). Here, we present detailed protocols for the synthesis and assay of STACs. We outline full syntheses for four structurally distinct classes of STACs, describe a general method for SIRT1 enzyme purification, and provide instructions on how to assay the effects of STACs on SIRT1 using two complementary activity assays (PNC1-OPT and RapidFire mass spectrometry).

2. MATERIALS
2.1 Synthesis of SIRT1-Activating Compounds

1. General organic chemistry glassware.
2. Reagent-grade chemicals outlined in the syntheses.
3. Equipment for Flash column chromatography (eg, ISCO CombiFlash Rf or similar system) with appropriate columns.
4. Analytical HPLC (eg, Agilent 1100 series), for compound verification.
5. Nuclear magnetic resonance (NMR) instrument (eg, Bruker Advance III), for compound verification.

6. High-resolution mass spectrometry (HRMS) (eg, Waters qTOF Premier Mass Spectrometer), for compound verification.

2.2 Expression and Purification of Recombinant His-Tagged SIRT1

1. pET-based His-tagged SIRT1 expression vector (or similar).
2. BL21 pLysS(DE3) or similar chemically competent bacteria.
3. Antibiotics (ampicillin or kanamycin, chloramphenicol).
4. LB media and LB-agar plates with appropriate antibiotics.
5. Large temperature-controlled bacterial incubator.
6. Centrifuge for spinning down large bacterial cultures (0.5–1 L volumes).
7. Protease inhibitor pellets (eg, Roche EDTA-free) (see Note 1).
8. Pipetteman with 10 or 25 mL serological pipettes.
9. Isopropyl β-D-1-thiogalactopyranoside (IPTG).
10. Ice buckets filled with ice.
11. Sonicator with wide-fitted head.
12. Ni-NTA agarose beads (eg, Qiagen).
13. Lysis buffer: 1% Triton X-100, 50 mM Tris pH 8.0, 150 mM NaCl, 20 mM imidazole, 3 mM β-mercaptoethanol.
14. Wash buffer: 1% Triton X-100, 50 mM Tris pH 8.0, 300 mM NaCl, 20 mM imidazole, 3 mM β-mercaptoethanol.
15. Elution buffer: 50 mM Tris pH 8.0, 250 mM imidazole, 3 mM β-mercaptoethanol.
16. Biorad Polyprep columns.
17. Spectrophotometer capable of performing absorbance measurements.

2.3 Assay of SIRT1 Activators Using the PNC1-OPT Assay

1. Purified recombinant SIRT1 enzyme (produced as described earlier).
2. Purified recombinant PNC1 enzyme (yeast PNC1 (yPNC1) may be purified using the same procedure described for SIRT1) (Howitz et al., 2003; Hubbard, Gomes, et al., 2013).
3. Assay buffer: Gibco phosphate-buffered saline (PBS) pH 7.4 (1 mM KH_2PO_4, 155 mM NaCl, 3 mM $Na_2HPO_4 \cdot 7H_2O$) supplemented with 1 mM dithiothreitol (DTT) (add fresh from a frozen stock).
4. Peptide substrate—typically between 5 and 15 amino acids (eg, Ac-RHKK(ac)W-NH_2), either synthesized in-house or purchased commercially (see Note 2).

5. β-Nicotinamide adenine dinucleotide (β-NAD)—prepare as 100 mM aliquots and store at −20°C.
6. OPT developer reagent: a 30% EtOH/70% PBS (pH 7.4) solution supplemented with 10 mM *ortho*-phthalaldehyde (OPT) and 10 mM DTT. This solution should be stored at −20°C until use and kept away from light (cover with aluminum foil).
7. Nicotinamide (NAM)—prepare 100 mM aliquots and store at −20°C.
8. 37°C incubator.
9. Orbital shaker.
10. 96-Well black opaque bottom plates for fluorometry.
11. Spectrophotometer with fluorescence capabilities and appropriate filters (excitation ∼413 nm and emission ∼476 nm).

2.4 Assay of SIRT1 Activators Using the RapidFire Mass Spectrometry Assay

1. Purified recombinant SIRT1 enzyme (produced as described earlier).
2. Bovine serum albumin (BSA).
3. Reaction buffer: 50 mM HEPES–NaOH, pH 7.5, 150 mM NaCl, 1 mM DTT, and 1 % DMSO.
4. β-NAD prepared as 100 mM aliquots and stored at −20°C.
5. Peptide substrate—typically between 5 and 15 amino acids (eg, Ac-RHKK(ac)W-NH$_2$) either synthesized in-house or purchased commercially.
6. Agilent RapidFire System.
7. SPE cartridges.
8. Mass spectrometer fitter with an electrospray ionization source (eg, ABSciex API 4000).
9. Stop Reagent: 10% formic acid and 50 mM nicotinamide.
10. 1:1 mixture of acetonitrile:methanol.
11. Buffer A: 90:10 acetonitrile:water with 0.1 mM ammonium acetate and 0.2% formic acid.
12. Buffer B: 60:40 acetonitrile:water supplemented with 1.6 mM ammonium acetate.

3. METHODS

3.1 Synthesis of SIRT1-Activating Compounds

A wide variety of natural molecules with the ability to increase SIRT1 activity have been reported (Hubbard & Sinclair, 2014). While these compounds comprise diverse structural scaffolds, including flavones, stilbenes, chalcones (Howitz et al., 2003), coumarins (Dao et al., 2012), and triterpenes (Yang et al., 2014), they all appear to activate SIRT1 through a common peptide K_m-lowering mechanism (Hubbard & Sinclair, 2014). In addition to natural product SIRT1 activators, Sirtris Pharmaceutics (GSK) has described the development of an expanding collection of synthetic SIRT1 activators with EC_{50} values in the nM range (hundreds of times more potent natural compounds) over the past 10 years (Dai et al., 2015, 2010; Hubbard, Gomes, et al., 2013; Milne et al., 2007). These activators constitute structurally distinct compounds including imidazothiazoles, thiazolopyridines, imidazopyridines, and bridged ureas (Dai et al., 2015, 2010; Hubbard, Gomes, et al., 2013; Milne et al., 2007). Below we present detailed synthetic routes for four STACs, consisting of one example from each structural class.

General points

1. Flash column chromatography should be performed using silica gel with a particle size of 60 Å, mesh of 230–400, using standard techniques. Unless otherwise indicated, chromatography refers to medium pressure chromatography performed on an ISCO CombiFlash Rf or similar system.
2. Following completion of each synthesis, final compound purities should be determined by analytical HPLC using the area percentage method on the UV trace recorded at a wavelength of 254 nm (see Note 3 for HPLC parameters). Expected purity is $\geq 95\%$ purity unless otherwise specified.
3. ^{1}H NMR spectra should be obtained to characterize each product. Example of NMR parameters may be found in Note 4. NMR spectral data reported in this protocol follow these conventions: chemical shift (δ) in ppm (multiplicity, coupling constants in Hertz, number of protons, assignment), s—singlet, d—doublet, t—triplet, q—quartet, m—multiplet, br—broad.
4. To further verify successful product synthesis, routine mass spectral (MS) analysis of each compound can be performed (eg, using an Agilent 1100 series spectrometer integrated into an Agilent 1200 series HPLC system). In addition, HRMS can be performed (see Note 5 for parameters). The HRMS acceptable error is 3 mDa or 5 ppm, although most analyses are observed within 0.5 mDa with isotope fits in good agreement with the proposed structures.

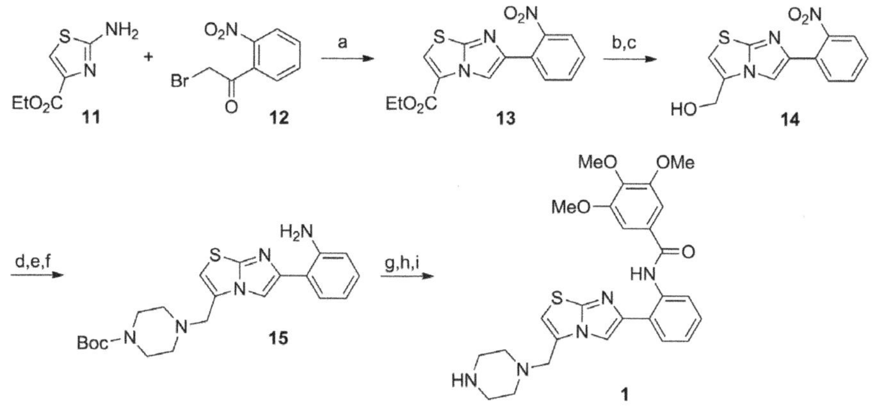

Fig. 1 Schematic outlining the synthesis of an imidazo[1,2-b]thiazole STAC. Compounds **11–15** represent intermediate products leading toward the production of the STAC 3,4,5-trimethoxy-N-(2-(3-(piperazin-1-ylmethyl)imidazo[2,1-b]thiazol-6-yl)phenyl) benzamide (**1**). Special reagents and conditions noted: (a) MEK, reflux (79%); (b) NaOH, 1:1 THF/H$_2$O, RT, then HCl; (c) iBuOCOCl, NMM, THF, 0°C, then NaBH$_4$, H$_2$O, 0°C (74% for 2 steps); (d) MsCl, Et$_3$N, CH$_2$Cl$_2$, 0°C; (e) Boc-piperazine, Et$_3$N, CH$_3$CN, RT; (f) NaSH, 6:1 MeOH/H$_2$O, reflux (60% for 3 steps); (g) 3,4,5-trimethoxybenzoyl chloride, pyridine; (h) TFA, CH$_2$Cl$_2$ (65%); (i) NaHCO$_3$, H$_2$O, then 3 N HCl, 1:1 CH$_3$CN/H$_2$O. *Adapted from Milne, J. C., Lambert, P. D., Schenk, S., Carney, D. P., Smith, J. J., Gagne, D. J., ..., Westphal, C. H. (2007). Small-molecule activators of SIRT1 as therapeutics for the treatment of type 2 diabetes.* Nature, 450(7170), 712–716.

3.1.1 Preparation of a Imidazo[1,2-b]thiazole STAC (Synthesis Adapted from Milne et al., 2007)

See Fig. 1.

6-(2-Nitrophenyl)-imidazo[2,1-b]thiazole-3-carboxylic acid ethyl ester (13)

1. Stir a solution of ethyl 2-aminothiazole-4-carboxylate (**11**) (2.1 g, 12 mmol) and 2-bromo-2′-nitroacetophenone (**12**) (3.0 g, 12 mmol) in methyl ethylketone (25 mL) under reflux for 18 h.
2. Cool the solution to room temperature and filter.
3. Concentrate the filtrate in vacuo to afford 3.10 g (79% yield) of 6-(2-nitrophenyl)-imidazo[2,1-b]thiazole-3-carboxylic acid ethyl ester (**13**). Characterization properties: ^{1}H NMR (300 MHz, DMSO-d_6): δ 8.39 (br s, 1 H), 8.31 (br s, 1 H), 7.92 (d, J=8 Hz, 1 H), 7.82 (d, J=7 Hz, 1 H), 7.3–7.7 (m, 2 H), 4.4 (q, J=7 Hz, 2 H), 1.37 (t, J=7 Hz, 3 H); MS m/z=318.0 (M+H)$^+$.

[6-(2-Nitrophenyl)imidazo[2,1-*b*]thiazol-3-yl]methanol (14)
4. Stir a solution of 6-(2-nitrophenyl)imidazo[2,1-*b*]thiazole-3-carboxylic acid ethyl ester (**13**) (14.50 g, 46 mmol) in THF (100 mL) and water (100 mL) containing NaOH (7.3 g, 4 equiv.) at room temperature for 18 h and then concentrate in vacuo.
5. Wash the aqueous layer once with CH_2Cl_2 and then acidify with 6 N HCl. Collect the solids by filtration and dry to afford 7.4 g of the acid intermediate.
6. Dissolve this material (7.4 g, 26 mmol) in anhydrous THF (200 mL) containing *N*-methylmorpholine (2.8 mL, 26 mmol) and cool to 0°C.
7. Add isobutyl chloroformate (3.35 mL, 26 mmol) and stir the reaction mixture in an ice bath for 3 h.
8. Add $NaBH_4$ (0.97 g, 25.6 mmol) in water (30 mL) and stir the mixture at 0°C for 45 min. Warm to room temperature and concentrate.
9. Extract the aqueous layer with CH_2Cl_2. Dry the combined organic layers (Na_2SO_4) and concentrate to afford the crude product.
10. Purify by chromatography to afford 5.20 g (74% yield) of [6-(2-nitrophenyl)imidazo[2,1-*b*]thiazol-3-yl]methanol (**14**). Characterization properties: ^1H NMR (300 MHz, DMSO-d_6): δ 8.14 (br s, 1 H), 7.2–7.9 (m, 4 H), 7.16 (br s, 1 H), 5.65 (t, *J*=7 Hz, 1 H), 4.6 (m, 2 H); MS *m/z* = 276.0 $(M+H)^+$.

4-[6-(2-Aminophenyl)imidazo[2,1-*b*]thiazol-3-ylmethyl]piperazine-1-carboxylic acid *tert*-butyl ester (15)
11. Treat a chilled (0°C) solution of [6-(2-nitrophenyl)imidazo[2,1-*b*]thiazol-3-yl]methanol (**14**) (1.0 g, 3.6 mmol) and Et_3N (0.51 mL, 3.64 mmol) in CH_2Cl_2 (100 mL) with methanesulfonyl chloride (0.28 mL, 3.7 mmol) and allow the resulting reaction to warm to room temperature.
12. Stir for 15 min.
13. Quench the reaction with brine and extract with CH_2Cl_2.
14. Dry (Na_2SO_4) the combined organic layers and concentrate in vacuo to afford the mesylate intermediate.
15. Dissolve this material in CH_3CN (4 mL) containing Et_3N (0.51 mL, 3.6 mmol) and Boc-piperazine (680 mg, 3.6 mmol) and stir the resulting solution at room temperature for 1 day.
16. Concentrate the reaction mixture and partition the resulting residue between CH_2Cl_2 and water.
17. Dry (Na_2SO_4) the organic layer and concentrate to afford the crude product. Dissolve this material in a solution of MeOH (6 mL) and water (1 mL) containing sodium hydrosulfide hydrate (200 mg).
18. Stir the resulting reaction mixture under reflux for 24 h, cool to room temperature, and concentrate. Dilute the residue with water

(2 mL) and extract with CH_2Cl_2. Dry (Na_2SO_4) the combined organic layers and concentrate to afford 0.90 g (60% yield) of 4-[6-(2-aminophenyl)-imidazo[2,1-b]thiazol-3-ylmethyl]-piperazine-1-carboxylic acid *tert*-butyl ester (**15**). Characterization properties: ^{1}H NMR (300 MHz, DMSO-d_6): δ 9.2 (br s, 1 H), 8.7 (br s, 1 H), 8.15 (s, 1 H), 8.10 (s, 1 H), 6.8–7.8 (m, 4 H), 6.16 (br s, 2 H), 3.72 (br s, 2 H), 1.39 (br s, 9 H); MS $m/z = 414.1$ $(M+H)^+$.

3,4,5-Trimethoxy-N-(2-(3-(piperazin-1-ylmethyl)imidazo[2,1-b] thiazol-6-yl)phenyl)benzamide (1)

19. Dissolve 4-[6-(2-aminophenyl)imidazo[2,1-b]thiazol-3-ylmethyl] piperazine-1-carboxylic acid *tert*-butyl ester (**15**) (300 mg, 0.73 mmol) in pyridine (5 mL) and treat with 3,4,5-trimethoxybenzoyl chloride (167 mg, 0.73 mmol).
20. Heat the reaction mixture in a microwave reactor (160°C × 10 min), cool to room temperature, and concentrate in vacuo.
21. Purify the resulting crude product by chromatography (gradient elution, CH_2Cl_2 to 95% CH_2Cl_2, 4% MeOH, and 1% Et_3N) and treat the purified product with a solution containing 25% TFA in CH_2Cl_2 (2 mL) for 2 h. Subsequently, concentrate and titrate the resulting residue with Et_2O to afford 335 mg (65% yield) of **1** as the TFA salt.
22. To prepare the corresponding HCl salt of **1**, dissolve the TFA salt in water and neutralize with $NaHCO_3$. Extract the resulting aqueous layer with CH_2Cl_2. Wash the combined organic layers with brine, dry (Na_2SO_4), and concentrate under reduced pressure.
23. Take up the resulting residue in 50% aqueous CH_3CN and add 1 mL of 3 N HCl. Lyophilize the mixture to obtain **1** as the HCl salt. Characterization properties: m.p.: 193.5°C (HCl salt). ^{1}H NMR (300 MHz, DMSO-d_6): δ 9.9 (br s, 1 H), 9.0 (br s, 1 H), 8.7–7.10 (m, 7 H), 8.5 (s, 1 H), 4.0 (br s, 9 H), 3.8 (m, 2 H), 3.2–2.8 (m, 8 H); ^{13}C NMR (100 MHz, DMSO-d_6): δ 47.56 50.01, 56.15, 60.16, 104.68, 111.52, 120.70, 120.93, 123.63, 126.89, 128.13, 130.13, 136.01, 140.54, 144.43, 147.75, 152.86, 164.09; HRMS calculated for $C_{26}H_{29}N_5O_4S$: 508.2018; found: 508.2039.

3.1.2 Preparation of a Thiazolopyridine STAC (Synthesis Adapted from Dai et al., 2010)

See Fig. 2.

N-(2-Chloro-5-methylpyridin-3-yl)-2-nitrobenzamide (22)

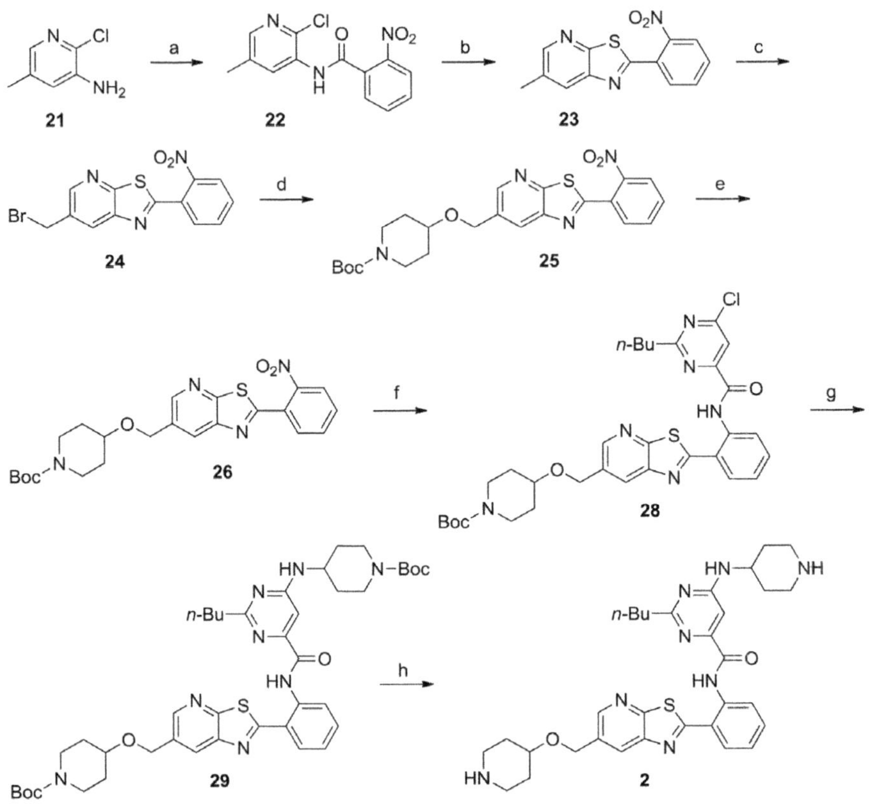

Fig. 2 Schematic outlining the synthesis of a thiazolopyridine STAC. Compounds 21–29 represent intermediate products leading toward the production of the STAC 2-butyl-6-(piperidin-4-ylamino)-N-(2-(6-((piperidin-4-yloxy)methyl)thiazolo[5,4-b]pyridin-2-yl)phenyl)pyrimidine-4-carboxamide (2). Special reagents and conditions noted: (a) 2-nitrobenzoyl chloride, pyridine 0°C to RT (91%); (b) P_2S_5, pyridine, p-xylene, 140°C (75%); (c) NBS, benzoyl peroxide, CCl_4; (d) tert-butyl 4-hydroxypiperidine-1-carboxylate, (n-Bu)$_4$N$^+$HSO$_4^-$, NaOH, toluene, H_2O (65%); (e) Fe, AcOH, THF, H_2O, 50°C (88%); (f) 2-butyl-6-chloropyrimidine-4-carbonyl chloride (27), Et_3N, CH_2Cl_2 (91%); (g) tert-butyl 4-aminopiperidine-1-carboxylate, (i-Pr)$_2$NEt, DMSO, 100°C (76%); (h) TFA, then NaHCO$_3$, H_2O (81%). Adapted from Dai, H., Kustigian, L., Carney, D., Case, A., Considine, T., Hubbard, B. P., ..., Stein, R. L. (2010). SIRT1 activation by small molecules: Kinetic and biophysical evidence for direct interaction of enzyme and activator. Journal of Biological Chemistry, 285(43), 32695–32703.

1. Add 2-nitrobenzoyl chloride (13.65 g, 73.6 mmol) dropwise to a solution of 5-amino-6-chloro-3-picoline (9.54 g, 66.9 mmol) (21) in pyridine (200 mL) at 0°C.
2. Stir the resulting mixture at room temperature for 18 h.
3. Dilute the dark mixture with water (1500 mL).
4. Add sat. NaHCO$_3$ solution until pH 8.

5. Collect the precipitate by filtration, rinse with water (3 × 30 mL), and dry in an oven to afford N-(2-chloro-5-methylpyridin-3-yl)-2-nitrobenzamide (17.70 g, 91%) as a pale solid.

6-Methyl-2-(2-nitrophenyl)thiazolo[5,4-*b*]pyridine (23)

6. Stir a mixture of N-(2-chloro-5-methylpyridin-3-yl)-2-nitrobenzamide (**22**; 5.0 g, 17.1 mmol) and P_2S_5 (7.6 g, 34.2 mmol) in pyridine (50 mL) and *p*-xylene (200 mL) at 140°C for 20 h.
7. Transfer the hot solution to another flask and remove the solvent in vacuo.
8. Purify the residue by recrystallization from EtOH to afford 6-methyl-2-(2-nitrophenyl)thiazolo[5,4-*b*]pyridine (3.5 g, 75%) as a yellow solid.

6-(Bromomethyl)-2-(2-nitrophenyl)thiazolo[5,4-*b*]pyridine (24)

9. Add 6-methyl-2-(2-nitrophenyl)thiazolo[5,4-*b*]pyridine (**23**; 2.9 g, 10.7 mmol), N-bromosuccinimide (NBS; 1.91 g, 10.7 mmol), CCl_4 (200 mL), and benzoyl peroxide (0.021 g) into a three-neck flask (500 mL) under argon.
10. Reflux the resulting yellow mixture for 2 h.
11. Add additional NBS (1.91 g) and benzoyl peroxide (0.021 g).
12. Add more NBS (0.95 g) and benzoyl peroxide (0.021 g) 2 h later and continually reflux the mixture for 3 h.
13. Cool the mixture to room temperature.
14. Transfer the solution into another flask and concentrate in vacuo to afford crude 6-(bromomethyl)-2-(2-nitrophenyl)thiazolo[5,4-*b*]pyridine (4.0 g).

***tert*-Butyl 4-((2-(2-nitrophenyl)thiazolo[5,4-*b*]pyridin-6-yl)methoxy)piperidine-1-carboxylate (25)**

15. Add 15 mL of toluene and 5.4 mL of water to a mixture of 2.73 g (13.56 mmol) of *tert*-butyl 4-hydroxypiperidine-1-carboxylate, 4.75 g (13.56 mmol) of 6-bromomethyl-2-(2-nitrophenyl)thiazolo[5,4-b]pyridine (**24**), 1 230 mg (0.678 mmol) of tetrabutylammonium bisulfate, and 5.42 g (135.6 mmol) of sodium hydroxide. Stir the reaction at ambient temperature at a rate sufficient to mix the two layers well.
16. After 15 h, dilute the reaction with 100 mL of 1 M HCl and then extract with ethyl acetate (3 × 25 mL).
17. Back-extract the combined ethyl acetate layers with water (1 × 25 mL), and brine (1 × 25 mL), dry over $MgSO_4$, filter, and concentrate to an orange oil. Purify this via medium pressure silica gel chromatography (240 g prepacked column), eluting with an isocratic mixture of 50% ethyl acetate: heptanes. It is expected that *tert*-butyl 4-hydroxypiperidine-1-carboxylate and the product coelute.

18. Pool and concentrate the product-containing fractions and take up the crude product in 20 mL of hot ethyl acetate and dilute with an equal volume of heptanes. The product should crystallize to give a yellow solid; the expected total yield is ~4.15 g (65%). Characterization properties: ^{1}H NMR (300 MHz, DMSO-d_6): δ 8.71 (d, J=1.9 Hz, 1 H), 8.41 (d, J=1.9 Hz, 1 H), 8.14 (dd, J=7.5, 2.3 Hz, 1 H), 8.03 (dd, J=7.5, 2.3 Hz, 1 H), 7.90 (m, 2 H), 4.75 (s, 2 H), 3.65 (m, 3 H), 3.06 (m, 2 H), 1.87 (m, 2 H), 1.46 (m, 2 H), 1.40 (s, 9 H); MS m/z=471 (M+H)$^+$.

tert-Butyl 4-((2-(2-aminophenyl)thiazolo[5,4-b]pyridin-6-yl)methoxy)piperidine-1-carboxylate (26)

19. To a solution of 5.00 g (10.6 mmol) of tert-butyl 4-((2-(2-nitrophenyl)thiazolo[5,4-b]pyridin-6-yl) methoxy)piperidine-1-carboxylate (25) in 50 mL of THF add 12.5 mL of water, then 3.00 g (53.5 mmol) of iron powder, and 3.2 g (53.3 mmol) of glacial acetic acid. Heat the mixture at 50°C for 2–4 h, monitoring it by HPLC to determine when the reaction is complete.
20. Next, filter the mixture through diatomaceous earth and wash the filter cake with CH$_2$Cl$_2$ (4 × 50 mL).
21. Separate the phases and then extract the organic phase with saturated NaHCO$_3$ (aq.) (1 × 50 mL) and brine (1 × 50 mL). Dry over Na$_2$SO$_4$, filter, and concentrate to 4.1 g (88%) of a yellow solid. Characterization properties: ^{1}H NMR (300 MHz, DMSO-d_6): δ 8.53 (d, J=1.9 Hz, 1 H), 8.30 (d, J=1.9 Hz, 1 H), 7.64 (dd, J=8.0, 1.3 Hz, 1 H), 7.41 (s, 2 H), 7.25 (m, 1 H), 6.90 (dd, J=8.3, 0.8 Hz, 1 H), 6.66 (m, 1 H), 4.70 (s, 2 H), 3.65 (m, 3 H), 3.07 (m, 2 H), 1.86 (m, 2 H), 1.46 (m, 2 H), 1.40 (s, 9 H); MS m/z=441 (M+H)$^+$.

2-Butyl-6-chloropyrimidine-4-carbonyl chloride (27)

22. To 21.0 g (100 mmol) of diethyl oxaloacetate sodium salt add 80 mL of water.
23. Add 16 mL (100 mmol) of 6.25 M NaOH over 1 min at ambient temperature to the stirred suspension, and stir the mixture for 10 min, until only traces of undissolved oxaloacetate ester remain, giving an orange solution.
24. Add a solution of 13.9 g (100 mmol) of n-pentanamidine hydrochloride in 20 mL of water to this mixture.
25. Monitor the reaction with a pH meter and add additional 6.25 M NaOH as necessary to keep the pH between 10 and 11.

26. After stirring at ambient temperature for 22 h, cool the mixture using an ice bath and then add 12 M HCl until pH 2.0, yielding a white precipitate. Filter this precipitate, wash with 50 mL of water, and then dry on the filter for 1 h.
27. Suspend the white solid in 50 mL of heptanes and distill at 1 bar with a Dean-Stark trap until no more water is collected in the trap.
28. Cool the suspension, filter, and dry on the filter to give 8.13 g (41%) of 2-butyl-6-hydroxypyrimidine-4-carboxylic acid as a white solid. Characterization properties: ^{1}H NMR (DMSO-d_6): δ 13.37 (br s, 1 H), 12.82 (br s, 1 H), 6.69 (s, 1 H), 2.56 (t, $J=7.7$ Hz, 2 H), 1.64 (m, 2 H), 1.31 (m, 2 H), 0.89 (t, $J=7.3$ Hz, 3 H); MS $m/z=197$ $(M+H)^+$.
29. Add 35 mL of phosphorus oxychloride to 5.00 g (25.48 mmol) of 2-butyl-6-hydroxypyrimidine-4-carboxylic acid. Stir the reaction at 105°C for 1 h and then concentrate in vacuo.
30. Suspend the dark residue in 50 mL of heptanes and then concentrate in vacuo to remove most of the remaining phosphorus oxychloride.
31. Next, suspend the residue in 100 mL of heptanes and extract with water (3 × 25 mL) and then brine (1 × 25 mL). Dry the organic layer over MgSO$_4$, filter, and concentrate in vacuo to give 5.1 g (86%) of acid chloride **27** as an amber oil.

tert-Butyl 4-((2-(2-(2-butyl-6-chloropyrimidine-4-carboxamido) phenyl)thiazolo[5,4-b]pyridin-6-yl)methoxy)piperidine-1-carboxylate (28)

32. To a solution of 795 mg (1.80 mmol) of *tert*-butyl 4-((2-(2-aminophenyl)thiazolo[5,4-b]pyridin-6-yl)methoxy)piperidine-1-carboxylate (**26**) in 8 mL of CH$_2$Cl$_2$ add 0.45 mL (3.2 mmol) of triethylamine, followed by a solution of 505 mg (2.17 mmol) of 2-butyl-6-chloropyrimidine-4-carbonyl chloride (**27**) in 2 mL of CH$_2$Cl$_2$.
33. After 1.25 h, dilute the reaction with 30 mL of methanol to give a crystalline precipitate.
34. Filter the precipitate, wash with 15 mL of methanol, and dry on the filter to give 1.04 g of **28** (91%) as colorless needles. Characterization properties: ^{1}H NMR (300 MHz, CDCl$_3$): δ 13.42 (s, 1 H), 8.98 (dd, $J=8.4$, 1.0 Hz, 1 H), 8.60 (d, $J=1.9$ Hz, 1 H), 8.35 (d, $J=1.9$ Hz, 1 H), 8.10 (s, 1 H), 7.93 (dd, $J=7.9$, 1.4 Hz, 1 H), 7.59 (m, 1 H), 7.30 (td, $J=1.0$, 7.6 Hz, 1 H), 4.75 (s, 2 H), 3.87 (m, 2 H), 3.66 (m, 1 H), 3.18 (t, $J=7.8$ Hz, 2 H), 3.08 (m, 2 H), 1.90 (m, 4 H), 1.60 (m, 2 H), 1.47 (s, 9 H), 1.41 (m, 2 H), 0.90 (t, $J=7.3$ Hz, 1 H).

***tert*-Butyl 4-((2-(2-(6-(1-(*tert*-butoxycarbonyl)piperidin-4-ylamino)-2-butylpyrimidine-4-carboxamido)phenyl)thiazolo [5,4-*b*]pyridin-6-yl)methoxy)piperidine-1-carboxylate (29)**

35. To a mixture of 709 mg (1.11 mmol) of *tert*-butyl 4-((2-(2-(2-butyl-6-chloropyrimidine-4-carboxamido) phenyl)thiazolo[5,4-*b*]pyridin-6-yl)methoxy)piperidine-1-carboxylate and 267 mg (1.33 mmol) of *tert*-butyl 4-aminopiperidine-1-carboxylate add 7 mL of DMSO and 0.40 mL (2.24 mmol) of *N*,*N*-diisopropyl-*N*-ethylamine. Heat the mixture at 100°C under N_2 for 4 h.

36. Monitor the reaction by ^{1}H NMR, taking an aliquot of the reaction and dissolving it in $CDCl_3$, then observing the resonances in the region downfield of 6 ppm. After the starting material has been consumed, dilute the reaction with 35 mL of water to give a granular precipitate.

37. Filter this precipitate and wash with 25 mL of water to give a light tan solid. Recrystallize the crude solid from 30 mL of isopropanol, and wash with 60 mL of cold isopropanol to give 674 mg (76%) of a light yellow solid. Characterization properties: ^{1}H NMR (300 MHz, $CDCl_3$): δ 13.3 (br s, 1 H), 8.98 (d, *J*=8.4 Hz, 1 H), 8.59 (d, *J*=1.7 Hz, 1 H), 8.38 (d, *J*=1.6 Hz, 1 H), 7.90 (dd, *J*=7.8, 1.2 Hz, 1 H), 7.56 (td, *J*=7.3, 1.2 Hz, 1 H), 7.26 (m, 1 H), 7.11 (s, 1 H), 5.10 (br s, 1 H), 4.74 (s, 2 H), 4.09 (m, 2 H), 3.83 (m, 2 H), 3.64 (m, 1 H), 3.12 (m, 2 H), 2.94 (m, 4 H), 2.06 (m, 2 H), 1.91 (m, 2 H), 1.82 (quintet, *J*=7.6 Hz, 2 H), 1.62 (m, 4 H), 1.47 (s, 18 H), 1.41 (m, 3 H), 0.90 (t, *J*=7.3 Hz, 3 H); MS *m/z*=802 $(M+H)^+$.

2-Butyl-6-(piperidin-4-ylamino)-N-(2-(6-((piperidin-4-yloxy) methyl)thiazolo[5,4-*b*]pyridin-2-yl)phenyl)pyrimidine-4-carboxamide (2):

38. To 600 mg (0.750 mmol) of *tert*-butyl 4-((2-(2-(6-(1-(*tert*-butoxycarbonyl)piperidin-4-ylamino)-2-butylpyrimidine-4-carboxamido) phenyl)thiazolo[5,4-*b*]pyridin-6-yl)methoxy)piperidine-1-carboxylate (29) slowly (to control the vigorous evolution of gas) add 5 mL of trifluoroacetic acid.

39. Stir the reaction for 10 min at ambient temperature and then remove the solvent in vacuo at 50°C.

40. Dilute this residue with 12 mL of saturated $NaHCO_3$, to give an oily suspension with pH 8, and then stir and heat at 80°C for 60 min, prior to cooling to ambient temperature, affording a pale yellow precipitate.

41. Filter the precipitate, then suspend in water, and filter again (3 × 20 mL). Dry the wet solid by suction on the filter for 18 h to give

364 mg (81%) of a pale yellow solid (**2**). Characterization properties: ^{1}H NMR (300 MHz, CDCl$_3$): δ 13.28 (s, 1 H), 8.98 (d, $J=8.3$ Hz, 1 H), 8.60 (d, $J=1.6$ Hz, 1 H), 8.40 (d, $J=1.6$ Hz, 1 H), 7.90 (dd, $J=7.8$, 1.1 Hz, 1 H), 7.56 (m, 1 H), 7.26 (m, 1 H [CHCl$_3$ overlap]), 7.11 (s, 1 H), 5.23 (br s, 1 H), 4.74 (s, 2 H), 3.81 (br s, 1 H), 3.56 (m, 1 H), 3.14 (m, 4 H), 2.93 (t, $J=7.7$ Hz, 2 H), 2.77 (m, 2 H), 2.65 (m, 2 H), 2.04 (m, 4 H), 1.82 (m, 4 H [H$_2$O overlap]), 1.50 (m, 6 H), 0.90 (t, $J=7.3$ Hz, 3 H); MS $m/z=301$ (M+2H)$^{2+}$, 601 (M+H)$^{+\cdot}$

3.1.3 Preparation of an Imidazo[4,5-c]pyridine STAC (Synthesis Adapted from Hubbard, Gomes, et al., 2013)

See Fig. 3.

4-Amino-5-nitro-nicotinic acid ethyl ester (32)

1. Add potassium nitrate (20.5 g, 200 mmol) to a stirred solution of **31** (27.6 g, 200 mmol) in concentrated H$_2$SO$_4$ (200 mL) at 0°C.
2. Stir the resulting mixture at 0°C for 30 min and then at 75°C for 3 h. Cool the reaction to ambient temperature and then add EtOH (540 mL).
3. Stir the resulting mixture at 60°C for 18 h and then slowly add it to an ice-cold solution of potassium acetate (800 g, in 1.5 L of water).

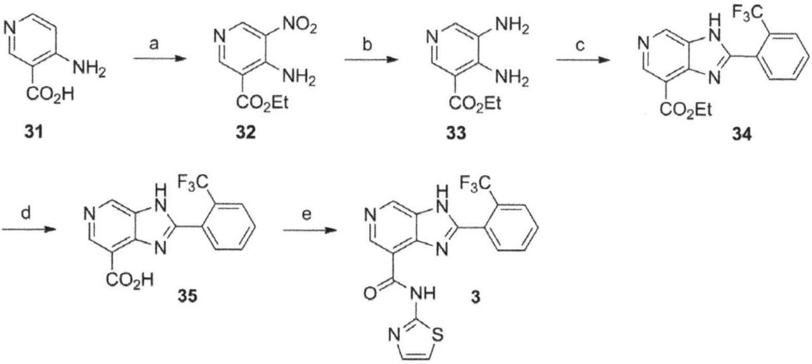

Fig. 3 Schematic outlining the synthesis of a imidazo[4,5-c]pyridine STAC. Compounds **31–35** represent intermediate products leading toward the production of the STAC *N*-(thiazol-2-yl)-2-(2-trifluoromethyl-phenyl)-3H-imidazo[4,5-c]pyridine-7-carboxamide (**3**). Special reagents and conditions noted: (a) KNO$_3$, H$_2$SO$_4$, 0–75°C, then add EtOH, RT to 60°C (35%); (b) H$_2$, 10% Pd/C, MeOH (80%); (c) 3-trifluoromethylbenzaldehyde, Na$_2$S$_2$O$_5$, DMF, 120°C (70%); (d) 10% aq. NaOH, EtOH, reflux, then 5 N HCl (85%); (e) thiazol-2-amine, HATU, (*i*-Pr)$_2$NEt, DMF, RT (12%). *Adapted from Hubbard, B. P., Gomes, A. P., Dai, H., Li, J., Case, A. W., Considine, T., ..., Sinclair, D. A. (2013). Evidence for a common mechanism of SIRT1 regulation by allosteric activators. Science, 339(6124), 1216–1219.*

4. Collect the resulting precipitate by filtration, wash with water, and dry over Na_2SO_4 to afford **32** (14.6 g, 35%), for direct use in step 5. Characterization properties: ^{1}H NMR (CDCl$_3$): δ 9.31(s, 1 H), 9.07 (s, 1 H), 9.05 (br, 1 H), 8.28 (br, 1 H), 4.42 (q, 2 H), 1.43 (t, 3 H); MS m/z = 211.92 (M+H)$^+$.

4,5-Diamino-nicotinic acid ethyl ester (33)

5. Stir a mixture of **32** (15 g, 71 mmol) and 10% Pd/C (500 mg) in MeOH (500 mL under 1 atm. of hydrogen at ambient temperature for 18 h).
6. Filter the resulting mixture through celite® and concentrate to afford **33** (10 g, 80%) to be used in the following step without additional purification. Characterization properties: ^{1}H NMR (CDCl$_3$): δ 8.62 (s, 1 H), 7.96 (s, 1 H), 6.14 (br, 2 H), 4.36 (q, 2 H), 3.15 (br, 2 H), 1.40 (t, 3 H).

2-(2-Trifluoromethyl-phenyl)-3H-imidazo[4,5-c]pyridine-7-carboxylic acid ethyl ester (34)

7. Stir a mixture of **33** (1.81 g, 10 mmol), trifluoromethylbenzaldehyde (1.9 g, 11 mmol), and $Na_2S_2O_5$ (9.5 g, 50 mmol) in DMF (50 mL) at 120°C for 18 h. Cool the resulting mixture to ambient temperature and pour into cold water (100 mL).
8. Filter the resulting solid, wash with water, and dry in vacuo to provide **34** (2.35 g, 70%).

2-(2-Trifluoromethyl-phenyl)-3H-imidazo[4,5-c]pyridine-7-carboxylic acid (35)

9. Heat at reflux for 30 min a solution of **34** (2.35 g, 7 mmol) in 10% aqueous NaOH (40 mL) and ethanol (20 mL). Allow the mixture to cool to ambient temperature and acidify with 5 N aqueous HCl.
10. Filter, wash with water, and dry the resulting yellow solid in vacuo to afford **35** (1.77 g, 85%). Characterization parameters: ^{1}H NMR (DMSO-d_6): δ 13.4 (br, 1 H), 9.18 (s, 1 H), 8.87 (s, 1 H), 7.95 (m, 1 H), 7.82 (m, 3 H); MS m/z = 308.0 (M+H)$^+$.

N-(Thiazol-2-yl)-2-(2-trifluoromethyl-phenyl)-3H-imidazo[4,5-c]pyridine-7-carboxamide (3)

11. Stir a mixture of **35** (64.0 mg, 0.21 mmol), HATU (160 mg, 0.42 mmol), DIPEA (70 µL, 0.42 mmol), and thiazol-2-amine (21.0 mg, 0.21 mmol) in DMF (2 mL) at room temperature for 18 h.
12. Remove the solvent in vacuo and purify the residue by chromatography (CH_2Cl_2/MeOH 50:1 to 5:1 gradient) to give **3** (10 mg, 12%) as a pale yellow solid. Characterization parameters: ^{1}H NMR (CH$_3$OD): δ 9.1(s, 2 H), 8.00 (d, 1 H), 7.94 (d, 1 H), 7.90–7.85 (m, 2 H), 7.49 (d, 1 H), 7.22 (d, 1 H); HRMS calculated for $C_{17}H_{11}N_5OSF_3$: 390.0634; found: 390.0635.

3.1.4 Preparation of a Bridged-Urea STAC (Synthesis Adapted from Dai et al., 2015)

See Fig. 4.

(S)-Dimethyl 2-((6-chloro-3-nitropyridin-2-yl)amino)succinate (43):

1. To a 2-L flask equipped with a thermometer, a reflux condenser, and a mechanical stirrer add 2,6-dichloro-3-nitropyridine (**41**; 100 g, 0.52 mol), (S)-aspartic acid dimethyl ester hydrochloride (**42**; 205 g, 1.04 mol), NaHCO$_3$ (174 g, 2.07 mol), and tetrahydrofuran (1 L).
2. Stir the reaction at 40°C for 16 h and monitor for the disappearance of 2,6-dichloropyridine by HPLC.
3. Following completion of the reaction, filter away solids and wash with ethyl acetate (3 × 300 mL). Concentrate the combined filtrate and washings to dryness, and take up the residue in 1 L of ethyl acetate.

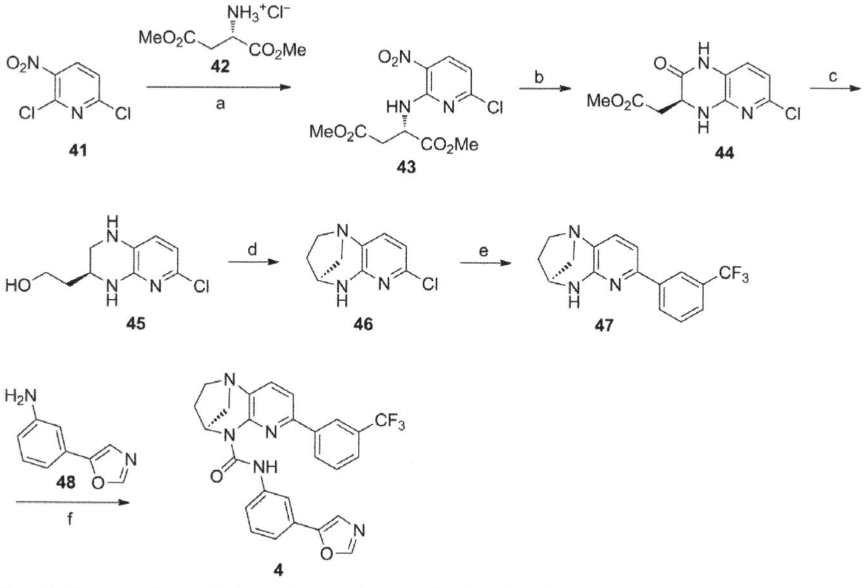

Fig. 4 Schematic outlining the synthesis of a bridged-urea STAC. Compounds 41–48 represent intermediate products leading toward the production of the STAC (4S)-N-(3-(oxazol-5-yl)phenyl)-7-(3-(trifluoromethyl)phenyl)-3,4-dihydro-1,4-methanopyrido[2,3-b][1,4]diazepine-5(2H)-carboxamide (**4**). Special reagents and conditions noted: (a) NaHCO$_3$, THF, 40°C (quantitative); (b) Fe, AcOH, i-PrOH, H$_2$O, 40–70°C (68%); (c) LiAlH$_4$, THF, 0°C to reflux (81%); (d) POCl$_3$, Et$_3$N, CH$_2$Cl$_2$, 0°C to RT (66%); (e) 3-trifluoromethylphenylboronic acid, Pd(OAc)$_2$, X-Phos, Cs$_2$CO$_3$, dioxane, H$_2$O, reflux (81%); (f) triphosgene, Et$_3$N, CH$_2$Cl$_2$, reflux, then add 3-(oxazol-5-yl)aniline (**48**), reflux (63%). *Adapted from Dai, H., Case, A. W., Riera, T. V., Considine, T., Lee, J. E., Hamuro, Y., ..., Ellis, J. L. (2015). Crystallographic structure of a small-molecule SIRT1 activator-enzyme complex. Nature Communications, 6, 7645.*

4. Stir the solution with charcoal (200 g) at ambient temperature for 2 h, filter away the charcoal, and wash with additional ethyl acetate (3 × 200 mL).
5. Concentrate the combined filtrate and washings in vacuo to obtain crude (S)-dimethyl 2-((6-chloro-3-nitropyridin-2-yl)amino)succinate (**43**; 180 g, >100%) as a yellow oil. Characterization properties: ^{1}H NMR (300 MHz, DMSO-d_6): δ 9.00 (d, $J=7.9$ Hz, 1 H), 8.50 (d, $J=8.6$ Hz, 1 H), 6.92 (d, $J=8.6$ Hz, 1 H), 5.23 (m, $J=5.7, 7.9$ Hz, 1 H), 3.67 (s, 3 H), 3.63 (s, 3 H), 3.06 (m, $J=5.8$ Hz, 2 H); ^{13}C NMR (APT) (75 MHz, DMSO-d_6): δ 170.93 (C), 170.65 (C), 154.65 (C), 150.59 (C), 138.82 (CH), 127.28 (C), 112.81 (CH), 52.23 (CH$_3$), 51.74 (CH$_3$), 50.20 (CH), 35.31 (CH$_2$); MS $m/z=318.0$ (M+H)$^+$; HRMS calculated for C$_{11}$H$_{13}$N$_3$O$_6$Cl: 318.0493; found: 318.0492.

(S)-Methyl 2-(6-chloro-2-oxo-1,2,3,4-tetrahydropyrido[2,3-b]pyrazin-3-yl)acetate (44)

6. Charge a 5-L three-necked flask equipped with a thermometer, a reflux condenser, and a mechanical stirrer with crude (S)-dimethyl 2-((6-chloro-3-nitropyridin-2-yl)amino)succinate (**43**; 180 g, 0.52 mol), iron powder (146 g, 2.59 mol), 2-propanol (2 L), and water (700 mL).
7. Stir the mixture at 40°C and then add acetic acid (15.5 g, 0.259 mmol) at a rate sufficient to keep the internal temperature below 70°C. Stir at 70°C for 30 min, or until HPLC indicates that the reaction is complete.
8. Cool the mixture to 40°C, then add Na$_2$CO$_3$ (165 g, 1.55 mol), and stir the mixture for 1 h. Filter the solids and wash with tetrahydrofuran (3 × 500 mL).
9. Concentrate the combined filtrate, wash in vacuo, and then stir the residue in ethanol (1 L) for 12 h.
10. Filter the solid, wash with cold ethanol, and dry in vacuo to obtain (S)-methyl 2-(6-chloro-2-oxo-1,2,3,4-tetrahydropyrido[2,3-b]pyrazin-3-yl)acetate as an off-white solid (**44**; 91 g, 68%). Characterization properties: ^{1}H NMR (300 MHz, DMSO-d_6): δ 10.55 (br s, 1 H), 7.35 (br s, 1 H), 6.92 (d, $J=7.9$ Hz, 1 H), 6.57 (d, $J=7.8$ Hz, 1 H), 4.43 (m, $J=1.4, 5.1$ Hz, 1 H), 3.57 (s, 3 H), 2.79 (m, $J=5.1, 16.4$ Hz, 2 H); ^{13}C NMR (APT) (75 MHz, DMSO-d_6): δ 170.32 (C), 164.96 (C), 146.13 (C), 140.32 (C), 122.41 (CH), 119.47 (C), 111.31 (CH), 51.81 (CH), 51.39 (CH$_3$), 37.01 (CH$_2$); MS $m/z=256.0$ (M+H)$^+$; HRMS calculated for C$_{10}$H$_{11}$N$_3$O$_3$Cl: 256.0489; found: 256.0487.

(S)-2-(6-Chloro-1,2,3,4-tetrahydropyrido[2,3-b]pyrazin-3-yl)ethanol (45)

11. Charge a 5-L three-necked flask equipped with a mechanical stirrer, a reflux condenser, and a nitrogen inlet with LiAlH$_4$ (60 g, 1.58 mol). Cool the flask with an ice bath and add tetrahydrofuran (500 mL).
12. Cool the stirred mixture to 0°C and then add a solution of (S)-methyl 2-(6-chloro-2-oxo-1,2,3,4-tetrahydropyrido[2,3-b]pyrazin-3-yl)acetate (44; 81 g, 0.32 mol) in tetrahydrofuran (2 L), while keeping the internal temperature below 5°C.
13. Heat the reaction at reflux for 16 h, while monitoring for the appearance of product by HPLC. The ester reduction should occur rapidly, while the lactam reduction may require longer for complete reduction.
14. Cool the reaction to 5°C and then add water (60 mL) while keeping the internal temperature below 10°C.
15. Stir the reaction for 15 min. Add 15% (w/w) aqueous NaOH (60 mL) while keeping the internal temperature below 5°C.
16. Stir the reaction for 15 min, then add water (180 mL), and stir at ambient temperature for 1 h.
17. Filter off and wash the solids with tetrahydrofuran (3 × 150 mL). Concentrate the filtrate, wash, in vacuo, and dry the solid residue in vacuo to obtain (S)-2-(6-chloro-1,2,3,4-tetrahydropyrido[2,3-b]pyrazin-3-yl)ethanol as a brown solid (45; 55 g, 81%). Characterization properties: ^{1}H NMR (300 MHz, DMSO-d_6): δ 6.60 (br s, 1 H), 6.58 (d, $J=7.8$ Hz, 1 H), 6.32 (d, $J=7.8$ Hz, 1 H), 5.69 (m, 1 H), 4.57 (t, $J=5.0$ Hz, 1 H), 3.56 (m, $J=5.8$ Hz, 2 H), 3.47 (m, 1 H), 3.22 (m, $J=2.7$, 11.1 Hz, 1 H), 2.84 (m, $J=1.6$, 6.7, 11.1 Hz, 1 H), 1.65 (m, $J=6.7$ Hz, 1 H), 1.54 (m, $J=6.3$ Hz, 1 H); ^{13}C NMR (APT) (75 MHz, DMSO-d_6): δ 146.75 (C), 134.44 (C), 128.20 (C), 118.97 (CH), 110.59 (CH), 57.97 (CH$_2$), 47.47 (CH), 43.99 (CH$_2$), 36.60 (CH$_2$); MS $m/z = 214.1$ (M+H)$^+$; HRMS calculated for C$_9$H$_{13}$N$_3$OCl: 214.0747: found: 214.0743.

(4S)-7-Chloro-2,3,4,5-tetrahydro-1,4-methanopyrido[2,3-b][1,4]diazepine (46):

18. Add triethylamine (95 g, 0.936 mol) to a solution of (S)-2-(6-chloro-1,2,3,4-tetrahydropyrido[2,3-b]pyrazin-3-yl)ethanol (45; 50 g, 0.234 mol) in CH$_2$Cl$_2$ (500 mL). Stir the mixture at ambient temperature until it is homogeneous, then cool to 0°C.

19. Add POCl₃ (54 g, 0.351 mol) dropwise to the reaction mixture while maintaining the temperature between 0 and 5°C. Remove cooling and stir the reaction at ambient temperature for 2 h while monitoring for the disappearance of the starting alcohol by HPLC.
20. After the reaction is complete, add 1.2 M aqueous NaHCO₃ (200 mL). Separate the layers and extract the aqueous layer with CH₂Cl₂.
21. Extract the combined CH₂Cl₂ layers with 1 M HCl (4 × 300 mL), adjust the pH of the combined HCl layers to 8 with solid NaHCO₃. Extract the resulting mixture with CH₂Cl₂ (4 × 300 mL) and dry (Na₂SO₄) this set of CH₂Cl₂ layers, filter, and treat with charcoal (50 g).
22. Stir the mixture at ambient temperature for 3 h, filter, and wash the charcoal with CH₂Cl₂ (200 mL). Concentrate the combined filtrate and wash solution to dryness, and dry the solid residue in vacuo to obtain (4S)-7-chloro-2,3,4,5-tetrahydro-1,4-methanopyrido[2,3-b][1,4]diazepine as an off-white crystalline solid (**46**; 30 g, 66%). Characterization properties: ¹H NMR (300 MHz, DMSO-d_6): δ 7.47 (br d, J=4.5 Hz, 1 H), 7.09 (d, J=7.7 Hz, 1 H), 6.39 (d, J=7.7 Hz, 1 H), 3.89 (m, J=5.0 Hz, 1 H), 2.95–3.13 (m, 2 H), 2.77 (m, 2 H), 1.98 (m, J=5.0 Hz, 1 H), 1.86 (m, J=6.9 Hz, 1 H); ¹³C NMR (APT) (75 MHz, DMSO-d_6): δ 153.45 (C), 144.50 (C), 134.32 (CH), 133.19 (C), 109.73 (CH), 59.88 (CH₂), 53.07 (CH₂), 50.08 (CH), 38.38 (CH₂); MS m/z=196.1 (M+H)⁺; HRMS calculated for C₉H₁₁N₃Cl: 196.0642; found: 196.0637.

(4S)-7-(3-(Trifluoromethyl)phenyl)-2,3,4,5-tetrahydro-1,4-methanopyrido[2,3-b][1,4]di-azepine (47)

23. Heat at reflux a solution of (4S)-7-chloro-2,3,4,5-tetrahydro-1, 4-methanopyrido[2,3-b][1,4]diazepine (**46**; 5.0 g, 25.6 mmol), 3-trifluoromethylphenylboronic acid (7.3 g, 38 mmol), Pd(OAc)₂ (0.14 g, 0.63 mmol), 2-dicyclohexylphosphino-2′,4′,6′-triisopropylbiphenyl (0.61 g, 1.3 mmol), and Cs₂CO₃ (24.9 g, 76.4 mmol) in a mixture of dioxane (100 mL) and water (10 mL) for 2.5 h and cool to room temperature.
24. Filter the reaction mixture through celite® and concentrate.
25. Dilute the residue with ethyl acetate, wash the organic layer with aqueous sat. NaHCO₃, water, and brine, then dry (Na₂SO₄), and concentrate to dryness.
26. Purify by silica gel chromatography (50–100% ethyl acetate gradient in pentane) to afford (4S)-7-(3-(trifluoromethyl)phenyl)-2,3,4,5-tetrahydro-1,4-methanopyrido[2,3-b][1,4]diazepine as a white solid

(**47**; 6.31 g, 81%). Characterization properties: ^{1}H NMR (300 MHz, DMSO-d_6): δ 8.27 (s, 1 H), 8.19 (d, $J=7.44$ Hz, 1 H), 7.69 (d, $J=7.9$ Hz, 1 H), 7.64 (t, $J=7.6$ Hz, 1 H), 7.27 (d, $J=4.5$ Hz, 1 H), 7.20 (d, $J=7.7$ Hz, 1 H), 7.09 (d, $J=7.7$ Hz, 1 H), 3.93 (m, $J=2.3$ Hz, 1 H), 3.00–3.17 (m, 2 H), 2.85 (d, $J=11.2$ Hz, 1 H), 2.80 (dd, $J=2.0$, 11.2 Hz, 1 H), 1.96–2.07 (m, 1 H), 1.85–1.96 (m, 1 H); ^{13}C NMR (APT) (75 MHz, DMSO-d_6): δ 153.31 (C), 149.41 (C), 140.10 (C), 134.64 (C), 132.42 (CH), 129.74 (CH), 129.54 (CH), 129.29 (q, $J_{CF}=31.5$ Hz, C), 124.42 (q, $J_{CF}=3.7$ Hz, CH), 124.33 (q, $J_{CF}=271.9$ Hz, C), 122.41 (q, $J_{CF}=3.9$ Hz, CH), 108.35 (CH), 59.85 (CH$_2$), 53.52 (CH$_2$), 50.31 (CH), 38.30 (CH$_2$); MS m/z = 305.1 (M+H)$^+$; HRMS calculated for C$_{16}$H$_{15}$N$_3$F$_3$: 306.1218; found: 306.1219.

(4S)-N-(3-(Oxazol-5-yl)phenyl)-7-(3-(trifluoromethyl)phenyl)-3,4-dihydro-1,4-methanopyrido[2,3-b][1,4]diazepine-5(2H)-carboxamide (4)

27. To a solution of (4S)-7-(3-(trifluoromethyl)phenyl)-2,3,4,5-tetrahydro-1,4-methanopyrido[2,3-b][1,4]diazepine (**47**; 3.05 g, 10 mmol) and triphosgene (2.37 g, 8.0 mmol) in CH$_2$Cl$_2$ (30 mL) add triethylamine (4.17 mL, 30 mmol).
28. Heat the solution at reflux for 1.5 h and then add 3-(oxazol-5-yl)aniline (**48**; 2.40 g, 15 mmol) as a solid. Then heat the reaction mixture at reflux for 1 h, cool to room temperature, and dilute with CH$_2$Cl$_2$.
29. Wash the resulting organic layer with sat. aqueous NaHCO$_3$, water, and brine, then dry (Na$_2$SO$_4$), and concentrate to dryness. During the aqueous workup the by-product (1,3-bis(3-(oxazol-5-yl)phenyl) urea) will likely form a rag layer, which can be removed by filtration.
30. Purify the crude product by silica gel chromatography (0–6% MeOH gradient in CH$_2$Cl$_2$) to obtain **4** as a foam. Sonicate the foam in pentane, concentrate, and dry under high vacuum to obtain (4S)-N-(3-(oxazol-5-yl)phenyl)-7-(3-(trifluoromethyl)phenyl)-3,4-dihydro-1,4-methanopyrido[2,3-b][1,4]diazepine-5(2H)-carboxamide as a free-flowing white solid (**4**; 3.08 g, 63%). Characterization properties: ^{1}H NMR (300 MHz, DMSO-d_6): δ 12.96 (s, 1 H), 8.46 (s, 1 H), 8.26 (d, $J=7.7$ Hz, 1 H), 8.20 (m, 1 H), 7.93 (m, 1 H), 7.90 (d, $J=7.9$ Hz, 1 H), 7.82 (t, $J=7.7$ Hz, 1 H), 7.72 (d, $J=7.9$ Hz, 1 H), 7.65 (d, $J=7.9$ Hz, 1 H), 7.61 (s, 1 H), 7.38–7.46 (m, 3 H), 5.51 (dd, $J=2.9$, 5.7 Hz, 1 H), 3.05–3.25 (m, 3 H), 2.98 (dd, $J=3.2$, 12.0 Hz, 1 H), 2.19–2.34 (m, 1 H), 1.91–2.00 (m, 1 H); ^{13}C NMR (APT)

(75 MHz, DMSO-d_6): δ 151.77 (C), 151.26 (CH assigned based on HSQC), 150.23 (C), 148.81 (C), 148.62 (C), 139.36 (C), 139.19 (C), 137.24 (C), 135.89 (CH), 130.65 (CH), 130.26 (CH), 129.91 (q, $J_{CF}=31.7$ Hz, C), 129.59 (CH), 128.05 (C), 125.69 (m, $J_{CF}=4.2$ Hz, CH), 123.97 (q, $J_{CF}=272.5$ Hz, C), 122.91 (m, $J_{CF}=3.9$ Hz, CH), 122.03 (CH), 119.28 (CH), 118.82 (CH), 115.59 (CH), 114.67 (CH), 58.73 (CH_2), 53.59 (CH_2), 51.66 (CH), 34.98 (CH_2); MS $m/z=491.9$ $(M+H)^+$; HRMS calculated for $C_{26}H_{21}N_5O_2F_3$: 492.1647; found: 492.1646.

3.2 Expression and Purification of Recombinant His-Tagged SIRT1

The SIRT1 expression and purification protocol below has been adapted from the previous works (Howitz et al., 2003; Hubbard, Gomes, et al., 2013; Schneider et al., 1998; Wood et al., 2004). While this procedure produces relatively crude SIRT1 protein (>75% purity), it does not require any specialized equipment (eg, fast-protein liquid chromatography (FPLC) machine), and it is both quick and cost-effective. Moreover, it can be readily adapted to incorporate FPLC or other chromatography-based approaches (starting at step #9), as previously described (Feldman, Baeza, & Denu, 2013).

Transformation of BL21 bacteria

1. Transform the pET His-SIRT1 plasmid into BL21 pLysS(DE3) bacteria according to the manufacturer's instructions and grow the transformed cells on LB-agar plates supplemented with the appropriate antibiotics overnight at 37°C.
2. Once colonies have formed (usually 14–16 h later), inoculate 10 mL of antibiotic-supplemented LB with a colony in a Falcon tube, and grow the culture overnight at 37°C.
3. The next day, inoculate a larger culture (eg, 2 L) with 1 mL of this culture and grow up in the presence of antibiotics. Take OD_{600} culture readings periodically using a spectrophotometer.
4. Once an OD of 0.6–0.8 has been reached, induce expression of His-SIRT1 by adding IPTG to a final concentration of 1 mM.
5. Immediately following addition of IPTG, transfer the flask into a shaker and grow the culture overnight (~16 h) at 16°C.

Purification of recombinant His-SIRT1

1. Supplement the lysis and wash buffers with protease inhibitor pellets (eg, Roche EDTA-free cocktail tablets—1 tablet/10 mL lysis buffer).

Approximately 15 mL of lysis buffer is needed for each liter of bacterial culture that was spun down. Allow the pellets to dissolve completely (eg, rotate for ~5 min at 4°C to dissolve) while keeping solutions chilled on ice.

2. Spin down cultures in a centrifuge capable of handling large volumes (eg, $2500 \times g$ for 20 min) at 4°C.
3. Discard the supernatant and keep the bacterial pellet on ice.
4. Add lysis buffer containing the protease inhibitors to the cell pellet and pipette the mixture up and down thoroughly to resuspend cells.
5. Allow the mixture to incubate on ice for ~30 min, until it becomes viscous. Viscosity can be monitored by drawing up the mixture in a Pasteur pipette.
6. Sonicate the mixture on ice for 30 s using a sonicator with a wide-fitted head (perform sonication in 50 mL Falcon tubes) set to 60%. Allow the sample to cool for 1 min and then repeat this process $4 \times$ (total of 5).
7. Transfer the sonicated lysate into centrifuge tubes and spin down at $27,000 \times g$ (16,000 rpm using a Sorvall SS34 rotor) for 30 min at 4°C to pellet cell debris.
8. While the sample is centrifuging, prepare the Ni-NTA resin. Aliquot approximately 1.5 mL of slurry (0.75 mL packed beads) per L of culture (see Note 6) into a 15-mL Falcon tube. Spin down the slurry (eg, $100 \times g$) and aspirate off the supernatant. Next, add 10 mL of cold lysis buffer (supplemented with protease inhibitor pellet) to the resin, invert to mix, spin down, and aspirate off the supernatant. Perform one additional wash.
9. Once centrifugation is complete, collect the sample supernatant into 50 mL Falcon tubes, add the appropriate amount of washed Ni-NTA resin, and rotate for 1 h at 4°C (eg, in a cold room).
10. Spin down the sample (eg, $100-200 \times g$) sufficiently to pellet the resin. Pipette or aspirate off the supernatant and discard.
11. Add twice the bead volume of ice-cold wash buffer (with protease inhib. pellet) to the resin, invert to resuspend, and centrifuge the resin ($100-200 \times g$). Pipette or aspirate off the supernatant and discard. Repeat this process at least four times.
12. Elute the bound protein by adding 1.5 times the bead volume of chilled elution buffer to the resin and allowing the mixture to rotate for 1 h at 4°C.
13. Spin down the beads ($100 \times g$ at 4°C) and transfer the supernatant to a Biorad Polyprep column. Allow the supernatant to filter through the column by gravity into a collector tube (eg, Falcon tube).

14. To reduce the concentration of imidazole for certain downstream applications, dialysis using Millipore Microcon columns may be performed according to the manufacturer's instructions.
15. Dilute the eluate 1:1 with glycerol (mix well), quantitate, and aliquot and store at $-20°C$.
16. To obtain a higher purity protein prep. (>90%), His-SIRT1 may be further purified by size exclusion chromatography in SEC buffer (50 mM Tris–HCl pH 7.5, 300 mM NaCl, 0.1 mM TCEP) using a Hi-load Superdex 200 16/60 column (GE LifeSciences, United States) via FPLC.

3.3 Assay of SIRT1 Activators Using the PNC1-OPT Assay

The PNC1-OPT assay affords a quick and reliable method to measure the deacetylase activity of SIRT1 without the need for any specialized equipment (Hubbard & Sinclair, 2013). Moreover, in contrast to previously developed SIRT1 assays which require the use of a fluorophore-conjugated peptide substrate, any custom peptide may be used in this assay (Hubbard & Sinclair, 2013). As outlined in Fig. 5, two distinct steps are involved in the measurement of SIRT1 activity using this assay. First, SIRT1 is incubated with acetylated peptide in the presence of β-NAD, reaction buffer, and saturating amounts of PNC1. As nicotinamide is produced from the

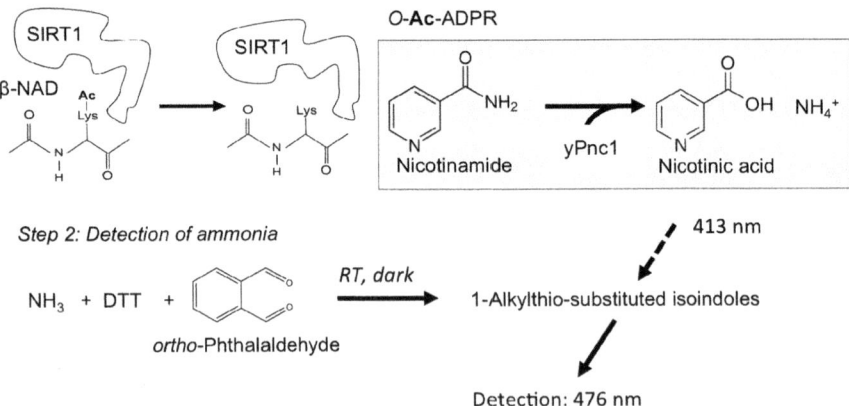

Fig. 5 Outline of the PNC1-OPT assay (Hubbard & Sinclair, 2013). In the first step, deacetylation of a custom peptide substrate by SIRT1 results in the production of nicotinamide (NAM), which is subsequently converted into nicotinic acid and ammonia by the nicotinamidase PNC1. In the second step, the reaction is quenched and ammonia is reacted with o-phthalaldehyde and dithiothreitol (DTT) (in the dark) to produce fluorescent adducts that are quantified using a spectrophotometer.

deacetylation reaction, it is converted into free ammonia (NH_3) by PNC1. Second, the reaction is stopped and the amount of ammonia present is quantified via a chemical reaction with OPT and DTT. This assay can be used to reliably study the effects of any STAC that does not interfere with the fluorescence signal or inhibit PNC1 (Hubbard & Sinclair, 2013).

Preparation of a nicotinamide standard curve

1. Thaw an aliquot of nicotinamide and perform a serial dilution to yield solutions that are $10 \times$ of the final reaction concentrations. A series of final concentrations that cover the dynamic range of the assay are 0, 5, 10, 20, 30, 40, and 50 μM. Pipette 10 μL of each of these solutions into appropriately labeled Eppendorf tubes.
2. Prepare a reaction mastermix corresponding to the total number of reactions to be performed (it is advisable to perform all reactions in triplicate). For each desired reaction, add 100 μL of assay buffer and 1 μg of PNC1 and mix the solution by gentle vortex.
3. Pipette 90 μL of the reaction mastermix into each tube containing the nicotinamide standards, mix by pipetting, and close the lid on each tube.
4. Place all of the sample tubes in a holder rack and shake using an orbital shaker at 37°C in an incubator for 1 h.
5. During the incubation period, thaw the OPT developer reagent by heating the stock at 42°C. Ensure that the solution is well mixed by vortexing, and that no DTT precipitate is present.
6. Remove the samples from the incubator, and under dim light, add 100 μL of the OPT developer reagent to each reaction as quickly as possible. Vortex all samples to mix on the highest setting for 5 s. Put the tubes back into a holder, cover all samples with aluminum foil, and incubate at room temperature on an orbital shaker for 1 h (see Note 7).
7. Transfer samples to a 96-well plate, under dim light, and read the fluorescence using a spectrophotometer with excitation and emission wavelengths set to 413 and 476 nm (Sugawara & Oyama, 1981), respectively (in practice an λ_{ex} of 420 ± 10 nm and $\lambda_{em} = 460 \pm 10$ nm work fine with a 0.1- or 1-s read time).
8. Subtract the background fluorescence (0 μM NAM) from all samples and plot a graph of normalized fluorescence vs concentration of NAM (standard curve).

Assay of SIRT1 activators

1. Thaw aliquots of β-NAD, peptide substrate, and PNC1 and SIRT1 enzymes on ice.

2. Arrange and label a series of Eppendorf tubes in a holder corresponding to *twice* the number of samples to be assayed: Label one set "+NAD" and the second set "−NAD." The latter set of samples will be used as background fluorescence control reactions (see Note 8). In addition, as stated earlier, it is advised that all samples be measured in triplicate (including the −NAD controls).
3. Pipette the various test compounds into each corresponding tube and include a vehicle control (eg, DMSO).
4. Prepare a reaction master mix corresponding to the total number of samples to be assayed (both +NAD and −NAD samples). For each reaction add the following (prepare on ice in a 15-mL Falcon tube): Reaction Buffer (100 μL minus the volume of other components), peptide substrate (typically 10–30 μM), PNC1 (1 μg), and SIRT1 purified as described in Section 3.1 (1 μg) (see Note 9). Mix by pipetting followed by gentle vortex (50% efficiency for 5 s).
5. Divide the mastermix into two Falcon tubes. To one tube add β-NAD to the appropriate final concentration (typically 100 μM) (+β-NAD mastermix), and to the second tube add an identical volume of water (for the no NAD control). Mix briefly.
6. Aliquot 100 μL of the +NAD mastermix to each experimental reaction tube, and 100 μL of the −NAD mastermix to each corresponding negative control reaction. Close the lids on each tube, and very gently vortex to mix (20% amplitude).
7. Incubate reactions at 37°C for 1 h.
9. During the incubation period, thaw the OPT developer reagent and incubate at 42°C for ∼15 min (keep covered in aluminum foil). Ensure that the solution is well mixed by vortexing, and that no DTT precipitate is observed. If particulates are visible, vortex and continue to heat.
8. Once the incubation period is complete, under dim light, quickly remove each sample tube from the holder and add 100 μL of OPT developer reagent. If available, a multichannel pipette may be used. It is imperative that the developer be added to each sample as quickly as possible to ensure consistency. Vortex all samples on the highest setting for 5 s.
9. Place the tubes back into the holder, cover all samples with aluminum foil (to prevent exposure to light), and incubate at room temperature on an orbital shaker for 1 h (see Note 7).
10. Following the development phase, remove the foil under dim light and transfer 150–200 μL from each tube into 1 well of a 96-well dark bottom plate.

11. Read the fluorescence using a spectrophotometer with excitation and emission wavelengths set to 413 and 476 nm (Sugawara & Oyama, 1981), respectively (as noted earlier, λ_{ex} of 420 ± 10 nm and $\lambda_{em} = 460 \pm 10$ nm work fine with a 0.1- or 1-s read time). Calculate the background fluorescence for each condition by taking the mean of the arbitrary fluorescence (AF) readings for the $-$NAD samples. Next, calculate the net fluorescence for each reaction condition by subtracting the mean background fluorescence from each reading, $F_{corrected} = F_{+NAD} - F_{-NADcontrol}$ (mean value). The resulting value is proportional to the amount of NAM produced during the deacetylation reaction.
12. AFU can be converted into amounts of NAM production using the linear equation obtained from the standard curve above.

3.4 Assay of SIRT1 Activators Using the RapidFire O-Ac-ADPR Detection Assay

RapidFire (Agilent) is a solid-phase extraction-based system that enables high sensitivity detection using mass spectrometry (MS) (Lim, Ozbal, & Kassel, 2010). SIRT1 activity can be assayed using RapidFire/MS technology by monitoring the production of either nicotinamide or 3′-O-Ac-ADPR following a deacetylation reaction. Since this assay does not rely on any type of fluorescence or require any additional coupling enzymes, it produces highly reliable results. Like the PNC1-OPT assay, the RapidFire detection assay can be performed using any custom native substrate, and it is amenable to medium to high-throughput compound screening. Below we outline a procedure for assaying SIRT1 activity using the RapidFire/MS system to detect production of O-Ac-ADPR.

Detection parameters and O–Ac–ADPR standard curve

1. Optimize MS parameters for the detection of O-Ac-ADPR (see Note 10).
2. Thaw an aliquot of O-Ac-ADPR and perform dilutions in Reaction buffer supplemented with 0.05% BSA and 4 mM β-NAD to yield 100 µL solutions that cover the dynamic range of the assay. All standards should be run in triplicate.
3. Mix 10 µL of each sample with 40 µL of a 1:1 acetonitrile:methanol solution.
4. Run the samples on the Agilent RapidFire system coupled to a mass spectrometer (eg, ABSciex API 4000) fitted with an electrospray ionization source: aspirate samples for 250 ms using the RapidFire system and

absorb 10 μL of each sample onto an SPE cartridge with Buffer A (3 s). Elute in Buffer B for 5 s.
5. Integrate peak data using the RapidFire Integrator software.
6. Subtract the background signal from all samples, plot a graph of peak area vs O-Ac-ADPR concentration, and fit the standard curve using linear regression.

Assay of SIRT1 activators
1. Thaw aliquots of β-NAD and peptide substrate, and SIRT1.
2. Dispense 1 μL of each test compound dissolved in DMSO into one well of a 96- or 384-well microtiter plate. In addition, prepare a vehicle-only control reaction (eg, DMSO) (in triplicate).
3. Add 50 μL of Reaction buffer supplemented with 0.05% BSA and SIRT1 (at a final concentration of 5 nM) to each well.
4. Incubate the enzyme-compound mixtures for 20 min at 25°C before starting the reaction with substrates.
5. Prepare a mastermix containing 2 × the final concentration of β-NAD and peptide substrate (see Note 11) in Reaction buffer supplemented with 0.05% BSA, and pipette 50 μL of this master mix into each reaction tube.
6. Allow the reaction to proceed for 30 min at 25°C (room temperature).
7. Add Stop reagent to each reaction (final concentrations should be ~1% formic acid and 5 mM NAM).
8. Dilute the quenched reactions fivefold with a solution of 1:1 acetonitrile:methanol.
9. Spin down samples at 5000 × g for 10 min to precipitate protein.
10. Run the samples on the Agilent RapidFire as described earlier: Aspirate samples for 250 ms and absorb 10 μL of each sample onto an SPE cartridge with Buffer A (3 s), then elute in Buffer B for 5 s.
11. Reequilibrate the system using Buffer A for 500 ms.
12. Use the standard curve produced above to calculate the amount of O-acetyl-adenosine diphosphate ribose (O-Ac-ADPR) produced.

4. NOTES

1. The protease inhibitor cocktail must be EDTA-free since EDTA is incompatible with Ni-NTA resin.
2. In order to minimize background due to peptide fluorescence, peptides that are shorter are preferred (aromatic groups also increase background fluorescence). Also, only certain peptides support STAC-mediated

SIRT1 activation, such as those with hydrophobic groups adjacent to the acetyl-lysine (Hubbard, Gomes, et al., 2013), so these are recommended for studying SIRT1 activation.

3. Analytical HPLC can be performed on an Agilent 1100 Series HPLC machine equipped with a 3.5 μm Eclipse XDB-C18 (4.6 mm × 100 mm) column using the following parameters: CH_3CN/H_2O, modified with 0.1% formic acid mobile phase, Gradient elution: 5% CH_3CN hold (2 min), 5–95% CH_3CN gradient (11 min), 95–5% CH_3CN gradient (0.3 min), 5% CH_3CN hold (2.7 min), 15 min total run time with a flow rate of 0.8 mL/min.

4. Proton NMR spectra may be obtained using any standard spectrometer (eg, Bruker Advance III 300 MHz spectrometer) and should be referenced to internal TMS (0.00 ppm), $CHCl_3$ (7.26 ppm), or DMSO (2.49 ppm).

5. HRMS can be completed on a Waters qTOF Premiere Mass Spectrometer operating in W mode positive ionization with a resolving power of approximately 15,000. A Waters nano-acquity LC may be used for flow injection.

6. To facilitate pipetting of the Ni-NTA resin, cut several mm off the pipette tip using scissors or a razor blade.

7. The length of the incubation period may need to be adjusted depending on signal strength. However, it is important to use a consistent time for all samples to minimize variance.

8. A background control reaction is performed to subtract out inherent substrate or small-molecule fluorescence and also NAD^+-independent deacetylase activity. Subtracting the AFU reading of the −NAD samples from the +NAD samples yields fluorescence values that are proportional to the amount of NAM produced ($F_{corrected} = F_{+NAD} - F_{-NADcontrol}$). Importantly, β-NAD also exhibits fluorescence in this assay at high concentrations (~200 μM). In these instances, control reactions lacking enzyme rather than NAD may be more appropriate. An alternative approach is to perform a parallel set of reactions using the corresponding nonacetylated peptide and use these as background controls.

9. The quantities of PNC1 and SIRT1 that are added to each reaction may vary depending on the specific activity of each preparation.

10. To optimize MS conditions use a direct infusion of O-Ac-ADPR in a 1:1 ethanol:water mixture containing 0.1% formic acid at a rate of 10 μL/min. MS parameters: curtain gas 20, probe temperature 550°

C, ion source gas 150, ion source gas 250, interface heater on, collision gas 10, ion source voltage −3800 V, declustering potential −85 V, entrance potential −10 V, collision energy −37 V, and collision exit potential −20 V. Negative MRM mode was used monitoring the transition 600.1/345.9 for the parent/daughter ion under low-resolution conditions.

11. Peptide substrate concentrations of approximately 1/10th of their K_m value should be used when assaying K_m-modulating activators.

ACKNOWLEDGMENTS

The authors thank Dr. W.H. Miller for critically reading the manuscript and providing suggestions for improvement. D.S. is supported by an NIA/NIH MERIT award 5R37AG028730-09 and generous support from the Glenn Foundation for Medical Research and Edward Schulak.

REFERENCES

Bhullar, K. S., & Hubbard, B. P. (2015). Lifespan and healthspan extension by resveratrol. *Biochimica et Biophysica Acta*, *1852*(6), 1209–1218. http://dx.doi.org/10.1016/j.bbadis.2015.01.012.

Borra, M. T., Smith, B. C., & Denu, J. M. (2005). Mechanism of human SIRT1 activation by resveratrol. *Journal of Biological Chemistry*, *280*(17), 17187–17195. http://dx.doi.org/10.1074/jbc.M501250200.

Cao, D., Wang, M., Qiu, X., Liu, D., Jiang, H., Yang, N., & Xu, R. M. (2015). Structural basis for allosteric, substrate-dependent stimulation of SIRT1 activity by resveratrol. *Genes & Development*, *29*(12), 1316–1325. http://dx.doi.org/10.1101/gad.265462.115.

Chung, J. H. (2012). Using PDE inhibitors to harness the benefits of calorie restriction: Lessons from resveratrol. *Aging (Albany, NY)*, *4*(3), 144–145.

Dai, H., Case, A. W., Riera, T. V., Considine, T., Lee, J. E., Hamuro, Y., ... Ellis, J. L. (2015). Crystallographic structure of a small molecule SIRT1 activator-enzyme complex. *Nature Communications*, *6*, 7645. http://dx.doi.org/10.1038/ncomms8645.

Dai, H., Kustigian, L., Carney, D., Case, A., Considine, T., Hubbard, B. P., ... Stein, R. L. (2010). SIRT1 activation by small molecules: Kinetic and biophysical evidence for direct interaction of enzyme and activator. *Journal of Biological Chemistry*, *285*(43), 32695–32703. http://dx.doi.org/10.1074/jbc.M110.133892.

Dao, T. T., Tran, T. L., Kim, J., Nguyen, P. H., Lee, E. H., Park, J., ... Oh, W. K. (2012). Terpenylated coumarins as SIRT1 activators isolated from Ailanthus altissima. *Journal of Natural Products*, *75*(7), 1332–1338. http://dx.doi.org/10.1021/np300258u.

Feige, J. N., Lagouge, M., Canto, C., Strehle, A., Houten, S. M., Milne, J. C., ... Auwerx, J. (2008). Specific SIRT1 activation mimics low energy levels and protects against diet-induced metabolic disorders by enhancing fat oxidation. *Cell Metabolism*, *8*(5), 347–358. http://dx.doi.org/10.1016/j.cmet.2008.08.017.

Feldman, J. L., Baeza, J., & Denu, J. M. (2013). Activation of the protein deacetylase SIRT6 by long-chain fatty acids and widespread deacylation by mammalian sirtuins. *Journal of Biological Chemistry*, *288*(43), 31350–31356. http://dx.doi.org/10.1074/jbc.C113.511261.

Firestein, R., Blander, G., Michan, S., Oberdoerffer, P., Ogino, S., Campbell, J., ... Sinclair, D. A. (2008). The SIRT1 deacetylase suppresses intestinal tumorigenesis

and colon cancer growth. *PLoS One*, *3*(4), e2020. http://dx.doi.org/10.1371/journal.pone.0002020.
Gertz, M., Nguyen, G. T., Fischer, F., Suenkel, B., Schlicker, C., Franzel, B., ... Steegborn, C. (2012). A molecular mechanism for direct sirtuin activation by resveratrol. *PLoS One*, *7*(11), e49761. http://dx.doi.org/10.1371/journal.pone.0049761.
Gomes, A. P., Price, N. L., Ling, A. J., Moslehi, J. J., Montgomery, M. K., Rajman, L., ... Sinclair, D. A. (2013). Declining NAD(+) induces a pseudohypoxic state disrupting nuclear-mitochondrial communication during aging. *Cell*, *155*(7), 1624–1638. http://dx.doi.org/10.1016/j.cell.2013.11.037.
Howitz, K. T., Bitterman, K. J., Cohen, H. Y., Lamming, D. W., Lavu, S., Wood, J. G., ... Sinclair, D. A. (2003). Small molecule activators of sirtuins extend Saccharomyces cerevisiae lifespan. *Nature*, *425*(6954), 191–196. http://dx.doi.org/10.1038/nature01960.
Hubbard, B. P., Gomes, A. P., Dai, H., Li, J., Case, A. W., Considine, T., ... Sinclair, D. A. (2013). Evidence for a common mechanism of SIRT1 regulation by allosteric activators. *Science*, *339*(6124), 1216–1219. http://dx.doi.org/10.1126/science.1231097.
Hubbard, B. P., Loh, C., Gomes, A. P., Li, J., Lu, Q., Doyle, T. L., ... Sinclair, D. A. (2013). Carboxamide SIRT1 inhibitors block DBC1 binding via an acetylation-independent mechanism. *Cell Cycle*, *12*(14), 2233–2240. http://dx.doi.org/10.4161/cc.25268.
Hubbard, B. P., & Sinclair, D. A. (2013). Measurement of sirtuin enzyme activity using a substrate-agnostic fluorometric nicotinamide assay. *Methods in Molecular Biology*, *1077*, 167–177. http://dx.doi.org/10.1007/978-1-62703-637-5_11.
Hubbard, B. P., & Sinclair, D. A. (2014). Small molecule SIRT1 activators for the treatment of aging and age-related diseases. *Trends in Pharmacological Sciences*, *35*(3), 146–154. http://dx.doi.org/10.1016/j.tips.2013.12.004.
Kaeberlein, M., McDonagh, T., Heltweg, B., Hixon, J., Westman, E. A., Caldwell, S. D., ... Kennedy, B. K. (2005). Substrate-specific activation of sirtuins by resveratrol. *Journal of Biological Chemistry*, *280*(17), 17038–17045. http://dx.doi.org/10.1074/jbc.M500655200.
Kim, D., Nguyen, M. D., Dobbin, M. M., Fischer, A., Sananbenesi, F., Rodgers, J. T., ... Tsai, L. H. (2007). SIRT1 deacetylase protects against neurodegeneration in models for Alzheimer's disease and amyotrophic lateral sclerosis. *EMBO Journal*, *26*(13), 3169–3179. http://dx.doi.org/10.1038/sj.emboj.7601758.
Lakshminarasimhan, M., Rauh, D., Schutkowski, M., & Steegborn, C. (2013). Sirt1 activation by resveratrol is substrate sequence-selective. *Aging (Albany NY)*, *5*(3), 151–154.
Lim, K. B., Ozbal, C. C., & Kassel, D. B. (2010). Development of a high-throughput online solid-phase extraction/tandem mass spectrometry method for cytochrome P450 inhibition screening. *Journal of Biomolecular Screening*, *15*(4), 447–452. http://dx.doi.org/10.1177/1087057110362581.
Milne, J. C., Lambert, P. D., Schenk, S., Carney, D. P., Smith, J. J., Gagne, D. J., ... Westphal, C. H. (2007). Small molecule activators of SIRT1 as therapeutics for the treatment of type 2 diabetes. *Nature*, *450*(7170), 712–716. http://dx.doi.org/10.1038/nature06261.
Morris, B. J. (2013). Seven sirtuins for seven deadly diseases of aging. *Free Radical Biology & Medicine*, *56*, 133–171. http://dx.doi.org/10.1016/j.freeradbiomed.2012.10.525.
Pacholec, M., Bleasdale, J. E., Chrunyk, B., Cunningham, D., Flynn, D., Garofalo, R. S., ... Ahn, K. (2010). SRT1720, SRT2183, SRT1460, and resveratrol are not direct activators of SIRT1. *Journal of Biological Chemistry*, *285*(11), 8340–8351. http://dx.doi.org/10.1074/jbc.M109.088682.
Price, N. L., Gomes, A. P., Ling, A. J., Duarte, F. V., Martin-Montalvo, A., North, B. J., ... Sinclair, D. A. (2012). SIRT1 is required for AMPK activation and the beneficial effects of resveratrol on mitochondrial function. *Cell Metabolism*, *15*(5), 675–690. http://dx.doi.org/10.1016/j.cmet.2012.04.003.

Satoh, A., Brace, C. S., Rensing, N., Cliften, P., Wozniak, D. F., Herzog, E. D., ... Imai, S. (2013). Sirt1 extends life span and delays aging in mice through the regulation of Nk2 homeobox 1 in the DMH and LH. *Cell Metabolism*, *18*(3), 416–430. http://dx.doi.org/10.1016/j.cmet.2013.07.013.

Sauve, A. A., Moir, R. D., Schramm, V. L., & Willis, I. M. (2005). Chemical activation of Sir2-dependent silencing by relief of nicotinamide inhibition. *Molecular Cell*, *17*(4), 595–601. http://dx.doi.org/10.1016/j.molcel.2004.12.032.

Schneider, A., Smith, R. W., Kautz, A. R., Weisshart, K., Grosse, F., & Nasheuer, H. P. (1998). Primase activity of human DNA polymerase alpha-primase. Divalent cations stabilize the enzyme activity of the p48 subunit. *Journal of Biological Chemistry*, *273*(34), 21608–21615.

Sugawara, K., & Oyama, F. (1981). Fluorogenic reaction and specific microdetermination of ammonia. *Journal of Biochemistry*, *89*(3), 771–774.

Wood, J. G., Rogina, B., Lavu, S., Howitz, K., Helfand, S. L., Tatar, M., & Sinclair, D. (2004). Sirtuin activators mimic caloric restriction and delay ageing in metazoans. *Nature*, *430*(7000), 686–689. http://dx.doi.org/10.1038/nature02789.

Yang, J. L., Ha, T. K., Dhodary, B., Kim, K. H., Park, J., Lee, C. H., ... Oh, W. K. (2014). Dammarane triterpenes as potential SIRT1 activators from the leaves of Panax ginseng. *Journal of Natural Products*, *77*(7), 1615–1623. http://dx.doi.org/10.1021/np5002303.

CHAPTER ELEVEN

Synthesis and Assays of Inhibitors of Methyltransferases

X.-C. Cai[1], K. Kapilashrami[1], M. Luo[2]

Memorial Sloan Kettering Cancer Center, New York, NY, United States
[2]Corresponding author: e-mail address: luom@mskcc.org

Contents

1. Introduction to Methyltransferases	246
2. Designing and Synthesizing Inhibitors of Methyltransferases	249
2.1 Overview of Methyltransferase Inhibitors	249
2.2 Pan-Inhibitors	250
2.3 Target-Selective Inhibitors of Methyltransferases	253
3. Evaluating Methyltransferase Inhibitors	283
3.1 Assay Formats	283
3.2 Radiometric Assays	283
3.3 Fluorescence-Based Detection of Methyltransferase Activity	289
3.4 Directly Monitoring SAH Formation	290
3.5 Mass Spectrometry-Based Detection of Methylated Product	291
3.6 Detailed Protocol for Radiometric Filter Paper Assay	292
4. Conclusion	296
References	298

Abstract

Epigenetic regulation requires site-specific modification of the genome and is involved in multiple physiological processes and disease etiology. Methyltransferases, which catalyze the transfer of a methyl group from S-adenosyl-L-methionine (SAM) to various substrates, are critical components of the epigenetic machinery. This group of enzymes can methylate diverse substrates including DNA, RNA, proteins, and small-molecule metabolites. Their dysregulation has also been implicated in multiple disease states such as cancer, neurological, and cardiovascular disorders. Developing potent and selective small-molecule inhibitors of methyltransferases is valuable not only for therapeutic intervention but also for investigating the roles of these enzymes in disease progression. In this chapter, we will discuss the strategies of designing and synthesizing methyltransferases inhibitors based on the SAM scaffold. Following the section of inhibitor design, we will briefly review representative assays that are available to evaluate the potency of these inhibitors along with a detailed description of the most commonly used radiometric assay.

---

[1] These authors made equal contribution.

1. INTRODUCTION TO METHYLTRANSFERASES

Epigenetics involves the study of heritable phenotypic changes without alterations in the DNA sequence. The regulation of the DNA topology and chromatin structure is tightly controlled by multiple factors such as the state of DNA methylation and various histone posttranslational modifications (PTMs such as methylation, acetylation, phosphorylation, and ubiquitination) (Kouzarides, 2007). Histone PTMs modulate gene expression by either affecting chromatin contexts or recruiting diverse cellular machinery (Kouzarides, 2007). These events are elegantly regulated by many histone modifiers including PTM writers, readers, and erasers. The PTM writers such as histone acetyltransferases and methyltransferases install chemical moieties at specific positions on their substrates, while the erasers such as histone deacetylases and demethylases are responsible for removal of these marks (Arrowsmith, Bountra, Fish, Lee, & Schapira, 2012; Kouzarides, 2007). Though critical for maintenance of a normal cellular stage, many chromatin modifiers, once deregulated, can lead to a variety of diseases including cancer, diabetes, and neurological disorders (Gnyszka, Jastrzebski, & Flis, 2013; Jones & Baylin, 2007; Kelly, De Carvalho, & Jones, 2010; Swierczynski et al., 2015; Urdinguio, Sanchez-Mut, & Esteller, 2009; Villeneuve & Natarajan, 2010; Wu, Sarkissyan, & Vadgama, 2015). Therefore, targeting these epigenetic modifiers can be a potential therapeutic strategy aimed at restoring the normal chromatin state of relevant genes (Arrowsmith et al., 2012).

Methyltransferases are a class of enzymes that catalyze the transfer of a methyl group from the methyl donor S-adenosyl-L-methionine (SAM) to their substrates. These enzymes are key players in epigenetic regulation through DNA methylation, RNA methylation, mRNA maturation, and histone modification. As part of the dynamic network of multiple PTMs, methylation is critical for cellular homeostasis and normal physiology (Arrowsmith et al., 2012; Copeland, Solomon, & Richon, 2009). For instance, hypomethylation of CpG-rich regions in the pericentromeric regions of the chromosome results in genome instability and cancer progression (Ehrlich, 2002; Goelz, Vogelstein, Hamilton, & Feinberg, 1985). On the other hand, promoter hypermethylation has been linked to aberrant silencing of tumor suppressors such as the retinoblastoma protein, p16, MLH1, and E-cadherin (Greger, Passarge, Hopping, Messmer, & Horsthemke, 1989; Laird & Jaenisch, 1994; Sakai et al., 1991; Santini, Kantarjian, & Issa, 2001). Hypermethylation often results from the dysregulated activity of DNA methyltransferases (DNMTs). DNMTs including DNMT1 and DNMT3A/3B catalyze the methylation of 5-cytosine and have been shown to be overexpressed in multiple cancers including colon, prostate, breast, liver, and hematopoietic cancers (Belinsky, Nikula, Baylin, & Issa, 1996; Eldeiry et al., 1991; Melki, Warnecke, Vincent, & Clark, 1998; Oh et al., 2007; Patra, Patra, Zhao, & Dahiya, 2002; Zhao et al., 2010). Nucleoside analogs such as 5-azacytidine (Vidaza) (Silverman et al., 2002) and 5-aza-2′-deoxycytidine (Decitabine) (Kantarjian et al., 2006) are the DNMT inhibitors approved by the FDA for the treatment of myelodysplastic syndromes. These small molecules are incorporated into DNA and thus sequester the activities of the DNMTs. There are other DNMT inhibitors documented in the literature including the EGCG tea polyphenol that has been shown to restore expression of tumor suppressors such as p16, RAR-β2, and MGMT in cellular contexts (Fang et al., 2003). Another DNMT inhibitor, hydralazine, has shown efficacy in early clinical evaluation in combination with valproic acid, a histone deacetylase inhibitor, against a wide range of tumors including ovarian, lung, breast, prostate, testicular, colon cancers, melanoma, and sarcomas (Chavez-Blanco et al., 2006; Duenas-Gonzalez et al., 2014; Graca et al., 2014; Song & Zhang, 2009).

There are two major classes of protein methyltransferases (PMTs): protein arginine methyltransferases (PRMTs) and protein lysine methyltransferases (PKMTs), which methylate arginine and lysine residues of their protein substrates, respectively. So far 10 PRMTs and over 50 PKMTs have been identified in the human genome (Luo, 2012). PRMTs share a characteristic THW loop, critical for SAM binding, along with four conserved motifs (I, Post I, II, and III) (Bedford & Clarke, 2009; Luo, 2012). In mammals, PRMTs methylate the ω-guanidino nitrogen and can be further classified into three distinct subtypes according to the final methylation states of their products (Bedford & Clarke, 2009). For example, type I PRMTs (PRMT1, 2, 3, 4, 6, and 8) catalyze the formation of monomethyl arginine (MMA) and asymmetric dimethyl arginine, while type II PRMTs (PRMT5 and 9) catalyze monomethylation and symmetric dimethylation of the arginine substrate. PRMT7 is classified as a type III PRMT because of its activity of MMA for the ending product of certain targets (Bedford & Clarke, 2009). The activity of PRMT10 has not been unambiguously characterized (Wei, Mundade, Lange, & Lu, 2014; Yang & Bedford, 2013). The sites and degrees of arginine methylation have been shown to render diverse functions in a cellular context (Bedford & Clarke, 2009). Dysregulation of PRMTs has been shown to be associated with multiple diseases including cancers, cardiovascular, and pulmonary disorders (Copeland et al., 2009; Wei et al., 2014). For instance, CARM1 (PRMT4) has been shown to be overexpressed in colorectal, breast, and prostate cancers. CARM1-mediated dimethylation of H3R17 at the E2F1 promoter results in transcriptional activation in parallel with treatment with estradiol in MCF-7 breast cancer cell lines (Frietze, Lupien, Silver, & Brown, 2008). PRMT1 and PRMT6 methylate H4R3 and H3R2, respectively, and are known to promote proliferation in lung and bladder cancers (Yang & Bedford, 2013; Yoshimatsu et al., 2011). Upregulation of PRMT5 activity has been associated with colorectal and lung cancers, lymphoma, and leukemia (Cho et al., 2012; Pal et al., 2007; Wang, Pal, & Sif, 2008; Wei et al., 2012).

Most PKMTs, with the exception of DOT1L, have a conserved catalytic SET (Su(var)3–9, enhancer of zeste, trithorax) domain, which is responsible for binding the SAM cofactor and catalyzing the methylation reactions (Kouzarides, 2002; Martin & Zhang, 2005). On the other hand, DOT1L, though methylating H3K79, is structurally similar to PRMTs (Krivtsov et al., 2008; Okada et al., 2005). PKMTs can mono-, di-, or trimethylate the ε-amino group of lysine residues. Similar to the situation of PRMTs, the sites and degrees of PKMT-mediated lysine methylation can lead to various downstream outcomes. For instance, polycomb repressive complex 2 (PRC2)-mediated trimethylation of H3K27 is typically associated with transcription repression, whereas a monomethyl mark at H3K27 has been shown to be associated with active promoters (Barski et al., 2007). While lysine methylation is well known to occur on H3 and H4, methylation on H1 and H2B has also been reported (Martin & Zhang, 2005). PKMTs are known to play critical roles in both normal physiology and disease settings. For example, EZH2 (the enzymatic subunit of the PRC2 complex), which catalyzes H3K27 methylation, has been implicated in prostate, bladder, colon, skin, lung, liver, and gastric cancers (Kleer et al., 2003; Simon & Lange, 2008). SETD2 catalyzes trimethylation of H3K36, a mark that has been associated with transcription elongation and mRNA splicing (Krogan et al., 2003; Luco et al., 2010). SETD2 has been shown to be a tumor suppressor with its dysregulation implicated in multiple cancer types including breast cancer and renal cell carcinoma (Copeland, 2013a; Dalgliesh et al., 2010; de Almeida et al., 2011; Duns et al., 2010; Kanu et al., 2015; Newbold & Mokbel, 2010). NSD2, a H3K36 dimethyltransferase, is deleted in patients with Wolf–Hirschhorn syndrome for developmental defects and mental retardation (Nimura et al., 2009;

Stec et al., 1998). Moreover, NSD2 is associated with t(4:14) translocation and the resultant overexpression in multiple myeloma (Hudlebusch et al., 2011; Kim et al., 2008; Kuo et al., 2011; Lauring et al., 2008). Overexpression of NSD2 in the t(4:14) + MM cell line leads to a global increase of H3K36 methylation, a mark for transcriptionally active loci, and a concomitant decrease in the global level of EZH2/PRC2-mediated H3K27 trimethylation, the characteristic mark for transcriptional repression (Martinez-Garcia et al., 2011; Wagner & Carpenter, 2012).

In addition to their roles in transcriptional control through histone modifications, PMTs have also been shown to methylate nonhistone targets and thus modulate their functions in the context of tumorigenesis (Hamamoto, Saloura, & Nakamura, 2015; Wei et al., 2014). For example, the Gozani Lab showed that SMYD3-mediated methylation of MAP3K2 (K260) upregulates oncogenic Ras signaling (Mazur et al., 2014). There is increasing evidence that methylation of cytoplasmic targets can also have a multitude of downstream effects not only through modulating protein–protein interactions as shown for SMYD3-mediated methylation of MAP3K2 but also through alteration in protein stability (Lee et al., 2012), protein localization (Cho et al., 2012; Chuikov et al., 2004), and cross talk of methylation with other PTMs (Kogure et al., 2013; Sone et al., 2014).

It is pertinent to point out that apart from protein and DNMTs, RNA and small-molecule methyltransferases have also been implicated in multiple pathological states (Blanco et al., 2014; Frye & Watt, 2006). For example, catechol-O-methyltransferase (COMT), a key enzyme involved in dopamine metabolism, has been a key therapeutic target in Parkinson's disease (Kurth & Adler, 1998). Recently, more evidence points to the role of COMT in multiple cancer types including colorectal, pancreatic, and breast cancers (Dawling, Roodi, Mernaugh, Wang, & Parl, 2001; Wu, Wu, Hong, Xiong, et al., 2015; Wu, Wu, Hong, Zhou, et al., 2015). While COMT does not fall within the class of epigenetic regulators, the strategy for designing SAM-based inhibitors targeting COMT activity is comparable to that employed for protein and DNMTs and will be briefly discussed in this chapter.

2. DESIGNING AND SYNTHESIZING INHIBITORS OF METHYLTRANSFERASES

2.1 Overview of Methyltransferase Inhibitors

Many PMTs consist of highly conserved SAM-binding pockets and less structured substrate-binding regions (Martin & McMillan, 2002). The

SAM-binding pocket of PKMTs and PRMTs is accessible from one face, and the substrate-binding regions account for the recruitment of lysine or arginine substrates at the opposite face. These two binding sites are often linked through a narrow channel to allow the cofactor (SAM, **1**) and substrate to approach in a proximity to allow the transfer of a methyl group from the cofactor to the substrate via an S_N2-type mechanism (Dalhoff & Weinhold, 2008; Fontecave, Atta, & Mulliez, 2004; Loenen, 2006; Smith & Denu, 2009; Struck, Thompson, Wong, & Micklefield, 2012). The nitrogen atom of lysine or arginine substrates donates their lone pair electrons to the sulfonium methyl group and likely forms a classical pentacoordinate carbon transition state, followed by the release of a methylated lysine or arginine substrate with S-adenosyl-L-homocysteine (SAH or AdoHcy, **2**) as a by-product (Fig. 1). Despite the homology of catalytic sites of many methyltransferases, the difference of SAM-binding modes, transitional state structures, and catalytic mechanisms has rendered it feasible to design target-selective inhibitors (Copeland, 2012; Ragno et al., 2007).

Over the last decade, tremendous efforts have been made to develop PMTs inhibitors as chemical probes and therapeutic reagents. These small-molecule inhibitors are of great values in interpreting biological functions and disease mechanisms of targeted enzymes. The DOT1L inhibitor EPZ-5676 has reached phase I clinical trials for the treatment of mixed linage rearranged leukemia, supporting the hypothesis that PMTs could be potential drug targets in cancer therapies (Daigle et al., 2013). Several other excellent reviews and book chapters have been published on the PMTs inhibitors (Bissinger, Heinke, Sippl, & Jung, 2010; Cole, 2008; Copeland et al., 2009; Kaniskan & Jin, 2015; Kaniskan, Konze, & Jin, 2015; Li et al., 2015, chapter 17; Zhang & Zheng, 2015) and DNMTs inhibitors (Gnyszka et al., 2013; Lyko & Brown, 2005; Martinet, Michel, Bertrand, & Benhida, 2012). In this chapter, we discuss the development of small-molecule inhibitors of methyltransferases with the primary focus on the design strategies and synthetic approaches to SAM/SAH-based analogs as PMTs inhibitors, which have not been extensively reviewed so far.

2.2 Pan-Inhibitors

As the by-product of SAM-dependent transmethylation reactions, SAH (or AdoHcy, **2**) and its mimics have long been recognized as pan-inhibitors of PMTs (Fig. 2). Structural mimics of SAH include methylthioadenosine (MTA, **3**) and 5′-alkylamine analog sinefungin (SNF, **4**).

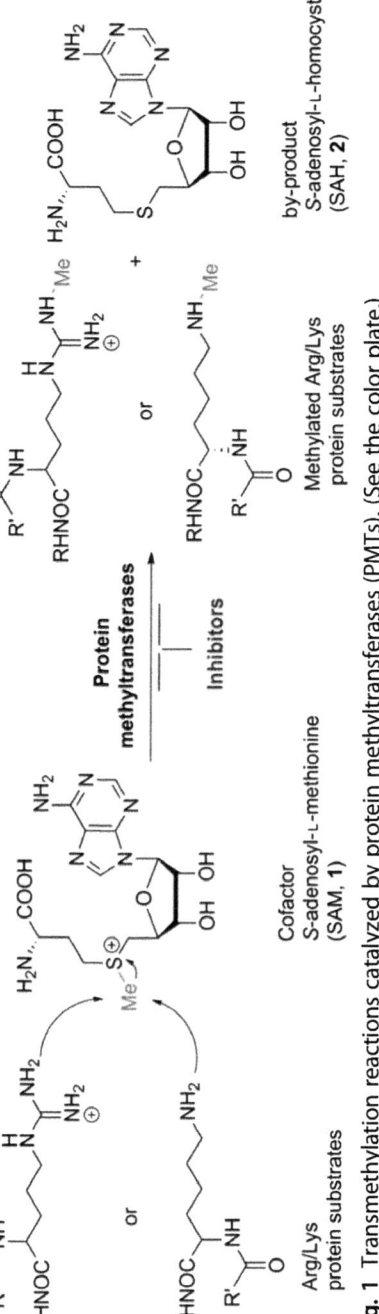

Fig. 1 Transmethylation reactions catalyzed by protein methyltransferases (PMTs). (See the color plate.)

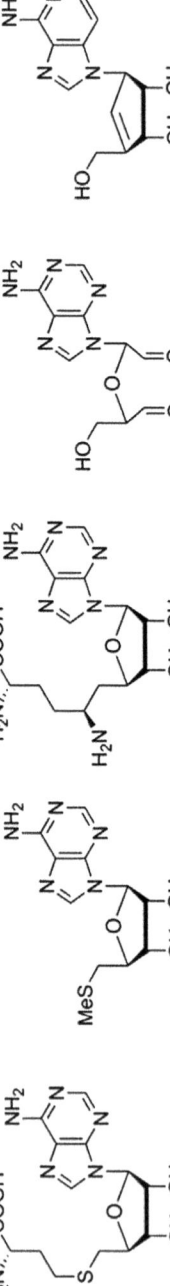

Fig. 2 Examples of pan-inhibitors targeting methyltransferases (**2–4**) and the inhibitors targeting SAH hydrolase (**5–6**).

SAH is a pan-inhibitor with IC_{50} values ranging from 0.1 to 20 μM against representative PMTs (Copeland et al., 2009; Richon et al., 2011). While SAH is active in a protein-based in vitro assay, it is inactive in a cellular context because it can be rapidly metabolized by SAH hydrolase to adenosine and homocysteine. Therefore, a series of SAH hydrolase inhibitors are used to indirectly modulate the activities of PMTs in a cellular context by increasing cellular accumulation of SAH. Among the frequently used SAH hydrolase inhibitors are periodate-oxidized adenosine (Adox, **5**) (Grant & Lerner, 1979; Patel-Thombre & Borchardt, 1985) and 3-deazaneplanocin (DZNep, **6**) (Fujiwara et al., 2014; Tseng et al., 1989).

MTA (**3**) is an abundant metabolite in the process of polyamine and methionine metabolism (Avila, Garcia-Trevijano, Lu, Corrales, & Mato, 2004). MTA has been reported to inhibit SAM-dependent methyltransferases in cellular settings, but the inhibitory mechanism has not been elucidated (Williams-Ashman, Seidenfeld, & Galletti, 1982). Both MTA (**3**) and Adox (**5**) are widely used as pan-inhibitors for cellular experiments.

(+)-Sinefungin (SNF, **4**), a nucleoside antibiotic, was isolated from cultures of *Streptomyces griseolus* (Hamil & Hoehn, 1973). (+)-Sinefungin (5′-aminoalkyl group) is structurally similar to SAH (5′-alkylthio group) with IC_{50} values ranging from 0.1 to 20 μM against representative PMTs (Copeland et al., 2009; Fuller & Nagarajan, 1978; McCammon & Parks, 1981). However, sinefungin shows poor cell membrane permeability, probably due to the polar methionine moiety. Moreover, toxicity in animal models precluded the potential use of sinefungin as a therapeutic agent (Zweygarth, Schillinger, Kaufmann, & Rottcher, 1986).

2.3 Target-Selective Inhibitors of Methyltransferases

Access to target-selective PMT inhibitors can facilitate biological characterization of PMTs. PMT inhibitors with pharmacological merits can be further pursued as drug candidates targeting PMT-implicated diseases. The combined efforts of industrial and academic laboratories have led to many encouraging findings. Small-molecule PMT inhibitors have been typically developed via three complementary approaches: rational design, high-throughput screening (HTS), and in silico screening. This chapter primarily focuses on the rational design and chemical synthesis of SAM mimics as PMTs inhibitors. We will provide general guidance to design and synthesize representative SAM mimics inhibitors. We will also underline the key structural motifs that are amenable for modification to improve potency and selectivity of these parent compounds.

On the basis of the characteristic substitutions at the 5′-position of the cofactor SAM (**1**), current SAM-based inhibitors can be classified into five categories (Fig. 3): 5′-alkylthio analogs (eg, DOT1L inhibitor BrSAH **16**), 5′-alkoxy analogs (eg, COMT inhibitor **36**), 5′-alkyl/alkenyl analogs (eg, PRMT inhibitors **48–49**), 5′-aminoalkyl (sinefungin) analogs (eg, SETD2 inhibitor PrSNF **63**), and 5′-amino analogs (eg, DOT1L inhibitor EPZ004777 **101**). In general, the types of 5′-substitutions are crucial for the potency and selectivity for inhibiting PMTs as demonstrated by the EZH2 inhibitor **20**, COMT inhibitor **41**, PRMT inhibitor **49**, and DOT1L inhibitor **102**. Here, we focus on the synthetic strategies to construct the diverse 5′-substitutions as well as nucleosides, amino acids, and 6′-amino group (for 5′-aminoalkyl analogs) of SAM mimics in the context of optimizing potency and selectivity against specific PMT targets. Meanwhile, we discuss their structure–activity relationships (SARs) and underline key modifications on these SAM scaffolds that render improved potency and specificity against specific enzymes.

2.3.1 5′-Alkylthio SAM Analogs as Protein and DNA Methyltransferase Inhibitors

The most common approach to prepare 5′-alkylthio SAM analogs is to convert the 5′-hydroxyl group of adenosine or its derivatives to a better leaving group such as chloride or tosylate, followed by nucleophilic substitution with homocysteine or other thiol-containing derivatives (Fig. 3). For instance, to implement the "bump-and-hole" chemical genetic approach (Alaimo, Shogren-Knaak, & Shokat, 2001) for allele-specific inhibition of Rmt1, Lin, Jiang, Schultz, and Gray (2001) relied on this synthetic strategy to access a set of N^6-substituted "bumped" SAH analogs (Fig. 4). In this study, a series of N^6-alkyl-substituted adenosine analogs similar to **8** were generated via aromatic substitution of 6-chloropurine ribonucleoside (**7**) with corresponding amines. Activation of the 5′-hydroxyl group of adenosine **8** afforded the 5′-chloro derivatives **9**, followed by nucleophilic substitution with homocysteine to yield N^6-alkyl SAH analogs. Among these "bumped" SAH analogs, N^6-benzyl SAH **10** and N^6-naphthylmethyl SAH **11** were characterized as allele-specific inhibitors of the Rmt1 E117G variant with K_i values of 4.4 and 5.0 μM, respectively, with a 20-fold selectivity over native Rmt1. However, these analogs showed poor potency

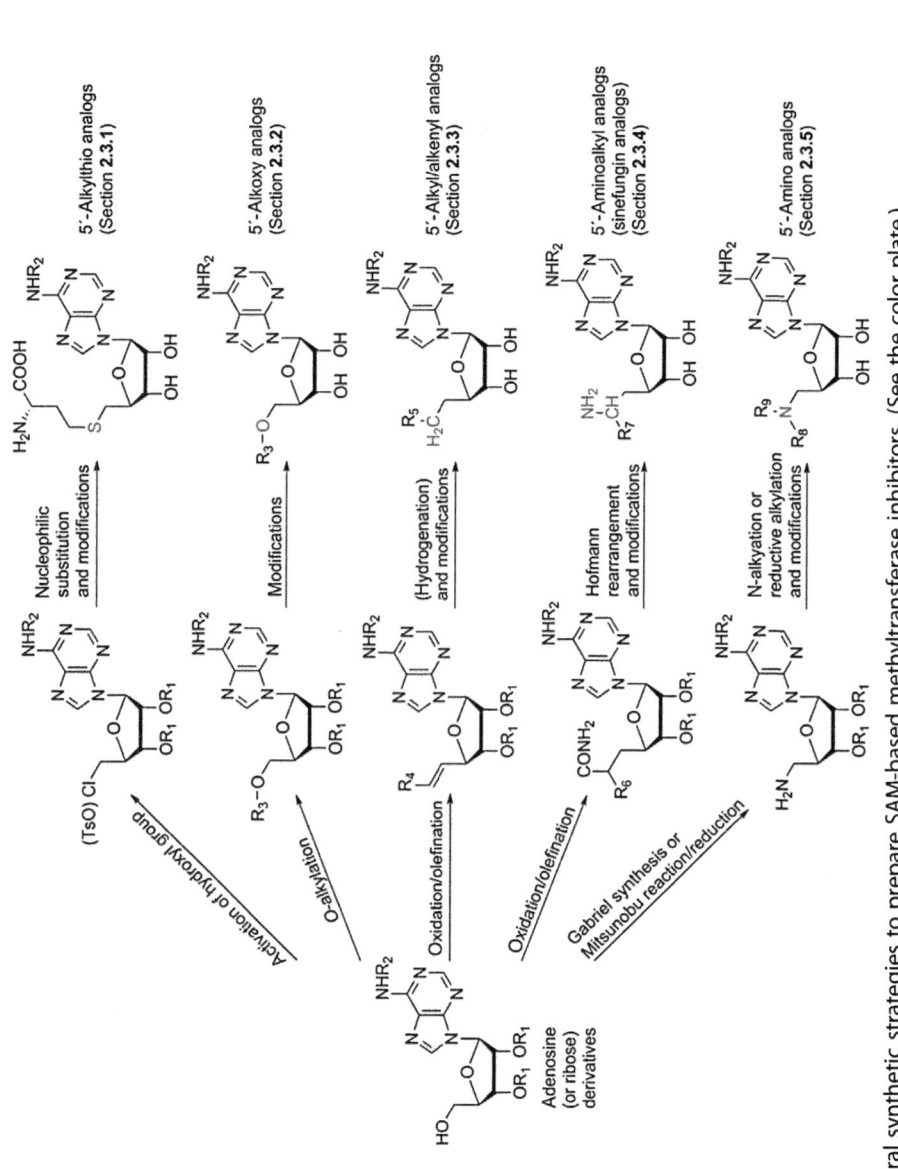

Fig. 3 General synthetic strategies to prepare SAM-based methyltransferase inhibitors. (See the color plate.)

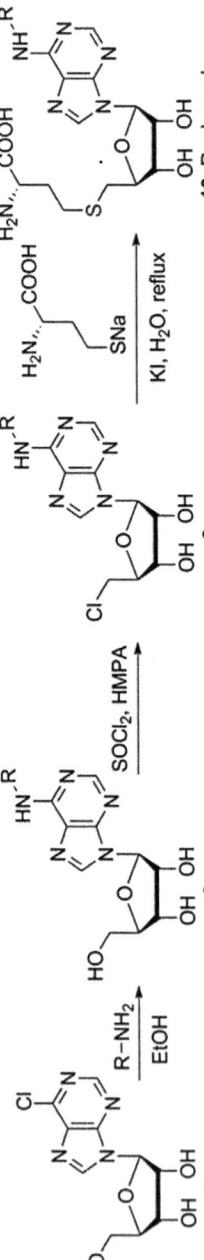

Fig. 4 Synthesis of allele-specific inhibitors of Rmt1 (yields were not reported).

($EC_{50} > 100$ μM) in a cellular context, due to the poor membrane permeability associated with the α-amino carboxylic acid moiety.

Yu et al. (2013) described the rationale and synthesis of 7-deaza-7-bromo SAH (BrSAH, **16**) as a potent and selective DOT1L inhibitor. DOT1L catalyzes the mono-, di-, and trimethylation of H3K79 and has been implicated in mixed lineage leukemia (MLL) (Okada et al., 2005). The modification of the 7-deaza-7-bromo moiety came through the screening of various adenosine analogs as potential DOT1L inhibitors. Given that the cofactors of PMTs (SAM) and kinases (ATP, adenosine triphosphate) share the same adenosine moiety, the authors examined a library of 3120 kinase inhibitors and identified 5-iodotubercidine (5ITC, **12**) as a modest DOT1L inhibitor with an IC_{50} value of 18.2 μM (Eswaran et al., 2009; Ugarkar et al., 2000). Structural docking of 5ITC into DOT1L's SAM-binding site suggested that the 7-iodo moiety of 5ITC occupies a distinct hydrophobic pocket in DOT1L (Pro133, Phe245, and Val249 in PDB 3QOX). The resulting van der Waals interaction was expected to facilitate the binding of 5ITC to DOT1L. This observation inspired Yu et al. to install hydrophobic moieties at the 7-position to improve the selectivity and affinity of the pan inhibitor SAH. In their synthetic design, BrSAH **16** was obtained from the 6,7-deaza-6-chloro-7-bromo adenosine derivative **13** (Fig. 5). Amination at the 6-position and deprotection generated 7-deaza-7-bromo adenosine (**14**). In a similar approach to prepare N^6-alkyl-substituted analogs **10–11** (Fig. 4), adenosine analog **14** was converted to its 5′-chloro derivative **15** and then treated with homocysteine under basic conditions to afford BrSAH (**16**). The structure of DOT1L in complex with BrSAH confirmed the binding interactions between the 7-bromo moiety and the hydrophobic cavity formed by the side chains of Pro133, Phe245, and Val249 residues (PDB 3SX0). Other SAH analogs with hydrophobic substitutions adjacent to SAH's 7-position (eg, 7-iodo and 6-methyl) were also prepared in a similar approach and were expected to engage DOT1L in a similar manner. It is worth pointing out that no cellular activity of these SAH analogs has been reported so far. The lack of such information is likely due to the poor membrane permeability of these SAH derivatives.

Given the poor membrane permeability caused by the α-amino carboxylic acid of SAH, the corresponding modifications have also been explored on this moiety to improve potency, specificity, and permeability. Kung et al. (2015) reported a series of SAM amide analogs, eg, compound **20**, as potent and selective inhibitors of EZH2. EZH2 is the enzymatic subunit of PRC2 and catalyzes the methylation of H3K27 (Cao et al., 2002; Czermin et al., 2002; Kirmizis et al., 2004). Several somatic activating mutations in the SET

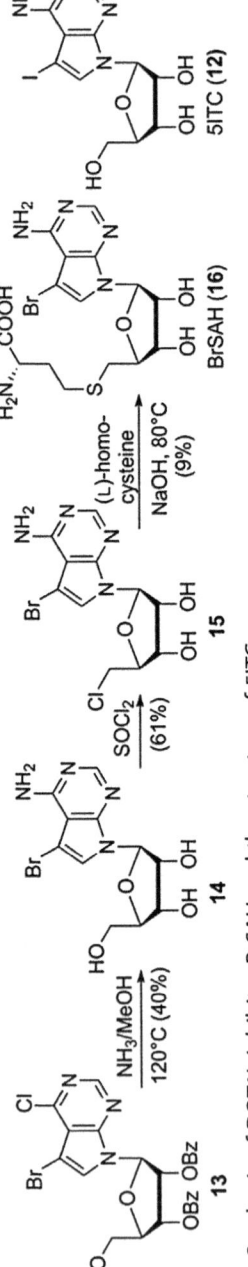

Fig. 5 Synthesis of DOT1L inhibitor BrSAH and the structure of 5ITC.

domain of EZH2, eg, Y641F/N and A677G, have also been observed in GCG and DLBCL (Morin et al., 2010). To increase the selectivity of SAH analogs targeting EZH2 and EZH2 mutants, the authors explored the feasibility of replacing the free carboxylic acid of SAH with various amides and systemically studied the SAR of N^6-subsitutitions (methyl, isopropyl, and benzyl), 5′-linker atoms (5′-S- and 5′-N-), and the stereochemistry (S and R) of the 9′-amino group. The synthesis started with the coupling between 5′-chloro adenosine derivative **17** and Boc-protected homocysteine **18** (Fig. 6). The resulting α-amino carboxylic acid **19** was subjected to coupling reactions with various amines and deprotection conditions to generate a series of amide analogs, such as compound **20**. While N^6-substitution, the 5′-sulfur atom, and S-amino group of SAH are conserved for its binding to EZH2, several amide replacements of the free carboxylic acid could be better accommodated by EZH2 providing more potent inhibitors than SAH. In particular, the most potent amide analog **20** showed IC_{50} values of 0.27 and 0.07 μM against wide-type and Y641N EZH2, respectively, as compared to 11 and 1.9 μM binding affinity of SAH. These compounds also demonstrated improved lipophilicity compared with SAH. However, up to a concentration of 100 μM the amide analog **20** did not show cellular activity as monitored by ablation of H3K27 methylation mark and by antiproliferation assay. Further modification of the physiochemical properties of compound **20** would be required to optimize the cellular activity of this set of compounds.

Among other examples of 5′-alkylthio SAM analogs as PMT inhibitors is a dual PRMT5–PRMT7 inhibitor DS-437 (**23**) disclosed by Smil et al. (2015). While PRMT5 carries out monomethylation and symmetric dimethylation on H4R3, H3R8, and Sm proteins, PRMT7 only carries out monomethylation on H3R2 and Sm proteins (Migliori et al., 2012; Zurita-Lopez, Sandberg, Kelly, & Clarke, 2012). PRMT5 has been reported to mediate p53 methylation and promote recruitment of DNMT3A, which results in CpG methylation and subsequent gene silencing (Jansson et al., 2008; Zhao et al., 2009). The structure of a human PRMT5–MEP50 complex bound to a cofactor mimic A9145C and the H4R3 peptide (PDB 4GQB) showed that the substrate's guanidinium moiety forms multiple hydrogen bonds with nearby Glu435 and Glu444 in the "double-E loop" which is known to be conserved in PRMTs (Antonysamy et al., 2012). The co-crystal structure of *Xenopus laevis* PRMT5–MEP50 complex with SAH but in the absence of the H4R3 peptide (PDB 4G56) revealed that the α-amino carboxylic acid moiety of SAH extends into the substrate-

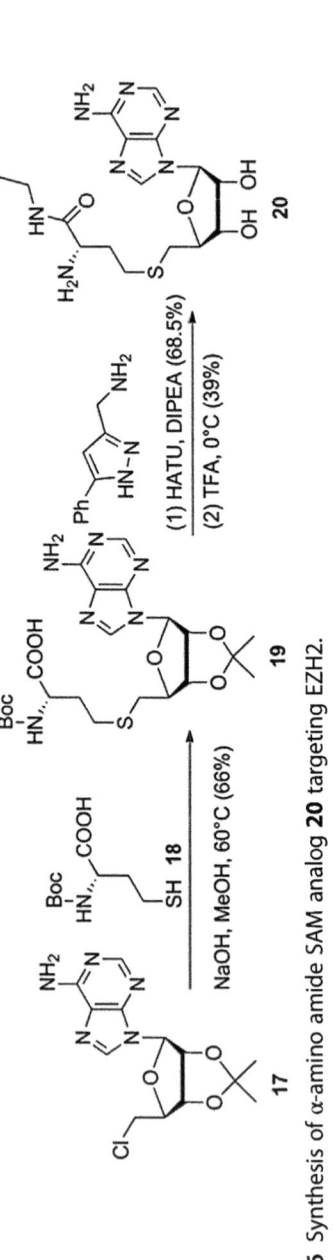

Fig. 6 Synthesis of α-amino amide SAM analog **20** targeting EZH2.

binding region that overlays with the aforementioned guanidinium in the hPRMT5–H4R3 complex (Ho et al., 2013). Based on this observation, Smil et al. designed SAH analog DS-437 (**23**) with a urea moiety in place of α-amino carboxylic acid of SAH. This urea analog **23** was expected to accommodate the cofactor-binding pocket by recapitulating the binding characters of SAH with the novel urea motif occupying the substrate-binding pocket via interaction with the "double-E loop" of PRMT5.

Synthesis of the urea SAH analog also started with the activated 5′-chloro-5′-deoxyadenosine (**21**), which was treated with cysteamine to yield amine derivative **22** (Fig. 7). Primary amine **22** was then reacted with ethyl isocyanate to give rise to DS-437 (**23**). DS-437 selectively inhibits PRMT5 and PRMT7 with both IC_{50} values of approximately 6 μM in a panel of 29 methyltransferases. Upon comparing the superimposed structures of PRMT5, PRMT7, and CARM1 (a Type I PRMT), the authors suggested that a YFxxY structural motif, which is conserved across Type I PRMTs but absent in Type II PRMTs, may prevent the binding of DS-437 (Schapira & Ferreira de Freitas, 2014). The cellular activity of DS-437 was further demonstrated by the dose-dependent reduction of the symmetrical dimethylation mark of SmD and SmB proteins that are known substrates of PRMT5. Simultaneously, steady-state kinetic studies confirmed the SAM-competitive nature of DS-437 binding. Collectively, replacing SAH's α-amino carboxylic acid moiety with the *N*-ethyl urea moiety not only allowed improvement in selectivity against PRMT5/7 but also improved membrane permeability for cellular activity.

Recently, Isakovic et al. (2009) developed a set of constrained SAH analogs **26–28** as inhibitors of DNMTs. DNMTs are established anticancer targets given their roles on epigenetic silencing of diverse sets of tumor suppressors (Bird, 2002; Siedlecki et al., 2003). 5-Azacytidine (Silverman et al., 2002) and 5-aza-2′-deoxycytidine (Kantarjian et al., 2006) are mechanism-based inhibitors targeting DNMTs that have been approved for the treatment of myelodysplastic syndromes. In contrast, these constrained SAH analogs **26–28** (Fig. 8) developed by Isakovic et al. are cofactor-competitive noncovalent inhibitors of DNMTs. To prepare analogs **26–28**, the tosyl compound **24** or its derivatives were coupled with various cyclized homocysteine derivatives, followed by deprotection to generate analogs **26–28** (Fig. 8) (Wahhab et al., 2006). Among the constrained SAH analogs, compound **26** inhibits DNMT1 and DNMT3b with IC_{50} values of 1.1 and 0.3 μM, respectively. SAR analysis further indicated that the 6′-stereochemistry, the ring size of cyclic amino acid,

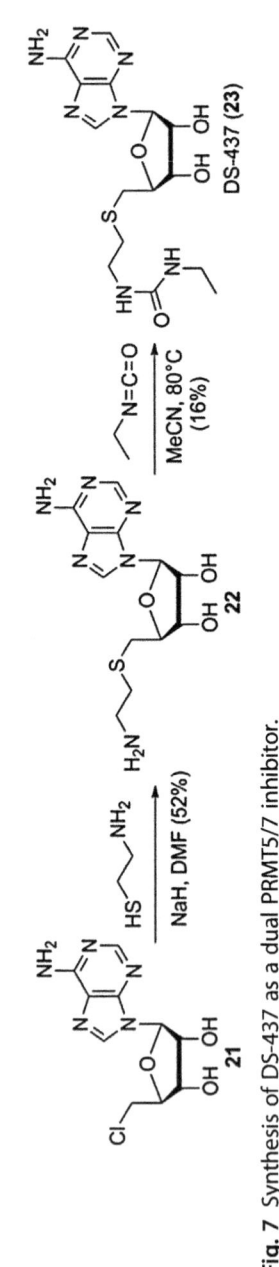

Fig. 7 Synthesis of DS-437 as a dual PRMT5/7 inhibitor.

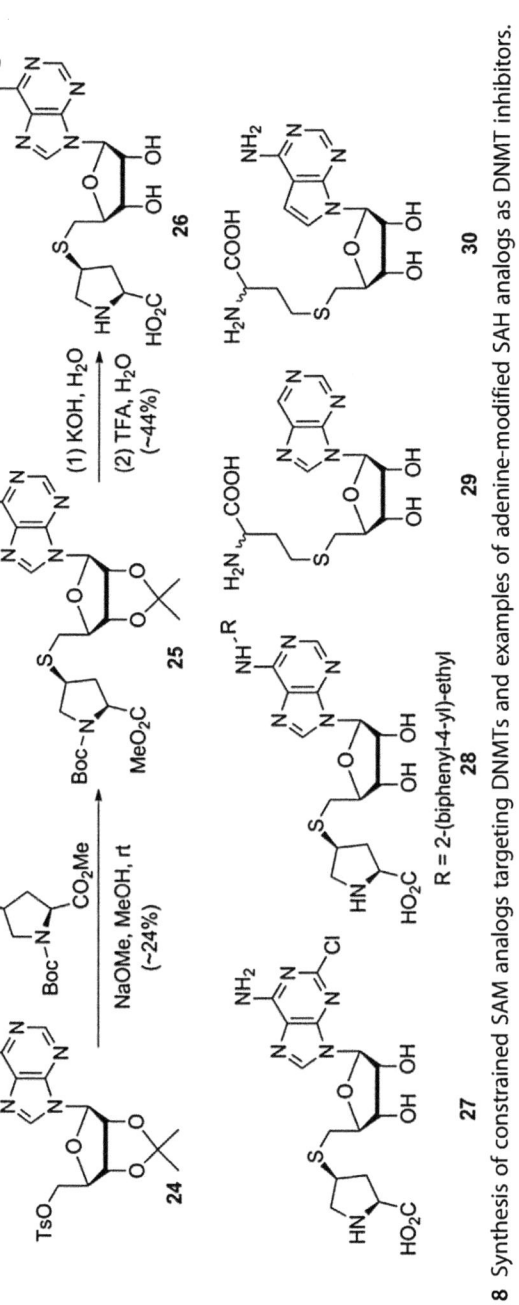

Fig. 8 Synthesis of constrained SAM analogs targeting DNMTs and examples of adenine-modified SAH analogs as DNMT inhibitors.

and the L-α-amino acid moiety of compound **26** are all critical for the potency. Chloro-substitution at the 2-postion (compound **27**) conserved the potency against DNMT1, while N^6-alkylation (compound **28**) maintained the activity toward DNMT3b2. In parallel, Saavedra et al. (2009) explored the modifications on the adenine moiety of SAH and demonstrated that the 6-NH_2 group of adenine (compound **29**, IC_{50} values of 300 and 28 μM) was critical for the affinity of DNMT1 and DNMT3b, while the 7-N (compound **30**, IC_{50} values of 1.5 and 0.3 μM) was dispensable (Fig. 8).

2.3.2 5′-Alkoxy SAM Analogs as COMT Inhibitors

Only a few 5′-alkoxy SAM analogs have been reported as methyltransferase inhibitors. The 5′-alkoxy moiety was typically introduced via O-alkylation of 5′-hydroxyl group of adenosine (Fig. 3). The Diederich group reported the synthesis of a bisubstrate inhibitor **36** of COMT (Masjost et al., 2000). Pharmacological inhibition of COMT has been used in combination therapy targeting Parkinson's disease through alleviating the COMT-mediated deprivation of L-dopa (Mannisto & Kaakkola, 1999; Mannisto et al., 1992). The design of inhibitor **36** was inspired by the X-ray structure of COMT in complex with the cofactor SAM and a competitive inhibitor 3,5-dinitrocatechol. The latter interacts with COMT via coordination with a catalytically essential Mg^{2+} ion (Jeffery & Roth, 1987; Vidgren, Svensson, & Liljas, 1994; Zheng & Bruice, 1997). The Diederich group proposed that the bisubstrate inhibitor **36** would occupy the binding pockets of both the substrate catechol moiety and SAM's adenosine moiety. Computational modeling was carried out to optimize the length of the linker, which was crucial for the synergistic engagement of the two binding sites. The synthesis of compound **36** involved the alkylation of the 5′-hydroxyl group of **31** followed by Fmoc protection to afford compound **32** (Fig. 9). Subsequent deprotection under acidic conditions and acetylation led to the triacetyl intermediate **33**. The adenosine derivative **34** was obtained upon treating **33** with silylated N^6-benzoyladenine in the presence of $SnCl_4$. It is worth noting that the use of TMSOTf instead of $SnCl_4$ led to exclusive ribosylation at adenine's 7-position, while the TMSOTf conditions afforded the desired regioisomers upon installation of cytosine or other bases onto the triacetyl intermediate **33**. The final compound **36** was obtained through Fmoc deprotection, amide formation, and global deprotection. Compound **36** showed an IC_{50} value of 2 μM against COMT. A detailed enzyme kinetic study indicated that **36** is a SAM-competitive and

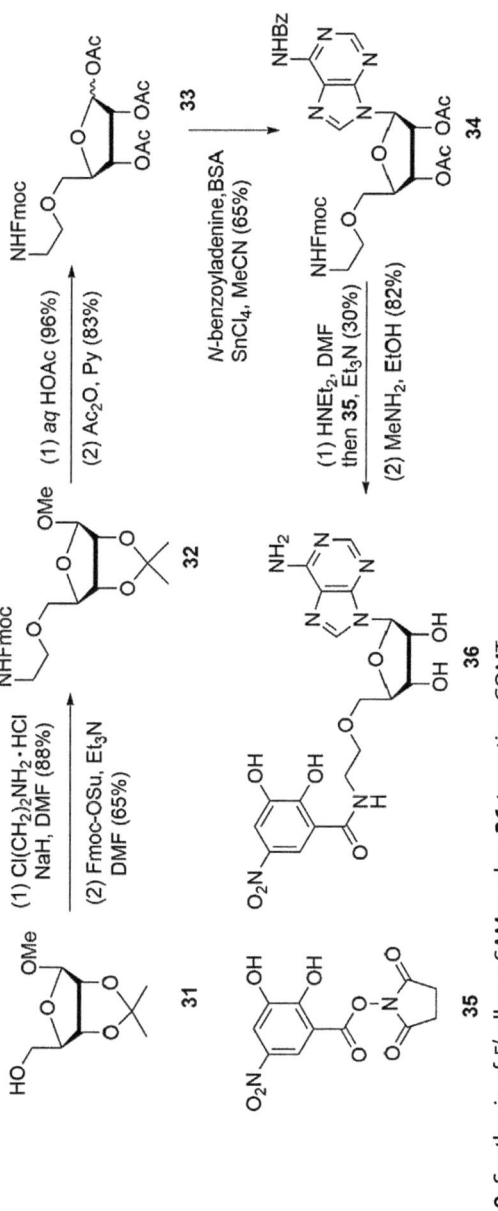

Fig. 9 Synthesis of 5′-alkoxy SAM analog **36** targeting COMT.

a catechol-noncompetitive inhibitor. In contrast, a 5′-alkenyl SAM analog **41** (Fig. 10) showed SAM-competitive and catechol-competitive characters as confirmed by the inhibitor-COMT binary structure. The bisubstrate-competitive binding mode of 5′-alkenyl analog **41** also significantly improved the IC_{50} value to 9 nM against COMT (Lerner et al., 2003).

2.3.3 5′-Alkyl/Alkenyl SAM Analogs as Methyltransferase Inhibitors

To improve the potency of **36** as a bisubstrate inhibitor of COMT, the Diederich group explored the linker atom by replacing the 5′-alkoxy with 5′-alkylthio or 5′-alkyl/alkenyl groups (Lerner et al., 2003). ^{1}H NMR analysis of a cytosine analog of **36** (Fig. 9) revealed a hydrophobic collapse conformation of the small molecule, which suggests that extra reorganizational energy would be needed for binding to COMT. This observation was consistent with the noncompetitive character of 5′-alkoxy analog **36** for the substrate-binding site (Masjost et al., 2000). Therefore, the authors proposed to replace the 5′-alkoxyl group with a 5′-alkyl/alkenyl group, which would increase the rigidity of the spacer to avoid the hydrophobic collapse. Bisubstrate analogs containing 5′-alkyl linkers of different lengths ($n = 1$–4, eg, compounds **42** and **43**) as well as 5′-*trans*-alkenyl compound **41** were synthesized (Fig. 10). For the corresponding synthesis, acetonide adenosine **37** was oxidized to aldehyde by IBX and subjected to Wittig reaction to afford α,β-unsaturated ethyl ester **38**. Reduction of **38** with DIBAL-H led to the allylic alcohol, followed by Mitsunobu reaction to yield phthalimide **39**. Removal of phthalimide and deprotection of acetonide afforded primary amine **40**. The amine intermediate was then coupled with catechol carbamate derivative **35** to yield compound **41**. 5′-Alkyl derivatives **42–43** were prepared in a similar approach with an extra hydrogenation step. The optimal spacers in analogs **42** (C-2 linker) and **43** (C-3 linker) may allow the adenosine and catechol moieties to occupy both the SAM-binding pocket and the substrate-binding pocket. 5′-Alkyl derivatives **42** and **43** inhibit COMT with IC_{50} values of 0.06 and 0.2 μM, respectively. The construction of the 5′-*trans*-alkene group in compound **41** was expected to further rigidify the linker and thus was responsible for the increased affinity with an IC_{50} value of 9 nM. Enzyme kinetic studies and the X-ray structure of COMT in complex with inhibitor **41** were consistent with the bisubstrate-competitive binding mode. To overcome the potential hepatotoxicity of the nitro group, the Diederich group further modified the catechol moiety of the bisubstrate inhibitor **41** (Fig. 10). Given the tight binding affinity afforded by the electron-withdrawing nitro group, a variety

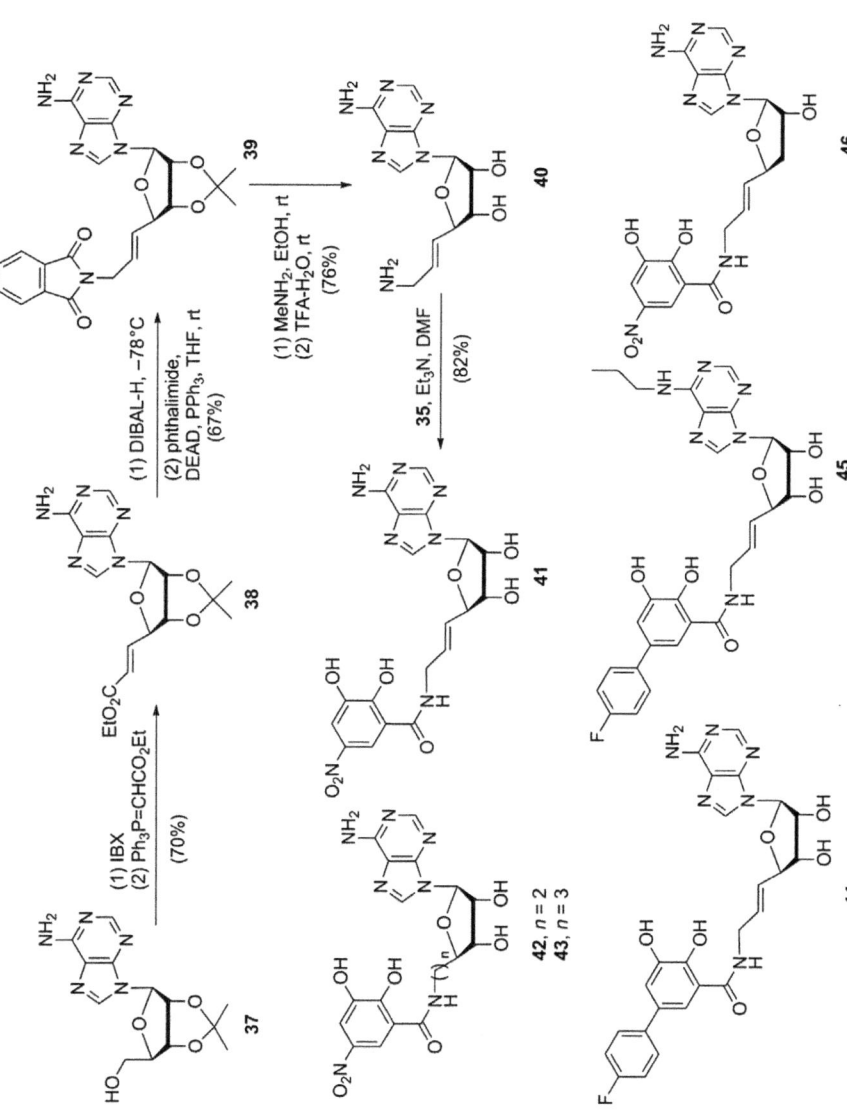

Fig. 10 Synthesis of 5′-alkyl/alkenyl SAM analogs targeting COMT and examples of further modified 5′-alkyl/alkenyl SAM COMT inhibitors.

of substitutions were examined and compound **44** was characterized to maintain high affinity to COMT with an IC_{50} value of 21 nM (Paulini et al., 2004). Besides the catechol motif, the adenine and the ribose moieties were also explored to afford N^6-n-Pr analog **45** and 3′-deoxyl analog **46** with IC_{50} values of 9 and 40 nM, respectively (Ellermann et al., 2009, 2011; Paulini et al., 2006).

Very recently, the Martin group reported the synthesis and characterization of 5′-alkyl/alkenyl SAM analogs **48–49** as PRMT inhibitors (van Haren, van Ufford, Moret, & Martin, 2015). Dysregulation of PRMTs has been reported to be associated with multiple diseases including cancer, cardiovascular, and pulmonary disorders (Copeland et al., 2009; Wei et al., 2014). Here the pan-PRMT inhibitors **48–49** were featured for its potential bisubstrate mode of inhibition by occupying both the SAM (adenosine) and the substrate-binding pockets. Different from the PRMT5 inhibitor DS-437 (**23**, Fig. 7) with a urea moiety to interact with PRMT5's E435/E444 residues, analogs **48–49** were expected to rely on a guanidine moiety to engage the interactions with E144/E153 in PRMT1. In addition, the Martin group systemically explored the linker between the adenosine and the guanidine moieties for the bisubstrate inhibitors, including 5′-alkylthio, 5′-alkoxy, 5′-alkyl/alkenyl, and 5′-amino groups. In the synthesis, the primary amine precursor **47** was treated with N,N'-di-Boc-N''-triflyl-guanidine to install the guanidine moiety, followed by deprotection to afford guanidine compound **48**. Analogs **49a** and **49b** were obtained in a similar approach with an extra hydrogenation step (Fig. 11). Analogs with a C2–C3 linker (either 5′-alkyl or 5′-alkenyl) are moderate inhibitors of PRMT1/4/6 with the range of IC_{50} values from 0.15 to 20 μM, but inert toward PKMT G9a with an IC_{50} value of >50 μM. In contrast, the derivatives containing 5′-alkylthio, 5′-alkoxy, and 5′-amino substitutions do not inhibit PRMT1, PRMT4, or PMRT6.

2.3.4 5′-Aminoalkyl SAM Analogs (Sinefungin Analogs) as PKMT Inhibitors

Although sinefungin, a 5′-aminoalkyl analog of SAH, is a pan-inhibitor against SAM-dependent methyltransferases, several selective inhibitors have been developed on the basis of the sinefungin scaffold. In general, 5′-alkylthio, 5′-alkoxy, 5′-alkyl/alkenyl, and 5′-amino SAM analogs can be readily prepared via modifications of 5′-hydroxyl group of adenosine or its derivatives (Fig. 3). However, the diastereoselective introduction of a 5′-aminoalkyl moiety in sinefungin analogs is synthetically more

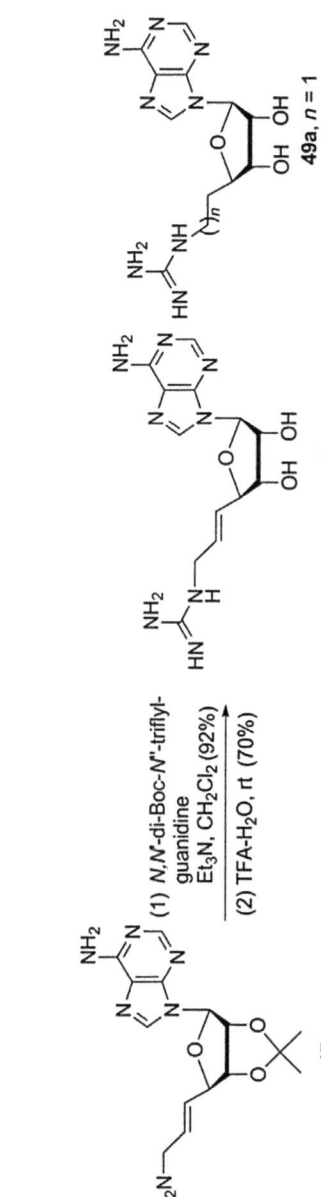

Fig. 11 Synthesis of 5′-alkyl/alkenyl SAM analogs targeting PRMTs.

challenging. In addition, installation of an L-amino acid moiety in sinefungin analogs, which is not available from the methionine building block, requires alternative synthetic strategies. Several sinefungin analogs have been reported, eg, 6′-epimer sinefungin (Geze, Blanchard, Fourrey, & Robert-Gero, 1983), carbocyclic sinefungin (Ghosh & Lv, 2014), 4′,5′-dehydro sinefungin (Blanchard, El Kortbi, Fourrey, Lawrence, & Robert-Gero, 1996), 9′-deamino sinefungin, and 5′-δ-lactam sinefungin, but here we will focus on the discussion of sinefungin analogs that show decent potency and selectivity against certain PKMTs.

SETD2 specifically trimethylates H3K36 and has been identified as a tumor suppressor in p53-dependent gene regulation, transcription elongation, and intron–exon splicing (Duns et al., 2010; Varela et al., 2011; Zhang et al., 2012). $N^{6\prime}$-Alkyl sinefungin derivatives **63–64** were first reported as selective SETD2 inhibitors by the Luo group (Zheng et al., 2012). The authors proposed that an extra alkyl chain at the 6′-amino position, upon occupying the lysine-binding pockets for the optimal interactions, can capture certain distinct transition state characteristics for specific PKMTs (Horowitz, Yesselman, Al-Hashimi, & Trievel, 2011; Zhang & Bruice, 2008). The synthesis of $N^{6\prime}$-alkyl sinefungin analogs started with adenosine derivative **50** was developed by (Ghosh and Liu (1996) (Fig. 12). The unsaturated ethyl ester **50** was hydrogenated and hydrolyzed to afford free carboxylic acid **51**. To construct the 6′-amine diastereoselectively, Evans oxazolidinone chiral auxiliary was used to carry out the asymmetric alkylation. Free carboxylic acid **51** was thus activated to a mixed anhydride and then coupled with chiral auxiliary **52** to generate carbamate **53**. Generation of Z-enolate and subsequent allylation yielded compound **54** as a single diastereomer. Hydrolysis of the chiral auxiliary gave rise to carboxylic acid **55**, followed by Curtius rearrangement to give 6′-amino compound **56** as a single diastereomer. Installation of benzyl groups (or methyl, ethyl, and n-propyl) via different approaches at the early stage afforded a set of $N^{6\prime}$-alkyl sinefungin analogs **57**. Oxidative cleavage of the allyl group, reduction of the resulting aldehyde to alcohol, and the Appel reaction afforded the iodide compound **58**. The chiral α-amino acid was introduced via Schöllkopf bis-lactam to generate compound **59**, followed by hydrolysis of the bis-lactam and protection to give the amino acid derivative **60**. Triacetyl derivative **61** was then obtained and treated with silylated N^6-benzoyladenine in the presence of TMSOTf to yield nucleoside **62**. Hydrogenolysis and deprotection under basic conditions yielded $N^{6\prime}$-n-propyl and benzyl sinefungin analogs (compounds **63–64**).

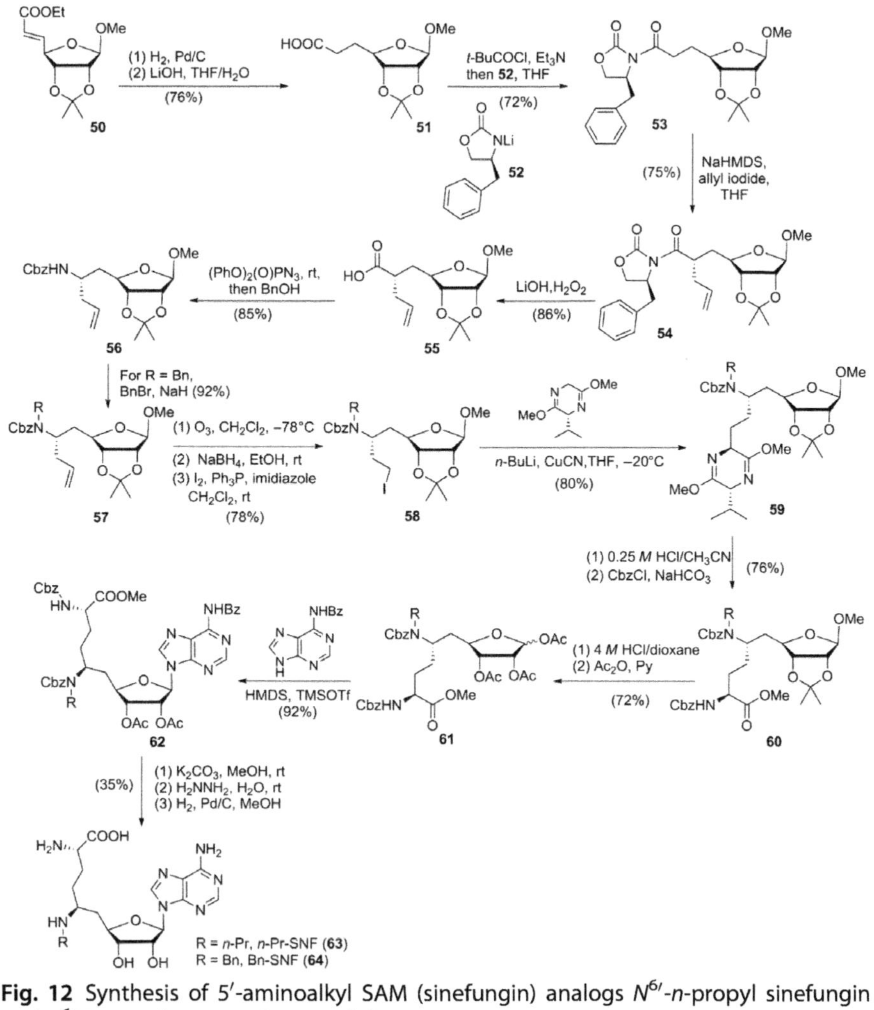

Fig. 12 Synthesis of 5'-aminoalkyl SAM (sinefungin) analogs $N^{6'}$-n-propyl sinefungin and $N^{6'}$-Bn-sinefungin as SETD2 inhibitors.

Their inhibition profiles were examined with a panel of methyltransferases including 10 SET-domain-containing PKMTs and 5 PRMTs. $N^{6'}$-n-Propyl sinefungin and $N^{6'}$-Bn-sinefungin were identified as selective inhibitors of SETD2 with IC_{50} values of 0.80 and 0.48 μM, respectively. SAR analysis revealed that SETD2 preferred sinefungin analogs with bulkier hydrophobic substitutions at the $N^{6'}$-position, such as n-propyl and benzyl groups. The co-crystal structure of SETD2 in complex with $N^{6'}$-n-propyl sinefungin indicated that the lysine-binding pocket of SETD2 was flexible to accommodate the $N^{6'}$-n-propyl group to render higher affinity and selectivity than sinefungin (4). More importantly, the co-crystal structure revealed that $N^{6'}$-n-propyl sinefungin preferentially binds to SETD2's catalytically active open conformer, which likely resulted in its selectivity against SETD2. The development of $N^{6'}$-n-propyl sinefungin thereby supports the feasibility of inhibiting closely related PMTs by targeting distinct conformations with structurally matched sinefungin analogs (Gutierrez et al., 2007; Schramm, 2011).

Recently, the Clausen group reported the synthesis and characterization of a 5′-aminoalkyl sinefungin analog 75 as a selective inhibitor of EHMT1 (GLP, G9a like proteins) and EHMT2 (G9a) (Devkota et al., 2014). EHMT1 and EHMT2 are the major PMTs for monomethylation and dimethylation of H3K9 and predominately exist as a heteromeric complex. The EHMT1/2 complex has been identified to regulate germ cell development, cell proliferation, and cancer cell formation (Shinkai & Tachibana, 2011). This series of inhibitors, eg, compound 75, was designed to simplify the scaffold of sinefungin by replacing the α-amino carboxylic acid moiety with alkyl groups. The synthesis of analog 75 started with a ribose derivative 65, which was treated with phosphonate 66 to generate α-nitrile phosphonate 67 (Fig. 13). Wittig olefination of 67 with a set of aldehydes or ketones allowed the installation of various alkyl chains in the position of α-amino carboxylic acid group of sinefungin. Hydrogenation of the resultant alkene 68 afforded a diastereomeric mixture of nitrile intermediate 69, which was converted to amide 70 using alkaline hydrogen peroxide. Triacetyl derivative 71 was then obtained and treated with silylated N^6-Bz-adenine in the presence of TMSOTf to afford 72 as a diastereomeric mixture. Deprotection and reinstallation of acetonide led to compound 73, which was subject to Hofmann rearrangement and Boc-protection to yield Boc-protected amine 74. Deprotection with TFA afforded a 1:1 mixture of 6′-diastereomers. The most potent compound 75 inhibits EHMT1/2 with IC_{50} values of 1.5 and 1.6 μM, respectively. The potency of these compounds against EHMT1/2 depends on the 5′-aminoalkyl substitutions with their IC_{50} values ranging from moderate (1.5 μM) to no inhibition. Additionally, the other set of analogs neglecting the 6′-amine group, eg, 5′-deoxy-5′-n-butyladenosine (76), is inert against EHMT1/2 with an IC_{50} value of >50 μM, which also indicated that the 6′-amino moiety was crucial for affinity. It is worth noting that although these compounds were designed to occupy the SAM-binding pocket, kinetic studies of inhibitor 75 showed that it is a noncompetitive inhibitor with respect to SAM. It would be interesting to further investigate the binding mode and

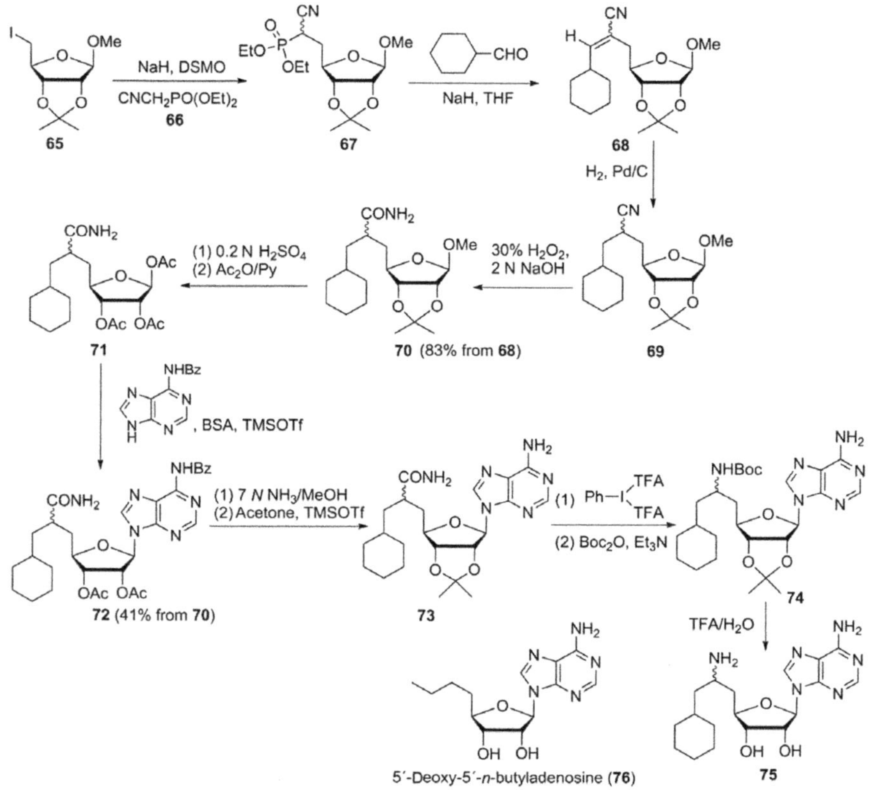

Fig. 13 Synthesis of 5′-aminoalkyl SAM (sinefungin) analog **75** targeting EHMT1/2 and the structure of compound **76**. The available yields are included as reported.

to evaluate the cellular activity of this series of inhibitors given that they do not contain an α-amino carboxylic acid moiety.

2.3.5 5′-Amino SAM Analogs as PMT Inhibitors

Another common strategy to develop SAM analogs is to replace the 5′-alkylthio group of SAM with a 5′-amino group (Fig. 3). At least three complementary strategies have been adopted to access structurally diverse 5′-amino SAM analogs as PMT inhibitors: in situ chemoenzymatic generation of inhibitors from an aziridine precursor, chemical synthesis of bisubstrate inhibitors with the 5′-amino group as a linker, and structure-guided rational design to explore novel binding modes with the 5′-amino moiety as an anchor.

The ring strain of aziridines renders high activity of aziridine toward various nucleophiles for ring opening (eg, compound **78** in Fig. 14). The utility

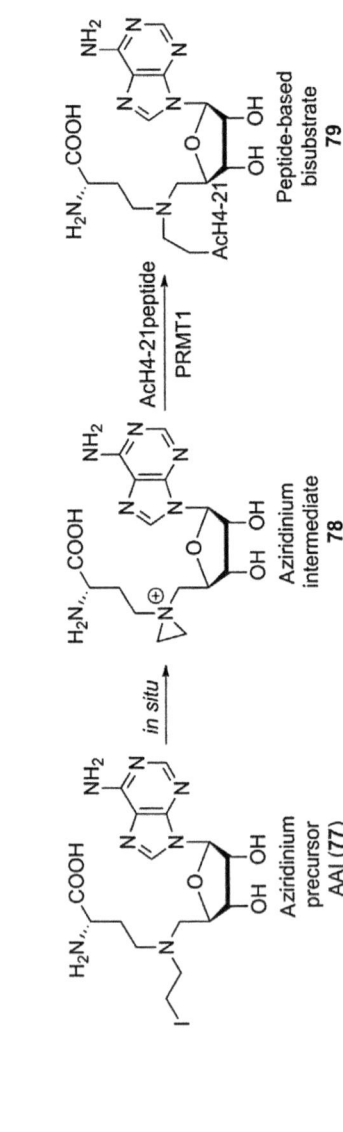

Fig. 14 Chemoenzymatic generation of PRMT1 inhibitor from an aziridine precursor AAI.

of aziridine as a warhead in SAM mimics was first reported by the Weinhold group as an enzymatic alkylation probe for DNMTs (Pignot, Siethoff, Linscheid, & Weinhold, 1998). Subsequently, the Thompson group reported the probe AAI (compound **77**) as an aziridinium precursor that can be converted to the bisubstrate derivative **79** in the presence of PRMT1 (Osborne, Roska, Rajski, & Thompson, 2008). PRMT1 has been demonstrated to be responsible for over 90% Arg methylation and have various substrates of important biological functions (Bedford & Clarke, 2009; Le Romancer et al., 2008). PRMT1 also has been shown to be critical for multiple diseases, including leukemia, breast, and colon cancer (Cheung, Chan, Thompson, Cleary, & So, 2007; Papadokostopoulou et al., 2009). The Thompson group demonstrated that PRMT1 promoted the regiospecific labeling of AcH4-21 peptide (a PRMT1 substrate) with AAI (compound **77**) through an in situ generated aziridinium intermediate **78** (Fig. 14). Bisubstrate analog AAI-AcH4-21 (compound **79**) thus inhibits PRMT1 with an IC_{50} value of 18.5 µM, which is comparable to that of SAH (8.3 µM). The probe AAI showed 4.4-fold selectivity of PMRT1 over CARM1. While no cellular activity was reported for the aziridine-based probe, the development of AAI proved the concept for the in situ chemoenzymatic production of 5′-amino SAM analog inhibitors through an aziridine precursor.

Recently, the Song group described the design and characterization of SAM-based aziridine precursor analogs **84** and **85** as DOT1L inhibitors (Yao et al., 2011). Upon analyzing the binary structure of the SAH–DOT1L complex (Min, Feng, Li, Zhang, & Xu, 2003), the authors noted that the 6-NH_2 of SAH, upon forming only one hydrogen bond with the side chain of Asp222, is in proximity to a unique hydrophobic pocket consisting of Phe223, Leu224, Val249, and Leu253. In contrast, 6-NH_2 of SAH forms two hydrogen bonds in PRMTs and SET-domain containing PKMTs that do not possess the hydrophobic pocket. This observation inspired the Song group to develop N^6-substituted SAM analogs **88** and **89**. The X-ray structure of DOT1L in complex with **88** revealed that the extra N^6-methyl group is accommodated by the hydrophobic cavity with the N^6-H forming one hydrogen bond with Asp222. Therefore, the Song group proposed that N^6-substitutions of the SAM analogs would afford selectivity against DOT1L. In the context of in situ generation of aziridine, the authors proposed compounds **84–87** as potential DOT1L inhibitors that could possibly form a covalent adduct with DOT1L's substrate and occupy the aforementioned hydrophobic pocket. In the synthesis of compounds **84** and **85**,

5′-amino adenosine derivative **81** was generated through Mitsunobu reaction and hydrolysis conditions with hydrazine from compound **80** (Fig. 15). Alkylation of **81** with ethyl bromoacetate and LAH-mediated reduction afforded ethanolamine **82**. Reductive amination of **82** with corresponding aldehydes allowed the installation of the protected amino acid chain to generate compound **83**. Final conversion of the hydroxyl group of **83** to iodide and deprotection yielded SAM-based aziridine precursor analogs **84** and **85**. SAR analysis of the series of inhibitors was of great interest. First, in comparison with SAH (**2**), analog **86** exhibits modest inhibition against DOT1L with an IC_{50} value of 15.7 μM likely because the C—N bonds in **86** are shorter than the C—S bonds in SAH. Adding one extra methylene group in the α-amino acid chain in compound **87** is expected to reflect the extended conformation of SAH and thus contributes to the increase in potency from 0.16 to 0.038 μM. Second, in comparison with **87**, the introduction of N^6-methyl (or Bn) group in **84** and **85** improves >100-fold selectivity against DOT1L over other PMTs. No cellular activity of these compounds was described in the work.

In the course of developing bisubstrate PMT inhibitors based on the SAM scaffold, the 5′-amino substitution in the place 5′-alkylthio group of SAH was preferred because the nitrogen atom can be used as a three-component anchor to link bioisosteres of the cofactor (adenosine and methionine moieties) as well as the substrate. In this context, the Dowden group reported the synthesis and characterization of a set of bisubstrate inhibitors **92–94** against PRMTs (Dowden, Hong, Parry, Pike, & Ward, 2010). Substrate binding in PRMTs is orchestrated by two conserved glutamate residues (Glu144 and Glu153 in PRMT1) that would readily engage interactions with guanidine or its bioisosteres of the bisubstrate inhibitors. The synthesis of these compounds started with 5′-amino adenosine derivative **90**, which went through reductive alkylation to install the protected α-amino acid group to afford SAH adenosine derivative **91** (Fig. 16). Subsequent reductive alkylation and global deprotection allowed the installation of guanidine side chain and afforded compounds **92–94**. These analogs inhibit PRMT1 with IC_{50} values of 3–6 μM, while these do not inhibit SET7 up to 100 μM. Interestingly, altering the distance between the guanidine and the 5′-amino position ($n=1$–3 in **92–94**) showed minimal effect on the potency against PRMT1.

Follow-up work by the Dowden group revealed that compounds **93** and **94** preferentially inhibit PRMT1 over CARM1 (Dowden et al., 2011). While **93–94** ($n=2$–3) inhibit PRMT1 with the IC_{50} values of 3–6 μM, compounds **93** and **94** demonstrated no inhibition against CARM1.

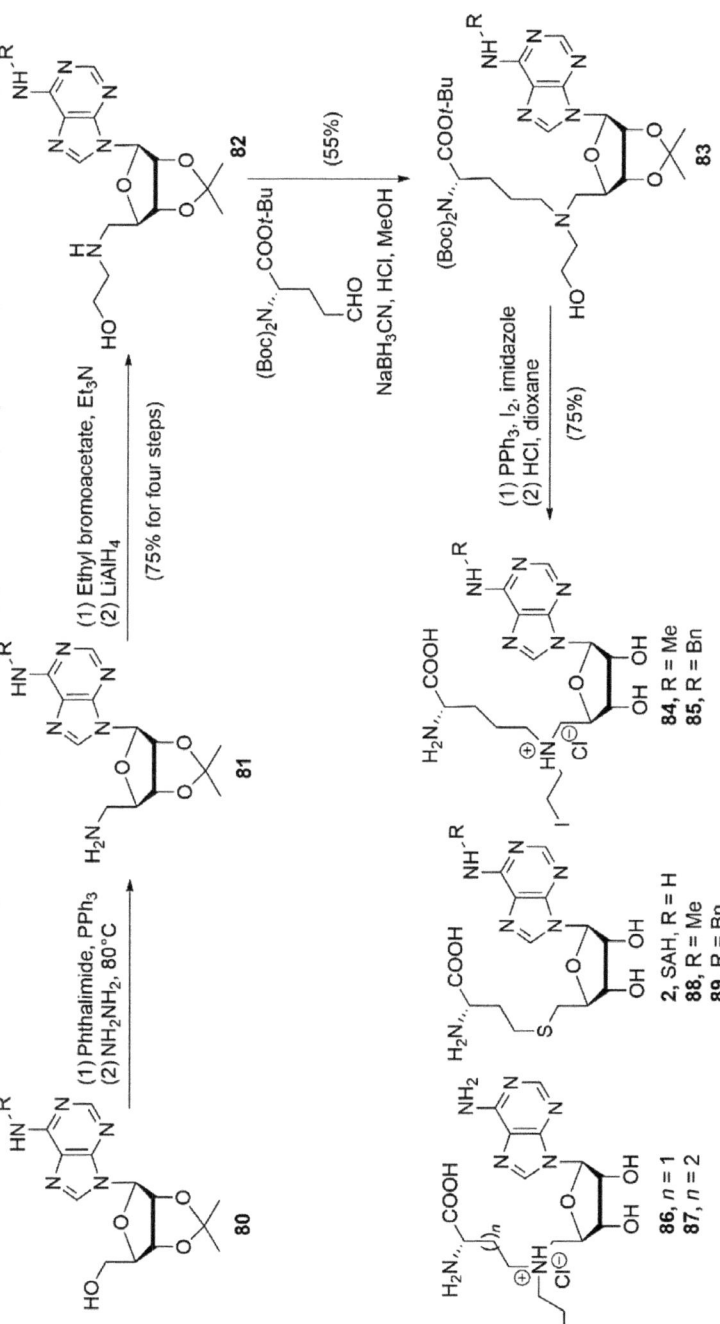

Fig. 15 Synthesis of 5′-amino SAM analogs targeting DOT1L and examples of SAM-based DOT1L inhibitors.

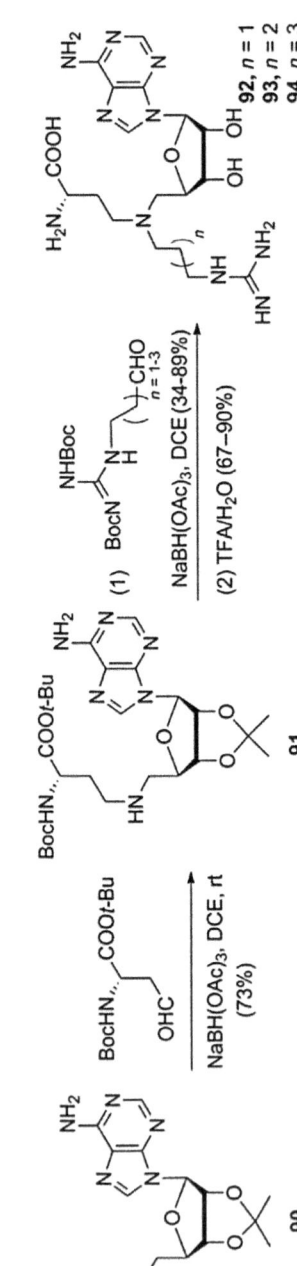

Fig. 16 Synthesis of 5′-amino SAM analogs targeting PRMT1.

Computational modeling suggested that Glu47 of PRMT1 is in a favored position to allow interaction with the guanidine moiety of **93** and **94**, while the corresponding Asn162 of CARM1 is not appropriately positioned. In addition, the longer spacers in **93** and **94** ($n=2$ and 3) also position their guanidinium motif beyond the range to interact with CARM1. In contrast, compound **92** ($n=1$) is a modest CARM1 inhibitor with an IC_{50} value of 13 μM, which is largely attributed to the optimal interaction between its guanidine moiety and CARM1's Glu258. While a structural characterization would provide more insight into the binding modes of these ligands, the current work demonstrated the feasibility of achieving selectivity by exploring sequence variations in the substrate-binding pockets of various PMTs.

5′-Amino SAM analogs were also widely reported based on structure-guided rational design to develop selective PMTs inhibitors. Daigle et al. (2011) reported a 5′-amino SAM analog as the first selective DOT1L inhibitor EPZ004777 (compound **101**). The synthesis of EPZ004777 and other derivatives started with 6-chloro tubercidin (**95**), which was treated with 1-(2,4-dimethylphenyl) methanamine to afford triol **96** (Fig. 17) (Basavapathruni et al., 2012). Protection of the *cis*-diol with acetonide and conversion of the primary alcohol to amine afforded 5′-amino analog **97**. Reductive amination allowed the installation of *iso*-propyl group to afford secondary amine **98**, which was subject to N-alkylation to afford the tertiary amine **99**. Cleavage of the *N*-phthaloyl group and trapping the resultant primary amine intermediate with isocyanate afforded the urea compound **100**, which was readily converted to EPZ004777 after deprotection. EPZ004777 demonstrated remarkable potency against DOT1L with an IC_{50} value of 0.4 nM and selectivity against DOT1L over a panel of PMTs (>1200-fold). Enzyme kinetics revealed that EPZ004777 is a SAM-competitive and substrate-noncompetitive inhibitor. The SAM-competitive mode of interaction of EPZ00477 with DOT1L was further confirmed by the binary structure of DOT1L in complex with EPZ004777. More importantly, the binary structure of DOT1L in complex with EPZ004777 suggested that DOT1L undergoes a conformational change upon binding the ligand, thereby contributing to the high affinity and selectivity of EPZ004777 against DOT1L vs other PMTs. The cellular assays demonstrated that EPZ004777 caused the dose-dependent reduction of the H3K79me2 mark, the characteristic methylation product of DOT1L. The 6–8-day treatment of EPZ004777 also suppressed the transcription of HOXA9 and MEISI, the target genes of DOT1L in the context

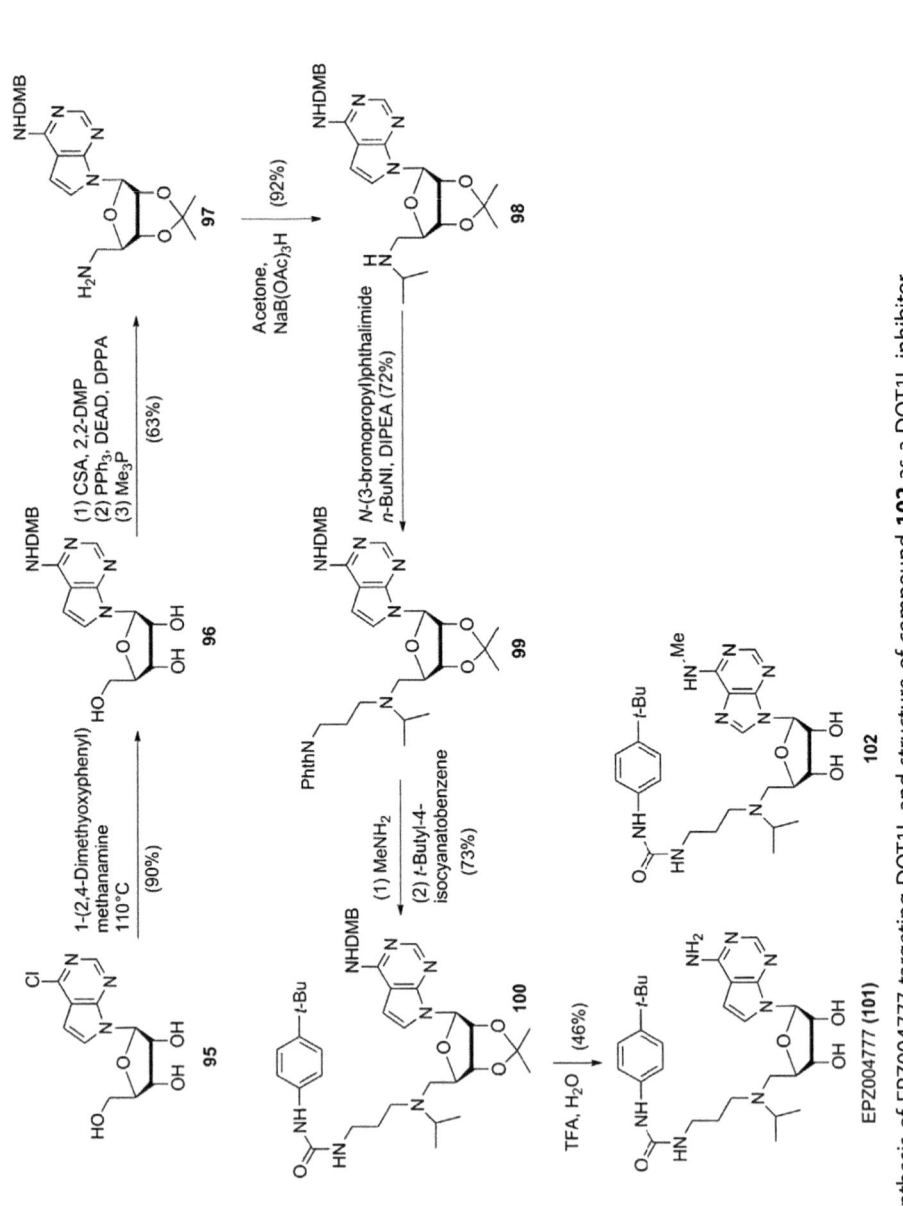

Fig. 17 Synthesis of EPZ004777 targeting DOT1L and structure of compound 102 as a DOT1L inhibitor.

of several MLL-rearranged leukemia cell lines such as MV4-11 and MOLM-13. In the mouse xenograft model of AF9-MLL-rearranged leukemia, the H3K79 methylation level was significantly reduced in the mice treated with EPZ004777, although the delivery was achieved through an implanted osmotic minipump because of its relatively poor pharmacokinetics.

The Song group reported a series of 5′-amino SAM analogs, including compound **102** (Fig. 17), as selective DOT1L inhibitors (Anglin et al., 2012). Inspired by their prior work on the development of DOT1L inhibitors **84** and **85** (Fig. 15), the authors systematically modified the substitutions of N^6-position, 5′-amino linker, and α-amino acid moiety. Their synthetic strategy of these compounds was similar to that of EPZ004777 except methyl acrylate rather than (3-bromopropyl)phthalimide was used to construct the 1,3-propanediamino moiety. SAR analysis of this series of inhibitors revealed that (i) the replacement of the α-amino acid moiety of SAH with a urea functional group led to several selective inhibitors of DOT1L; (ii) the replacement of 5′-thiol with 5′-*iso*-proply amine further increased the affinity to DOT1L by a 100-fold; and (iii) the N^6-modification with small alkyl chains (eg, methyl) could be well tolerated as compound **102** inhibits DOT1L with an IC_{50} value of 0.76 nM.

Yu et al. (2012) reported the design and characterization of SGC0946 (compound **108**) as a more potent DOT1L inhibitor. The X-ray structures of DOT1L in complex with several urea-based SAM analogs revealed the conformational changes in the DOTL1 catalytic site upon binding as well as the existence of a hydrophobic cavity adjacent to the N-7 position of these ligands. The authors thus argued that hydrophobic modification of N-7 would further enhance the affinity and selectivity of urea-based 5′-amino SAM analogs against DOT1L. Here the authors developed a different synthetic route to access SGC0946, in which both the *iso*-propyl and urea moieties were installed via reductive alkylation. The synthesis started with the generation of nucleoside **104** from 1′-chloro-ribose derivative **103** (Fig. 18). Deprotection of the silyl group of **104**, followed by the conversion of the resulting alcohol as well as 6′-chloride to amines, afforded 5′-amino derivative **105**. Primary amine **105** was then subjected to sequential reductive alkylations to afford bis-alkylated product **107**. SGC0946 (compound **108**) was then readily obtained after deprotection of acetonide. SGC0946 showed improved potency in comparison with EPZ004777 (K_d values of 0.06 vs 0.25 nM). Similar to EPZ004777, SGC0946 is highly selective against DOT1L over other PMTs. The improved potency of SGC0946 is also consistent with its better ability to abolish H3K79 methylation levels

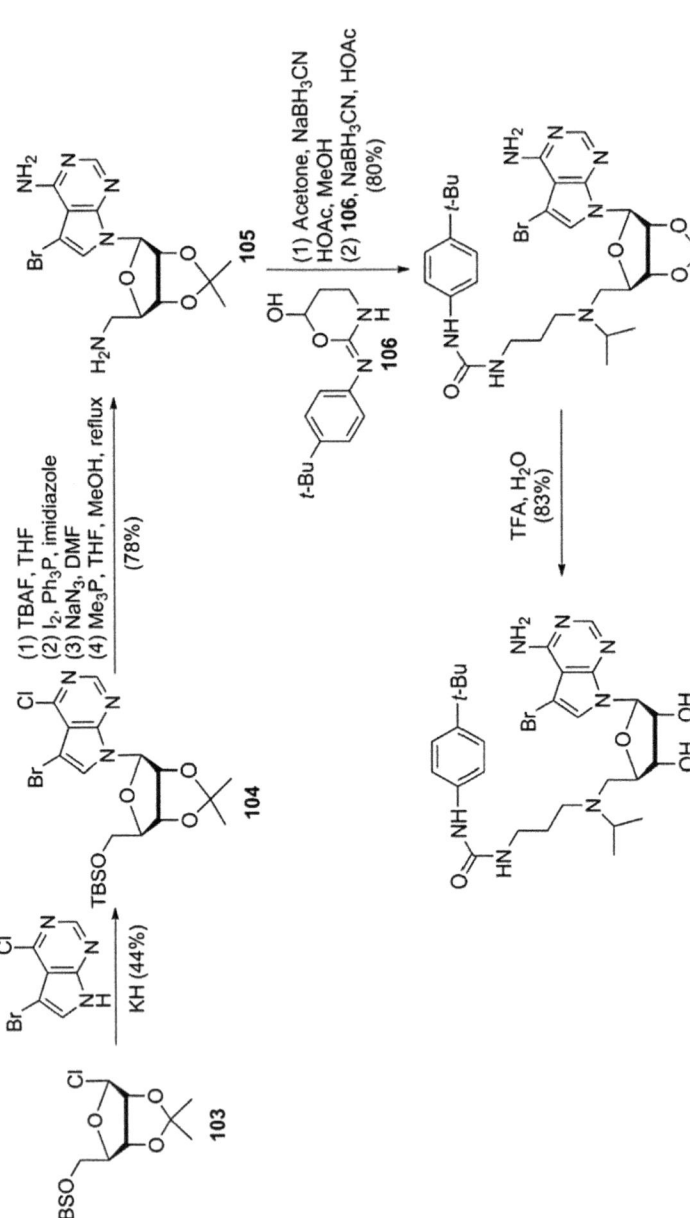

Fig. 18 Synthesis of SGC0946 as a DOT1L inhibitor.

and to reduce the viability of the cord blood cells transformed with the MLL-AF9 fusion protein.

Very recently, Daigle et al. (2013) reported another 5′-amino SAM analog EPZ-5676 (compound **115**) as a clinical DOT1L inhibitor. To synthesize EPZ-5676, 5′-amino derivative **109** went through reductive alkylation with cyclobutanone **110** and N-alkylation with *i*-PrI to afford the bisalkylated amine **111** (Fig. 19) (Olhava et al., 2012). DIBAL reduction, Horner–Wadsworth–Emmons olefination, hydrogenation, and saponification gave rise to the free carboxylic acid **113**. This intermediate was then coupled with diamine **114**, followed by acid-mediated benzimidazole cyclization and deprotection to afford EPZ-5676 (**115**). EPZ-5676 was shown to be a better DOT1L inhibitor than the parent compound EPZ004777 in several aspects. First, it inhibits DOT1L with a K_i value of 0.08 nM in comparison with a K_i value of 0.3 nM of EPZ004777. Second, the residence time of EPZ-5676 was prolonged to >24 h in comparison with 1 h of EPZ004777. Additionally, treating MV4-11 and HL-60 cells with EPZ-5676 for 3–4 days resulted in the selective reduction of the H3K79 methylation mark with EC_{50} of 3 nM. More importantly, pharmacokinetics of EPZ-5676 was much improved compared with that of EPZ004777.

3. EVALUATING METHYLTRANSFERASE INHIBITORS

3.1 Assay Formats

Methyltransferases utilize SAM as a methyl donor, regardless of their substrates as proteins, nucleic acids, or small-molecule substrates, and with the formation of SAH as a by-product (Carmel, Jacobsen, & Hajjar, 2001). Multiple approaches have been developed to assay the activities of methyltransferases by monitoring the accumulation of the methylated products of SAH rather than the disappearance of the substrates because of the slow enzymatic turnover of many methyltransferases (Fig. 20) (Luo, 2012). We describe several assays that are commonly used to measure kinetics of the methylation reaction and evaluate the in vitro potency of methyltransferase inhibitors. While we focus on PMTs, most of these assays can be used to examine DNA, RNA, and small-molecule methyltransferases.

3.2 Radiometric Assays

Detection of $^{14}C/^{3}H$-labeled substrate has thus far been the gold standard for monitoring methyltransferase activity. The assay involves the enzymatic

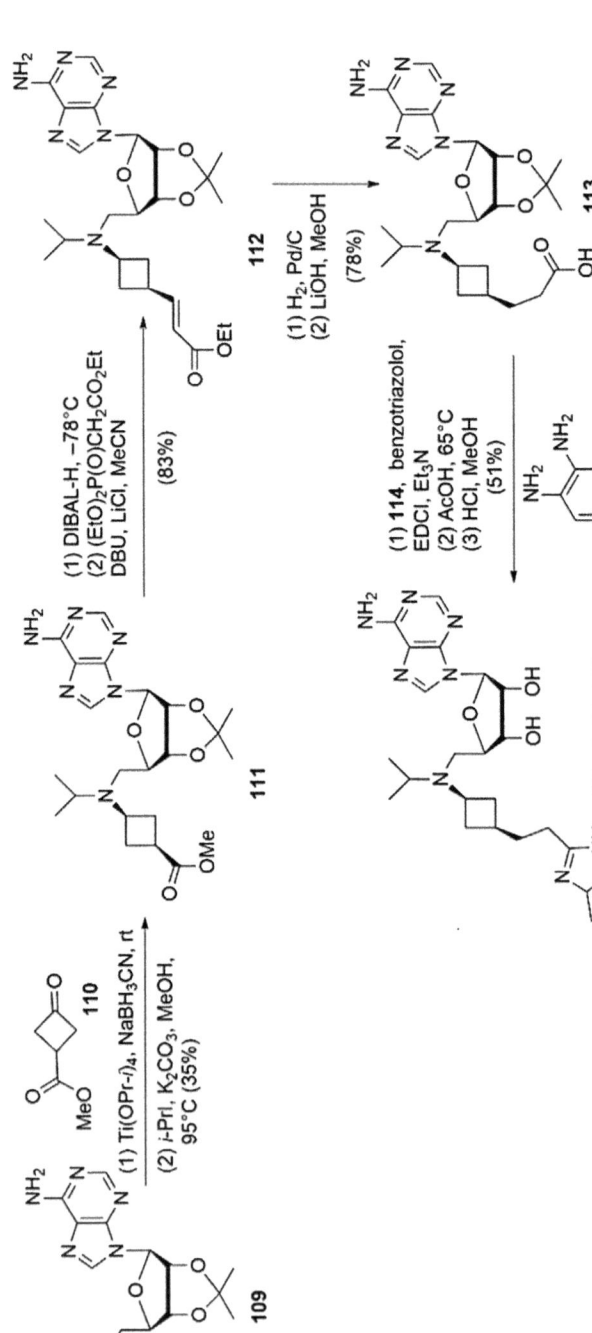

Fig. 19 Synthesis of EPZ-5676 as a clinical DOT1L inhibitor.

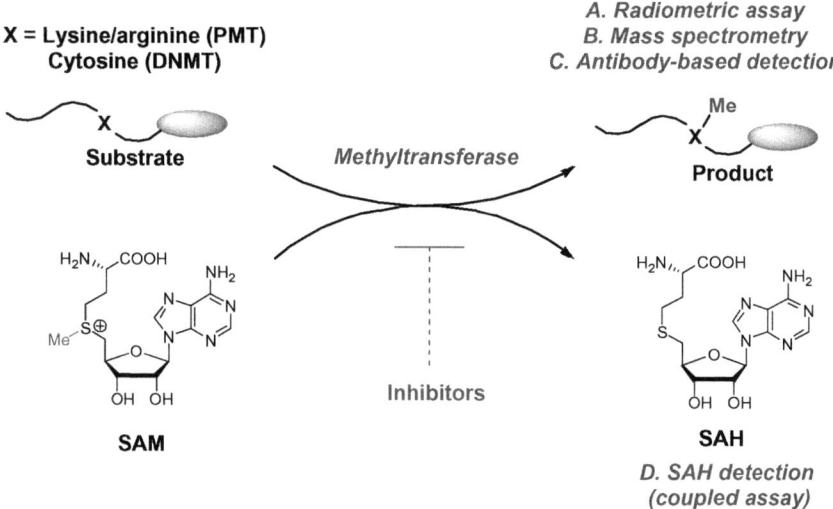

Fig. 20 The reaction catalyzed by methyltransferases and the different assay formats used for evaluating inhibitors. The methyl group is transferred from S-adenosyl-L-methionine to the substrate (ε-nitrogen on Lys; ω-guanidino nitrogen on Arg for protein methyltransferases; C^5-cytosine for DNA methyltransferases). The methylated product is accompanied by formation of SAH. The methylated product can be quantified using: (A) radiometric detection using [^{3}H]-SAM; (B) mass spectrometric quantitation of peptides and protein substrates; (C) antibody-based detection of the methylated products or the by-product SAH; (D) spectrophotometric and luminescent detection using coupled enzymes that detect SAH formation.

transfer of a ^{3}H- or ^{14}C-labeled methyl group to the substrate. The reaction is typically carried out in a 10–20 μL format with [^{3}H]-SAM (or [^{14}C]-SAM) in an appropriate reaction buffer, followed by separation of the labeled product from the unreacted SAM for scintillation counting or autoradiography (Fig. 20). Since this assay is typically performed in an end point format, it requires knowing the kinetic parameters of the enzyme reaction, in particular the region for the linear progression, so that the reaction can be performed in this time window. Moreover, a negative control (either 100% inhibition of enzyme with known inhibitors or no enzyme) is often carried out in parallel for background correction. To implement the assay to screen inhibitors, it is also advised to measure the K_m for the substrates and then perform the assay under "balanced conditions" (the concentrations of the SAM cofactor and the substrate roughly around their K_m values) (Copeland, 2003). The balanced conditions are expected to identify the inhibitors with diverse modes of interaction (eg, competitive, uncompetitive, or noncompetitive with SAM or substrate).

3.2.1 Filter Paper-Based Scintillation Assay

The format of the filter paper-based scintillation assay takes advantage of the ability of cation exchange filters (phospho-cellulose or anion glass fiber) to bind peptides or protein products with greater affinity than free SAM. The reaction is typically initiated by adding an equal volume of a substrate–cofactor mixture (peptide substrate and [^{3}H]-SAM) into the enzyme–inhibitor complex. It is critical to preincubate the enzyme and the inhibitor before initiating the reaction in order to characterize the inhibitors that act on the target in a time-dependent manner. The reaction is allowed to proceed for a predetermined amount of time within the time window of the linear progression and then quenched by either acidifying the reaction or adding an excess of unlabeled SAM. The reaction mixture is then spotted on individual filter paper or filters assembled in a 96-well plate format. Subsequently, the unreacted [^{3}H]-SAM is washed away using 50 mM sodium bicarbonate pH 9.2 buffer. The immobilized [^{3}H]-labeled peptide can be quantified by a scintillation counter. This method has been extensively used to identify and characterize inhibitors (compounds **16**, **20**, **63**, **77**, **101**, **108**, and **115**) targeting multiple PMTs including DOT1L, EZH1/2, SETD2, SETD8, SET7/9, and G9a/GLP (Basavapathruni et al., 2012; Daigle et al., 2011; Horiuchi et al., 2013; Kung et al., 2015; McCabe et al., 2012; Yao et al., 2011; Yu et al., 2012, 2013; Zheng et al., 2012). While several obvious advantages of this method are its high sensitivity, a straightforward protocol, and easily accessible reagents, the assay is laborious and time consuming especially if the labeled substrate has to be separated from SAM using SDS electrophoresis (compounds **48** and **94**) (Dowden et al., 2010; Luo, 2012; Suh-Lailam & Hevel, 2010). The SDS separation is then followed by excising the protein band directly from the gel or from a nitrocellulose membrane after a transfer step (Dowden et al., 2010, 2011). In order to expedite the process time of the assay and reduce the radioactive waste, Hevel et al. utilized ZipTip$_{c4}$ pipette tips to separate the radiolabeled products from unreacted [^{3}H]-SAM. The latter method allowed the total process time to be reduced from several hours to a maximum of 15 min (Luo, 2012; Suh-Lailam & Hevel, 2010).

The filter paper assay can be easily adapted to study DNMT inhibition by using DEAE/DE81 filter paper in comparison with phospho-cellulose filter paper for protein substrates, followed by sequential washing of the unreacted [^{3}H]-SAM with 0.2 M ammonium bicarbonate, water, and ethanol. This radiometric assay has been used to measure the kinetic parameters of DNMTs and characterize resveratrol and SAH mimics (compound **26**) as

DNMT inhibitors (Aldawsari et al., in press; Hemeon, Gutierrez, Ho, & Schramm, 2011; Isakovic et al., 2009). Due to the straightforward nature of this assay, it has also been adapted to evaluate the in vitro potency of bisubstrate inhibitors (**36, 41**, and **44–46**) of COMT (Lerner et al., 2003; Masjost et al., 2000).

3.2.2 Scintillation Proximity Assay

The Jeltsch and Zheng groups, with an effort to minimize the washing and pipetting steps, developed the scintillation proximity assay (SPA) to monitor the activities of PMTs (Rathert, Cheng, & Jeltsch, 2007; Wu, Xie, Feng, & Zheng, 2012). This assay involves the use of biotinylated peptides as substrates for the methylation reaction along with [^{3}H]-SAM as the cofactor. Subsequent to the methylation reaction, the [^{3}H]-labeled peptide is immobilized on streptavidin-coated microscopic capsid beads containing the scintillation fluid (Fig. 21). The captured [^{3}H]-labeled peptide on the beads creates the proximity between the [^{3}H] β-emitter and the scintillant fluid, resulting in luminescence. Since the β emission from the unreacted [^{3}H]-SAM is relatively remote to the scintillation beads and thus quenched by the solvent, this method obviates the need for separating the unreacted cofactor and then allows rapid quantitation of the reaction product. The SPA has been used to study enzyme kinetics, conduct HTS of PMT inhibitors, and characterize lead compounds for G9a (Dhayalan, Dimitrova, Rathert, & Jeltsch, 2009), PRMT5/7 (Smil et al., 2015), Set7/9, and SETD2 (Ibanez et al., 2012; Zheng et al., 2012). For instance, Smil et al. (2015) used the SPA to aid their structure-based design of PRMT5 inhibitors. They also successfully implemented the SPA to screen the inhibitory activity of DS-437 (**23**) against a panel of methyltransferases, to measure the IC$_{50}$ values, and to characterize the cofactor-competitive inhibition of **23** against PRMT5 and PRMT7. This compound showed low micromolar IC$_{50}$ values against PRMT5-MEP50 and PRMT7 (5.9 and 6 μM, respectively) and demonstrated SAM-competitive binding for their lead compound. Ibanez et al. (2012) implemented the SPA in an HTS format to screen a compound library (~6–7K) against SETD8, GLP, SET7/9, and SETD2. Optimization of the enzyme concentrations, reaction conditions, and the enzyme tolerance test for DMSO was performed using a standard radiometric filter paper assay. The optimized assay conditions were then transformed to a high-throughput SPA platform to calculate the Z' score for the PMTs. The hits were then identified with the SAP assay in an HTS format and confirmed with the filter paper assay as a secondary assay (Ibanez et al., 2012).

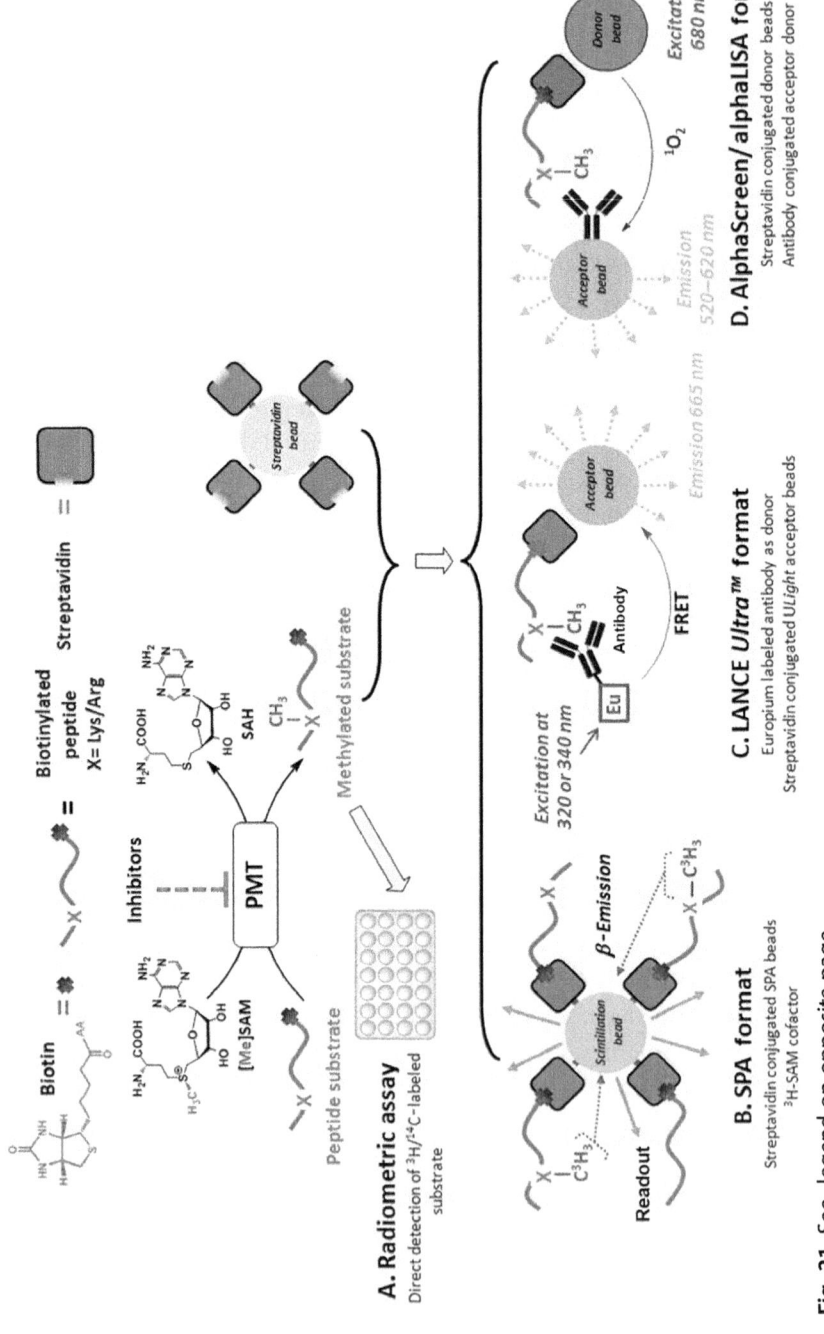

Fig. 21 See legend on opposite page.

3.3 Fluorescence-Based Detection of Methyltransferase Activity

Fluorescence polarization (FP)-based detection methods have been utilized by several groups for monitoring the activities and screening inhibitors of methyltransferases (Cheng et al., 2004; Feng, Xie, Wu, Yang, & Zheng, 2009; Graves, Zhang, & Scott, 2008; Liu et al., 2009; Yi et al., 2015). Fluorescence polarization or anisotropy is based on the principle that the degree of polarization of a fluorescent probe is inversely related to its molecular rotation. The binding of a fluorophore to methyltransferases results in the increase of FP signal, while the binding of a competing ligand results in the release of the bound fluorophore and thus decreases the FP signal. Fluorescein-labeled peptide substrates can thus be used as an FP probe to monitor the binding of substrate-competitive inhibitors. This technology has been implemented to characterize peptide-competitive inhibitors AIM-1 (PRMT1) and UNC0224 (G9a) (Cheng et al., 2004; Feng et al., 2009; Liu et al., 2009). A similar approach can be used to characterize SAM-competitive inhibitors as exemplified by commercial SAM-Screener™ kits tailored for specific PMTs (eg, MLL1, SET7/9, and G9a/GLP). Bradner et al. relied on the FP assay, in conjunction with an AlphaScreen assay and a cellular assay, to identify and characterize selective DOT1L inhibitors (Yi et al., 2015).

Anisotropy-based fluorescence assays are sensitive, easily adaptable to an HTS format, and free of radioactive materials. However, a direct FP assay requires a high amount of the enzyme for a robust signal and therefore

Fig. 21 Assay formats to evaluate inhibitors targeting methyltransferases. (A) Radiometric filter paper/plate assay is based on the transfer of the radiolabeled methyl group from [^{3}H]-SAM to the substrate. β-Emission from the radiolabel to the scintillant fluid results in luminescence signal as a readout. (B) Scintillation proximity assay is a variant of the radiometric assay featured by the capture of the [^{3}H]-labeled substrate on streptavidin-conjugated scintillant beads. (C) LANCE *Ultra*™ involves the capture of the biotinylated peptide on streptavidin-coated acceptor beads. The methylated peptide is recognized by lanthanide (Eu)-labeled antibodies. Förster resonance energy transfer between the lanthanide donor ($\lambda_{ex}=320–340$ nm) and the acceptor beads gives a fluorescence emission at 665 nm. (D) AlphaScreen(AlphaLISA) format involves the capture of the methylated, biotin-tagged peptide product on streptavidin-coated donor beads. The methylated peptide is recognized by antimethyllysine antibody-conjugated on acceptor beads. Excitation of the donor beads at 680 nm results in emission of a singlet oxygen ($^1O_2$), which excites the acceptor beads causing an emission at 520–620 nm (AlphaScreen) or at 615 nm (AlphaLISA). *Adapted from Luo, M. (2012). Current chemical biology approaches to interrogate protein methyltransferases. ACS Chemical Biology, 7(3), 443–463.* (See the color plate.)

has limitations for a large-scale screening. With the commercially available anti-SAH antibody in combination with SAH-fluorescein conjugate, Scott et al. developed the comparable FP methodology to monitor the accumulation of the by-product SAH generated during the COMT-catalyzed methylation reaction (Graves et al., 2008). While the assay is extremely sensitive with a detection threshold of 0.15 pmol SAH (5 nM for a 30-μL reaction mixture), the cross-reactivity of the antibody with SAM sets an upper limit for the amount of SAM that can be used in the assay (3 μM for a 30-μL reaction mixture). Recently, multiple fluorescence-based HTS strategies have been developed wherein the emission of the acceptor fluorophore is excited either through Förster resonance energy transfer signal from a nearby lanthanide-labeled donor (LANCE *Ultra* to characterize the compound **75**) (Devkota et al., 2014; Gauthier et al., 2012) or through a singlet oxygen species generated from nearby Alpha donor beads (AlphaScreen/AlphaLISA) (Gauthier et al., 2012; Yi et al., 2015) (Fig. 21). The acceptor–donor proximity in the two cases is achieved by the biotinylated methylated linker peptides that connect the antimethyllysine antibodies (the donor for LANCE *Ultra* and the acceptor for AlphaScreen/AlphaLISA) and the streptavidin-coated florescent beads (the acceptor bead for LANCE *Ultra* and the donor bead for AlphaScreen/AlphaLISA).

3.4 Directly Monitoring SAH Formation

SAH is known to be a feedback inhibitor of methyltransferases (Deguchi & Barchas, 1971). One merit of the assays coupled with SAH degradation is to eliminate the potential inhibitory effect of SAH as a by-product of methylation reactions. SAH can be readily hydrolyzed by MTA/SAH-nucleosidase (MTAN) into adenine and S-ribosylhomocysteine (Luo, 2012). Adenine can then be processed by adenine deaminase to product hypoxanthine, a process that can be monitored by the decrease of the absorbance at 265 nm (Dorgan et al., 2006). While this SAH-based assay can provide a convenient continuous readout, it may suffer from a high background because of the potential overlap with the absorbance spectra of protein substrates and inhibitors. One way of minimizing the overlap with other assay components is to use other coupled enzymes such as xanthine oxidase, which can further convert hypoxanthine into fluorogenic xanthine and xanthine derivatives (Luo, 2012). Alternatively, SAH can be converted into adenosine and homocysteine by SAH hydrolase for quantifying SAH's production via the free thiol residue of homocysteine (Hendricks, Ross,

Pichersky, Noel, & Zhou, 2004; Wang, Leffler, Thompson, & Hrycyna, 2005). A similar outcome can also be achieved by coupling the methylation reaction with MTAN and S-ribosylhomocysteinase (LuxS) (Hendricks et al., 2004). The free thiol of homocysteine released by SAH hydrolase or LuxS can then be quantified using Ellman's reagent or other thiol-reactive dyes such as Thio-Glo1 (Collazo, Couture, Bulfer, & Trievel, 2005) and 7-diethyl-3-(4′-maleimidylphenyl)-4-methylcoumarin (Ibanez, McBean, Astudillo, & Luo, 2010). While this assay avoids the use of expensive radioactive materials and can tolerate a wide range of the concentrations of substrates and the SAM cofactor, it has the limitation because of its high background of interfering cysteine residues of enzymes and substrates. However, such issue can be addressed by a luminescence-based detection method. In this scenario, the adenine generated from SAH by MTAN can be further coupled with adenine phosphoribosyl transferase and pyruvate orthophosphate dikinase to generate ATP. The resultant ATP can then be quantified using the highly sensitive luciferase/luciferin assay kits (ATPlite). The luciferase/luciferin-based assay is extremely sensitive with a threshold to detect 0.3 pmol SAH and shows a broad linear response to the concentrations of SAH (Ibanez et al., 2010).

These aforementioned coupled assays are excellent for kinetic measurement of methyltransferases in general because of their feasibility to examine a wide range of the concentrations of substrates and the SAM cofactor, and their adaptability to different types of substrates. Furthermore, because of their adaptability into an HTS format, these assays are thus amenable to inhibitor screening and subsequent characterization of hit compounds. In contrast, a limitation of these assays for HTS, as with all coupling assays, is the potential of high false positive rates because of the likely cross-inhibition against the coupling enzymes.

3.5 Mass Spectrometry-Based Detection of Methylated Product

Despite being used extensively, the previously mentioned assays are not suitable to determine the sites and states of methylation. Mass spectrometry (MS) is therefore a complimentary method in this aspect. Mono-, di-, and trimethylation states of lysine methylation can be readily distinguished by a +14, +28, or +42 mass shifts, respectively. Moreover, since the addition of a methyl group does not significantly alter the size and the charge of substrates, the MS peak ratios of the methylated peptides can be used directly to quantify the progression of the different states of methylation reaction (Chang et al.,

2009; Islam, Zheng, Yu, Deng, & Luo, 2011; Patel, Dharmarajan, Vought, & Cosgrove, 2009). As an earlier example, this assay format was utilized by Lin et al. (2001) to assist their rational probe design to target the E117G variant of *Saccharomyces cerevisiae* Rmt1 (compounds **10** and **11**). Moreover, a top-down MS approach allows for mapping the site of methylation, especially when working with protein substrates of decent molecular weights.

3.6 Detailed Protocol for Radiometric Filter Paper Assay

In this section, the protocol of the radiometric assay will be documented. This assay, conducted by monitoring the reaction with a radioactive (^{3}H/^{14}C) tracer, is extremely straightforward and therefore stands as a gold standard. While the focus of this protocol is the filter paper-based assay as a representative example, the method can be easily coupled with other detection methods such as autoradiography for ^{3}H/^{14}C-SAM and MS for nonradioactive SAM.

3.6.1 Materials
- Methyltransferase of interest.
- [^{3}H]-SAM (Perkin Elmer)—5–15 Ci/mmol or 55–85 Ci/mmol.
- Protein or peptide substrate of interest—peptide substrates are preferred if the assay is expected to be adapted into a SPA format.
- 100 nM MTAN—to alleviate potential SAH inhibition.
- Optimized reaction buffer: There are multiple buffer conditions that can be used for assaying methyltransferase activity including 50 mM HEPES pH 8.0, 50 mM Tris–HCl pH 8.0. It is important to note that SAM is unstable at alkaline pH (Parks & Schlenk, 1958), so one should avoid performing the SAM-dependent reactions at a pH higher than 8.5. Additional components to the reaction buffer include a reducing agent (1 mM TCEP or 1 mM DTT), MgCl$_2$ (5–10 mM), bovine serum albumin (BSA) (0.0005%), and detergents (0.005% Tween). It is advisable to make the concentrated stocks of BSA and the reducing agents in advance and store them at -20°C until use.

Note
1. Optimization of the buffer and enzyme conditions requires a few trials.
2. BSA and the detergent are required to stabilize some enzymes. Their use and optimal concentrations can vary on the basis of the stability of the enzymes and the temperatures at which the reactions are conducted.

3.6.2 Preliminary Test for Methyltransferase Activity

For preliminary experiments, an enzyme concentration of 1 μM, a substrate concentration of 1 μM, and a SAM cofactor concentration of 0.5 μM are good starting points.

1. The reactions are initiated by adding 10 μL of the substrate–cofactor mix into 10 μL of the enzyme solution. The methylation reaction is allowed to proceed for approximately 12–15 h at ambient temperature. The experiment without the enzyme is set up in parallel as a negative control.
2. It is prudent to set up another reaction containing MTAN to obviate the potential low readout that may arise by SAH inhibition.
3. The reaction is quenched by either acidification of the reaction mixture (1–2% trifluoroacetic acid) or adding a micromolar excess of unlabeled SAM as a cold carrier.
4. Subsequently, 6 μL of the reaction mixture is spotted on a 1.2 cm × 1.2 cm grid of phospho-cellulose (P-81) filter paper (P-81 filter paper plate/fiber glass filter plate as alternatives). Typically, the reaction mixture is spotted three times as technical replicates.
5. The filters are allowed to dry for 0.5–2 h and then washed five times with 50 mM sodium bicarbonate (pH 9.2) buffer to remove unreacted [^{3}H]-SAM.
6. The filters are allowed to dry for ~0.5 h and then immersed in the scintillant fluid (7 mL scintillant/0.5 mL doubly deionized water or 50–100 μL scintillant fluid/well if a filter plate is used). The samples are allowed to incubate for ~1 h and read on a scintillation counter (eg, Perkin Elmer; Liquid Scintillation Analyzer Tri-Carb 2910 TR, or MicroBeta2 Plate Counter).
7. An approximately 10-fold or higher difference between the positive and the negative control is considered acceptable.
8. Once methyltransferase activity is confirmed with the biochemical assay, the buffer components and enzyme concentrations can be further optimized for optimal signal-to-noise ratio.

3.6.3 Measurement of Steady-State Kinetic Parameters

1. After finalizing the assay conditions, a time course measurement will be performed.
2. Set up the reaction under the optimized conditions, albeit in a larger volume.
3. A negative control is not required for the time course experiment as the time zero data set can be used as the blank.

4. Monitor the increase of the methylation signal as the function of time. If necessary, adjust the concentrations of the SAM cofactor, substrate, and enzyme to assure a linear increase in the signal for at least ~30 min without significantly compromising signal-to-noise ratios. Note that the slope of the linear region is proportional to the final concentration of the methyltransferase used in the assay.
5. Measure the initial velocity of the reaction as a function of the concentrations of the substrate and the SAM cofactor to obtain their respective K_m values and the k_{cat} (Segel, 1976).

3.6.4 Evaluating Potency of Inhibitors In Vitro

The first step of characterizing a lead inhibitor is to obtain the concentration required to achieve 50% inhibition of the target enzyme. As a rough starting point, the enzyme is preincubated with a range of concentrations of the inhibitor for 20–30 min. This reaction is initiated by adding the aforementioned substrate–cofactor mix. After waiting for a specific period of time (a linear region of enzyme activity), the reaction is quenched and processed with at least two to three technical replicates. Percentage of inhibition is calculated using Eq. (1):

$$\% \text{ Inhibition} = \frac{(R_{pos} - R_{[I]})}{(R_{pos} - R_{neg})} * 100 \qquad (1)$$

where R_{pos} and R_{neg} are the signal of the positive control (without inhibitor) and the negative control (without enzyme), and $R_{[I]}$ is the radiometric signal in the presence of a given concentration of the inhibitor. The data are then fit to a nonlinear regression (Eq. [2]) to obtain the IC_{50} value (see Fig. 22 as an example):

$$\frac{v_i}{v_o} = 1 + \frac{[I]}{IC_{50}} \qquad (2)$$

where the term of v_i/v_o is the inhibition percentage at a given concentration of the inhibitor [I]. After obtaining the IC_{50} value, it is critical to investigate the mode of inhibition. These following experiments include measuring the dependence of the IC_{50} values on the concentrations of the substrate and the SAM cofactor to obtain the K_i value for the inhibitor (Copeland, 2013b, 2013c). For the compounds that exhibit slow-onset inhibition mechanism (time-dependent inactivation of enzyme), k_{off}, k_{on}, K_i (K_i^*),

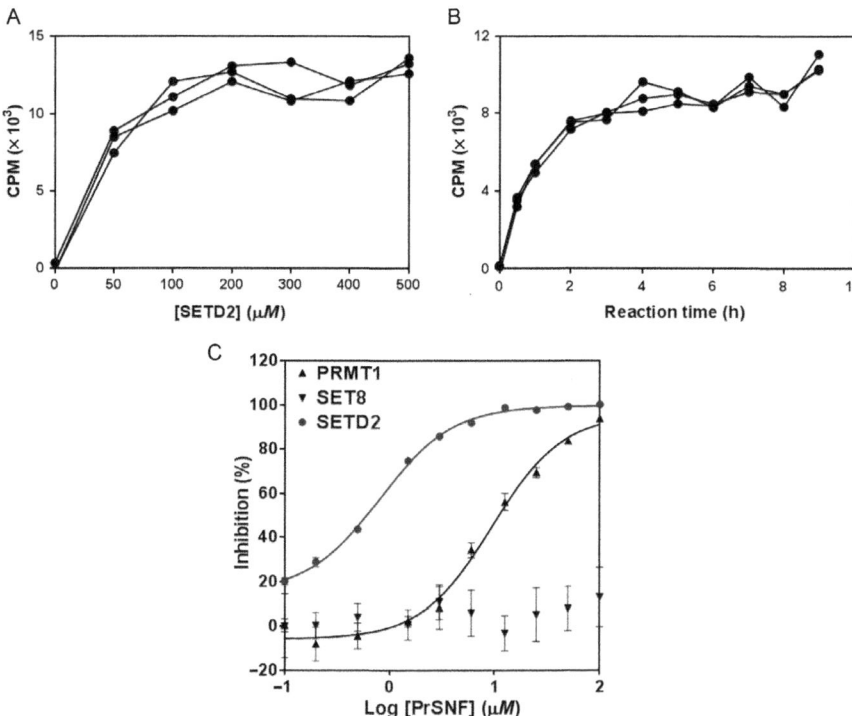

Fig. 22 Representative data for assay optimization and IC$_{50}$ measurement with the radiometric assay. (A) Three replicates of the radioactivity of methylated production after an overnight reaction as a function of SETD2 concentration. (B) Time course of SETD2-mediated methylation with 250 nM enzyme (Ibanez et al., 2012). (C) Dose–response curves of PrSNF against PMTs including SETD2 •, SET8 ▼, PRMT1 ▲ (Zheng et al., 2012).

and k_{inact}/K_i, if applicable, should be characterized to further evaluate these inhibitors (Copeland, 2013a, 2013d).

The aforementioned strategy was previously demonstrated by the Luo group to characterize the inhibition of 5′-aminopropyl sinefungin (compound **63**) and other sinefungin derivatives against SETD2 as well as a panel of PMTs (Zheng et al., 2012). These analogs were designed through a structure-based rational approach and were identified as potent and selective SETD2 inhibitors. The assay conditions were optimized using the previously described radiometric assay (Section 3.6). Subsequently, the dose–response experiments were conducted using the optimized conditions (250 nM SETD2). After measuring the IC$_{50}$ values of the inhibitor against a panel of methyltransferases (Fig. 22), the authors investigated the mode of inhibition followed by a structural analysis of SETD2 in complex with **63**.

4. CONCLUSION

In this chapter, we have discussed different strategies to synthesize and evaluate SAM-based methyltransferase inhibitors. Dysregulation of methyltransferases has been shown to have multiple outcomes including, but not limited to, developmental disorders and tumorigenesis (Gnyszka et al., 2013; Jones et al., 2007; Kelly et al., 2010; Swierczynski et al., 2015; Urdinguio et al., 2009; Villeneuve & Natarajan, 2010; Wu, Sarkissyan, et al., 2015). Due to the complicated roles of methyltransferases in normal human physiology and disease states, it is extremely valuable to develop potent and selective inhibitors not only for therapeutic intervention but also as probes to elucidate the biology of these targets. As discussed in this chapter, the sinefungin/SAM structural motif provides a valuable starting point to develop pan- and selective methyltransferase inhibitors. As a privileged scaffold, the sinefungin/SAM motif also enables one to probe the potential differences of transition states across a panel of closely related methyltransferases. Toward this goal, a structural understanding of the apo enzyme and the enzyme bound to either substrate(s) or the inhibitor will provide information on the reaction coordinate along the pathway of the enzyme-catalyzed reaction and likely present specific conformational states that are amenable for selective inhibition by small-molecule inhibitors. For example, the structure of SETD2 in complex with PrSNF (**63**) enabled the Luo group to probe the specific conformational states for SETD2's catalysis and thus provide mechanistic insight for designing a selective inhibitor (Zheng et al., 2012). Yu et al. (2013) demonstrated the utility of the structure-based rational design to develop bromo-deaza-SAH analog (**16**) as a selective inhibitor of DOT1L. The implementation of the SAM/sinefungin scaffold further lies in developing high-quality molecular probes to elucidate the biological roles of the target enzymes. It is worth noting that there are still a large number of methyltransferases whose biological functions are unclear (Hamamoto et al., 2015; Lucio-Eterovic & Carpenter, 2011). Development of selective inhibitors is useful toward conducting cell-based experiments to delineate the functional roles of these targets.

Stepwise evaluation of inhibitors is critical to guide the optimization of lead compounds. Given that the choices of available assays vary from case to case, one must be cautious upon selecting appropriate assays. For instance, enzyme-coupled assays that detect the formation of the SAH by-product are excellent tools for monitoring methyltransferase activity. However, these assays are often associated with a high rate of false positives in the context of an HTS format. Monitoring changes of fluorescence polarization signals using a fluorescently labeled substrate or cofactor mimic is another sensitive way to monitor methyltransferase activities (Cheng et al., 2004; Feng et al., 2009; Liu et al., 2009). However, it may require prohibitive amounts of the target enzymes. The amount of the enzyme required can be reduced when the SAH production can be qualified with a fluorescein-SAH conjugate as a probe in combination with anti-SAH antibody (Graves et al., 2008). While the SAH-based fluorescence polarization assay is highly sensitive, the cross-reactivity of the anti-SAH antibody to SAM makes it difficult to examine multiple methyltransferases with intrinsic high $K_{m,SAM}$ values (1–25 μM) (Graves et al., 2008). Given the robustness of the radiometric assay, we detailed this assay to allow the reader to design their own experiments with more convenience. The advantage of this assay lies not only in its straightforward feature and readily accessible reagents but also in its high sensitivity and low false positive rates. It is essential to know that, prior to evaluating inhibitors, enough time and effort should be spent to carefully characterize the enzyme and the reaction conditions to ensure a robust, reproducible assay with decent signal-to-noise ratios.

The high degree of homology of the cofactor-binding pockets of PMTs can make it challenging to design selective inhibitors for a specific target. Following the identification of a lead compound or scaffold, a structural understanding of the lead–target interaction is essential for further optimization. In order to improve selectivity of lead compounds, the team requires a nuanced understanding of the enzyme's active site as well as be equipped with medicinal chemistry prowess. A comprehensive analysis of SAR will eventually help correlate the SAR in the relevant biological setting. A recent example of such an accomplishment is the development of EPZ004777 (**101**) as a potent and selective SAM-competitive inhibitor of DOT1L (Daigle et al., 2011). The potency, selectivity, and pharmacokinetic properties of EPZ004777 were further optimized to lead the development of EPZ5676 (**115**), a picomolar inhibitor of DOT1L with >37,000-fold selectivity against other PMTs (Daigle et al., 2013).

Despite recent advances in developing potent SAM mimic inhibitors, there are a few limitations to be considered before embarking on a rational design project to further pursue this scaffold. While SAM is expected to a privileged scaffold against methyltransferases, most SAM-based probes show very poor cell permeability. This limitation therefore prevents many SAM analog inhibitors from being used in cellular assays. However, multiple strategies can be employed to overcome the issue of membrane permeability and thus improve the pharmacokinetic properties of this scaffold. For example, despite limited bioisosteres of the amino acid moiety for SAM analog inhibitors, certain modifications (eg, the urea in EPZ004777 against DOT1L) can be tolerated and show improved pharmacokinetic properties. Another potential strategy is to develop cell permeable prodrug species that release the SAM analog inhibitors inside the cells.

REFERENCES

Alaimo, P. J., Shogren-Knaak, M. A., & Shokat, K. M. (2001). Chemical genetic approaches for the elucidation of signaling pathways. *Current Opinion in Chemical Biology, 5*(4), 360–367.

Aldawsari, F. S., Aguayo-Ortiz, R., Kapilashrami, K., Yoo, J., Luo, M., Medina-Franco, J. L., et al. (in press). Resveratrol-salicylate derivatives as selective DNMT3 inhibitors and anticancer agents. *Journal of Enzyme Inhibition and Medicinal Chemistry*, 1–9. PMID: 26118420.

Anglin, J. L., Deng, L., Yao, Y., Cai, G., Liu, Z., Jiang, H., et al. (2012). Synthesis and structure-activity relationship investigation of adenosine-containing inhibitors of histone methyltransferase DOT1L. *Journal of Medicinal Chemistry, 55*(18), 8066–8074.

Antonysamy, S., Bonday, Z., Campbell, R. M., Doyle, B., Druzina, Z., Gheyi, T., et al. (2012). Crystal structure of the human PRMT5:MEP50 complex. *Proceedings of the National Academy of Sciences of the United States of America, 109*(44), 17960–17965.

Arrowsmith, C. H., Bountra, C., Fish, P. V., Lee, K., & Schapira, M. (2012). Epigenetic protein families: A new frontier for drug discovery. *Nature Reviews Drug Discovery, 11*(5), 384–400.

Avila, M. A., Garcia-Trevijano, E. R., Lu, S. C., Corrales, F. J., & Mato, J. M. (2004). Methylthioadenosine. *The International Journal of Biochemistry & Cell Biology, 36*(11), 2125–2130.

Barski, A., Cuddapah, S., Cui, K., Roh, T. Y., Schones, D. E., Wang, Z., et al. (2007). High-resolution profiling of histone methylations in the human genome. *Cell, 129*(4), 823–837.

Basavapathruni, A., Jin, L., Daigle, S. R., Majer, C. R., Therkelsen, C. A., Wigle, T. J., et al. (2012). Conformational adaptation drives potent, selective and durable inhibition of the human protein methyltransferase DOT1L. *Chemical Biology & Drug Design, 80*(6), 971–980.

Bedford, M. T., & Clarke, S. G. (2009). Protein arginine methylation in mammals: Who, what, and why. *Molecular Cell, 33*(1), 1–13.

Belinsky, S. A., Nikula, K. J., Baylin, S. B., & Issa, J. P. J. (1996). Increased cytosine DNA-methyltransferase activity is target-cell-specific and an early event in lung cancer. *Proceedings of the National Academy of Sciences of the United States of America, 93*(9), 4045–4050.

Bird, A. (2002). DNA methylation patterns and epigenetic memory. *Genes & Development, 16*(1), 6–21.

Bissinger, E.-M., Heinke, R., Sippl, W., & Jung, M. (2010). Targeting epigenetic modifiers: Inhibitors of histone methyltransferases. *Medicinal Chemistry Communications, 1*(2), 114–124.

Blanchard, P., El Kortbi, M. S., Fourrey, J. L., Lawrence, F., & Robert-Gero, M. (1996). Relationship between chemical structure and antileishmanial effect of sinefungine analogues. *Nucleosides & Nucleotides, 15*(6), 1121–1135.

Blanco, S., Dietmann, S., Flores, J. V., Hussain, S., Kutter, C., Humphreys, P., et al. (2014). Aberrant methylation of tRNAs links cellular stress to neuro-developmental disorders. *The EMBO Journal, 33*(18), 2020–2039.

Cao, R., Wang, L., Wang, H., Xia, L., Erdjument-Bromage, H., Tempst, P., et al. (2002). Role of histone H3 lysine 27 methylation in polycomb-group silencing. *Science*, *298*(5595), 1039–1043.

Carmel, R., Jacobsen, D. W., & Hajjar, K. A. (2001). *Homocysteine in health and disease*. Cambridge, NY: Cambridge University Press.

Chang, Y. Q., Zhang, X., Horton, J. R., Upadhyay, A. K., Spannhoff, A., Liu, J., et al. (2009). Structural basis for G9a-like protein lysine methyltransferase inhibition by BIX-01294. *Nature Structural & Molecular Biology*, *16*(3), 312–317.

Chavez-Blanco, A., Perez-Plasencia, C., Perez-Cardenas, E., Carrasco-Legleu, C., Rangel-Lopez, E., Segura-Pacheco, B., et al. (2006). Antineoplastic effects of the DNA methylation inhibitor hydralazine and the histone deacetylase inhibitor valproic acid in cancer cell lines. *Cancer Cell International*, *6*, 2.

Cheng, D., Yadav, N., King, R. W., Swanson, M. S., Weinstein, E. J., & Bedford, M. T. (2004). Small molecule regulators of protein arginine methyltransferases. *The Journal of Biological Chemistry*, *279*(23), 23892–23899.

Cheung, N., Chan, L. C., Thompson, A., Cleary, M. L., & So, C. W. (2007). Protein arginine-methyltransferase-dependent oncogenesis. *Nature Cell Biology*, *9*(10), 1208–1215.

Cho, H. S., Shimazu, T., Toyokawa, G., Daigo, Y., Maehara, Y., Hayami, S., et al. (2012). Enhanced HSP70 lysine methylation promotes proliferation of cancer cells through activation of Aurora kinase B. *Nature Communications*, *3*, 1072.

Cho, E. C., Zheng, S., Munro, S., Liu, G., Carr, S. M., Moehlenbrink, J., et al. (2012). Arginine methylation controls growth regulation by E2F-1. *The EMBO Journal*, *31*(7), 1785–1797.

Chuikov, S., Kurash, J. K., Wilson, J. R., Xiao, B., Justin, N., Ivanov, G. S., et al. (2004). Regulation of p53 activity through lysine methylation. *Nature*, *432*(7015), 353–360.

Cole, P. A. (2008). Chemical probes for histone-modifying enzymes. *Nature Chemical Biology*, *4*(10), 590–597.

Collazo, E., Couture, J. F., Bulfer, S., & Trievel, R. C. (2005). A coupled fluorescent assay for histone methyltransferases. *Analytical Biochemistry*, *342*(1), 86–92.

Copeland, R. A. (2003). Mechanistic considerations in high-throughput screening. *Analytical Biochemistry*, *320*(1), 1–12.

Copeland, R. A. (2012). Protein methyltransferase inhibitors as personalized cancer therapeutics. *Drug Discovery Today: Therapeutic Strategies*, *9*(2–3), e83–e90.

Copeland, R. A. (2013a). Irreversible enzyme inactivators. Evaluation of enzyme inhibitors in drug discovery (pp. 345–382). New Jersey: John Wiley & Sons, Inc.

Copeland, R. A. (2013b). *Quantitative biochemistry in the pharmacological evaluation of drugs. Evaluation of enzyme inhibitors in drug discovery*. New Jersey: John Wiley & Sons, Inc., pp. 383–469

Copeland, R. A. (2013c). *Reversible modes of inhibitor interactions with enzymes. Evaluation of enzyme inhibitors in drug discovery*. New Jersey: John Wiley & Sons, Inc., pp. 57–121

Copeland, R. A. (2013d). *Slow binding inhibitors. Evaluation of enzyme inhibitors in drug discovery*. New Jersey: John Wiley & Sons, Inc., pp. 203–244

Copeland, R. A., Solomon, M. E., & Richon, V. M. (2009). Protein methyltransferases as a target class for drug discovery. *Nature Reviews Drug Discovery*, *8*(9), 724–732.

Czermin, B., Melfi, R., McCabe, D., Seitz, V., Imhof, A., & Pirrotta, V. (2002). Drosophila enhancer of Zeste/ESC complexes have a histone H3 methyltransferase activity that marks chromosomal polycomb sites. *Cell*, *111*(2), 185–196.

Daigle, S. R., Olhava, E. J., Therkelsen, C. A., Basavapathruni, A., Jin, L., Boriack-Sjodin, P. A., et al. (2013). Potent inhibition of DOT1L as treatment of MLL-fusion leukemia. *Blood*, *122*(6), 1017–1025.

Daigle, S. R., Olhava, E. J., Therkelsen, C. A., Majer, C. R., Sneeringer, C. J., Song, J., et al. (2011). Selective killing of mixed lineage leukemia cells by a potent small-molecule DOT1L inhibitor. *Cancer Cell, 20*(1), 53–65.

Dalgliesh, G. L., Furge, K., Greenman, C., Chen, L., Bignell, G., Butler, A., et al. (2010). Systematic sequencing of renal carcinoma reveals inactivation of histone modifying genes. *Nature, 463*(7279), 360–363.

Dalhoff, C., & Weinhold, E. (2008). S-Adenosyl-L-methionine and related compounds. In P. Herdewijn (Ed.), *Modified nucleosides: In biochemistry, biotechnology and medicine* (pp. 223–247). KGaA, Weinheim: Wiley-VCH Verlag GmbH & Co.

Dawling, S., Roodi, N., Mernaugh, R. L., Wang, X., & Parl, F. F. (2001). Catechol-O-methyltransferase (COMT)-mediated metabolism of catechol estrogens: Comparison of wild-type and variant COMT isoforms. *Cancer Research, 61*(18), 6716–6722.

de Almeida, S. F., Grosso, A. R., Koch, F., Fenouil, R., Carvalho, S., Andrade, J., et al. (2011). Splicing enhances recruitment of methyltransferase HYPB/Setd2 and methylation of histone H3 Lys36. *Nature Structural & Molecular Biology, 18*(9), 977–983.

Deguchi, T., & Barchas, J. (1971). Inhibition of transmethylations of biogenic amines by S-adenosylhomocysteine—Enhancement of transmethylation by adenosylhomocysteinase. *Journal of Biological Chemistry, 246*(10), 3175–3181.

Devkota, K., Lohse, B., Liu, Q., Wang, M. W., Staerk, D., Berthelsen, J., et al. (2014). Analogues of the natural product sinefungin as inhibitors of EHMT1 and EHMT2. *ACS Medicinal Chemistry Letters, 5*(4), 293–297.

Dhayalan, A., Dimitrova, E., Rathert, P., & Jeltsch, A. (2009). A continuous protein methyltransferase (G9a) assay for enzyme activity measurement and inhibitor screening. *Journal of Biomolecular Screening, 14*(9), 1129–1133.

Dorgan, K. M., Wooderchak, W. L., Wynn, D. P., Karschner, E. L., Alfaro, J. F., Cui, Y. Q., et al. (2006). An enzyme-coupled continuous spectrophotometric assay for S-adenosylmethionine-dependent methyltransferases. *Analytical Biochemistry, 350*(2), 249–255.

Dowden, J., Hong, W., Parry, R. V., Pike, R. A., & Ward, S. G. (2010). Toward the development of potent and selective bisubstrate inhibitors of protein arginine methyltransferases. *Bioorganic & Medicinal Chemistry Letters, 20*(7), 2103–2105.

Dowden, J., Pike, R. A., Parry, R. V., Hong, W., Muhsen, U. A., & Ward, S. G. (2011). Small molecule inhibitors that discriminate between protein arginine N-methyltransferases PRMT1 and CARM1. *Organic & Biomolecular Chemistry, 9*(22), 7814–7821.

Duenas-Gonzalez, A., Coronel, J., Cetina, L., Gonzalez-Fierro, A., Chavez-Blanco, A., & Taja-Chayeb, L. (2014). Hydralazine-valproate: A repositioned drug combination for the epigenetic therapy of cancer. *Expert Opinion on Drug Metabolism & Toxicology, 10*(10), 1433–1444.

Duns, G., van den Berg, E., van Duivenbode, I., Osinga, J., Hollema, H., Hofstra, R. M., et al. (2010). Histone methyltransferase gene SETD2 is a novel tumor suppressor gene in clear cell renal cell carcinoma. *Cancer Research, 70*(11), 4287–4291.

Ehrlich, M. (2002). DNA hypomethylation, cancer, the immunodeficiency, centromeric region instability, facial anomalies syndrome and chromosomal rearrangements. *The Journal of Nutrition, 132*(8 Suppl.), 2424S–2429S.

Eldeiry, W. S., Nelkin, B. D., Celano, P., Yen, R. W. C., Falco, J. P., Hamilton, S. R., et al. (1991). High expression of the DNA methyltransferase gene characterizes human neoplastic-cells and progression stages of colon cancer. *Proceedings of the National Academy of Sciences of the United States of America, 88*(8), 3470–3474.

Ellermann, M., Jakob-Roetne, R., Lerner, C., Borroni, E., Schlatter, D., Roth, D., et al. (2009). Molecular recognition at the active site of catechol-o-methyltransferase: Energetically favorable replacement of a water molecule imported by a bisubstrate inhibitor. *Angewandte Chemie International Edition, 48*(48), 9092–9096.

Ellermann, M., Paulini, R., Jakob-Roetne, R., Lerner, C., Borroni, E., Roth, D., et al. (2011). Molecular recognition at the active site of catechol-O-methyltransferase (COMT): Adenine replacements in bisubstrate inhibitors. *Chemistry, 17*(23), 6369–6381.

Eswaran, J., Patnaik, D., Filippakopoulos, P., Wang, F., Stein, R. L., Murray, J. W., et al. (2009). Structure and functional characterization of the atypical human kinase haspin. *Proceedings of the National Academy of Sciences of the United States of America, 106*(48), 20198–20203.

Fang, M. Z., Wang, Y., Ai, N., Hou, Z., Sun, Y., Lu, H., et al. (2003). Tea polyphenol (−)-epigallocatechin-3-gallate inhibits DNA methyltransferase and reactivates methylation-silenced genes in cancer cell lines. *Cancer Research, 63*(22), 7563–7570.

Feng, Y., Xie, N., Wu, J., Yang, C., & Zheng, Y. G. (2009). Inhibitory study of protein arginine methyltransferase 1 using a fluorescent approach. *Biochemical and Biophysical Research Communications, 379*(2), 567–572.

Fontecave, M., Atta, M., & Mulliez, E. (2004). S-Adenosylmethionine: Nothing goes to waste. *Trends in Biochemical Sciences, 29*(5), 243–249.

Frietze, S., Lupien, M., Silver, P. A., & Brown, M. (2008). CARM1 regulates estrogen-stimulated breast cancer growth through up-regulation of E2F1. *Cancer Research, 68*(1), 301–306.

Frye, M., & Watt, F. M. (2006). The RNA methyltransferase Misu (NSun2) mediates Myc-induced proliferation and is upregulated in tumors. *Current Biology, 16*(10), 971–981.

Fujiwara, T., Saitoh, H., Inoue, A., Kobayashi, M., Okitsu, Y., Katsuoka, Y., et al. (2014). 3-Deazaneplanocin A (DZNep), an inhibitor of S-adenosylmethionine-dependent methyltransferase, promotes erythroid differentiation. *The Journal of Biological Chemistry, 289*(12), 8121–8134.

Fuller, R. W., & Nagarajan, R. (1978). Inhibition of methyltransferases by some new analogs of S-adenosylhomocysteine. *Biochemical Pharmacology, 27*(15), 1981–1983.

Gauthier, N., Caron, M., Pedro, L., Arcand, M., Blouin, J., Labonte, A., et al. (2012). Development of homogeneous nonradioactive methyltransferase and demethylase assays targeting histone H3 lysine 4. *Journal of Biomolecular Screening, 17*(1), 49–58.

Geze, M., Blanchard, P., Fourrey, J. L., & Robert-Gero, M. (1983). Synthesis of sinefungin and its C-6′ epimer. *Journal of the American Chemical Society, 105*(26), 7638–7640.

Ghosh, A. K., & Liu, W. (1996). Total synthesis of (+)-sinefungin. *The Journal of Organic Chemistry, 61*(18), 6175–6182.

Ghosh, A. K., & Lv, K. (2014). A convergent synthesis of carbocyclic sinefungin and its C5 epimer. *European Journal of Organic Chemistry, 2014*(30), 6761–6768.

Gnyszka, A., Jastrzebski, Z., & Flis, S. (2013). DNA methyltransferase inhibitors and their emerging role in epigenetic therapy of cancer. *Anticancer Research, 33*(8), 2989–2996.

Goelz, S. E., Vogelstein, B., Hamilton, S. R., & Feinberg, A. P. (1985). Hypomethylation of DNA from benign and malignant human colon neoplasms. *Science, 228*(4696), 187–190.

Graca, I., Sousa, E. J., Costa-Pinheiro, P., Vieira, F. Q., Torres-Ferreira, J., Martins, M. G., et al. (2014). Anti-neoplastic properties of hydralazine in prostate cancer. *Oncotarget, 5*(15), 5950–5964.

Grant, A. J., & Lerner, L. M. (1979). Dialdehydes derived from adenine nucleosides as substrates and inhibitors of adenosine aminohydrolase. *Biochemistry, 18*(13), 2838–2842.

Graves, T. L., Zhang, Y., & Scott, J. E. (2008). A universal competitive fluorescence polarization activity assay for S-adenosylmethionine utilizing methyltransferases. *Analytical Biochemistry, 373*(2), 296–306.

Greger, V., Passarge, E., Hopping, W., Messmer, E., & Horsthemke, B. (1989). Epigenetic changes may contribute to the formation and spontaneous regression of retinoblastoma. *Human Genetics, 83*(2), 155–158.

Gutierrez, J. A., Luo, M., Singh, V., Li, L., Brown, R. L., Norris, G. E., et al. (2007). Picomolar inhibitors as transition-state probes of 5′-methylthioadenosine nucleosidases. *ACS Chemical Biology, 2*(11), 725–734.

Hamamoto, R., Saloura, V., & Nakamura, Y. (2015). Critical roles of non-histone protein lysine methylation in human tumorigenesis. *Nature Reviews Cancer, 15*(2), 110–124.

Hamil, R. L., & Hoehn, M. M. (1973). A9145, a new adenine-containing antifungal antibiotic. I. Discovery and isolation. *The Journal of Antibiotics, 26*(8), 463–465.

Hemeon, I., Gutierrez, J. A., Ho, M. C., & Schramm, V. L. (2011). Characterizing DNA methyltransferases with an ultrasensitive luciferase-linked continuous assay. *Analytical Chemistry, 83*(12), 4996–5004.

Hendricks, C. L., Ross, J. R., Pichersky, E., Noel, J. P., & Zhou, Z. S. (2004). An enzyme-coupled colorimetric assay for S-adenosylmethionine-dependent methyltransferases. *Analytical Biochemistry, 326*(1), 100–105.

Ho, M. C., Wilczek, C., Bonanno, J. B., Xing, L., Seznec, J., Matsui, T., et al. (2013). Structure of the arginine methyltransferase PRMT5-MEP50 reveals a mechanism for substrate specificity. *PLoS One, 8*(2), e57008.

Horiuchi, K. Y., Eason, M. M., Ferry, J. J., Planck, J. L., Walsh, C. P., Smith, R. F., et al. (2013). Assay development for histone methyltransferases. *ASSAY and Drug Development Technologies, 11*(4), 227–236.

Horowitz, S., Yesselman, J. D., Al-Hashimi, H. M., & Trievel, R. C. (2011). Direct evidence for methyl group coordination by carbon-oxygen hydrogen bonds in the lysine methyltransferase SET7/9. *The Journal of Biological Chemistry, 286*(21), 18658–18663.

Hudlebusch, H. R., Santoni-Rugiu, E., Simon, R., Ralfkiaer, E., Rossing, H. H., Johansen, J. V., et al. (2011). The histone methyltransferase and putative oncoprotein MMSET is overexpressed in a large variety of human tumors. *Clinical Cancer Research, 17*(9), 2919–2933.

Ibanez, G., McBean, J. L., Astudillo, Y. M., & Luo, M. (2010). An enzyme-coupled ultrasensitive luminescence assay for protein methyltransferases. *Analytical Biochemistry, 401*(2), 203–210.

Ibanez, G., Shum, D., Blum, G., Bhinder, B., Radu, C., Antczak, C., et al. (2012). A high throughput scintillation proximity imaging assay for protein methyltransferases. *Combinatorial Chemistry & High Throughput Screening, 15*(5), 359–371.

Isakovic, L., Saavedra, O. M., Llewellyn, D. B., Claridge, S., Zhan, L., Bernstein, N., et al. (2009). Constrained (L-)-S-adenosyl-l-homocysteine (SAH) analogues as DNA methyltransferase inhibitors. *Bioorganic & Medicinal Chemistry Letters, 19*(10), 2742–2746.

Islam, K., Zheng, W. H., Yu, H. Q., Deng, H. T., & Luo, M. K. (2011). Expanding cofactor repertoire of protein lysine methyltransferase for substrate labeling. *ACS Chemical Biology, 6*(7), 679–684.

Jansson, M., Durant, S. T., Cho, E. C., Sheahan, S., Edelmann, M., Kessler, B., et al. (2008). Arginine methylation regulates the p53 response. *Nature Cell Biology, 10*(12), 1431–1439.

Jeffery, D. R., & Roth, J. A. (1987). Kinetic reaction mechanism for magnesium binding to membrane-bound and soluble catechol O-methyltransferase. *Biochemistry, 26*(10), 2955–2958.

Jones, P. A., & Baylin, S. B. (2007). The epigenomics of cancer. *Cell, 128*(4), 683–692.

Kaniskan, H. U., & Jin, J. (2015). Chemical probes of histone lysine methyltransferases. *ACS Chemical Biology, 10*(1), 40–50.

Kaniskan, H. U., Konze, K. D., & Jin, J. (2015). Selective inhibitors of protein methyltransferases. *Journal of Medicinal Chemistry, 58*(4), 1596–1629.

Kantarjian, H., Issa, J. P., Rosenfeld, C. S., Bennett, J. M., Albitar, M., DiPersio, J., et al. (2006). Decitabine improves patient outcomes in myelodysplastic syndromes: Results of a phase III randomized study. *Cancer, 106*(8), 1794–1803.

Kanu, N., Gronroos, E., Martinez, P., Burrell, R. A., Yi Goh, X., Bartkova, J., et al. (2015). SETD2 loss-of-function promotes renal cancer branched evolution through replication stress and impaired DNA repair. *Oncogene, 34*(46), 5699–5708.

Kelly, T. K., De Carvalho, D. D., & Jones, P. A. (2010). Epigenetic modifications as therapeutic targets. *Nature Biotechnology, 28*(10), 1069–1078.

Kim, J. Y., Kee, H. J., Choe, N. W., Kim, S. M., Eom, G. H., Baek, H. J., et al. (2008). Multiple-myeloma-related WHSC1/MMSET isoform RE-IIBP is a histone methyltransferase with transcriptional repression activity. *Molecular and Cellular Biology, 28*(6), 2023–2034.

Kirmizis, A., Bartley, S. M., Kuzmichev, A., Margueron, R., Reinberg, D., Green, R., et al. (2004). Silencing of human polycomb target genes is associated with methylation of histone H3 Lys 27. *Genes & Development, 18*(13), 1592–1605.

Kleer, C. G., Cao, Q., Varambally, S., Shen, R., Ota, I., Tomlins, S. A., et al. (2003). EZH2 is a marker of aggressive breast cancer and promotes neoplastic transformation of breast epithelial cells. *Proceedings of the National Academy of Sciences of the United States of America, 100*(20), 11606–11611.

Kogure, M., Takawa, M., Saloura, V., Sone, K., Piao, L. H., Ueda, K., et al. (2013). The oncogenic polycomb histone methyltransferase EZH2 methylates lysine 120 on histone H2B and competes ubiquitination. *Neoplasia, 15*(11), 1251–1261.

Kouzarides, T. (2002). Histone methylation in transcriptional control. *Current Opinion in Genetics & Development, 12*(2), 198–209.

Kouzarides, T. (2007). Chromatin modifications and their function. *Cell, 128*(4), 693–705.

Krivtsov, A. V., Feng, Z., Lemieux, M. E., Faber, J., Vempati, S., Sinha, A. U., et al. (2008). H3K79 methylation profiles define murine and human MLL-AF4 leukemias. *Cancer Cell, 14*(5), 355–368.

Krogan, N. J., Kim, M., Tong, A., Golshani, A., Cagney, G., Canadien, V., et al. (2003). Methylation of histone H3 by Set2 in *Saccharomyces cerevisiae* is linked to transcriptional elongation by RNA polymerase II. *Molecular and Cellular Biology, 23*(12), 4207–4218.

Kung, P. P., Huang, B., Zehnder, L., Tatlock, J., Bingham, P., Krivacic, C., et al. (2015). SAH derived potent and selective EZH2 inhibitors. *Bioorganic & Medicinal Chemistry Letters, 25*(7), 1532–1537.

Kuo, A. J., Cheung, P., Chen, K. F., Zee, B. M., Kioi, M., Lauring, J., et al. (2011). NSD2 links dimethylation of histone H3 at lysine 36 to oncogenic programming. *Molecular Cell, 44*(4), 609–620.

Kurth, M. C., & Adler, C. H. (1998). COMT inhibition: A new treatment strategy for Parkinson's disease. *Neurology, 50*(5 Suppl. 5), S3–S14.

Laird, P. W., & Jaenisch, R. (1994). DNA methylation and cancer. *Human Molecular Genetics, 3*(Spec No.), 1487–1495.

Lauring, J., Abukhdeir, A. M., Konishi, H., Garay, J. P., Gustin, J. P., Wang, Q., et al. (2008). The multiple myeloma associated MMSET gene contributes to cellular adhesion, clonogenic growth, and tumorigenicity. *Blood, 111*(2), 856–864.

Lee, J. M., Lee, J. S., Kim, H., Kim, K., Park, H., Kim, J. Y., et al. (2012). EZH2 generates a methyl degron that is recognized by the DCAF1/DDB1/CUL4 E3 ubiquitin ligase complex. *Molecular Cell, 48*(4), 572–586.

Lerner, C., Masjost, B., Ruf, A., Gramlich, V., Jakob-Roetne, R., Zurcher, G., et al. (2003). Bisubstrate inhibitors for the enzyme catechol-O-methyltransferase (COMT): Influence of inhibitor preorganisation and linker length between the two substrate moieties on binding affinity. *Organic & Biomolecular Chemistry, 1*(1), 42–49.

Le Romancer, M., Treilleux, I., Leconte, N., Robin-Lespinasse, Y., Sentis, S., Bouchekioua-Bouzaghou, K., et al. (2008). Regulation of estrogen rapid signaling through arginine methylation by PRMT1. *Molecular Cell, 31*(2), 212–221.

Li, K. K., Huang, K., Kondengaden, S., Wooten, J., Reyhanfard, H., Qing, Z., et al. (2015). Histone methyltransferase inhibitors for cancer therapy. In Y. G. Zheng (Ed.), *Epigenetic technological applications* (pp. 363–395). Boston: Academic Press.

Lin, Q., Jiang, F., Schultz, P. G., & Gray, N. S. (2001). Design of allele-specific protein methyltransferase inhibitors. *Journal of the American Chemical Society, 123*(47), 11608–11613.

Liu, F., Chen, X., Allali-Hassani, A., Quinn, A. M., Wasney, G. A., Dong, A. P., et al. (2009). Discovery of a 2,4-diamino-7-aminoalkoxyquinazoline as a potent and selective inhibitor of histone lysine methyltransferase G9a. *Journal of Medicinal Chemistry, 52*(24), 7950–7953.

Loenen, W. A. (2006). S-adenosylmethionine: Jack of all trades and master of everything? *Biochemical Society Transactions, 34*(Pt. 2), 330–333.

Lucio-Eterovic, A. K., & Carpenter, P. B. (2011). An open and shut case for the role of NSD proteins as oncogenes. *Transcription, 2*(4), 158–161.

Luco, R. F., Pan, Q., Tominaga, K., Blencowe, B. J., Pereira-Smith, O. M., & Misteli, T. (2010). Regulation of alternative splicing by histone modifications. *Science, 327*(5968), 996–1000.

Luo, M. (2012). Current chemical biology approaches to interrogate protein methyltransferases. *ACS Chemical Biology, 7*(3), 443–463.

Lyko, F., & Brown, R. (2005). DNA methyltransferase inhibitors and the development of epigenetic cancer therapies. *Journal of the National Cancer Institute, 97*(20), 1498–1506.

Mannisto, P. T., & Kaakkola, S. (1999). Catechol-O-methyltransferase (COMT): Biochemistry, molecular biology, pharmacology, and clinical efficacy of the new selective COMT inhibitors. *Pharmacological Reviews, 51*(4), 593–628.

Mannisto, P. T., Ulmanen, I., Lundstrom, K., Taskinen, J., Tenhunen, J., Tilgmann, C., et al. (1992). Characteristics of catechol O-methyl-transferase (COMT) and properties of selective COMT inhibitors. *Progress in Drug Research, 39*, 291–350.

Martin, J. L., & McMillan, F. M. (2002). SAM (dependent) I AM: The S-adenosylmethionine-dependent methyltransferase fold. *Current Opinion in Structural Biology, 12*(6), 783–793.

Martin, C., & Zhang, Y. (2005). The diverse functions of histone lysine methylation. *Nature Reviews Molecular Cell Biology, 6*(11), 838–849.

Martinet, N., Michel, B. Y., Bertrand, P., & Benhida, R. (2012). Small molecules DNA methyltransferases inhibitors. *Medicinal Chemistry Communications, 3*(3), 263–273.

Martinez-Garcia, E., Popovic, R., Min, D. J., Sweet, S. M., Thomas, P. M., Zamdborg, L., et al. (2011). The MMSET histone methyl transferase switches global histone methylation and alters gene expression in t(4;14) multiple myeloma cells. *Blood, 117*(1), 211–220.

Masjost, B., Ballmer, P., Borroni, E., Zurcher, G., Winkler, F. K., Jakob-Roetne, R., et al. (2000). Structure-based design, synthesis, and in vitro evaluation of bisubstrate inhibitors for catechol O-methyltransferase (COMT). *Chemistry, 6*(6), 971–982.

Mazur, P. K., Reynoird, N., Khatri, P., Jansen, P. W., Wilkinson, A. W., Liu, S., et al. (2014). SMYD3 links lysine methylation of MAP3K2 to Ras-driven cancer. *Nature, 510*(7504), 283–287.

McCabe, M. T., Ott, H. M., Ganji, G., Korenchuk, S., Thompson, C., Van Aller, G. S., et al. (2012). EZH2 inhibition as a therapeutic strategy for lymphoma with EZH2-activating mutations. *Nature, 492*(7427), 108–112.

McCammon, M. T., & Parks, L. W. (1981). Inhibition of sterol transmethylation by S-adenosylhomocysteine analogs. *Journal of Bacteriology, 145*(1), 106–112.

Melki, J. R., Warnecke, P., Vincent, P. C., & Clark, S. J. (1998). Increased DNA methyltransferase expression in leukaemia. *Leukemia, 12*(3), 311–316.

Migliori, V., Muller, J., Phalke, S., Low, D., Bezzi, M., Mok, W. C., et al. (2012). Symmetric dimethylation of H3R2 is a newly identified histone mark that supports euchromatin maintenance. *Nature Structural & Molecular Biology, 19*(2), 136–144.

Min, J., Feng, Q., Li, Z., Zhang, Y., & Xu, R. M. (2003). Structure of the catalytic domain of human DOT1L, a non-SET domain nucleosomal histone methyltransferase. *Cell*, *112*(5), 711–723.
Morin, R. D., Johnson, N. A., Severson, T. M., Mungall, A. J., An, J., Goya, R., et al. (2010). Somatic mutations altering EZH2 (Tyr641) in follicular and diffuse large B-cell lymphomas of germinal-center origin. *Nature Genetics*, *42*(2), 181–185.
Newbold, R. F., & Mokbel, K. (2010). Evidence for a tumour suppressor function of SETD2 in human breast cancer: A new hypothesis. *Anticancer Research*, *30*(9), 3309–3311.
Nimura, K., Ura, K., Shiratori, H., Ikawa, M., Okabe, M., Schwartz, R. J., et al. (2009). A histone H3 lysine 36 trimethyltransferase links Nkx2-5 to Wolf-Hirschhorn syndrome. *Nature*, *460*(7252), 287–291.
Oh, B. K., Kim, H., Park, H. J., Shim, Y. H., Choi, J., Park, C., et al. (2007). DNA methyltransferase expression and DNA methylation in human hepatocellular carcinoma and their clinicopathological correlation. *International Journal of Molecular Medicine*, *20*(1), 65–73.
Okada, Y., Feng, Q., Lin, Y., Jiang, Q., Li, Y., Coffield, V. M., et al. (2005). hDOT1L links histone methylation to leukemogenesis. *Cell*, *121*(2), 167–178.
Olhava, E. J., Chesworth, R., Kuntz, K. W., Richon, V. M., Pollock, R. M., & Daigle, S. R. (2012). Substituted purine and 7-deazapurine compounds as modulators of epigenetic enzymes. *PCT Int. Appl.* WO 2012075381.
Osborne, T., Roska, R. L., Rajski, S. R., & Thompson, P. R. (2008). In situ generation of a bisubstrate analogue for protein arginine methyltransferase 1. *Journal of the American Chemical Society*, *130*(14), 4574–4575.
Pal, S., Baiocchi, R. A., Byrd, J. C., Grever, M. R., Jacob, S. T., & Sif, S. (2007). Low levels of miR-92b/96 induce PRMT5 translation and H3R8/H4R3 methylation in mantle cell lymphoma. *The EMBO Journal*, *26*(15), 3558–3569.
Papadokostopoulou, A., Mathioudaki, K., Scorilas, A., Xynopoulos, D., Ardavanis, A., Kouroumalis, E., et al. (2009). Colon cancer and protein arginine methyltransferase 1 gene expression. *Anticancer Research*, *29*(4), 1361–1366.
Parks, L. W., & Schlenk, F. (1958). The stability and hydrolysis of S-adenosylmethionine; isolation of S-ribosylmethionine. *The Journal of Biological Chemistry*, *230*(1), 295–305.
Patel, A., Dharmarajan, V., Vought, V. E., & Cosgrove, M. S. (2009). On the mechanism of multiple lysine methylation by the human mixed lineage leukemia protein-1 (MLL1) core complex. *Journal of Biological Chemistry*, *284*(36), 24242–24256.
Patel-Thombre, U., & Borchardt, R. T. (1985). Adenine nucleoside dialdehydes: Potent inhibitors of bovine liver S-adenosylhomocysteine hydrolase. *Biochemistry*, *24*(5), 1130–1136.
Patra, S. K., Patra, A., Zhao, H., & Dahiya, R. (2002). DNA methyltransferase and demethylase in human prostate cancer. *Molecular Carcinogenesis*, *33*(3), 163–171.
Paulini, R., Lerner, C., Jakob-Roetne, R., Zurcher, G., Borroni, E., & Diederich, F. (2004). Bisubstrate inhibitors of the enzyme catechol O-methyltransferase (COMT): Efficient inhibition despite the lack of a nitro group. *ChemBioChem*, *5*(9), 1270–1274.
Paulini, R., Trindler, C., Lerner, C., Brandli, L., Schweizer, W. B., Jakob-Roetne, R., et al. (2006). Bisubstrate inhibitors of catechol O-methyltransferase (COMT): The crucial role of the ribose structural unit for inhibitor binding affinity. *ChemMedChem*, *1*(3), 340–357.
Pignot, M., Siethoff, C., Linscheid, M., & Weinhold, E. (1998). Coupling of a nucleoside with DNA by a methyltransferase. *Angewandte Chemie International Edition*, *37*(20), 2888–2891.
Ragno, R., Simeoni, S., Castellano, S., Vicidomini, C., Mai, A., Caroli, A., et al. (2007). Small molecule inhibitors of histone arginine methyltransferases: Homology modeling, molecular docking, binding mode analysis, and biological evaluations. *Journal of Medicinal Chemistry*, *50*(6), 1241–1253.

Rathert, P., Cheng, X., & Jeltsch, A. (2007). Continuous enzymatic assay for histone lysine methyltransferases. *Biotechniques, 43*(5), 602–608.

Richon, V. M., Johnston, D., Sneeringer, C. J., Jin, L., Majer, C. R., Elliston, K., et al. (2011). Chemogenetic analysis of human protein methyltransferases. *Chemical Biology & Drug Design, 78*(2), 199–210.

Saavedra, O. M., Isakovic, L., Llewellyn, D. B., Zhan, L., Bernstein, N., Claridge, S., et al. (2009). SAR around (l)-S-adenosyl-l-homocysteine, an inhibitor of human DNA methyltransferase (DNMT) enzymes. *Bioorganic & Medicinal Chemistry Letters, 19*(10), 2747–2751.

Sakai, T., Toguchida, J., Ohtani, N., Yandell, D. W., Rapaport, J. M., & Dryja, T. P. (1991). Allele-specific hypermethylation of the retinoblastoma tumor-suppressor gene. *American Journal of Human Genetics, 48*(5), 880–888.

Santini, V., Kantarjian, H. M., & Issa, J. P. (2001). Changes in DNA methylation in neoplasia: Pathophysiology and therapeutic implications. *Annals of Internal Medicine, 134*(7), 573–586.

Schapira, M., & Ferreira de Freitas, R. (2014). Structural biology and chemistry of protein arginine methyltransferases. *Medicinal Chemistry Communications, 5*(12), 1779–1788.

Schramm, V. L. (2011). Enzymatic transition states, transition-state analogs, dynamics, thermodynamics, and lifetimes. *Annual Review of Biochemistry, 80*, 703–732.

Segel, I. H. (1976). Enzyme-kinetics. *Bioscience, 26*(7), 425–426.

Shinkai, Y., & Tachibana, M. (2011). H3K9 methyltransferase G9a and the related molecule GLP. *Genes & Development, 25*(8), 781–788.

Siedlecki, P., Garcia Boy, R., Comagic, S., Schirrmacher, R., Wiessler, M., Zielenkiewicz, P., et al. (2003). Establishment and functional validation of a structural homology model for human DNA methyltransferase 1. *Biochemical and Biophysical Research Communications, 306*(2), 558–563.

Silverman, L. R., Demakos, E. P., Peterson, B. L., Kornblith, A. B., Holland, J. C., Odchimar-Reissig, R., et al. (2002). Randomized controlled trial of azacitidine in patients with the myelodysplastic syndrome: A study of the cancer and leukemia group B. *Journal of Clinical Oncology, 20*(10), 2429–2440.

Simon, J. A., & Lange, C. A. (2008). Roles of the EZH2 histone methyltransferase in cancer epigenetics. *Mutation Research, 647*(1–2), 21–29.

Smil, D., Eram, M. S., Li, F., Kennedy, S., Szewczyk, M. M., Brown, P. J., et al. (2015). Discovery of a dual PRMT5-PRMT7 inhibitor. *ACS Medicinal Chemistry Letters, 6*(4), 408–412.

Smith, B. C., & Denu, J. M. (2009). Chemical mechanisms of histone lysine and arginine modifications. *Biochimica et Biophysica Acta, 1789*(1), 45–57.

Sone, K., Piao, L., Nakakido, M., Ueda, K., Jenuwein, T., Nakamura, Y., et al. (2014). Critical role of lysine 134 methylation on histone H2AX for gamma-H2AX production and DNA repair. *Nature Communications, 5*, 5691.

Song, Y., & Zhang, C. (2009). Hydralazine inhibits human cervical cancer cell growth in vitro in association with APC demethylation and re-expression. *Cancer Chemotherapy and Pharmacology, 63*(4), 605–613.

Stec, I., Wright, T. J., van Ommen, G. J., de Boer, P. A., van Haeringen, A., Moorman, A. F., et al. (1998). WHSC1, a 90 kb SET domain-containing gene, expressed in early development and homologous to a *Drosophila* dysmorphy gene maps in the Wolf-Hirschhorn syndrome critical region and is fused to IgH in t(4;14) multiple myeloma. *Human Molecular Genetics, 7*(7), 1071–1082.

Struck, A. W., Thompson, M. L., Wong, L. S., & Micklefield, J. (2012). S-adenosyl-methionine-dependent methyltransferases: Highly versatile enzymes in biocatalysis, biosynthesis and other biotechnological applications. *ChemBioChem, 13*(18), 2642–2655.

Suh-Lailam, B. B., & Hevel, J. M. (2010). A fast and efficient method for quantitative measurement of S-adenosyl-L-methionine-dependent methyltransferase activity with protein substrates. *Analytical Biochemistry, 398*(2), 218–224.

Swierczynski, S., Klieser, E., Illig, R., Alinger-Scharinger, B., Kiesslich, T., & Neureiter, D. (2015). Histone deacetylation meets miRNA: Epigenetics and post-transcriptional regulation in cancer and chronic diseases. *Expert Opinion on Biological Therapy, 15*(5), 651–664.

Tseng, C. K., Marquez, V. E., Fuller, R. W., Goldstein, B. M., Haines, D. R., McPherson, H., et al. (1989). Synthesis of 3-deazaneplanocin A, a powerful inhibitor of S-adenosylhomocysteine hydrolase with potent and selective in vitro and in vivo antiviral activities. *Journal of Medicinal Chemistry, 32*(7), 1442–1446.

Ugarkar, B. G., DaRe, J. M., Kopcho, J. J., Browne, C. E., 3rd, Schanzer, J. M., Wiesner, J. B., et al. (2000). Adenosine kinase inhibitors. 1. Synthesis, enzyme inhibition, and antiseizure activity of 5-iodotubercidin analogues. *Journal of Medicinal Chemistry, 43*(15), 2883–2893.

Urdinguio, R. G., Sanchez-Mut, J. V., & Esteller, M. (2009). Epigenetic mechanisms in neurological diseases: Genes, syndromes, and therapies. *The Lancet Neurology, 8*(11), 1056–1072.

van Haren, M., van Ufford, L. Q., Moret, E. E., & Martin, N. I. (2015). Synthesis and evaluation of protein arginine N-methyltransferase inhibitors designed to simultaneously occupy both substrate binding sites. *Organic & Biomolecular Chemistry, 13*(2), 549–560.

Varela, I., Tarpey, P., Raine, K., Huang, D., Ong, C. K., Stephens, P., et al. (2011). Exome sequencing identifies frequent mutation of the SWI/SNF complex gene PBRM1 in renal carcinoma. *Nature, 469*(7331), 539–542.

Vidgren, J., Svensson, L. A., & Liljas, A. (1994). Crystal structure of catechol O-methyltransferase. *Nature, 368*(6469), 354–358.

Villeneuve, L. M., & Natarajan, R. (2010). The role of epigenetics in the pathology of diabetic complications. *American Journal of Physiology-Renal Physiology, 299*(1), F14–F25.

Wagner, E. J., & Carpenter, P. B. (2012). Understanding the language of Lys36 methylation at histone H3. *Nature Reviews Molecular Cell Biology, 13*(2), 115–126.

Wahhab, A., Besterman, J. M., Delorme, D., Isakovic, L., Rahil, J., Saavedra, O., et al. (2006). Inhibitors of DNA methyltransferase. *PCT Int. Appl.* WO 2006078752.

Wang, C. H., Leffler, S., Thompson, D. H., & Hrycyna, C. A. (2005). A general fluorescence-based coupled assay for S-adenosylmethionine-dependent methyltransferases. *Biochemical and Biophysical Research Communications, 331*(1), 351–356.

Wang, L., Pal, S., & Sif, S. (2008). Protein arginine methyltransferase 5 suppresses the transcription of the RB family of tumor suppressors in leukemia and lymphoma cells. *Molecular and Cellular Biology, 28*(20), 6262–6277.

Wei, T. Y., Juan, C. C., Hisa, J. Y., Su, L. J., Lee, Y. C., Chou, H. Y., et al. (2012). Protein arginine methyltransferase 5 is a potential oncoprotein that upregulates G1 cyclins/cyclin-dependent kinases and the phosphoinositide 3-kinase/AKT signaling cascade. *Cancer Science, 103*(9), 1640–1650.

Wei, H., Mundade, R., Lange, K. C., & Lu, T. (2014). Protein arginine methylation of non-histone proteins and its role in diseases. *Cell Cycle, 13*(1), 32–41.

Williams-Ashman, H. G., Seidenfeld, J., & Galletti, P. (1982). Trends in the biochemical pharmacology of 5′-deoxy-5′-methylthioadenosine. *Biochemical Pharmacology, 31*(3), 277–288.

Wu, Y. Y., Sarkissyan, M., & Vadgama, J. V. (2015). Epigenetics in breast and prostate cancer. *Methods in Molecular Biology, 1238*, 425–466.

Wu, W., Wu, Q., Hong, X., Xiong, G., Xiao, Y., Zhou, J., et al. (2015a). Catechol-O-methyltransferase inhibits colorectal cancer cell proliferation and invasion. *Archives of Medical Research, 46*(1), 17–23.

Wu, W., Wu, Q., Hong, X., Zhou, L., Zhang, J., You, L., et al. (2015b). Catechol-O-methyltransferase, a new target for pancreatic cancer therapy. *Cancer Science, 106*(5), 576–583.

Wu, J., Xie, N., Feng, Y., & Zheng, Y. G. (2012). Scintillation proximity assay of arginine methylation. *Journal of Biomolecular Screening, 17*(2), 237–244.

Yang, Y., & Bedford, M. T. (2013). Protein arginine methyltransferases and cancer. *Nature Reviews Cancer, 13*(1), 37–50.

Yao, Y., Chen, P., Diao, J., Cheng, G., Deng, L., Anglin, J. L., et al. (2011). Selective inhibitors of histone methyltransferase DOT1L: Design, synthesis, and crystallographic studies. *Journal of the American Chemical Society, 133*(42), 16746–16749.

Yi, J. S., Federation, A. J., Qi, J., Dhe-Paganon, S., Hadler, M., Xu, X., et al. (2015). Structure-guided DOT1L probe optimization by label-free ligand displacement. *ACS Chemical Biology, 10*(3), 667–674.

Yoshimatsu, M., Toyokawa, G., Hayami, S., Unoki, M., Tsunoda, T., Field, H. I., et al. (2011). Dysregulation of PRMT1 and PRMT6, type I arginine methyltransferases, is involved in various types of human cancers. *International Journal of Cancer, 128*(3), 562–573.

Yu, W., Chory, E. J., Wernimont, A. K., Tempel, W., Scopton, A., Federation, A., et al. (2012). Catalytic site remodelling of the DOT1L methyltransferase by selective inhibitors. *Nature Communications, 3*, 1288.

Yu, W., Smil, D., Li, F., Tempel, W., Fedorov, O., Nguyen, K. T., et al. (2013). Bromo-deaza-SAH: A potent and selective DOT1L inhibitor. *Bioorganic & Medicinal Chemistry, 21*(7), 1787–1794.

Zhang, X., & Bruice, T. C. (2008). Mechanism of product specificity of AdoMet methylation catalyzed by lysine methyltransferases: Transcriptional factor p53 methylation by histone lysine methyltransferase SET7/9. *Biochemistry, 47*(9), 2743–2748.

Zhang, J., Ding, L., Holmfeldt, L., Wu, G., Heatley, S. L., Payne-Turner, D., et al. (2012). The genetic basis of early T-cell precursor acute lymphoblastic leukaemia. *Nature, 481*(7380), 157–163.

Zhang, J., & Zheng, Y. G. (2015). SAM/SAH analogs as versatile tools for SAM-dependent methyltransferases. *ACS Chemical Biology.* in press.

Zhao, Q., Rank, G., Tan, Y. T., Li, H., Moritz, R. L., Simpson, R. J., et al. (2009). PRMT5-mediated methylation of histone H4R3 recruits DNMT3A, coupling histone and DNA methylation in gene silencing. *Nature Structural & Molecular Biology, 16*(3), 304–311.

Zhao, Z. J., Wu, Q. X., Cheng, J. A., Qiu, X. M., Zhang, J. Q., & Fan, H. (2010). Depletion of DNMT3A suppressed cell proliferation and restored PTEN in Hepatocellular carcinoma cell. *Journal of Biomedicine and Biotechnology, 2010*, 737535.

Zheng, Y.-J., & Bruice, T. C. (1997). A theoretical examination of the factors controlling the catalytic efficiency of a transmethylation enzyme: Catechol O-methyltransferase. *Journal of the American Chemical Society, 119*(35), 8137–8145.

Zheng, W. H., Ibanez, G., Wu, H., Blum, G., Zeng, H., Dong, A. P., et al. (2012). Sinefungin derivatives as inhibitors and structure probes of protein lysine methyltransferase SETD2. *Journal of the American Chemical Society, 134*(43), 18004–18014.

Zurita-Lopez, C. I., Sandberg, T., Kelly, R., & Clarke, S. G. (2012). Human protein arginine methyltransferase 7 (PRMT7) is a type III enzyme forming omega-NG-monomethylated arginine residues. *The Journal of Biological Chemistry, 287*(11), 7859–7870.

Zweygarth, E., Schillinger, D., Kaufmann, W., & Rottcher, D. (1986). Evaluation of sinefungin for the treatment of Trypanosoma (Nannomonas) congolense infections in goats. *Tropical Medicine and Parasitology, 37*(3), 255–257.

PART III

Epigenetics and Biological Connections

CHAPTER TWELVE

Exploring the Dynamic Relationship Between Cellular Metabolism and Chromatin Structure Using SILAC-Mass Spec and ChIP-Sequencing

P. Mews, S.L. Berger[1]

Perelman School of Medicine, University of Pennsylvania, Philadelphia, PA, United States
[1]Corresponding author: e-mail address: bergers@mail.med.upenn.edu

Contents

1. Introduction 312
2. Analyzing the Turnover Dynamics of Histone Modifications 314
   2.1 SILAC-Mass Spec Methodology 314
   2.2 Methyl-SILAC 315
   2.3 Analysis of Histone Methylation Dynamics Using Methyl-SILAC 315
   2.4 Acetyl-SILAC 317
   2.5 Limitations of SILAC-Mass Spec 318
3. Genome-Wide Mapping of Histone Modifications 320
   3.1 Charting the Epigenetic Landscape by ChIP-seq 320
   3.2 ChIP—Materials and Buffer Recipes 320
   3.3 ChIP—Cell Harvest and Cross-linking 321
   3.4 ChIP Day 1 | Bead Preparation, Cell Lysis, and Sonication, Setting Up the IP 322
   3.5 ChIP Day 2 | Washes, Elution, and Reverse Cross-Linking 323
   3.6 ChIP Day 3 | DNA Purification 324
   3.7 ChIP-seq Library Preparation 325
   3.8 Computational Pipeline and Considerations in ChIP-seq 326
4. Summary 327
References 327

Abstract

Metabolic state and chromatin structure are tightly linked, enabling adaptation of gene expression to changing environment and metabolism. The bioenergetic pathways and enzymes that provide metabolic cofactors for histone modification have recently emerged as central regulators of chromatin. Current research therefore focuses on the dynamic interface of cellular metabolism and chromatin structure. Here, we provide an adaptable approach to examine broadly in changing physiological states, how chromatin structure is dynamically modulated by metabolic activity. We employ two complementary methods: high-throughput sequencing to establish the location of epigenetic changes, and stable isotope tracing using mass spectrometry to evaluate chromatin modification dynamics. Our two-pronged approach is of particular advantage when interrogating how metabolic and oncogenic mutations influence the dynamic relationship between metabolism, nutritional environment, and chromatin regulation.

1. INTRODUCTION

The intimate relationship between metabolic state and chromatin regulation is gaining major interest in the field of epigenetics. Cells must continually adapt to metabolic needs and changing environment, and thus require an acutely responsive system that adjusts chromatin structure and gene transcription to maintain homeostasis. In recent years, a new paradigm of transcriptional regulation has emerged, based on direct sensing of intermediary metabolites by chromatin-modifying enzymes (Kaelin & McKnight, 2013; Katada, Imhof, & Sassone-Corsi, 2012). Key metabolites, beside their role in central biochemical pathways as intermediates, also participate as cofactors in chromatin modification, and can therefore dynamically adapt global chromatin structure and gene expression to metabolic change (Gut & Verdin, 2013). Advancing our insight into how metabolic state influences chromatin structure thus holds promise for new intervention points in the treatment of diseases that are characterized by epigenetic aberration, including cancer and neurodegenerative disorders. Hence, metabolic pathways and the enzymes that provide metabolites used in chromatin modification have now moved into the limelight of epigenetic studies.

In particular, the temporal turnover dynamics of histone modifications and their dysregulation upon nutritional and metabolic disruption are of increasing interest. While our knowledge of the location of chromatin aberrations that accompany disease expanded with the advent of chromatin immunoprecipitation sequencing (ChIP-seq) technology (Baylin & Jones, 2011; Rivera & Ren, 2013; Shen & Laird, 2013), this now routine method fails to capture important dynamic aspects in epigenetic regulation. Here, we provide a highly adaptable approach that employs two complementary methods to examine how cellular metabolism regulates dynamic change in chromatin structure, by making use of both high-throughput sequencing data to establish the location of changes and stable isotope tracing using mass spectrometry to evaluate their dynamics. First, we describe the use of stable isotope tracing in cell culture to track metabolic flux into histone modifications and their time-associated turnover. We then describe how to chart the epigenetic landscape by performing ChIP-seq in order to generate genome-wide maps of individual histone modifications. We discuss computational tools that have been developed to build such maps, and that allow investigation of genomic patterns of chromatin modifications and how their distribution responds to changing environment. Both approaches are of great value when interrogating how metabolic and oncogenic mutations influence the dynamic relationship between metabolism, nutritional environment, and chromatin regulation.

In one example of the use of these methods, we applied these technologies to study metabolically controlled gene regulation in the context of nutrient-induced growth and cell cycle reentry from quiescence in the yeast *Saccharomyces cerevisiae* (Mews et al., 2014). Reports previous to our study showed that histone acetylation is exquisitely responsive to the nutritional state of cells (Cai, Sutter, Li, & Tu, 2011; Friis et al., 2009; Wellen et al., 2009). Our findings from Stable Isotope Labeling using Amino acids in Cell culture (SILAC) and ChIP-seq provided unequivocal evidence that nutrients induce an immediate histone acetylation response, driven by external glucose. Interestingly, our application of both technologies demonstrated that histone methylation—in contrast to highly dynamic histone acetylation—remains exceptionally stable and uncoupled from rapid metabolic and transcriptional reprogramming as cells exit quiescence. Instead, SILAC-mass spec and ChIP-seq revealed that the methylation landscape is extensively restructured genome-wide only at a later time, in conjunction with genome replication and ensuing cell division (Mews et al., 2014). Notably, as global histone methylation levels do not change immediately in the exit from quiescence, these dynamic changes in genome-wide methylation patterns would have remained obscured when analyzed by Western blot alone. Thus the combination of SILAC-mass spec with ChIP-seq provided distinct interpretations of the dynamics of histone acetylation compared to histone methylation during reentry into the cell cycle.

Another context in which a combined approach of SILAC and ChIP-seq is poised to yield novel insight are studies of oncogenic lesions and the metabolic adaptations they precipitate, which consequently alter chromatin structure and the epigenome. For instance, in many cancers, deregulated metabolism resulting in aberrant histone methylation has been implicated in tumorigenesis (Chi, Allis, & Wang, 2010; Greer & Shi, 2012; Maddocks, Labuschagne, Adams, & Vousden, 2015). Mutations in the metabolic enzymes isocitrate dehydrogenase 1 and 2 are found in a variety of tumors and cause neomorphic catalytic activities that generate 2-hydroxyglutarate, an oncometabolite that alters the histone methylation landscape and disrupts regulated gene expression (Lu et al., 2012). Similarly, a large body of evidence has shown that acetyl–CoA metabolism and histone acetylation are frequently dysregulated in cancer (Carrer & Wellen, 2014; Comerford et al., 2014). Therefore, in order to develop novel therapeutic approaches that target cellular metabolism to alter the tumor epigenome, it is key to understand how chromatin is regulated by metabolic state in health and disease.

2. ANALYZING THE TURNOVER DYNAMICS OF HISTONE MODIFICATIONS

2.1 SILAC-Mass Spec Methodology

Quantitative proteomics using SILAC is a popular quantitative proteomics approach applicable to most experimental workflows. The SILAC approach relies on the endogenous translational machinery of living cells to incorporate tagged chemical analogs of molecular building blocks into chromatin (Ong & Mann, 2006; Ong et al., 2002). In the classical SILAC experiment, two cell populations are compared (Ong & Mann, 2006). These populations are grown in different media containing distinct forms of amino acids that are either of natural isotopic abundance or heavy-labeled stable isotopes. Incorporation of these light and heavy amino acids into de novo synthesized proteins creates proteomes that can be discerned by their specific mass differences. Mass spectrometry is used to measure the relative abundance of light and heavy peptides in pooled samples, which allows for the detection and quantification of differential changes in protein expression.

In order to study temporal turnover dynamics of chromatin modifications, the classic SILAC approach has been adapted to analyze site-specific dynamics of histone acetylation and methylation (Zee, Levin, Dimaggio, & Garcia, 2010). Our labeling approach provides an effective marker to distinguish preexisting from newly catalyzed chromatin modifications, and permits the monitoring over time for dynamic turnover, when coupled with advanced quantitative liquid chromatography–mass spec technology (Mews et al., 2014; Zee et al., 2010). Here we describe how SILAC and the development of quantitative models characterizing the steady-state kinetics of global methylation and acetylation provide a sensitive metric to study gene regulatory mechanisms in different physiological and metabolic states.

2.2 Methyl-SILAC

Histone methylation is dependent on the metabolic intermediate S-adenosyl methionine (SAM), the sole donor of the methyl group. The SAM precursor methionine is an essential amino acid that mammal cells cannot synthesize de novo. Thus, when heavy stable isotope form of methionine, L-methionine-methyl-^{13}C, D$_3$, is used in the cell culture medium, heavy-methyl groups will be introduced by methyltransferases into lysine and arginine residues on histone proteins. This labeling technology coupled with quantitative mass spectrometry technology facilitates the study of site-specific methylation dynamics on histone proteins (Cao, Zee, & Garcia, 2013; Zee et al., 2010). To this end, cells are introduced into the heavy methionine-labeled medium, after which they are harvested at a series of time points. Histone proteins are then purified by nuclei isolation and acid extraction, and then sequentially subjected to propionylation, trypsin digestion, and liquid chromatography-tandem mass spectrometry (LC-MS/MS) analysis. These steps are discussed in detail by Karch et al. in "Identification and Quantification of Histone PTMs Using High-Resolution Mass Spectrometry". Finally, the abundances of light, heavy, and intermediate forms of particular methylation sites can be extracted, and normalized to construct kinetic models toward investigating their turnover dynamics (Cao et al., 2013; Zee, Levin, et al., 2010). The experimental scheme of SILAC-mass spectrometry is shown in Fig. 1.

2.3 Analysis of Histone Methylation Dynamics Using Methyl-SILAC

2.3.1 Heavy-Labeled Medium Preparation

1. 1000× Heavy methionine (L-methionine-methyl-^{13}C, D$_3$; Sigma-Aldrich) stock solution is prepared in DPBS (15.4 mg/mL).
2. Stock solution is filtered using 0.22-μm syringe filter and store at −20°C.
3. Weigh the required amounts of all of the cell culture medium components except methionine and add them to 800 mL MilliQ water. Mix until completely dissolved. Components of homemade medium (all from Sigma-Aldrich): L-arginine monohydrochloride; L-cysteine dihydrochloride; L-histidine dihydrochloride; L-isoleucine; L-leucine; L-lysine monohydrochloride; L-phenylalanine; L-threonine; L-tryptophan; L-tyrosine; L-valine; choline chloride; folic acid; myo-inositol; niacinamide; D-pantothenic acid hemicalcium salt; pyridoxal hydrochloride; riboflavin; thiamine hydrochloride; MgCl$_2$; KCl; NaCl; Na$_2$HPO$_4$; glucose; phenol red sodium salt; NaHCO$_3$. Heavy methionine: L-methionine-methyl-^{13}C, D$_3$ (Sigma-Aldrich).

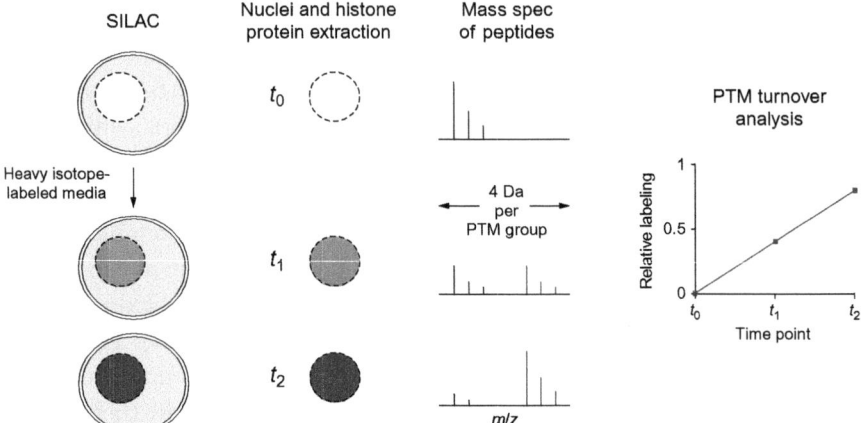

Figure 1 Methodology of SILAC-based quantitative mass spectrometry analysis of histone methylation dynamics. Cells are transferred to medium containing heavy [^{13}C]D$_3$-methionine-labeled medium, and isotope incorporation (*orange* (*gray* in the print version)) into histone methylation pools is monitored over time by tandem mass spectrometry analysis. Specific shifts in the mass-to-charge (*m/z*) ratio are used to distinguish unlabeled from labeled modifications and reveal methylation turnover dynamics when measured over time (eg, 4-Da shift per methyl group). *Adapted from Mews, P., et al. (2014). Histone methylation has dynamics distinct from those of histone acetylation in cell cycle reentry from quiescence. Molecular and Cellular Biology 34, 3968–3980.*

4. Adjust pH to 7.2 and supplement MilliQ water to 900 mL.
5. Filter the solution with a 0.22-μm membrane filter. Store at 4°C.
6. Add dialyzed FBS, heavy methionine stock solution, and supplements to the medium to their working concentration when use.

2.3.2 Mammalian Cell Culture and Methyl-SILAC Labeling

HeLa S3 cells are maintained at 37°C in 0.2 LPM CO$_2$ in minimum essential Joklik modified medium (SAFC Biosciences), supplemented with 10% newborn calf serum, penicillin, streptomycin, and 1% GlutaMAX (Invitrogen).

1. Pellet cells by centrifugation (80 rcf); aspirate unlabeled media.
2. Resuspend cells with labeled media (Joklik media depleted of unlabeled methionine, supplemented with L-methionine-methyl-^{13}CD$_3$; 5% dialyzed fetal bovine serum, penicillin, streptomycin, and 1% GlutaMAX).
3. Aliquots are taken at various time points, depending on experimental design and question, eg, daily over the course of 1 week.
4. Media should be replenished every day to maintain a cell density of 2–6 × 10^5 cells/mL throughout the experiment.

5. Cells can be pelleted at 600 rcf and washed in phosphate-buffered saline.
6. Flash-freeze pelleted cells and stored at −80°C.

2.3.3 Methyl-SILAC Peptide Quantification and Kinetics Modeling

For detailed instruction on how to isolate nuclei and acid-extract histone proteins, we refer the user to see chapter "Identification and Quantification of Histone PTMs Using High-Resolution Mass Spectrometry" by Karch et al. This chapter also discusses how to prepare histone peptides for LC-MS/MS analysis by propionylation and trypsin digestion. The obtained MS and MS/MS spectra can be analyzed with Qual Browser (Thermo Scientific), and peptide abundances are obtained by chromatographic peak integration. Methylated peptide species are commonly denoted for each methylation residue with X:Y, where X is the number of total methyl groups and Y is the number of labeled methyl groups. For instance, H3K36me1:0 refers to unlabeled monomethylated H3K36, and H3K36me1:1 refers to labeled monomethylated H3K36. In order to reveal the distribution of differently labeled species within a methylation state, their relative abundance is calculated after all methylation intermediates for a specific methylation state of a particular residue have been determined. For instance, the abundance of H3K36me1:0 would be determined with respect to H3K36me1:0 and me1:1. This calculation is based on the logic that growth conditions should not change with $^{12}CH_3$- or $^{13}CD_3$-methionine, and that the total abundance of each methylated species does not change (Zee et al., 2010). The relative distribution can then be used to quantify half-maximal times of the formation of labeled peptides ($t\frac{1}{2}$). The half-maximal time provides a measure of how quickly a particular species is formed, and thus is predictive of the overall rate of methylation for that species. For a detailed description of this approach and how to characterize the dynamics of methylated residues in an idealized system using kinetics modeling, we refer the reader to Zee et al. (2010).

2.4 Acetyl-SILAC

Metabolic production of acetyl-CoA has been linked to histone acetylation and gene regulation. Acetyl-CoA is the key metabolite used as a cofactor in all protein acetylation, and altered intracellular pools of acetyl-CoA can readily manipulate histone acetylation (Albaugh, Arnold, & Denu, 2010; Wellen et al., 2009). Accordingly, metabolic enzymes that provide nuclear pools of acetyl-CoA have been postulated to directly control chromatin acetylation (Pietrocola, Galluzzi, Bravo-San Pedro, Madeo, & Kroemer, 2015).

SILAC technology offers novel insight into how nutrient abundance and metabolic state influence histone acetylation dynamics, in particular when utilized to investigate how mutations in metabolic pathways affect this relationship.

Glucose-dependent histone acetylation can be studied utilizing heavy glucose SILAC (Mews et al., 2014). Similar to Methyl-SILAC, this method relies on mass spectrometry to detect and quantify heavy carbon labeling of histone acetylation at several time points, resulting from the glycolytic breakdown of heavy $[^{13}C_6]$-glucose in the labeling media. Histone proteins are purified by nuclei isolation and acid extraction and sequentially subjected to propionylation, trypsin digestion, and LC-MS/MS analysis, as outlined in the section on methyl-SILAC above. Acetyl-SILAC has been successfully used to investigate the temporal dynamics of the intimate relationship between glucose uptake, glycolytic metabolism, and regulated histone acetylation (Mews et al., 2014).

2.5 Limitations of SILAC-Mass Spec

SILAC-based mass spec provides insight into the residue-specific turnover dynamics of histone modifications. These turnover dynamics remain obscured in classic Western blot analysis that measures only the total relative abundance of specific histone modifications. SILAC therefore presents a valuable tool to reveal whether particular modification pools acutely respond to altered metabolic flux. Importantly, neither approach provides insight into the genome-wide distribution of the examined histone modifications. Genomic locations can be investigated with modern ChIP-sequencing technology, which measures genome-wide enrichment of histone modifications and allows for the monitoring of epigenomic changes over time. Such insight into the epigenetic landscape is critical when examining chromatin-based mechanisms of gene regulation.

For instance, changes in gene expression can be triggered by metabolic activities that result in the altered turnover of particular histone modifications. SILAC-mass spec may indicate whether particular histone modifications are created de novo, but ChIP-seq is needed to gain insight into their genomic location in order to distinguish (1) localized dynamic turnover from (2) epigenomic restructuring that involves regional gains and losses, both of which may be coupled to altered gene expression, as outlined in Fig. 2. Likewise, increases in global histone acetylation and methylation may be detected by either Western blot or SILAC analysis, but only ChIP-sequencing provides information regarding their genomic location, and whether such change is linked to altered gene expression (Fig. 2E

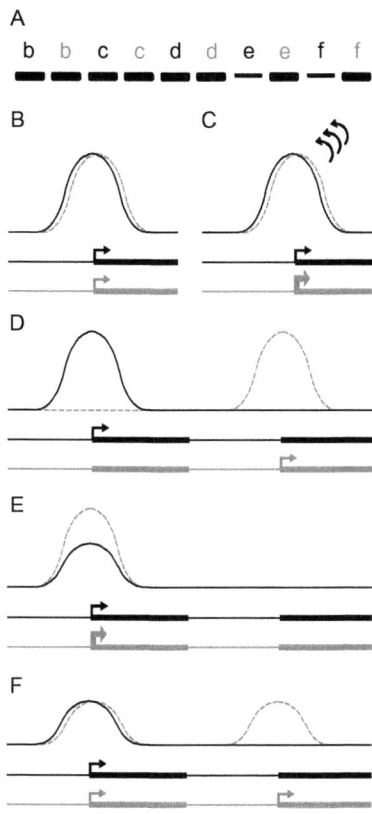

Figure 2 SILAC-mass spec and ChIP-seq work synergistically to capture dynamic aspects of chromatin-mediated gene regulation. (A) Hypothetical Western blot analysis of global histone modification levels for theoretical chromatin state alterations illustrated in (B)–(F) (from *black* to gray), demonstrating the inherent limitations of Western method when a dynamic change in chromatin structure is analyzed. (B–D) Enrichment of a histone modification over the promoter region of a gene. Shifts in metabolic state may alter the turnover dynamics of particular histone modifications, coupled to changes in gene expression. Whereas SILAC-mass spec analysis indicates whether or not histone modifications are created de novo (example B vs C, D), this method on its own is inadequate to distinguish localized dynamic turnover (example C) from epigenomic restructuring involving both regional gains and losses of enrichment (example D). These outcomes can be distinguished by ChIP-seq, which provides important information on the genomic location of histone modifications. Conversely, ChIP-seq does not yield insight into their turnover dynamics that may respond to metabolic state (example B vs C). Notably, none of these changes in chromatin structure can be detected by classic Western blot analysis (in A, a–d are identical levels). (E and F) Increases in total abundance of histone modification may be detected with either Western or SILAC analysis. Yet, only ChIP-seq technology may determine the location of such epigenetic change, and whether it is associated with altered gene transcription.

and F). Therefore SILAC-mass spec and ChIP-seq technologies are especially powerful when applied together, and work in concert to afford novel insight into chromatin-mediated mechanisms of transcriptional control by metabolic state.

3. GENOME-WIDE MAPPING OF HISTONE MODIFICATIONS

3.1 Charting the Epigenetic Landscape by ChIP-seq

Since SILAC-mass spec methods cannot provide information regarding the genomic distribution of histone modifications, next-generation sequencing technology is applied to generate maps of the epigenetic landscape. In the past decade, epigenomic analyses by ChIP-seq have become the prevalent method to interrogate the genome-wide binding pattern of epigenetic enzymes and their cognate chromatin modifications. Our protocol is based on the standard ChIP methodology originally published by the Young Lab (Lee, Johnstone, & Young, 2006), which requires further optimization for cell type and experimental question. We discuss experimental design considerations to prevent common sources of bias in ChIP-seq experiments, and emphasize important steps to ensure high-quality next-generation sequencing data. We note that our two-pronged approach to study the relationship between metabolism and histone modification dynamics also involves SILAC-mass spec, a technology that demands higher cell numbers than needed in recent low-cell ChIP-seq protocols (Lara-Astiaso et al., 2014; Rotem et al., 2015; van Galen et al., 2016), which are more relevant to other experimental questions.

3.2 ChIP—Materials and Buffer Recipes

Note on ChIP-seq antibody choice: It is critically important to determine the efficiency and specificity of the antibody that is to be used in ChIP-seq. Antibodies are key to success, as they frequently are not site specific and hence cross-react with similar modifications on different residues. Off-target antibody binding results in contaminating DNA in the final material, yielding biologically inaccurate ChIP-seq results. We strongly encourage the user to test antibody specificity and to consult resources in the public domain that are devoted to this concern, eg, the Histone Antibody Specificity Database (http://histoneantibodies.com).

Block Solution (make fresh each time, store at 4°C)
   1× PBS, 0.5% BSA (powdered stock at 4°C)
Lysis Buffer 1 (filter and store at 4°C, add PI before use)
   50 mM HEPES–KOH, pH 7.5, 140 mM NaCl, 1 mM EDTA, 10% glycerol, 0.5% NP-40, 0.25% Triton X-100, 1× protease inhibitors—add fresh before using
Lysis Buffer 2 (filter and store at 4°C, add PI before use)
   10 mM Tris–HCl, pH 8.0, 200 mM NaCl, 1 mM EDTA, 0.5 mM EGTA, 1× protease inhibitors—add fresh before using
Lysis Buffer 3 (filter and store at 4°C, add PI before use)
   10 mM Tris–HCl, pH 8.0, 100 mM NaCl, 1 mM EDTA, 0.5 mM EGTA, 0.1% Na-deoxycholate, 0.5% N-lauroylsarcosine, 1× protease inhibitors—add fresh before using
RIPA Wash Buffer (filter and store at 4°C)
   50 mM HEPES–KOH, pH 7.5, 500 mM LiCl, 1 mM EDTA, 1% NP-40, 0.7% Na-deoxycholate
ChIP Final Wash Buffer (filter and store at 4°C)
   1× TE, 50 mM NaCl
ChIP Elution Buffer (filter and store at room temperature)
   50 mM Tris–HCl, pH 8.0, 10 mM EDTA, 1% SDS

3.3 ChIP—Cell Harvest and Cross-linking

Note: This protocol is optimized for about 5.5×10^6 cells, the typical yield of a single 100-mm-cell culture dish with a density of 1×10^5 cells/cm^2.

1. Add formaldehyde (37% w/v) to final concentration of 1% and rotate tubes for 10 min at room temperature to cross-link.
2. Incubate at 37°C for 10 min.
3. Add glycine (2.5 M) to final concentration of 125 mM and incubate for an additional 5 min at room temperature to quench cross-linking.
4. Scrape cells off plate using disposable cell lifter.
5. Collect cells in 15-mL conical tube.
   Note: Keep samples on ice after this step.
6. Centrifuge cells at 1200 rpm for 5 min at 4°C.
7. Aspirate supernatant and wash tissue pieces once in 10 mL cold PBS.
8. Spin as in step 7, aspirate supernatant, and freeze cross-linked cells (in conical tubes) indefinitely at −80°C or proceed directly with ChIP.

3.4 ChIP Day 1 | Bead Preparation, Cell Lysis, and Sonication, Setting Up the IP

3.4.1 Bead Preparation

1. Early in the day (at least 6 h before setting up IPs), aliquot 30 µL Protein G Magnetic Beads (Dynabeads) into 1.5-mL microcentrifuge tube; set up 1 tube/IP.
2. Collect beads on magnet, put tubes on magnet, allow beads to separate for 30 s to 1 min, and remove supernatant by aspiration.
3. Resuspend beads in 1 mL Block Solution and mix by inverting tube.
4. Repeat Steps 2 and 3.
5. Repeat Step 2. Resuspend beads in 250 µL Block Solution.
6. Add 4 µg antibody (or IgG/no antibody for control) per IP tube.
7. Rotate at 4°C for a minimum of 6 h.
8. After 6 h incubation, quick spin tubes containing the beads–antibody mixture to remove any liquid and beads from the lid.

3.4.2 Cell Lysis

Note: It is recommended to begin cell lysis about 2 h before setting up IPs.

1. Remove frozen cross-linked cell pellets from −80°C and put on ice.
2. Resuspend cell pellet in cold 10 mL Lysis Buffer 1 (plus protease inhibitors) and rock on platform rocker for 10 min at 4°C.
3. Spin cells at 3000 rpm at 4°C for 5 min and aspirate supernatant.
4. Resuspend cell pellet in cold 10 mL Lysis Buffer 2 and rock on platform rocker for 10 min at room temperature.
5. Spin and process pellet as in Step 3.
6. Resuspend pellet in 1 mL cold Lysis Buffer 3 and shear chromatin by sonication.

    Note: Sonication conditions need to be optimized for sonication system, cell type, and desired fragment size. This is critical as fragment size is a key determinant of both ChIP efficiency and the quality of prepared ChIP libraries.
7. Add 1/10 volume 10% Triton X-100 to sonicated lysate.
8. Spin at full speed in microcentrifuge for 10 min at 4°C.
9. Transfer supernatants to new tubes.

    Note: Lysates can be combined for same samples if performing ChIP for low-abundance histone modifications.
10. Measure approximate protein or DNA concentration in each sample.

    Note: 500 µg lysate/IP, but yield may vary depending on the cell number and type, and the specific cross-linking conditions.

3.4.3 Setting Up the IP
Prepare beads
1. Quick spin tubes containing beads/antibody mixture to remove any liquid and beads from the lid.
2. Collect beads on magnet; put tubes on magnet, allow beads to separate for 30 s to 1 min, and remove supernatant by aspiration.
3. Resuspend beads in 1 mL Block Solution and mix by inverting tube.
4. Repeat Steps 2 and 3 twice more (ie, total of three washes).
5. Following final wash, resuspend each aliquot of beads in 50 μL Block Solution; beads are then ready to use.

Combine lysate and beads
1. Remove 10% (v/v) aliquot from each lysate for input control, combine with TES so that final volume is 300 μL, and store at −20°C.
2. In low-adhesion microcentrifuge tube, aliquot lysate corresponding to 500 μg protein, or to 1–2 μg DNA.
   Note: If lysate volume is greater than 800 μL, set up IPs in conical tubes, to allow maximum rotation of beads within the tube.
3. Add 50 μL bead mixture to tube containing lysate.
   Note: Final IP mix contains 500 μg lysate, 30 μL beads (initial volume), and 2 μg antibody (prebound to beads). If yield is not enough for 500 μg lysate protein/IP, adjust all lysate amounts so that the same amount of protein is being used in every IP. All ChIPs must contain the same amount of input protein/IP to be comparable.
4. Rotate IP tubes overnight at 4°C.

3.5 ChIP Day 2 | Washes, Elution, and Reverse Cross-Linking
3.5.1 Washes
1. Prechill microcentrifuge tubes on ice before beginning washes; set up 1 tube per IP.
2. Collect beads using magnet, put tubes on magnet, allow beads to separate for 30 s, and remove supernatant by aspiration.
3. Save the supernatant from an IgG control IP to check sonication and assess whether equal DNA has been added to each IP.
4. Add 1 mL cold RIPA Wash Buffer, mix tube by inversion, and incubate briefly on ice (1–2 min).
5. Collect beads using magnet, put tubes on magnet, allow beads to separate for 30 s, and remove supernatant by aspiration.
6. Repeat Steps 5 and 6 four times, resulting in a total of five RIPA washes.

7. Resuspend beads in 1 mL Final ChIP Wash Buffer.
8. Collect beads as in Step 6.
9. Spin tubes at 1500 rpm for 3 min in microcentrifuge at 4°C.
10. Collect beads on magnet as in Step 6 to remove any residual buffer from beads.

3.5.2 Elution
1. Resuspend beads in 210 μL Elution Buffer.
2. Incubate tubes at 65°C for 30 min.

    Note: Use a Thermomixer with gentle agitation (approx. 400 rpm), or flick or invert tubes every 5 min to keep beads in suspension.
3. Spin tubes at top speed for 1 min in microcentrifuge at room temperature.
4. Collect beads using magnet; put tubes on magnet, allow beads to separate for 30 s to 1 min.

    Note: Do not aspirate supernatant! This contains the eluted ChIP-DNA!
5. Remove 200 μL supernatant to fresh tube (remaining 10 μL left behind to prevent transfer of beads).

3.5.3 Reverse Cross-linking
1. Thaw tubes containing input DNA.
2. Add 150 μL Elution Buffer to input DNA tubes (3 volumes more than input volume; adjust accordingly if using greater than 50 μL input lysate).
3. Incubate all tubes (input and IP) at 65°C overnight to reverse cross-linking.

    Note: Incubate for at least 12 h and not more than 18 h to maximize DNA content, but minimize background signal resulting from excess reverse cross-linking.

3.6 ChIP Day 3 | DNA Purification

Note: These steps are optimized for 200 μL samples; if using greater than 200 μL volume for input sample, adjust reagent volumes accordingly.
1. Add 200 μL 1× TE to each tube.
2. Add 8 μL 10 mg/mL RNaseA (0.2 mg/mL final concentration) to each tube and incubate at 37°C for 2 h.
3. Add 4 μL 20 mg/mL Proteinase K (0.2 mg/mL final concentration) to each tube and incubate at 55°C for 2 h.

Note: Incomplete digestion may result in DNA-binding proteins carrying a fraction of DNA into the phenol phase, which results in uneven genome coverage owing to chromatin effects.
4. Add 400 μL phenol:chloroform:isoamyl alcohol (P:C:IA) and vortex.
5. Prespin 2 mL Phase Lock Heavy Gel tubes for 1 min at top speed in microcentrifuge.
6. Pipet eluate/P:C:IA mixture onto the gel in the Phase Lock tubes and spin for 15 min at top speed in microcentrifuge.
7. Transfer aqueous layer into fresh 1.5-mL microcentrifuge tube containing 16 μL 5 MNaCl and 1.5 μL 20 μg/μL glycogen (final 30 μg).
8. Add 800 μL cold 100% EtOH and mix well by inversion.
9. Incubate tubes at −20°C for 30 min to precipitate DNA.
10. Spin samples at top speed for 10 min in microcentrifuge at 4°C; note orientation of tube in microcentrifuge, as pellets may be difficult to see.
11. Carefully decant/aspirate supernatant (pellets may be loose).
12. Add 500 μL 80% EtOH to wash DNA (pellets may become dislodged).
13. Spin as in Step 10.
14. Remove all EtOH by pipet/aspiration.
15. Allow pellets to air dry for 10–15 min at room temperature.
16. Resuspend pellet in 60 μL 10 mM Tris–HCl, pH 8.0.

Note: This volume is adjustable; 60 μL is used here, to allow for part of the sample to be analyzed by qPCR and the remainder of the sample to be used for ChIP-seq. Volumes can be adjusted depending on the amount of DNA that is immunoprecipitated.

3.7 ChIP-seq Library Preparation

All sequencing platforms have guidelines for optimal ChIP-DNA size in order to ensure efficient bridge amplification and sequencing. The sonication conditions need to be optimized accordingly to generate DNA material for optimal sequencing output. Based on various aspects of the library preparation procedure, such as adapter ligation and PCR amplification, most library preparation kits suggest a particular range of starting material to ensure high library quality. It is recommended to follow these guidelines, and to start with ChIP-DNA devoid of contamination and fragment size deviations. A major concern for low starting material is the need for additional DNA amplification, which may result in reduced complexity of the sequencing sample.

Before sequencing, the prepared ChIP-seq libraries must be assessed for both quality and quantity. Typically, library quality and size distribution are evaluated using the BioAnalyzer (Agilent Technologies), a nanofluidics device that performs size fractionation and quantification of small DNA samples. Libraries with adapter contamination, large DNA fragments, or unexpected size distributions are considered low quality and may result in inefficient or failed sequencing. In certain cases, improvement of the quality of prepared libraries is attainable by performing a new size selection using AMPure beads (Beckman Coulter). Refinement of techniques that make ChIP-seq possible with minimal material for library preparation is the focus of many NGS experiments and is expected to advance the field dramatically.

3.8 Computational Pipeline and Considerations in ChIP-seq

The raw sequencing data that are produced in NGS runs are many millions of reads per sample that correspond to the genomic regions bound by histone proteins that bear the particular modification of interest. These short sequencing reads are mapped onto a reference genome before subsequent computational analysis. Such sequence alignment to a reference genome is achieved with short read aligners like Bowtie or Burrows-Wheeler Alignment (Langmead, Trapnell, Pop, & Salzberg, 2009; Li & Durbin, 2009). Initial quality control of sequencing data can be further accomplished using FastQC. For more detailed information regarding quality control and useful computational tools, we refer the user to in-depth reviews and valuable repositories online (Landt et al., 2012; Maze et al., 2014; Schones & Zhao, 2008; Sims, Sudbery, Ilott, Heger, & Ponting, 2014).

A key task in ChIP-seq data analysis is to identify genomic regions that show significant above-background enrichment for the histone modification subjected to the ChIP experiment. Peak calling is typically done using Model-based Analysis of ChIP-Seq (MACS) (Zhang et al., 2008) and Hypergeometric Optimization of Motif Enrichment (HOMER) (Heinz et al., 2010). Regional annotation is useful to determine the typical location of the detected peaks across the genomes, for instance genic, nongenic, or pericentromeric. Analysis tools such as the diffRep package (Shen et al., 2013) or ChIPpeakAnn package (Zhu et al., 2010) can be used to this end.

Differential analysis of ChIP-seq data to determine significant enrichment changes between conditions is often frustrated by confounding biases that are difficult to mitigate, such as batch effects, PCR amplification bias, background noise, and sequencing depth (Sims et al., 2014). One approach is to extract regions of significant differential enrichment uses a sliding window-based strategy that scans and continuously scores the entire genome, which can be done with the Perl-based program diffReps (Shen et al., 2013). Further bias may be mitigated by using unsupervised techniques, such as surrogate variable analysis that aims to remove systematic effects of unknown origin. For further considerations of bias we point to excellent reviews that discuss detailed issues regarding ChIP-seq peak detection, peak deconvolution, domain calling, and differential identification using an array of computational tools (Maze et al., 2014; Sims et al., 2014). NGS technology and the associated tools and methods are evolving rapidly, and their implementation in modular computational packages is expected to make future ChIP-seq analyses more flexible, effective, and less biased.

4. SUMMARY

Epigenetic enzymes have critical roles for chromatin regulatory events in normal cellular development and plasticity. We provide a powerful and highly adaptable approach to investigate how nutritional status and metabolic state impact the genome-wide pattern of individual histone modifications, and their turnover. ChIP-seq is a powerful technology that enables the study of a broad range of chromatin phenomena on a genome-wide scale and, in particular, provides position- and gene-specific mapping. However, ChIP-seq does not capture the temporal dynamics of histone modification turnover. Acetyl-SILAC and methyl-SILAC, in contrast, are formidable techniques that permit detection and quantification of relative differential changes in histone modifications over time. SILAC-mass spec has now been successfully applied to interrogate the reciprocal relationship between glucose uptake, glycolytic metabolism, and regulated histone acetylation. In combination, ChIP-seq and SILAC-mass spec further allow comprehensive profiling of chromatin during changing nutritional and metabolic environments. Used together, these technologies yield superior insight that each approach on its own does not provide. Taking a synergistic approach to utilize ChIP-seq and SILAC-mass spectrometry technology will benefit future studies that aim for a deeper understanding of how chromatin structure and the epigenome are dynamically regulated by metabolic state.

REFERENCES

Albaugh, B. N., Arnold, K. M., & Denu, J. M. (2010). KAT(ching) metabolism by the tail: Insight into the links between lysine acetyltransferases and metabolism. *ChemBioChem*, *53706*, 290–298.

Baylin, S. B., & Jones, P. A. (2011). A decade of exploring the cancer epigenome—Biological and translational implications. *Nature Reviews. Cancer*, *11*, 726–734.

Cai, L., Sutter, B. M., Li, B., & Tu, B. P. (2011). Acetyl-CoA induces cell growth and proliferation by promoting the acetylation of histones at growth genes. *Molecular Cell, 42*, 426–437.
Cao, X. J., Zee, B. M., & Garcia, B. A. (2013). Heavy methyl-SILAC labeling coupled with liquid chromatography and high-resolution mass spectrometry to study the dynamics of site-specific histone methylation. *Methods in Molecular Biology, 977*, 299–313.
Carrer, A., & Wellen, K. E. (2014). Metabolism and epigenetics: A link cancer cells exploit. *Current Opinion in Biotechnology, 34*, 23–29.
Chi, P., Allis, C. D., & Wang, G. G. (2010). Covalent histone modifications—Miswritten, misinterpreted and mis-erased in human cancers. *Nature Reviews. Cancer, 10*, 457–469.
Comerford, S. A., et al. (2014). Acetate dependence of tumors. *Cell, 159*, 1591–1602.
Friis, R. M. N., et al. (2009). A glycolytic burst drives glucose induction of global histone acetylation by picNuA4 and SAGA. *Nucleic Acids Research, 37*, 3969–3980.
Greer, E. L., & Shi, Y. (2012). Histone methylation: A dynamic mark in health, disease and inheritance. *Nature Reviews. Genetics, 13*, 343–357.
Gut, P., & Verdin, E. (2013). The nexus of chromatin regulation and intermediary metabolism. *Nature, 502*, 489–498.
Heinz, S., et al. (2010). Simple combinations of lineage-determining transcription factors prime cis-regulatory elements required for macrophage and B cell identities. *Molecular Cell, 38*, 576–589.
Kaelin, W. G., & McKnight, S. L. (2013). Influence of metabolism on epigenetics and disease. *Cell, 153*, 56–69.
Katada, S., Imhof, A., & Sassone-Corsi, P. (2012). Connecting threads: Epigenetics and metabolism. *Cell, 148*, 24–28.
Landt, S. G., et al. (2012). ChIP-seq guidelines and practices of the ENCODE and modENCODE consortia. *Genome Research, 22*, 1813–1831.
Langmead, B., Trapnell, C., Pop, M., & Salzberg, S. L. (2009). Ultrafast and memory-efficient alignment of short DNA sequences to the human genome. *Genome Biology, 10*, R25. http://dx.doi.org/10.1186/gb-2009-10-3-r25.
Lara-Astiaso, D., et al. (2014). Chromatin state dynamics during blood formation. *Science, 345*, 943–949.
Lee, T. I., Johnstone, S. E., & Young, R. A. (2006). Chromatin immunoprecipitation and microarray-based analysis of protein location. *Nature Protocols, 1*, 729–748.
Li, H., & Durbin, R. (2009). Fast and accurate short read alignment with Burrows-Wheeler transform. *Bioinformatics, 25*, 1754–1760.
Lu, C., et al. (2012). IDH mutation impairs histone demethylation and results in a block to cell differentiation. *Nature, 483*, 474–478.
Maddocks, O. D. K., Labuschagne, C. F., Adams, P. D., & Vousden, K. H. (2015). Serine metabolism supports the methionine cycle and DNA/RNA methylation through de novo ATP synthesis in cancer cells. *Molecular Cell, 61*, 210–221. http://dx.doi.org/10.1016/j.molcel.2015.12.014.
Maze, I., et al. (2014). Analytical tools and current challenges in the modern era of neuroepigenomics. *Nature Neuroscience, 17*, 1476–1490.
Mews, P., et al. (2014). Histone methylation has dynamics distinct from those of histone acetylation in cell cycle reentry from quiescence. *Molecular and Cellular Biology, 34*, 3968–3980.
Ong, S.-E., & Mann, M. (2006). A practical recipe for stable isotope labeling by amino acids in cell culture (SILAC). *Nature Protocols, 1*, 2650–2660.
Ong, S. E., et al. (2002). Stable isotope labeling by amino acids in cell culture, SILAC, as a simple and accurate approach to expression proteomics. *Molecular & Cellular Proteomics, 1*, 376–386.

Pietrocola, F., Galluzzi, L., Bravo-San Pedro, J. M., Madeo, F., & Kroemer, G. (2015). Acetyl coenzyme A: A central metabolite and second messenger. *Cell Metabolism, 21*, 805–821.

Rivera, C. M., & Ren, B. (2013). Mapping human epigenomes. *Cell, 155*, 39–55.

Rotem, A., et al. (2015). Single-cell ChIP-seq reveals cell subpopulations defined by chromatin state. *Nature Biotechnology, 33*, 1165–1172.

Schones, D. E., & Zhao, K. (2008). Genome-wide approaches to studying chromatin modifications. *Nature Reviews. Genetics, 9*, 179–191.

Shen, H., & Laird, P. W. (2013). Interplay between the cancer genome and epigenome. *Cell, 153*, 38–55.

Shen, L., et al. (2013). diffReps: Detecting differential chromatin modification sites from ChIP-seq data with biological replicates. *PLoS One, 8*, e65598.

Sims, D., Sudbery, I., Ilott, N. E., Heger, A., & Ponting, C. P. (2014). Sequencing depth and coverage: Key considerations in genomic analyses. *Nature Reviews. Genetics, 15*, 121–132.

van Galen, P., et al. (2016). A multiplexed system for quantitative comparisons of chromatin landscapes. *Molecular Cell, 61*, 170–180.

Wellen, K. E., et al. (2009). ATP-citrate lyase links cellular metabolism to histone acetylation. *Science, 324*, 1076–1080.

Zee, B. M., Levin, R. S., Dimaggio, P. A., & Garcia, B. A. (2010). Global turnover of histone post-translational modifications and variants in human cells. *Epigenetics & Chromatin, 3*, 22.

Zee, B. M., et al. (2010). In vivo residue-specific histone methylation dynamics. *The Journal of Biological Chemistry, 285*, 3341–3350.

Zhang, Y., et al. (2008). Model-based analysis of ChIP-Seq (MACS). *Genome Biology, 9*, R137.

Zhu, L. J., et al. (2010). ChIPpeakAnno: A bioconductor package to annotate ChIP-seq and ChIP-chip data. *BMC Bioinformatics, 11*, 237.

CHAPTER THIRTEEN

Current Proteomic Methods to Investigate the Dynamics of Histone Turnover in the Central Nervous System

L.A. Farrelly*, B.D. Dill[†], H. Molina[†], M.R. Birtwistle*, I. Maze*,[1]
*Icahn School of Medicine at Mount Sinai, New York, NY, United States
[†]The Rockefeller University Proteomics Resource Center, The Rockefeller University, New York, NY, United States
[1]Corresponding author: e-mail address: ian.maze@mssm.edu

Contents

1. Introduction	332
2. Early Methods to Study Histone Turnover in Brain	334
2.1 The Introduction of Radioactive Tracers	335
2.2 Limitations of Radioactive Labeling	336
3. Current Proteomic Methods to Study Histone Turnover in Brain	337
3.1 Mass Spectrometry: A Brief Introduction	337
3.2 Sample Preparation Considerations	338
3.3 Label Free vs Stable Isotope Incorporation	339
3.4 Acquisition Types	340
3.5 Mass Spectrometry-Based Methods to Study Histone Turnover in Brain	341
4. Retrospective Birth Dating of Histones in Human Postmortem Brain	342
4.1 "Bomb Pulse Labeling" Coupled to Accelerator Mass Spectrometry	342
4.2 Bomb Pulse Labeling: Analytical Considerations	345
5. Conclusion	347
6. Methodology: Preparing Chromatin from Neurons for Mass Spectrometry Analysis of Histone Variants and Turnover	348
Acknowledgments	352
References	352

Abstract

Characterizing the dynamic behavior of nucleosomes in the central nervous system is vital to our understanding of brain-specific chromatin-templated processes and their roles in transcriptional plasticity. Histone turnover—the complete loss of old, and replacement by new, nucleosomal histones—is one such phenomenon that has recently been shown to be critical for cell-type-specific transcription in brain, synaptic plasticity, and cognition. Such revelations that histones, long believed to static proteins in postmitotic cells, are highly dynamic in neurons were only possible owing to significant advances in analytical chemistry-based techniques, which now provide a platform for investigations of histone dynamics in both healthy and diseased tissues. Here, we discuss both past and present proteomic methods (eg, mass spectrometry, human "bomb pulse labeling") for investigating histone turnover in brain with the hope that such information may stimulate future investigations of both adaptive and aberrant forms of "neuroepigenetic" plasticity.

1. INTRODUCTION

Eukaryotic gene transcription is a highly complex and dynamic process mediated by several critical and coordinated mechanisms that function to govern cellular diversity and plasticity. These mechanisms, and their respective timings, are critical during organismal development and are ultimately responsible for appropriate patterns of lineage specification, preservation of cellular identity, and phenotypic variation. Patterns of gene transcription, however, can be modified by a large variety of environmental exposures, thereby inducing both short- and long-term changes in gene expression, which can then act to alter the trajectory of previously "defined" cellular states.

Neuroplasticity refers to the brain's ability to adapt to changing internal and external environmental stimuli leading to changes in neuronal function, circuit formation, structural morphology, and behavior, all of which are directed, at least in part, by altered patterns of gene expression. These gene–environment interactions are multifaceted and involve several critical mediators and intrinsic mechanisms. Some of these processes are referred to as "epigenetic" and involve histone–DNA interactions that are mediated by dynamic posttranslational modifications (PTMs) (both on histones and DNA), histone variant exchange, and nucleosomal remodeling (Maze, Noh, & Allis, 2013).

Chromatin, the DNA–protein complex that functions as the defining substrate for processes regulating cellular gene expression in eukaryotes, is comprised of both genomic DNA and core basic histone proteins. The nucleosome exists as the essential repeating subunit of chromatin and consists of an octamer of highly conserved core histone proteins (H2A, H2B, H3, and H4, or variants thereof) wrapped around ~147 bp of superhelical DNA (Luger, Mader, Richmond, Sargent, & Richmond, 1997). Modulating the accessibility of genes to the transcriptional machinery via alterations in chromatin structure has explicit implications for gene expression in brain and has been consistently linked to neuroplasticity and cognition, as well as aberrant adaptations (Maze et al., 2013). Investigating the mechanisms of epigenetic plasticity in the central nervous system is a challenging feat; however, studies of chromatin function in brain have been increasing at an exponential rate over the last decade, effectively identifying many novel "players" and mechanisms involved in the regulation of neuroepigenetic states. Such advances have largely been due to an ever-growing interest in the role of such processes in human health and neurological disease (Cramer et al., 2011).

Histone variant proteins, which vary in primary amino acid sequence from their canonical counterparts (eg, H3.3 vs H3.1 and H3.2), play a pivotal role in cellular development, lineage commitment, and transcriptional potential (Maze, Noh, Soshnev, & Allis, 2014). H3 variants are generally less diverse than those arising from the H2A and H2B families, which are more commonly linked to the direct regulation of nucleosomal stability and are accompanied by substantial variation in their amino and carboxy-terminal "tail" regions.

In recent years, it has become clear that histone variants play critical roles in the regulation of gene transcription serving to alter the PTM landscape on chromatin or to impact nucleosomal structure via (1) recruitment of distinct chromatin effector complexes or (2) sequence-based structural effects that lead to octameric instability. Interestingly, however, in postmitotic cells, canonical histones (eg, H3.1 and H3.2) are not able to incorporate into chromatin (ie, they are considered to be replication dependent) resulting in an imbalance between chromatin-associated levels of "variant" histones (eg, H3.3), which are typically incorporated into chromatin in a replication-independent manner, and canonical isoforms (Maze et al., 2014).

Recent work from our laboratory demonstrated that H3.3, but not H3.1/2, turnover in brain is extraordinarily high during early phases of neurodevelopment and remains constitutive, albeit at lower rates, for the remainder of life following rapid accumulation of H3.3 from late embryogenesis to mid-adolescence (Maze et al., 2015). We later showed, using many of the recent methods described throughout this chapter, that these histone turnover events are activity dependent, are critical for cell-type-specific transcription, are necessary and sufficient to drive both normal and aberrant patterns of synaptic connectivity/plasticity, and are essential for mammalian cognition (Fig. 1). Here, we provide a comprehensive overview of the proteomic methods, both old and new, that can be utilized to

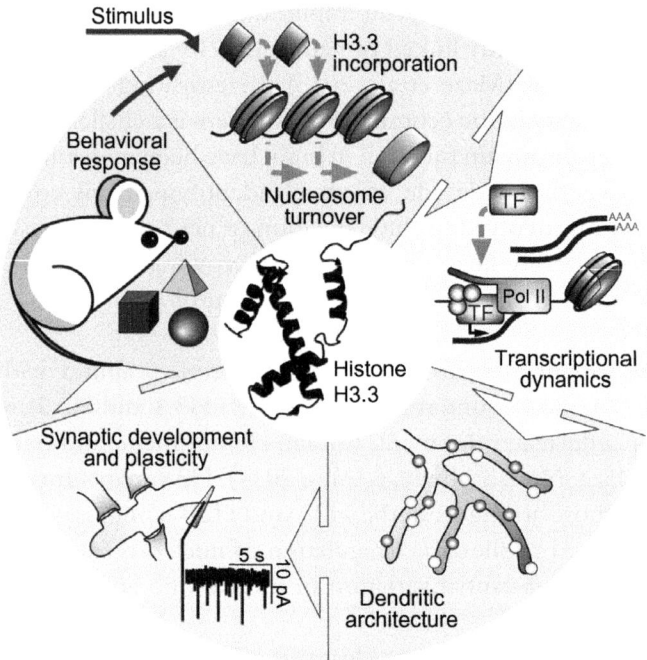

Fig. 1 Histone turnover is a critical mediator of neurological plasticity. In the CNS, during periods of heightened neuronal activity and cellular plasticity, H3.3-associated nucleosomal turnover is increased to allow for activity-dependent transcriptional plasticity that results in alterations in dendritic architecture, synaptic connectivity, and behavioral plasticity (eg, cognition). (See the color plate.)

investigate histone turnover in brain, future investigations of which will surely provide great insights into our understanding of human neurological disease and psychiatric disorders.

2. EARLY METHODS TO STUDY HISTONE TURNOVER IN BRAIN

There is a strong basis for monitoring protein synthesis and degradation in mammalian systems stemming from a basic interest in understanding cellular dynamics and their relationships to human disease. Measurements of protein turnover often require sophisticated methods that are capable of determining dynamic, and oftentimes subtle, changes resulting from the synthesis and degradation of specific polypeptides extracted from heterogeneous mixtures of both stable and dynamic molecules. Since the early 1970s, a range of methodologies have been developed. Herein, we will briefly discuss some of the past methods used in these pioneering studies aimed at addressing the function of histone turnover.

2.1 The Introduction of Radioactive Tracers

The act of "labeling" proteins in vivo to measure protein turnover originally served to enhance the sensitivity of turnover detection and to increase quantitative measurements of protein–protein interactions. Protein labeling, in its earliest sense, was based on the incorporation of tracers into live cells. Early studies utilizing this method often took advantage of radioactivity and employed radiolabeled precursors, such as ^{35}S-labeled methionines or cysteines. Rates of protein synthesis could therefore be measured by determining the incorporation of a given label into a given cell type being investigated. In turn, protein turnover (ie, synthesis vs degradation) was then assessed through direct measurements of the incorporation and loss of the label. Radioactive labeling with ^{35}S-methionine was frequently used to label newly synthesized proteins and was therefore compatible with the study of protein turnover. An interest in investigating histone turnover commenced in the early 1970s with the overarching goal of elucidating whether histone dynamics occur independently from cellular proliferation to contribute to biological functions. Radioactive labeling, as described earlier, was used in these pioneering studies, which then "paved the way" for safer and more high-throughput methods that are utilized today. In these initial studies, Gurley and Walters radiolabeled Chinese hamster cells in tissue culture in an attempt to correlate histone turnover rates with phosphorylation events (Gurley & Walters, 1971). In doing so, they found that reductions in histone turnover correlated strongly with decreases in histone phosphorylation, suggesting that these two processes may be linked. Bondy and colleagues later measured turnover of cerebral histones using 4,5-^{3}H-leucine in the developing chick embryo and demonstrated that although all "bulk" histone fractions appeared to decay at similar rates, initially with a half-life of approximately 5 days and later with a delayed half-life of ~19 days, DNA itself was significantly more stable. They therefore concluded that histone turnover occurs independently from that of DNA (Bondy, 1971), suggesting that histone dynamics may themselves provide biological functionality in the absence of cellular proliferation.

Although informative, some of the assumptions made using these more antiquated labeling approaches were later proved to be incorrect using more sophisticated methods. For example, by labeling mammalian cells with

^{14}C-methionine, followed by hydrolysis and gas chromatography, Byvoet and colleagues originally concluded that histone methylation was irreversible following comparisons between the half-lives of histones and their methyl lysine and/or arginine components (Byvoet, Shepherd, Hardin, & Noland, 1972). Following decades of debate regarding the reversibility of histone methylation, however, Shi and colleagues identified the first histone demethylases (Shi et al., 2004), thereby refuting these earlier assumptions and changing our understanding of the stability of chromatin-templated processes.

In later studies of mouse kidney and liver cells, which are nonproliferating and terminally differentiated, specific histone variants were found to undergo continuous replacement throughout life as assessed by the incorporation of radioactive-labeled amino acids (Djondjurov, Yancheva, & Ivanova, 1983). Other classical studies used tritiated water intake in rodents to induce radiolabeling for analysis of protein turnover in rodent brain demonstrating that canonical histones (eg, H3.1 and H3.2) are remarkably stable throughout the lifetime of an animal (Commerford, Carsten, & Cronkite, 1982). Such labeling approaches work on the principle that ^{2}H atoms rapidly equilibrate across pools of water in the body following intake, whereby C—H bonds of free glycines and/or alanines, as well as several other nonessential amino acids, approach isotopic equilibrium. Protein biosynthesis then becomes the rate-limiting step for ^{2}H incorporation (De Riva, Deery, McDonald, Lund, & Busch, 2010), thus enabling extended labeling protocols for quantification of proteins exhibiting defined turnover rates (Grove & Zweidler, 1984).

2.2 Limitations of Radioactive Labeling

As discussed, protein synthesis and decay have historically been measured by incorporation and clearance of radiolabeled amino acids. Such approaches, however, are difficult to employ in animal models and are not well suited for high-throughput screening. Furthermore, many labeling procedures require the utilization of in vitro culture systems, which often do not mimic in vivo biology, especially with respect to neurons, which function as highly coordinated circuits in brain. Also, in comparison to tissue culture cells, where the amount of incorporated tracer can easily be monitored, incorporation rates in multicellular organisms are often confounded by tissue/cell-type-specific rates of protein turnover. Lastly, the use of substantial amounts of radiation carries with it obvious health and safety concerns.

3. CURRENT PROTEOMIC METHODS TO STUDY HISTONE TURNOVER IN BRAIN

Advances in the field of neuroproteomics have equipped modern investigators with a series of high-throughput methods that now allow for efficient investigations of histone turnover in brain. These techniques require novel methodologies of sample preparation coupled to mass spectrometry to allow for powerful analyses when combined with sophisticated bioinformatics tools.

3.1 Mass Spectrometry: A Brief Introduction

Modern high mass accuracy mass spectrometers coupled to nanoflow liquid chromatography (LC) systems have developed into a capable platform enabling sensitive, fast, and high-confidence identifications for quantifying histone isoform levels and modifications (Garcia, Shabanowitz, & Hunt, 2007). Mass spectrometry data can be acquired by a multitude of methodologies, but fundamentally consist of MS spectra with precursor ion mass-to-charge (m/z) signals and/or MS/MS spectra resulting from the gas-phase fragmentation of isolated, preferably individual, ions (Hunt, Yates, Shabanowitz, Winston, & Hauer, 1986). MS and MS/MS can either be used together or individually for peptide/protein identifications and quantification. In the most common workflows, peptides are generated by digestion of the sample protein using a sequence-specific cleavage enzyme, such as trypsin, which allows matching of proteolytic peptide fragments to predicted in silico digestion products corresponding to parent proteins in a sequence database (Aebersold & Goodlett, 2001; Dhingra, Gupta, Andacht, & Fu, 2005; Domon & Aebersold, 2006). Mass analyzers provide high resolution, high mass accuracy, and broad dynamic range. Their sensitivity and resolving power ensure that high-performance analyses of complex mixtures and their relatively low cost make them attractive instruments for a variety of laboratory settings.

Nano LC separates peptides/proteins, typically by C_{18} reversed-phase chromatography based on hydrophobicity, prior to being introduced to the mass spectrometer by nanoelectrospray (Vanhoutte et al., 1997) in a process termed "shotgun proteomics" (Wolters, Washburn, & Yates, 2001). Quantitative profiling can be further enhanced by the use of stable isotope labeling (eg, SILAC), through in situ labeling, subsequent chemical labeling, or spiked-in references, in advanced quantitation informatics pipelines.

It is without doubt that continued improvements in scan speed, dynamic range, and mass resolution will enable mass spectrometry to quantitate histone dynamics in a truly unbiased manner and become the forefront player in the field of chromatin studies. Sections 3.1.1–3.5 will review mass spectrometry-based methods to study histone turnover in brain ranging from label-free approaches to stable isotope incorporation.

3.1.1 Bottom Up vs Top Down

Proteomic analyses can be categorized into two broad categories: "top-down" analysis of intact proteins and protein complexes, or "bottom-up" analysis of proteins proteolytically digested (most commonly with trypsin) to provide smaller, more manageable peptides. While top-down approaches have definitive advantages in protein analytics including the avoidance of loss of information regarding the protein context of modifications and/or isoform identities, significantly more technological development in the areas of chromatographic separation and mass spectrometry instrumentation is required to properly resolve intact proteins given their inherently high multiple charge states during routine analyses. Bottom-up approaches, on the other hand, allow for robust characterizations, albeit via peptide-centric data.

3.2 Sample Preparation Considerations

The ultimate success of proteomic analyses begins with, and relies upon, appropriate and efficient sample processing. Histone core purifications from acid-extracted chromatin limit the dynamic range and complexity of the sample enabling higher sensitivity and detection of substoichiometric modifications without complicating downstream detection. Proteins can be digested directly in-solution or in-gel following SDS-PAGE separation, which allows for further separation from other contaminating proteins if necessary. Due to the significantly higher proportion of lysine and arginine residues in histones, posttranslational modifications are identified most successfully through propionylation blocking of lysines via propionic anhydride (Garcia, Mollah, et al., 2007); this step is often necessary to achieve reasonable sequence coverage with typical workflows. Not only does this process block trypsin digestion at modified lysines, thereby preventing the production of peptides too small to measure effectively in the optimal range for most mass spectrometers, it reduces the positive charge on peptides to reduce the dominant charge state of most peptides toward +2 or +3, which produce the most interpretable fragmentation patterns in collision induced dissociation mass spectra (MS/MS).

3.3 Label Free vs Stable Isotope Incorporation

Numerous quantitation strategies have historically been used in proteomics and can be generally separated into label-free vs stable isotope-based approaches. In label-free quantitation, peptides are quantified and interpreted without referencing heavy-isotope versions. Spectral counting is the most straightforward label-free quantification technique, in which proteins can be quantified by the number of MS/MS events that can be attributed to them relying on the stochastic nature of MS/MS sampling. Otherwise, the ion intensity signals of precursor ions can be used to compare the relative abundance of a given analyte in subsequent LC–MS runs, either through maximal intensity or through chromatographic elution profiles. Neither method is based on absolute measurements but can be used to compare signals for a given peptide between subsequent analyses. Additionally, as ionization efficiency differs among peptide sequence and modification states, and since coeluting analytes can cause signal suppression, different peptides are not easily compared to one another directly. However, it is possible to compare relative signals of differing peptides to one another within samples, such as looking for a change in modification state or isoform stoichiometry via comparisons of the relative degree of modifications against an unmodified version.

In isotope-based analyses, the introduction of heavy nitrogen or carbon isotopes can be used to provide highly accurate quantitation measurements between heavy and light channels, thus reducing systematic and nonsystematic variations. Isotope labeling can also remove sample preparation biases, such as differences in digestion efficiency, fractionation patterns, and sample loss during cleanup steps, such as solid-phase extraction. Isotope labeling can be accomplished through metabolic labeling with heavy isotopes of nonradioactive amino acids, such as with SILAC (Ong et al., 2002) or ^{15}N (Wu, MacCoss, Howell, Matthews, & Yates, 2004), as well as through chemical labeling approaches, such as dimethyl labeling (Arnaudo, Molden, & Garcia, 2011), isotope-coded affinity tags (eg, ICAT; Gygi et al., 1999), or isobaric tags, such as iTRAQ (Ross et al., 2004) or TMT (Thompson et al., 2003). Isobaric tags differ from aforementioned approaches due to an ability to massively multiplex samples through the mixing of isobarically labeled peptides that can be later differentiated by specific reporter ions in MS/MS scans.

Metabolic isotope labeling has an additional application: pulse labeling to evaluate time-resolved information pertaining to net turnover: translation, degradation, and translocation. By adding heavy media for a selected period of time, "pulse" experiments can trace histone dynamics over space and time (Schwanhausser, Gossen, Dittmar, & Selbach, 2009).

In some experimental scenarios, it is desirable to measure the absolute, rather than relative, abundance of proteins. While absolute quantitative data cannot be directly determined from quantitative values, since peptides have different physical properties that dictate ionization efficiency, absolute quantitation can be derived by comparisons to spiked-in internal standards of defined abundance (Gerber, Rush, Stemman, Kirschner, & Gygi, 2003).

3.4 Acquisition Types

Aside from quantitative schemes for measuring abundance of peptides/proteins, another important consideration is the methodology of data acquisition. In classical LC–MS/MS experiments, peptides were selected for fragmentation to derive peptide identity based on selection of the most abundant ions at a given time during the LC timescale, termed data-dependent acquisition. Particularly in analyzing complex samples, a primary limitation to peptide sampling is the duty cycle of the mass spectrometer measurements, as this occurs on a chromatographic timescale. While faster mass spectrometers have continually been developed, low-abundance proteins and modifications are still prone to under sampling.

Targeted acquisition resolves issues of undersampling and improves sensitivity limitations through *a priori* determination of masses corresponding to analytes of interest. This can be accomplished through the use of triple quadrupole mass spectrometers by selected reaction monitoring (Lange, Picotti, Domon, & Aebersold, 2008), which measure individual fragment ions after fragmenting selected precursor masses, or by mass spectrometers more routinely utilized for data-dependent acquisition, such as an Orbitrap (ThermoScientific, San Jose, USA) or Q-TOF (Sciex, Toronto, Canada) by high mass accuracy monitoring of full fragmentation patterns of selected ions, termed parallel reaction monitoring (Peterson, Russell, Bailey, Westphall, & Coon, 2012). By selecting peptides of interest, in particular peptides that discern histone isotypes (eg, H3.1 vs H3.2 vs H3.3) or represent modification sites, reproducibility and sensitivity of detection can be maximized, thereby limiting noise and irrelevant signals, as well as facilitating the most accurate possible quantifications. The primary drawback to targeted analysis is that subsequent reanalysis of a dataset for previously unknown analytes is hampered, since data for these would not necessarily be collected, as would be the case for data-dependent approaches. An additional acquisition approach, termed data-independent analysis, is possible for "archival" data collection through an intentionally wide isolation protocol prior to fragmentation; however, this comes at the greatest expense of sensitivity and discernment of related analytes (Gillet et al., 2012).

3.5 Mass Spectrometry-Based Methods to Study Histone Turnover in Brain

The introduction of high-throughput MS-based proteomic techniques in the investigation of chromatin dynamics has provided us with a new understanding and appreciation of the nucleosome. The earliest insights arose in the early 2000s and introduced a game-changing dimension to what we know about histone proteins and their respective functions. This was, with the aid of MS, the realization of new PTMs including O-linked β-N-acetyl glucosamine (O-GlycNac) and lysine modifications such as propionylation, crotonylation, succinylation, and malonylation (Sakabe, Wang, & Hart, 2010; Tan et al., 2011; Xie et al., 2012). And most recently, MS was a key player in the groundbreaking finding of histone turnover as a previously undocumented regulator of synaptic formation and memory (Maze et al., 2015). Mass spectrometry can be used to study the dynamics of PTMs in the total histone pool by briefly labeling newly synthesized histones with a heavy isotope, such as ^{15}N or ^{13}C. Histone modifications can then be determined by MS for both the old and new histones based on mass differences between these two pools. Incorporation of heavy amino acids into a peptide leads to a known mass shift compared to the peptide that contains a natural (ie, "light") version of the amino acid. In contrast to early labeling studies, the isotopes of SILAC amino acids are, as the name indicates, stable, thereby differing dramatically from radioactive approaches (Mann, 2006). SILAC studies of histone turnover by Zee and colleagues have revealed a similar turnover rate for core histones, except for those of the H2A variant family (Zee, Levin, Dimaggio, & Garcia, 2010). Moreover, acetylated histones were recently shown to display faster turnover rates in comparison to unacetylated or methylated histone proteins. Additionally, histone modifications relating to active gene transcription report higher turnover rates in comparison to their silent counterparts.

Our latest pioneering study using SILAC in the investigation of histone turnover in postreplicative neurons (Fig. 2) found that in mice, approximately one-third of the total H3.3 pool was replaced over a 4-week period, even during times in which the overall amount of H3.3 remained unaltered. This demonstrated, contrary to dogma, that H3.3 undergoes constitutive turnover in neurons, while H3.1 and H3.2 remain static (Zovkic & Sweatt, 2015). In sum, mass spectrometry-based methods have allowed for groundbreaking strides in our understanding of neuroepigenetics, methods that can now be applied to both basic and translational investigations.

4. RETROSPECTIVE BIRTH DATING OF HISTONES IN HUMAN POSTMORTEM BRAIN

4.1 "Bomb Pulse Labeling" Coupled to Accelerator Mass Spectrometry

Until approximately 10 years ago, our understanding of cell turnover in human brain was greatly restricted. This was due to a general lack of methods available to study organic synthesis and decay in human tissues, where pulse-chase-labeling techniques were generally not feasible. The technique referred to as "bomb pulse labeling," however, revolutionized our ability to resolve these complicated phenomena and will serve as the focus of Sections 4.1 and 4.2.

High levels of open-air nuclear weapons testing from the mid 1950s to the early 1960s resulted in a dramatic increase in global levels of radioactive ^{14}C in the atmosphere. Elevated levels of ^{14}C resulting from this so-called bomb pulse effectively provided us with a novel means for monitoring organic synthesis and decay in human tissues, as ^{14}C levels have remained in exponential decay since the cessation of open-air bomb testing in the early 1960s. Thus, the principle of "bomb pulse labeling" assumes that most molecules within a given cell, with the exception of genomic DNA, are in a constant stage of dynamic regulation. In line with this assumption, molecules of ^{14}C that become integrated into genomic DNA should therefore reflect the level of ^{14}C in the atmosphere at the time of incorporation (Spalding, Bhardwaj, Buchholz, Druid, & Frisen, 2005). In other words, if a new neuron is born at time "X" and ^{14}C levels in the atmosphere at time "X" = "Y," then one can assume that if that cell persists until a later time of "Z" (as would be presumed to be the case for many nondividing cells), then levels of ^{14}C in the DNA of this cell should remain at Y over the course of time

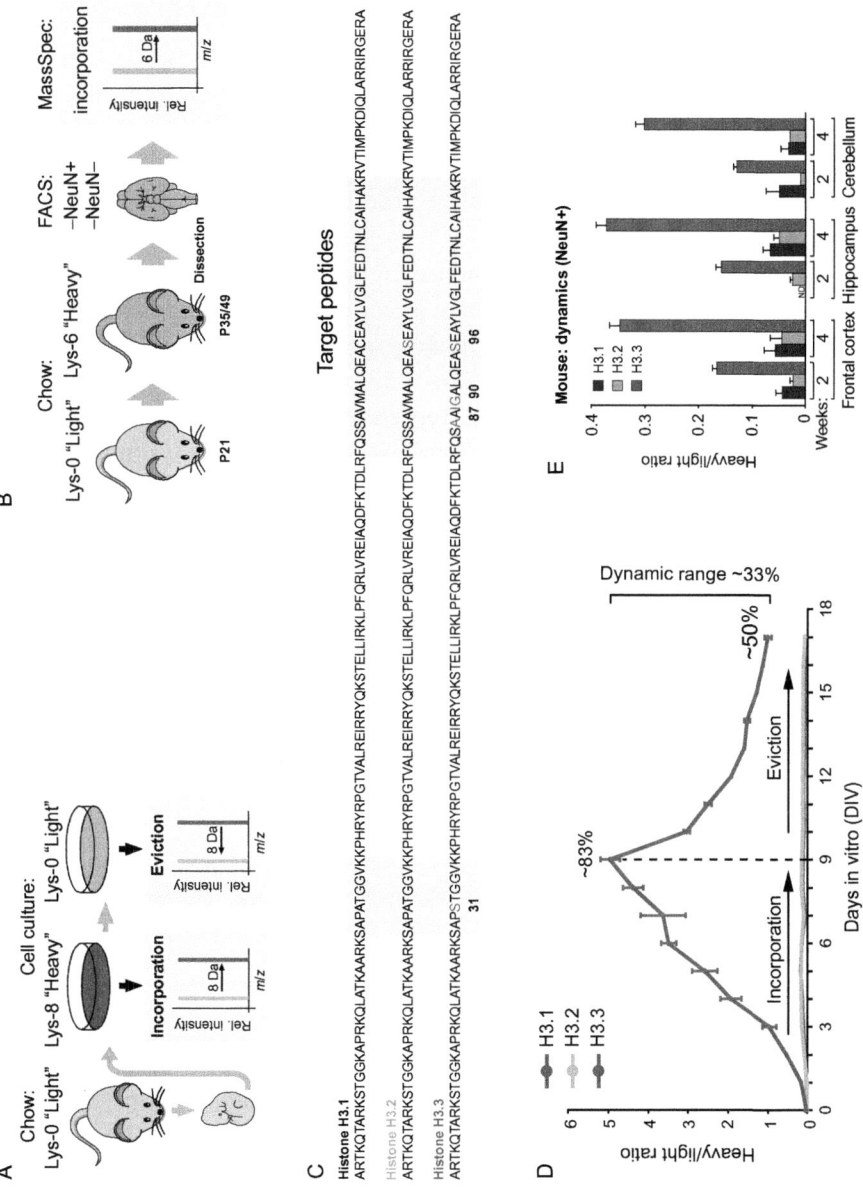

Fig. 2 See legend on next page.

approaching that of "Z". In line with such assumptions, pioneering studies by Frisen and colleagues were the first to demonstrate that levels of ^{14}C in human neuronal genomic DNA (isolated from postmortem tissues) closely correspond with atmospheric levels and can consequently be exploited to establish the timeline in which DNA is synthesized and new neurons are born (Spalding et al., 2013, 2005). By determining the age of DNA, which is reflected in the number of radioactive carbon atoms incorporated vs atmospheric ^{14}C levels, one can determine exactly when a cell is created utilizing accelerator mass spectrometry (AMS) to examine carbon turnover.

Until recently, dogma in the neuroscience field had long postulated that the adult human brain did not create new neurons, although several groups utilizing histological and/or autoradiography-based techniques had firmly established that adult neurogenesis, within specific regions of the brain (eg, hippocampal dentate gyrus), represents a phenomenon that persists throughout life (Altman & Das, 1965; Eriksson et al., 1998), similar to findings in rodents (Kuhn, Dickinson-Anson, & Gage, 1996). Using AMS, Spalding et al. were the first to conclusively determine the age of neurons in adult human hippocampus from postmortem neuronal DNA and concretely established the existence of adult human neurogenesis.

Although much of the recent work utilizing AMS had focused exclusively on cellular turnover in human brain (ie, DNA measurements), we recently set out to investigate whether protein dynamics, specifically within the context of histone turnover, could similarly be assessed using this method. To do so, we performed high-performance liquid chromatography (HPLC) to purify

---

Fig. 2 Mass spectrometry-based assessments of histone turnover in neurons using SILAC. (A) Schematic of SILAC to assess chromatin-associated H3.x incorporation and eviction from cultured neurons. (B) Schematic describing the SILAC mouse model to assess chromatin-associated H3.x incorporation in adult neurons. (C) Amino acid sequences for H3.x proteins highlighting differences between H3.1 vs H3.2 and H3.3 (H3.3, *red*). Putative target peptides for mass spectrometric analysis in the N-terminal tail (*yellow*) or histone core (*blue*) are indicated. Target peptides in *blue* are used for all subsequent analyses. (D) SILAC time course of H3.x chromatin incorporation and eviction in mouse embryonic neurons over the course of 17 d in vitro. Percentages reflect H3.3 peptide labeling by SILAC. (E) SILAC LC–MS/MS analysis of H3.1/2 vs H3.3 in NeuN+ mouse chromatin from multiple brain structures after 2 or 4 weeks of feeding on a heavy lysine (6 Da) diet. *Panels (D) and (E) displayed with permission from Maze, I., Wenderski, W., Noh, K. M., Bagot, R. C., Tzavaras, N., Purushothaman, I., et al. (2015). Critical role of histone turnover in neuronal transcription and plasticity. Neuron, 87(1), 77–94.* (See the color plate.)

chromatin-associated H3.3 from postmortem human brain, followed by AMS to examine H3.3 synthesis and decay rates throughout development and into adulthood. In doing so, we demonstrated that not only does H3.3 remain dynamic in neuronal chromatin throughout the lifetime of an individual, but also that rates of turnover differ dramatically during different stages of development, in line with a newly established role for these turnover events in synaptic connectivity and plasticity (Maze et al., 2015).

4.2 Bomb Pulse Labeling: Analytical Considerations

Open-air bomb testing resulted in essentially a global ^{14}C pulse-chase experiment from which one might infer turnover properties of various biological quantities. For example, depressed ^{14}C ratios in individuals born after the pulse initiation, or increased ^{14}C ratios in individuals born before the pulse initiation, suggest turnover. However, such conclusions still leave many important questions. For example, how quick is the turnover? Is turnover restricted to a subset of cells or a particular pool of molecules within a cell? Do turnover rates change with age? How does accumulation or depletion of the analyte (eg, due to cell growth/death, protein level increases/decreases, or isoform switching) affect these ratios and conclusions?

Drawing more precise inferences from ^{14}C ratios to gain insight into such questions is not straightforward for several reasons, many of which we detail later. These reasons to be outlined arguably apply not only to histone proteins, but most analytes, although here we focus on how to resolve them in the context of histone turnover. This analysis task is facilitated by interpreting the ^{14}C ratios in the context of a mathematical model that formalizes particular assumptions related to these (and other) questions posed earlier. The general approach is to write systems of differential equations that account for rates of ^{14}C uptake and removal due to the biological mechanisms deemed important for the questions at hand. These equations can be solved to predict how ^{14}C levels change over time in an individual with certain characteristics, as defined by the translation of biological assumptions into equations. Coherence of such equation-based predictions with observed ^{14}C levels from an experimental cohort suggests that the assumptions are consistent with reality (although such coherence never conclusively proves mechanism). Alternatively, incoherence strongly suggests that the assumptions do not reflect reality. Inherent in these equations are rate constants (and other parameters) whose estimated values will allow for quantitative and precise insight about turnover.

One of the potentially largest confounds in such analyses is that histone turnover can arise from cell birth/death or from subcellular histone turnover (ie, from transcriptional regulation). Simply having a ^{14}C ratio indicative of turnover does not on its own tell us which of these mechanisms may be responsible, and moreover, to what extent. If one has measurements of ^{14}C turnover in DNA, however, then this can be used to constrain the contribution of cell birth/death independently. Inherent in this analysis is the reasonably fixed ratio of histones/DNA carbons (Maze et al., 2015). This analysis requires the assumptions that (i) subcellular DNA turnover (ie, due to damage and repair) is negligible compared to that of cell turnover—which fortunately is typically the case (but may not always be), and also (ii) the overall cell number within the tissue of interest is essentially constant over a timescale of years. This is reasonably the case in many brain regions, but may not be in others or in different tissues (or disease contexts), so caution must be taken to evaluate this assumption critically. If the cell number in the tissue of interest is not constant over an individual's time span, the situation is not unworkable, but rather requires more information to constrain and formalize how cell birth and death rates change over time. For example, knowing the dynamics of cell number over time, and how the balance of cell proliferation and cell death dictate those dynamics, would be sufficient.

Another significant issue is accumulation (or depletion) of the histone of interest over time. In our work on H3.3, we found that its levels increase approximately threefold from birth to adolescence. Thus, one would expect changes in ^{14}C levels simply due to such accumulation, regardless of turnover. Like above with cell turnover, though, knowledge of such accumulation/depletion dynamics allows one to account for its contribution on observed ^{14}C levels (Fig. 3).

Given data on only a single subject, limited information can be inferred from ^{14}C data. In our experience, data from multiple subjects born both prebomb pulse and postbomb pulse, with a variety of birth and death years, significantly enhance one's ability to draw precise conclusions. In the case of H3.3, we found that data from individuals born prebomb pulse helped us to constrain the turnover rates later in life, whereas data from individuals born postbomb pulse helped us to constrain turnover rates early in life. Thus, identifying potential changes in turnover rates as a function of aging depends on picking individuals with a variety of birth and death ages pre and postbomb pulse. Although not yet performed, we postulate that such analyses

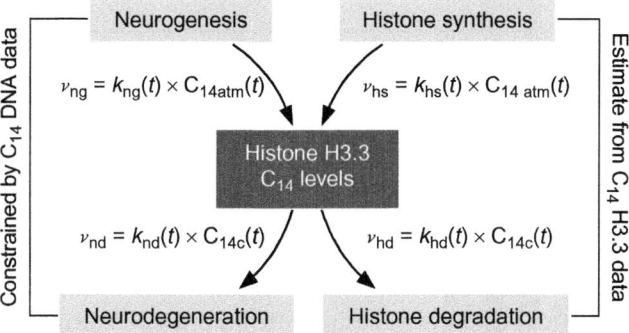

Fig. 3 Modeling nucleosomal turnover in human brain with AMS. The rate of fluctuation of ^{14}C levels in histone H3.3 pools from human brain tissue depends on the rates of (1) cellular (neuronal and glial) proliferation and death (eg, neurogenesis vs neurodegeneration), (2) histone H3.3 synthesis and accumulation in brain chromatin, and (3) degradation, which can be directly monitored using "bomb pulse labeling." With this information, one can effectively model H3.3 turnover rates as proportional to the current atmospheric ^{14}C/^{12}C levels and degradation rates as proportional to current cellular ^{14}C/^{12}C levels. *Displayed with permission from Maze, I., Wenderski, W., Noh, K. M., Bagot, R. C., Tzavaras, N., Purushothaman, I., et al. (2015). Critical role of histone turnover in neuronal transcription and plasticity. Neuron, 87(1), 77–94.*

can be applied to future investigations aimed at uncovering the role of histone dynamics in neurological disease states and may open the door to a new understanding of the role of histone regulation in brain.

5. CONCLUSION

Here, we have attempted to provide researchers with an updated "toolkit" of current proteomic methods for use in the dissection of histone turnover events in brain with the hopes that such information will stimulate deeper investigations into the underlying mechanisms of neuroplasticity and neurological disease. It is without doubt that the field of neuroepigenetics is progressing rapidly toward significantly enhanced and higher throughput methods to monitor the rates and functions of histone turnover in brain. It is our hope that by introducing these principles and protocols that researchers across diverse disciplines may now apply these tools in their specific research programs to address critical mechanistic questions relating to the role of histone turnover in brain development and human health.

6. METHODOLOGY: PREPARING CHROMATIN FROM NEURONS FOR MASS SPECTROMETRY ANALYSIS OF HISTONE VARIANTS AND TURNOVER

See Fig. 4 for schematic of analysis pipeline.

Note: The nucleus and the cytoplasm are highly distinct with respect to their macromolecular composition, and separation of these fractions allows for efficient and sensitive downstream proteomic analyses. To extract chromatin from various cell types in brain (or from primary cultures), the following stepwise protocol is ideal and is compatible with SILAC-based experiments.

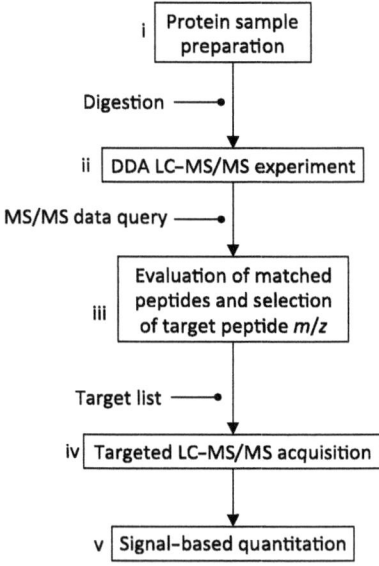

Fig. 4 Typical workflow for LC–MS-based quantitation of histone proteins. (i) High-quality histone-enriched protein samples, including nuclear fractions and chromatin, can be generated so that downstream fractionations are unnecessary (eg, for typical histone enrichment protocols, such as histone acid extraction (Shechter, Dormann, Allis, & Hake, 2007), histone signals are approximately 10-fold higher than the most abundant "nonhistone" protein in the sample). (ii) Digested samples are analyzed in data-dependent acquisition (DDA) experiments, (iii) which allows assessment of samples, including digestion efficiency and modification types, to select appropriate mass-to-charge ratios for peptides to be (iv) targeted in quantitative analyses. Preferably, samples are analyzed in biological and technical replicates to enable statistical analyses. (v) Target peptide signals can further be extracted and processed using appropriate software.

If one aims to isolate chromatin from intact brain, additional steps aimed at isolating neurons from nonneuronal cells are necessary due to the heterogeneity of brain tissues. To do so, fluorescence-activated cell sorting, otherwise referred to as "FACS," can be used. In short, this involves preextraction of nuclei in hypotonic lysis buffer, followed by ultracentrifugation and immunotagging with the neuronal specific marker NeuN using antibody-based methodologies that have been previously described (Matevossian & Akbarian, 2008).

Step 1: Chromatin isolation from neurons

Primary neurons

- First, cells are washed $2\times$ in phosphate-buffered saline, collected using a sterile cell scraper, and then spun in a table top centrifuge at $1500 \times g$ for 5 min.
- Pelleted cells are then resuspended in 200 μL of buffer A [10 mM HEPES, pH 7.9, 10 mM KCl, 1.5 mM MgCl$_2$, 0.34 M sucrose, 10% glycerol, 0.5 mM phenylmethylsulfonyl fluoride (PMSF), 0.1% Triton-X, 10 mM sodium butyrate, plus EDTA-free protease inhibitors (Roche, Basel, Switzerland), and PhosSTOP phosphatase inhibitors (Roche)].
- Resuspension of the cell pellet can be achieved in several ways. We recommend utilizing a Dounce homogenizer, or alternatively, they can be broken up and resuspended by drawing the cell suspension/buffer solution through a 22-gauge needle and 1-mL syringe.
- Homogenized samples are then incubated on ice for approximately 30 min, briefly vortexing the samples every 5 min.
- At this point, the sample is then ready to undergo initial crude fractionations. Samples are centrifuged at $1300 \times g$ for 5 min to separate the supernatants (crude cytosol*) from the pellets (nuclei).

  *Isolation of pure cytosol involves an additional centrifugation step at $20,000 \times g$ for 5 min to remove insoluble materials.

Both primary and FACS (see note earlier) purified neuronal nuclei

- Nuclear pellets are then washed once more with 200 μL of buffer A (without Triton-X) and then resuspend in 100 μL of buffer B [3 mM EDTA, 0.2 mM EGTA, 0.2 mM PMSF, 10 mM sodium butyrate, plus EDTA-free protease inhibitors, and PhosSTOP phosphatase inhibitors] (note: pelleted nuclei should appear white and opaque).
- Nuclear samples in buffer B are then incubated on ice again for 30 min, briefly vortexing the tube every 5 min.

- Following incubation, nuclear samples are centrifuged at 1700 × g for 5 min at 4°C to further fractionate the nuclear sample into a supernatant (soluble nuclear fraction) and final pellet (chromatin).
- Chromatin pellets can then be washed in 200 µL of buffer B prior to further processing into purified histones for mass spectrometry (note: chromatin pellets should appear glassy and translucent).

Step 2: Acid extraction of histones from chromatin
- Deviating slightly from published protocols (Shechter et al., 2007), chromatin pellets are then resuspended in 400 µL of 0.4 N sulfuric acid. This addition acts to solubilize histone proteins. Samples are then drawn through a 22-gauge needle and 1-mL syringe to remove any clumps that may be present.
- Samples are then incubated on a rotator for 30 min at 4°C. This step can be extended to overnight if desired, but we find 30 min to be sufficient to recover high yields of histone for subsequent processing.
- Following incubation, samples are centrifuged at 16,000 × g for 10 min at 4°C. This step acts to remove nuclear debris and nonacid soluble proteins.
- Histone containing supernatant is the carefully transferred into fresh 1.5-mL tubes.
- Trichloroacetic acid is then added to the supernatants to a final concentration of 33%, followed by gentle inversions to mix. Samples are then incubated overnight at 4°C. This is done to precipitate histone proteins from the acid solution.
- The next day, histones are pelleted by centrifugation at 16,000 × g for 10 min at 4°C, followed by subsequent removal of the supernatant. Take caution not to disrupt the histone pellet, which will likely appear as a smear along the side of the tube.
- Ice-cold acetone is then used to wash (2 ×) the histone pellets, typically using 1 mL/wash, followed by and centrifugation at 16,000 × g for 5 min at 4°C.
- Following the final wash, all supernatants are removed and histone pellets are allowed to air-dry for 15–20 min at room temperature.
- Pellets are then solubilized in up to 100 µL of water, taking note to wash the sides of the inner tube (note: if sufficient amounts of histones are present, resuspension in water should cause sample bubbling during pipetting. Once bubbles are present, the samples are now ready for subsequent processing).

Step 3: In solution digestion for mass spectrometry analyses

- Resuspended samples are then prepared with 8 M urea in 0.1 M ammonium bicarbonate (ABC). This serves to solubilize and denature proteins, making them more susceptible to subsequent enzymatic cleavage.
- *Reduction*: Dithiothreitol is added to a final concentration of 5 mM for 30 min at room temperature (note: it is very important to never heat with urea present, in particular with histone samples containing high contents of lysines that can be carbamylated) to disrupt disulfide bonds within and between proteins.
- *Alkylation*: Add iodoacetamide (IAA) to a final concentration of 15 mM and incubate for 30 min at room temperature in the dark. Alkylation with IAA increases the mass of a peptide by 57.021464 Da for each cysteine present.
- Samples are then diluted with 0.1 M ABC to reduce urea concentration to <4 M.
- *First digestion*: LysC is added at 1:50. This endoproteinase specifically hydrolyzes proteins at the carboxyl side of lysine, and efficient protein digestion can be completed in 6 h at room temperature.
- Dilute sample to <2 M urea with 0.1 M ABC.
- *Second digestion*: Trypsin (1:50) is added to the sample to further digest the proteins, also hydrolyzing following lysine residues but also arginine, and is incubated overnight at room temperature shaking.
- Trypsin digestion is stopped by adding trifluoroacetic acid (TFA) (0.5%, v/v). By lowering the pH of the solution, trypsin can no longer actively digest proteins.
- Samples are next desalted and concentrated using a C_{18} Stop And Go Extraction (StAGE) tip (Rappsilber, Ishihama, & Mann, 2003).
- Samples are then dried down in a speed vacuum centrifuge before resuspension in a mass spectrometry compatible buffer for analysis, such as 0.4% acetic acid (v/v) and 5% acetonitrile (v/v) in HPLC grade H_2O.

Step 4: Mass spectrometry analysis for identification of neuronal histone peptides

- Peptides of interest can be analyzed by a high-performance LC–MS/MS setup such as a Q-Exactive coupled to a Dionex NCP3200RS HPLC setup (ThermoScientific, San Jose, USA).

- *Data acquisition*: Depending on LC conditions including gradient, up to the 20 precursor ions are subjected to MS/MS per MS survey scan.
- *Analysis of data*: Software, such as "ProteomeDiscoverer/MASCOT," can be used to interrogate raw data which included querying MS/MS spectra against Uniprot's complete human or mouse proteome databases concatenated with common known contaminants (Bunkenborg, Garcia, Paz, Andersen, & Molina, 2010).
- For label-free quantitation experiments, the area of the target peptides is used, whereas for SILAC experiments, ratios can be calculated as heavy over light as previously described (Maze et al., 2015).

ACKNOWLEDGMENTS

We would like to thank members of the Maze laboratory for critical readings of the manuscript. We would also like to thank Dr. Alexey Soshnev for help with illustrations.

REFERENCES

Aebersold, R., & Goodlett, D. R. (2001). Mass spectrometry in proteomics. *Chemical Reviews*, *101*(2), 269–295.

Altman, J., & Das, G. D. (1965). Autoradiographic and histological evidence of postnatal hippocampal neurogenesis in rats. *The Journal of Comparative Neurology*, *124*(3), 319–335.

Arnaudo, A. M., Molden, R. C., & Garcia, B. A. (2011). Revealing histone variant induced changes via quantitative proteomics. *Critical Reviews in Biochemistry and Molecular Biology*, *46*(4), 284–294.

Bondy, S. C. (1971). The synthesis and decay of histone fractions and of deoxyribonucleic acid in the developing avian brain. *The Biochemical Journal*, *123*(3), 465–469.

Bunkenborg, J., Garcia, G. E., Paz, M. I., Andersen, J. S., & Molina, H. (2010). The minotaur proteome: Avoiding cross-species identifications deriving from bovine serum in cell culture models. *Proteomics*, *10*(16), 3040–3044.

Byvoet, P., Shepherd, G. R., Hardin, J. M., & Noland, B. J. (1972). The distribution and turnover of labeled methyl groups in histone fractions of cultured mammalian cells. *Archives of Biochemistry and Biophysics*, *148*(2), 558–567.

Commerford, S. L., Carsten, A. L., & Cronkite, E. P. (1982). Histone turnover within nonproliferating cells. *Proceedings of the National Academy of Sciences of the United States of America*, *79*(4), 1163–1165.

Cramer, S. C., Sur, M., Dobkin, B. H., O'Brien, C., Sanger, T. D., Trojanowski, J. Q., et al. (2011). Harnessing neuroplasticity for clinical applications. *Brain*, *134*(Pt. 6), 1591–1609.

De Riva, A., Deery, M. J., McDonald, S., Lund, T., & Busch, R. (2010). Measurement of protein synthesis using heavy water labeling and peptide mass spectrometry: Discrimination between major histocompatibility complex allotypes. *Analytical Biochemistry*, *403*(1–2), 1–12.

Dhingra, V., Gupta, M., Andacht, T., & Fu, Z. F. (2005). New frontiers in proteomics research: A perspective. *International Journal of Pharmaceutics*, *299*(1–2), 1–18.

Djondjurov, L. P., Yancheva, N. Y., & Ivanova, E. C. (1983). Histones of terminally differentiated cells undergo continuous turnover. *Biochemistry*, *22*(17), 4095–4102.

Domon, B., & Aebersold, R. (2006). Mass spectrometry and protein analysis. *Science*, *312*(5771), 212–217.

Eriksson, P. S., Perfilieva, E., Bjork-Eriksson, T., Alborn, A. M., Nordborg, C., Peterson, D. A., et al. (1998). Neurogenesis in the adult human hippocampus. *Nature Medicine*, *4*(11), 1313–1317.

Garcia, B. A., Mollah, S., Ueberheide, B. M., Busby, S. A., Muratore, T. L., Shabanowitz, J., et al. (2007). Chemical derivatization of histones for facilitated analysis by mass spectrometry. *Nature Protocols*, *2*(4), 933–938.

Garcia, B. A., Shabanowitz, J., & Hunt, D. F. (2007). Characterization of histones and their post-translational modifications by mass spectrometry. *Current Opinion in Chemical Biology*, *11*(1), 66–73.

Gerber, S. A., Rush, J., Stemman, O., Kirschner, M. W., & Gygi, S. P. (2003). Absolute quantification of proteins and phosphoproteins from cell lysates by tandem MS. *Proceedings of the National Academy of Sciences of the United States of America*, *100*(12), 6940–6945.

Gillet, L. C., Navarro, P., Tate, S., Rost, H., Selevsek, N., Reiter, L., et al. (2012). Targeted data extraction of the MS/MS spectra generated by data-independent acquisition: A new concept for consistent and accurate proteome analysis. *Molecular & Cellular Proteomics*, *11*(6). O111.016717.

Grove, G. W., & Zweidler, A. (1984). Regulation of nucleosomal core histone variant levels in differentiating murine erythroleukemia cells. *Biochemistry*, *23*(19), 4436–4443.

Gurley, L. R., & Walters, R. A. (1971). Response of histone turnover and phosphorylation to X irradiation. *Biochemistry*, *10*(9), 1588–1593.

Gygi, S. P., Rist, B., Gerber, S. A., Turecek, F., Gelb, M. H., & Aebersold, R. (1999). Quantitative analysis of complex protein mixtures using isotope-coded affinity tags. *Nature Biotechnology*, *17*(10), 994–999.

Hunt, D. F., Yates, J. R., 3rd., Shabanowitz, J., Winston, S., & Hauer, C. R. (1986). Protein sequencing by tandem mass spectrometry. *Proceedings of the National Academy of Sciences of the United States of America*, *83*(17), 6233–6237.

Kuhn, H. G., Dickinson-Anson, H., & Gage, F. H. (1996). Neurogenesis in the dentate gyrus of the adult rat: Age-related decrease of neuronal progenitor proliferation. *The Journal of Neuroscience*, *16*(6), 2027–2033.

Lange, V., Picotti, P., Domon, B., & Aebersold, R. (2008). Selected reaction monitoring for quantitative proteomics: A tutorial. *Molecular Systems Biology*, *4*, 222.

Luger, K., Mader, A. W., Richmond, R. K., Sargent, D. F., & Richmond, T. J. (1997). Crystal structure of the nucleosome core particle at 2.8 Å resolution. *Nature*, *389*(6648), 251–260.

Mann, M. (2006). Functional and quantitative proteomics using SILAC. *Nature Reviews. Molecular Cell Biology*, *7*(12), 952–958.

Matevossian, A., & Akbarian, S. (2008). Neuronal nuclei isolation from human postmortem brain tissue. *Journal of Visualized Experiments*, *20*.

Maze, I., Noh, K. M., & Allis, C. D. (2013). Histone regulation in the CNS: Basic principles of epigenetic plasticity. *Neuropsychopharmacology*, *38*(1), 3–22.

Maze, I., Noh, K. M., Soshnev, A. A., & Allis, C. D. (2014). Every amino acid matters: Essential contributions of histone variants to mammalian development and disease. *Nature Reviews. Genetics*, *15*(4), 259–271.

Maze, I., Wenderski, W., Noh, K. M., Bagot, R. C., Tzavaras, N., Purushothaman, I., et al. (2015). Critical role of histone turnover in neuronal transcription and plasticity. *Neuron*, *87*(1), 77–94.

Ong, S. E., Blagoev, B., Kratchmarova, I., Kristensen, D. B., Steen, H., Pandey, A., et al. (2002). Stable isotope labeling by amino acids in cell culture, SILAC, as a simple and accurate approach to expression proteomics. *Molecular & Cellular Proteomics*, *1*(5), 376–386.

Peterson, A. C., Russell, J. D., Bailey, D. J., Westphall, M. S., & Coon, J. J. (2012). Parallel reaction monitoring for high resolution and high mass accuracy quantitative, targeted proteomics. *Molecular & Cellular Proteomics*, *11*(11), 1475–1488.

Rappsilber, J., Ishihama, Y., & Mann, M. (2003). Stop and go extraction tips for matrix-assisted laser desorption/ionization, nanoelectrospray, and LC/MS sample pretreatment in proteomics. *Analytical Chemistry*, *75*(3), 663–670.

Ross, P. L., Huang, Y. N., Marchese, J. N., Williamson, B., Parker, K., Hattan, S., et al. (2004). Multiplexed protein quantitation in Saccharomyces cerevisiae using amine-reactive isobaric tagging reagents. *Molecular & Cellular Proteomics*, *3*(12), 1154–1169.

Sakabe, K., Wang, Z., & Hart, G. W. (2010). Beta-N-acetylglucosamine (O-GlcNAc) is part of the histone code. *Proceedings of the National Academy of Sciences of the United States of America*, *107*(46), 19915–19920.

Schwanhausser, B., Gossen, M., Dittmar, G., & Selbach, M. (2009). Global analysis of cellular protein translation by pulsed SILAC. *Proteomics*, *9*(1), 205–209.

Shechter, D., Dormann, H. L., Allis, C. D., & Hake, S. B. (2007). Extraction, purification and analysis of histones. *Nature Protocols*, *2*(6), 1445–1457.

Shi, Y., Lan, F., Matson, C., Mulligan, P., Whetstine, J. R., Cole, P. A., et al. (2004). Histone demethylation mediated by the nuclear amine oxidase homolog LSD1. *Cell*, *119*(7), 941–953.

Spalding, K. L., Bergmann, O., Alkass, K., Bernard, S., Salehpour, M., Huttner, H. B., et al. (2013). Dynamics of hippocampal neurogenesis in adult humans. *Cell*, *153*(6), 1219–1227.

Spalding, K. L., Bhardwaj, R. D., Buchholz, B. A., Druid, H., & Frisen, J. (2005). Retrospective birth dating of cells in humans. *Cell*, *122*(1), 133–143.

Tan, M., Luo, H., Lee, S., Jin, F., Yang, J. S., Montellier, E., et al. (2011). Identification of 67 histone marks and histone lysine crotonylation as a new type of histone modification. *Cell*, *146*(6), 1016–1028.

Thompson, A., Schafer, J., Kuhn, K., Kienle, S., Schwarz, J., Schmidt, G., et al. (2003). Tandem mass tags: A novel quantification strategy for comparative analysis of complex protein mixtures by MS/MS. *Analytical Chemistry*, *75*(8), 1895–1904.

Vanhoutte, K., Van Dongen, W., Hoes, I., Lemiere, F., Esmans, E. L., Van Onckelen, H., et al. (1997). Development of a nanoscale liquid chromatography/electrospray mass spectrometry methodology for the detection and identification of DNA adducts. *Analytical Chemistry*, *69*(16), 3161–3168.

Wolters, D. A., Washburn, M. P., & Yates, J. R., 3rd. (2001). An automated multidimensional protein identification technology for shotgun proteomics. *Analytical Chemistry*, *73*(23), 5683–5690.

Wu, C. C., MacCoss, M. J., Howell, K. E., Matthews, D. E., & Yates, J. R., 3rd. (2004). Metabolic labeling of mammalian organisms with stable isotopes for quantitative proteomic analysis. *Analytical Chemistry*, *76*(17), 4951–4959.

Xie, Z., Dai, J., Dai, L., Tan, M., Cheng, Z., Wu, Y., et al. (2012). Lysine succinylation and lysine malonylation in histones. *Molecular & Cellular Proteomics*, *11*(5), 100–107.

Zee, B. M., Levin, R. S., Dimaggio, P. A., & Garcia, B. A. (2010). Global turnover of histone post-translational modifications and variants in human cells. *Epigenetics & Chromatin*, *3*(1), 22.

Zovkic, I. B., & Sweatt, J. D. (2015). Memory-associated dynamic regulation of the "stable" core of the chromatin particle. *Neuron*, *87*(1), 1–4.

CHAPTER FOURTEEN

ChIP-Sequencing to Map the Epigenome of Senescent Cells Using Benzonase Endonuclease

T.S. Rai*,[1], P.D. Adams[†],[1]
*Institute of Biomedical and Environmental Health Research, University of the West of Scotland, Paisley, United Kingdom
[†]CR-UK Beatson Labs, Institute of Cancer Sciences, University of Glasgow, Glasgow, United Kingdom
[1]Corresponding authors: e-mail address: taranjitsingh.rai@uws.ac.uk; p.adams@beatson.gla.ac.uk

Contents

1. Introduction	356
2. Buffer Compositions	358
3. Protocol	359
3.1 Antibody Bead Preparation	359
3.2 Cell Culture, Lysis, and ChIP	359
4. Conclusion	362
References	363

Abstract

Cellular senescence is a state of stable cell cycle arrest triggered by diverse stresses. Establishment of senescence occurs in conjunction with a multitude of chromatin changes, which are just beginning to be studied. These chromatin changes are hypothesized to be causative for senescence. Currently, a preferred method to study such changes is chromatin immunoprecipitation followed by sequencing (ChIP-Seq). This is usually done by cross-linking the cells with formaldehyde and then generating chromatin fragments between 150 and 300 bp by sonication. The DNA replication-independent histone chaperone HIRA plays an important role in control of chromatin in nonproliferating senescent cells. While investigating the role of HIRA in senescence, we found conventional ChIP protocols to be problematic, routinely yielding too low amounts of DNA for sequencing. To overcome these problems we adapted and optimized an alternative ChIP method that does not rely on cross-linking and sonication for chromatin fragmentation, and is able to easily isolate chromatin from senescent cells ready for immunoprecipitation. This method uses Benzonase endonuclease for solubilization of uncross-linked chromatin by digestion of DNA and RNA, in the absence of proteolytic activity. Using this protocol, we were easily able to immunoprecipitate HIRA with sufficient DNA for Illumina sequencing.

1. INTRODUCTION

Cellular senescence is a state of stable cell cycle arrest associated with an inflammatory secretory phenotype. Cellular senescence can be triggered by a variety of stresses such as oncogenic stress, DNA damage, drug treatment, and others. Since senescence puts a brake on cell proliferation, it is thought to be tumor suppressive (Braig et al., 2005; Chen et al., 2005; Collado et al., 2005; Michaloglou et al., 2005). However, long-term senescence is also considered to have detrimental consequences due to its chronic inflammatory secretory phenotype (Campisi, 2005). One such detrimental consequence is thought to be acceleration of tissue aging (Baker et al., 2013, 2011).

Establishment of senescence is accompanied by large-scale changes in chromatin structure and function. An important chromatin regulator in senescent cells is histone chaperone, HIRA. HIRA specifically deposits a histone variant H3.3 into nucleosomes in a DNA replication-independent manner (Ray-Gallet et al., 2002; Tagami, Ray-Gallet, Almouzni, & Nakatani, 2004). HIRA and its complex partners, CABIN1, UBN1, and ASF1a, relocalize to PML nuclear bodies in different models of senescence and in various cell types (Banumathy et al., 2009; Rai et al., 2014, 2011; Zhang et al., 2005). This relocalization is thought to reflect trafficking of histones into chromatin of senescent cells (Rai et al., 2014). In addition to changes in the HIRA complex, several other chromatin changes have been reported in senescence. For example, in certain cell types there is formation of senescence-associated heterochromatic foci (SAHF), which are enriched for various heterochromatic marks, such as H3K9Me3 (Chicas et al., 2012; Narita et al., 2006, 2003). Some histone variants accumulate in senescent chromatin, including macroH2A (Zhang et al., 2005) and histone H3.3 (Rogakou & Sekeri-Pataryas, 1999). Further reports have indicated that other nonhistone proteins such as HP1 and HMGA can also accumulate at SAHF (Narita et al., 2006, 2003; Zhang et al., 2005).

Understanding these and other chromatin changes in senescence can be beneficial to our understanding of both aging and cancer. A widely used method to study such changes is chromatin immunoprecipitation followed by sequencing (ChIP-seq) (Barski et al., 2007; Johnson, Mortazavi, Myers, & Wold, 2007; Robertson et al., 2007). Typically, this involves cross-linking the cells with formaldehyde and then generating chromatin fragments containing less than 500 DNA base pairs by sonication. However, this method can generate inconsistent results between different experiments under seemingly identical conditions. In addition, formaldehyde cross-linking and sonication can denature protein epitopes, thus decreasing the efficiency of immunoprecipitation. In fact, while studying the role of HIRA in senescence, we found conventional ChIP protocols to be problematic, routinely yielding too low amounts of DNA for sequencing, even though such protocols previously worked well in proliferating HeLa cells (Pchelintsev et al., 2013). In addition to the aforementioned problems associated with cross-linking and sonication, this might in part also be due to other inherent problems associated with senescent cells, for example, large-scale changes in chromatin conformation and packaging, a large cytoplasmic to nuclear ratio, extensive vacuolization, and an extended cytoskeletal structure, all of which perhaps decreases the quality and quantity of chromatin obtained from senescent cells by conventional protocols.

To overcome these problems we adapted and optimized an alternative ChIP method that does not rely on cross-linking and sonication for chromatin fragmentation and is able to easily isolate chromatin from senescent cells ready for immunoprecipitation. Conventional "native ChIP" typically uses micrococcal nuclease (MNase) digestion, instead of sonication, of uncross-linked chromatin to fragment and solubilize the chromatin and has advantages and disadvantages over conventional ChIP (O'Neill & Turner, 2003). Our method uses Benzonase endonuclease for solubilization of uncross-linked chromatin by digestion of DNA and RNA. Benzonase is a nonspecific endonuclease, isolated from *Serratia marcescens* (Eaves & Jeffries, 1963; Meiss, Friedhoff, Hahn, Gimadutdinow, & Pingoud, 1995), that can degrade both DNA and RNA (Miller, Tanner, Alpaugh, Benedik, & Krause, 1994; Yonemura, Matsumoto, & Maeda, 1983). Importantly for ChIP, Benzonase is specific for nucleic acids with no detectable proteolytic activity (Yonemura et al., 1983). Benzonase does not differentiate between DNA, RNA, strandedness, and linear or circular nucleic acids (Meiss et al., 1995). Unlike MNase, Benzonase maintains its activity over a wide range of reaction conditions, from pH 6.0 to pH 10.0, temperature from 0 to 42°C, and monovalent cations, such as Na^+ and K^+, from 0 to 100 mM (Eaves & Jeffries, 1963). Benzonase can tolerate a variety of detergents such as Triton X-100 (upto 0.4%), SDS (upto 0.1%), and sodium deoxycholate (upto 0.4%). Further, Benzonase requires the presence of only one cation (Mg^{2+} at 1–2 mM), as opposed to two cations required by MNase (Mg^{2+} and Ca^{2+}). These properties make it an ideal enzyme to be used in ChIP for generating pure mono- and dinucleosome-enriched chromatin fractions.

Using Benzonase, there is efficient and uniform chromatin solubilization and minimal damage to epitopes on proteins of interest to be immunoprecipitated. Further, in this modified ChIP method described, we have also significantly decreased the total time to carry out the ChIP protocol to just 1 day. Using this protocol, we were easily able to immunoprecipitate HIRA from senescent cells with sufficient DNA for sequencing (Rai et al., 2014).

2. BUFFER COMPOSITIONS

Note 1. All buffers should be made in pyrogen-free, DNase-free, RNase-free ultrapure water (Millipore Milli-Q Synthesis A10, Cat: QGARD00R1) and then filtered using a VWR bottle-top sterile filtration unit (Cat: 514-0297).

Note 2. On the day of ChIP, all working solutions (except wash buffers) should be supplemented with protease (Sigma, P8340) and phosphatase (P5726) inhibitors. 100 mM final sodium butyrate can also be added to working solutions if ChIP is for acetylated histones and or proteins.

Note 3. After preparation, all buffers should be stored at 4°C.

Buffer 1: *Mild lysis buffer (MLB)*: 50 mM Tris, pH 7.5, 150 mM NaCl, 0.5% NP40, and 15 mM MgCl$_2$.

Note: This buffer is a stock buffer and should be diluted 1:4 using Milli-Q water on the day of ChIP.

Buffer 2: *Benzonase buffer (BB)*: 50 mM Tris, pH 7.5, 300 mM NaCl, 0.5% NP40, and 2.5 mM MgCl$_2$.

Note: Benzonase requires only Mg^{2+} for its endonuclease activity.

Buffer 3: *Dilution buffer (DB)*: 50 mM Tris, pH 7.5, 300 mM NaCl, 0.5% NP40, and 15 mM EDTA.

Note: This buffer has 15 mM EDTA that chelates Mg^{2+} ions and stops Benzonase activity.

Buffer 4: *Wash buffer 1 (WB1)*: 50 mM Tris, 150 mM NaCl, 0.5% NP40, and 5 mM EDTA.

Buffer 5: *Wash buffer 2 (WB2)*: 50 mM Tris, 5 mM EDTA, 0.5% NP40, and 300 mM NaCl.

Note: Salt concentration in this buffer could be increased to 500 mM–1 M for more stringent washes.

3. PROTOCOL

3.1 Antibody Bead Preparation

Step 1:
At the start of the day/experiment, pipette out 150 μL of dynabeads per IP (Invitrogen M-280 Cat: 112.02D for mouse IgG or Cat: 112.04D for rabbit IgG) in 1.5-mL eppendorf tubes on a magnetic rack for 1 min. Aspirate supernatant and wash the beads 3 × with 1 × phosphate buffered saline (PBS) + 0.5% bovine serum albumin (BSA).
Step 2:
Add 1 mL of 1× PBS + 0.5% BSA to the washed beads.
Step 3:
Add 6 μg of specific antibody to the beads per IP (depending upon number of ChIP experiments).
Note: As a control ChIP, always have same amount of species matched and class matched antibody + beads. If possible we also recommend having a CRISPR/siRNA/shRNA knockout cell line for the specific protein to be ChIPed in addition to control IgG.
Step 4:
Incubate beads + antibody mix on a slow rotating wheel at room temperature for at least 1 h.
Note: This mixture can be kept longer on a rotating wheel if the chromatin preparation steps take longer (as discussed later).
Step 5:
Wash the beads + antibody mix 3 × with WB1. Beads + antibody mix is now ready to be added to chromatin preparations.
Note: Any unbound or loosely bound antibody to the beads will be washed away in this step.

3.2 Cell Culture, Lysis, and ChIP

Having good quality cells/tissues is a prerequisite to a good ChIP-seq experiment. In the case of senescent cells, they should be quality controlled for markers of senescence, such as the presence of SA-β-gal staining, lack of EdU incorporation, and expression of p16 (Rai et al., 2014). When comparing proliferating and senescent cells, it is equally important to ensuring proliferating cells are also of very good quality and stress free.

Step 1: Grow cells to 80–95% confluency. Volumes in this protocol are for 1 × 15-cm plate. The number of cells per 15 cm varies; for example, this amounts to approximately 8×10^{-6} proliferating cells, 5×10^{-6} oncogene (HRasG12V)-induced senescent (OIS) cells, and 2×10^{-6} replicative senescent (RS) cells. For histone chaperone ChIP, we recommend pooling four to five 15-cm plates to yield greater than 10 ng of ChIP DNA.

Step 2: Aspirate media, wash cells once with PBS (room temperature).

Step 3: Scrape cells into 2 mL of MLB:H_2O = 1:4, supplemented with protease and phosphatase inhibitors (see note under MLB preparation). Transfer cells + buffer to 15-mL falcon tubes (prechilled to 4°C on ice).

Note: Scrape cells quickly and efficiently. If working with a small number of 15-cm dishes, we recommend scraping on ice at the bench. If there is a large number of 15-cm dishes, we recommend scraping cells in the cold room.

Step 4: Incubate 15-mL falcons containing cells + buffer on ice for 10 min and resuspend the mixture occasionally.

Step 5: Centrifuge cells + buffer for 3 min at 300 × g, +4°C.

Note: This mild lysis and centrifugation step separates cytoplasmic contents from nuclear contents. An aliquot of supernatant can also be saved for WB to confirm fractionation efficiency.

Step 6: Aspirate supernatant and recover the pellet.

Step 7: Resuspend pellet (nuclei) in 400 µL of BB with inhibitors (see earlier) and add 25 U/100 µL of Benzonase (Sigma, E1014).

Step 8: Incubate nuclei + BB on ice for 30 min; resuspend the mixture occasionally. An aliquot of supernatant can also be saved prior to addition of Benzonase ($t=0$) to check DNA digestion efficiency. DNA can be isolated from this aliquot using standard DNA isolation kits such as Qiagen Cat: 80004. Isolated DNA should be run on an agarose gel to check digestion efficiency of Benzonase. We recommend starting with 25 U/mL of Benzonase and testing nucleosome fractions over a period of time. Fig. 1A and B illustrates generation of mono-, di-, and trinucleosomes over a period of time in RKO cells (low exposure left and high exposure right). Generally, incubation with 25 U/mL of Benzonase for 25–30 min generates mononucleosome-rich chromatin preps in multiple cell lines (Fig. 1B). Similar conditions can also be applied to chromatin preparations for senescent cells. Fig. 1C shows digestion patterns over several independent replicates in two different

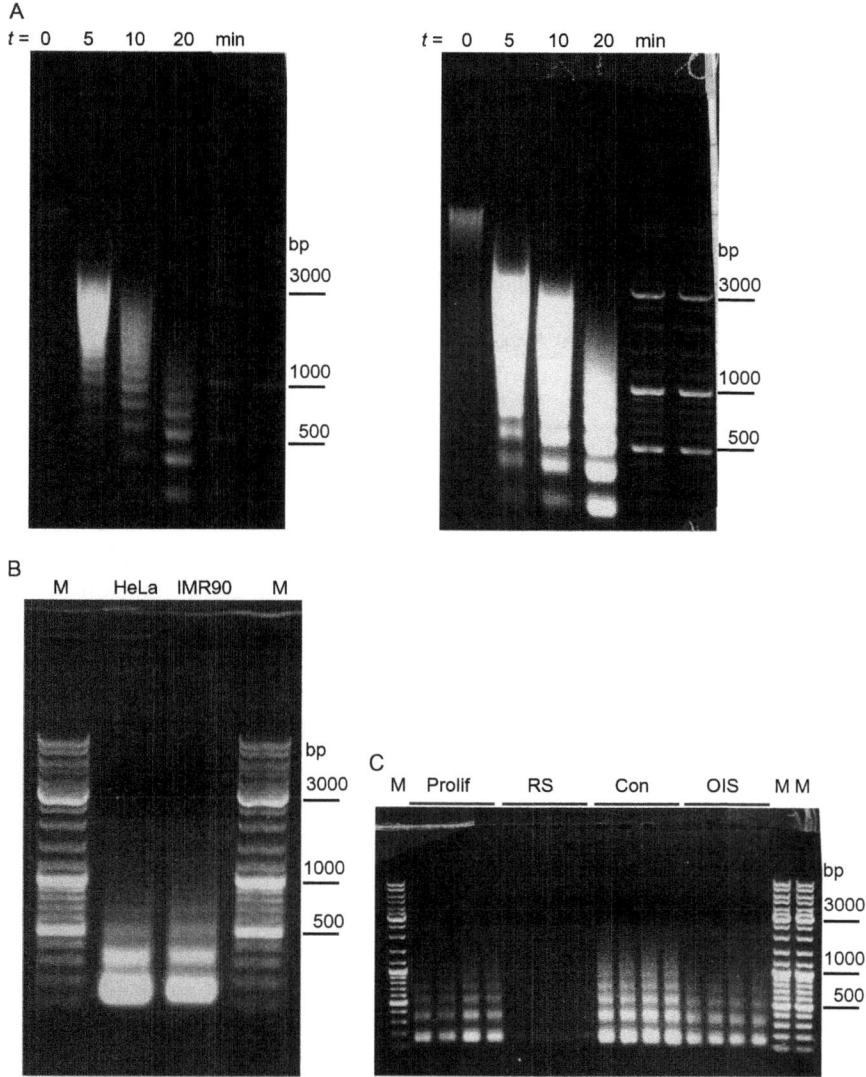

Fig. 1 Optimization of Benzonase digestion in various cell lines and digestion patterns analyzed on 1.0% agarose gels. (A) Comparison of chromatin digestion by Benzonase in RKO cells over a period of 20 min. *Left image*: lower exposure, *Right image*: higher exposure. (B) Comparison of chromatin digestion by Benzonase in HeLa and IMR90 cells, showing predominantly mono- and dinucleosomal fractions. (C) Comparison of chromatin digestion by Benzonase in RS and OIS models. Each lane is independent Benzonase-digested cell samples.

models of senescence. RS and OIS (low exposure left and high exposure right).

Note: This step is very important and should be optimized for each cell line tested. Enrichment of mono- and dinucleosomes is important for a good immunoprecipitation, just as in the case of the cross-linked ultrasound sonication protocols.

Step 9: Centrifuge the mixture 3 min at $300 \times g$, +4°C.

Step 10: Recover supernatant—it contains solubilized chromatin and other nuclear proteins.

Step 11: Dilute supernatant twofold by adding 400 µL of DB with inhibitors. EDTA in the buffer will inhibit Benzonase activity, which relies on Mg^{2+}.

Step 12: Use this diluted supernatant for ChIP. Take out 50 µL of this diluted supernatant as input to be processed for DNA isolation later (see step 16). Determine the protein concentration of the supernatant by Bradford assay. The amount of DNA in the diluted sample can also be measured using spectrophotometric methods. Using dilution buffer, equilibrate to equal amount of protein or DNA in each IP.

Step 13: Add preprepared beads (from step 5) to diluted chromatin and incubate 2–6 h in cold room.

Note: We recommend incubation times of 6 h or less, as higher or O/N incubation will result in higher background.

Step 14: Wash $3 \times$ with WB1.

Step 15: Wash $2-3 \times$ with WB2.

Step 16: For protein analysis, elute beads into SB by boiling for 5 min and proceed with Western blot. For DNA analysis, elute beads into 600 µL RLT buffer + 6 µL beta mercaptoethanol (Qiagen Cat: 80004) and apply to DNA isolation column (AllPrep DNA Spin Column). Purify according to the manual. Using this protocol we have consistently recovered greater than 10 ng of ChIP DNA for histone chaperone ChIPs. Recovered DNA was used for library preparation using NEBNext Ultra DNA library prep Kit.

4. CONCLUSION

Formaldehyde cross-linking and sonication can potentially degrade proteins and denature protein epitopes. Degradation of proteins is likely detrimental to efficient chromatin immunoprecipitation. Moreover, chromatin preparation from senescent cells can also be problematic by conventional protocols. To overcome these limitations, we describe a ChIP method that uses Benzonase digestion. The protocol just takes 1 day and is applicable to a wide range of cell lines and is reproducible across multiple experiments. Using this protocol, we have successfully mapped the genome-wide distribution of HIRA histone chaperone not only in senescent cells but also in proliferating cells (Rai et al., 2014).

REFERENCES

Baker, D. J., Dawlaty, M. M., Wijshake, T., Jeganathan, K. B., Malureanu, L., van Ree, J. H., ... van Deursen, J. M. (2013). Increased expression of BubR1 protects against aneuploidy and cancer and extends healthy lifespan. *Nature Cell Biology, 15*(1), 96–102. http://dx.doi.org/10.1038/ncb2643.

Baker, D. J., Wijshake, T., Tchkonia, T., LeBrasseur, N. K., Childs, B. G., van de Sluis, B., ... van Deursen, J. M. (2011). Clearance of p16Ink4a-positive senescent cells delays ageing-associated disorders. *Nature, 479*(7372), 232–236. http://dx.doi.org/10.1038/nature10600. nature10600 [pii].

Banumathy, G., Somaiah, N., Zhang, R., Tang, Y., Hoffmann, J., Andrake, M., ... Adams, P. D. (2009). Human UBN1 is an ortholog of yeast Hpc2p and has an essential role in the HIRA/ASF1a chromatin-remodeling pathway in senescent cells. *Molecular and Cellular Biology, 29*(3), 758–770. http://dx.doi.org/10.1128/MCB.01047-08. MCB.01047-08 [pii].

Barski, A., Cuddapah, S., Cui, K., Roh, T. Y., Schones, D. E., Wang, Z., ... Zhao, K. (2007). High-resolution profiling of histone methylations in the human genome. *Cell, 129*(4), 823–837. http://dx.doi.org/10.1016/j.cell.2007.05.009. S0092-8674(07) 00600-9 [pii].

Braig, M., Lee, S., Loddenkemper, C., Rudolph, C., Peters, A. H., Schlegelberger, B., ... Schmitt, C. A. (2005). Oncogene-induced senescence as an initial barrier in lymphoma development. *Nature, 436*(7051), 660–665.

Campisi, J. (2005). Senescent cells, tumor suppression, and organismal aging: Good citizens, bad neighbors. *Cell, 120*(4), 513–522.

Chen, Z., Trotman, L. C., Shaffer, D., Lin, H. K., Dotan, Z. A., Niki, M., ... Pandolfi, P. P. (2005). Crucial role of p53-dependent cellular senescence in suppression of Pten-deficient tumorigenesis. *Nature, 436*(7051), 725–730.

Chicas, A., Kapoor, A., Wang, X., Aksoy, O., Evertts, A. G., Zhang, M. Q., ... Lowe, S. W. (2012). H3K4 demethylation by Jarid1a and Jarid1b contributes to retinoblastoma-mediated gene silencing during cellular senescence. *Proceedings of the National Academy of Sciences of the United States of America, 109*(23), 8971–8976. http://dx.doi.org/10.1073/pnas.1119836109. 1119836109 [pii].

Collado, M., Gil, J., Efeyan, A., Guerra, C., Schuhmacher, A. J., Barradas, M., ... Serrano, M. (2005). Tumour biology: Senescence in premalignant tumours. *Nature, 436*(7051), 642.

Eaves, G. N., & Jeffries, C. D. (1963). Isolation and properties of an exocellular nuclease of Serratia marcescens. *Journal of Bacteriology, 85*(2), 273–278.

Johnson, D. S., Mortazavi, A., Myers, R. M., & Wold, B. (2007). Genome-wide mapping of in vivo protein-DNA interactions. *Science, 316*(5830), 1497–1502. http://dx.doi.org/10.1126/science.1141319.

Meiss, G., Friedhoff, P., Hahn, M., Gimadutdinow, O., & Pingoud, A. (1995). Sequence preferences in cleavage of dsDNA and ssDNA by the extracellular Serratia marcescens endonuclease. *Biochemistry, 34*(37), 11979–11988.

Michaloglou, C., Vredeveld, L. C., Soengas, M. S., Denoyelle, C., Kuilman, T., van der Horst, C. M., ... Peeper, D. S. (2005). BRAFE600-associated senescence-like cell cycle arrest of human naevi. *Nature, 436*(7051), 720–724.

Miller, M. D., Tanner, J., Alpaugh, M., Benedik, M. J., & Krause, K. L. (1994). 2.1 A structure of Serratia endonuclease suggests a mechanism for binding to double-stranded DNA. *Nature Structural Biology, 1*(7), 461–468.

Narita, M., Narita, M., Krizhanovsky, V., Nunez, S., Chicas, A., Hearn, S. A., ... Lowe, S. W. (2006). A novel role for high-mobility group a proteins in cellular senescence and heterochromatin formation. *Cell, 126*(3), 503–514.

Narita, M., Nunez, S., Heard, E., Lin, A. W., Hearn, S. A., Spector, D. L., ... Lowe, S. W. (2003). Rb-mediated heterochromatin formation and silencing of E2F target genes during cellular senescence. *Cell, 113*(6), 703–716.

O'Neill, L. P., & Turner, B. M. (2003). Immunoprecipitation of native chromatin: NChIP. *Methods, 31*(1), 76–82. Comparative Study.

Pchelintsev, N. A., McBryan, T., Rai, T. S., van Tuyn, J., Ray-Gallet, D., Almouzni, G., & Adams, P. D. (2013). Placing the HIRA histone chaperone complex in the chromatin landscape. *Cell Reports, 3*(4), 1012–1019. http://dx.doi.org/10.1016/j.celrep.2013.03.026.

Rai, T. S., Cole, J. J., Nelson, D. M., Dikovskaya, D., Faller, W. J., Vizioli, M. G., ... Adams, P. D. (2014). HIRA orchestrates a dynamic chromatin landscape in senescence and is required for suppression of neoplasia. *Genes & Development, 28*(24), 2712–2725. http://dx.doi.org/10.1101/gad.247528.114.

Rai, T. S., Puri, A., McBryan, T., Hoffman, J., Tang, Y., Pchelintsev, N. A., ... Adams, P. D. (2011). Human CABIN1 is a functional member of the human HIRA/UBN1/ASF1a histone H3.3 chaperone complex. *Molecular and Cellular Biology, 31*(19), 4107–4118. http://dx.doi.org/10.1128/MCB.05546-11. MCB.05546-11 [pii].

Ray-Gallet, D., Quivy, J.-P., Scamps, C., Martini, E. M., Lipinski, M., & Almouzni, G. (2002). HIRA is critical for a nucleosome assembly pathway independent of DNA synthesis. *Molecular Cell, 9*, 1091–1100.

Robertson, G., Hirst, M., Bainbridge, M., Bilenky, M., Zhao, Y., Zeng, T., ... Jones, S. (2007). Genome-wide profiles of STAT1 DNA association using chromatin immunoprecipitation and massively parallel sequencing. *Nature Methods, 4*(8), 651–657. http://dx.doi.org/10.1038/nmeth1068.

Rogakou, E. P., & Sekeri-Pataryas, K. E. (1999). Histone variants of H2A and H3 families are regulated during in vitro aging in the same manner as during differentiation. *Experimental Gerontology, 34*(6), 741–754.

Tagami, H., Ray-Gallet, D., Almouzni, G., & Nakatani, Y. (2004). Histone H3.1 and H3.3 complexes mediate nucleosome assembly pathways dependent or independent of DNA synthesis. *Cell, 116*(1), 51–61.

Yonemura, K., Matsumoto, K., & Maeda, H. (1983). Isolation and characterization of nucleases from a clinical isolate of Serratia marcescens kums 3958. *Journal of Biochemistry, 93*(5), 1287–1295.

Zhang, R., Poustovoitov, M. V., Ye, X., Santos, H. A., Chen, W., Daganzo, S. M., ... Adams, P. D. (2005). Formation of MacroH2A-containing senescence-associated heterochromatin foci and senescence driven by ASF1a and HIRA. *Developmental Cell, 8*(1), 19–30.

CHAPTER FIFTEEN

Exploiting Chromatin Biology to Understand Immunology

J.L. Johnson, G. Vahedi[1]

Institute for Immunology, Pereleman School of Medicine, University of Pennsylvania, Philadelphia, PA, United States
[1]Corresponding author: e-mail address: vahedi@mail.med.upenn.edu

Contents

1. Introduction: Design Principle of Immune Responses	366
2. Epigenome: Our Software	366
3. Mapping DNA Methylation	367
4. Mapping Histone Modifications by Chromatin Immunoprecipitation and Sequencing	368
5. Limitations of ChIP-seq	369
6. MNase-seq for Nucleosome Positioning	370
7. DNase-seq and ATAC-seq for DNA Accessibility	371
8. Analysis of NGS Data	372
9. Painting an Enhancer Landscape	376
10. Chromatin Biology to Understand Gene Regulation in Immune Cells	377
11. Role of Intrinsic and Extrinsic Signals on Enhancer Formation	378
12. Conclusions	379
Acknowledgments	380
References	380

Abstract

The activation of immune responses relies on variations of a common rule where immune cells that are able to sense infections produce one set of cytokines to induce lymphocytes to produce another set of cytokines, which in turn activate the appropriate effector responses. This multitiered immune response is in fact a remarkable showcase of different ways the same genome can be used to facilitate cellular communications. Here, we review next-generation sequencing methods enabling us to map the differential usage of our genome in primary immune cells.

1. INTRODUCTION: DESIGN PRINCIPLE OF IMMUNE RESPONSES

Multitiered immune responses, as recently reviewed by Iwasaki and Medzhitov (2015), are orchestrated by three groups of cells. The first tier relates to cells that function as *sensors*, detect pathogens, and secrete *level 1* cytokines. The second tier relates to the tissue-resident lymphocytes that respond to level 1 cytokines and lead to the secretion of *level 2* cytokines. And finally, we have the *effector* cells that respond to level 2 cytokines to carry out the appropriate effector functions to eliminate the pathogen. Dendritic cells and macrophages serve as sensors for the type 1 immune response and epithelial cells and mast cells for the type 2 immune response. Level 1 cytokines act on terminally differentiated lymphocytes $CD4^+$ T helper cells. In response to level 1 cytokines, these lymphocytes produce level 2 cytokines. The level 2 cytokines act on the effector cells of the immune response. These include macrophages, neutrophils, epithelial cells, eosinophils, and basophils, B cells as well as sensory neurons, endothelium, and smooth muscle cells. These cells perform diverse effector functions, including barrier defenses, killing, and expulsion of pathogens, in addition to antibody production.

In this multitiered design, cytokines are simply "signals" acting as the major building blocks of communication between cells. To permit this communication without undue interference, it is expected that signals create distinct impacts on cells. Overwhelming evidence indicates that the chromatin structure represents a vehicle for specifying cellular regulatory states and is an attractive template for transmitting and recording immunological events (Ciofani et al., 2012; Gosselin et al., 2015; Ostuni et al., 2013; Vahedi et al., 2015, 2012). In this chapter, we briefly review genomic methods that map various aspects of the epigenome. Although DNA methylation is an important aspect of epigenomics, we will mostly focus on methodologies mapping histone modifications and chromatin accessibility. We present computational workflow required to analyze epigenomic measurements and demonstrate studies in immune cells that help advance our understanding of gene regulation.

2. EPIGENOME: OUR SOFTWARE

The DNA sequence that makes up our genome can be seen as our body's "hardware." Sitting on top of the genomic hardware is the

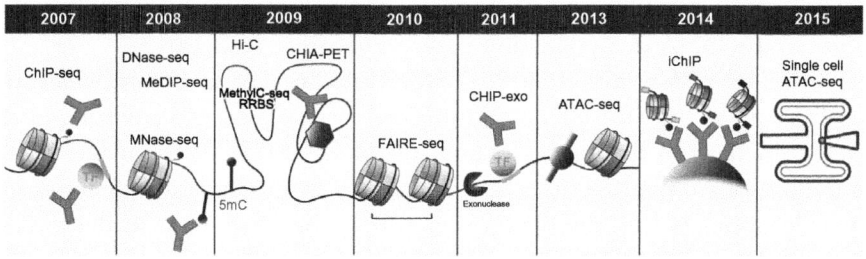

Fig. 1 Timeline of sequencing-based technologies for mapping epigenomes. *Adapted from Rivera, C. M., & Ren, B. (2013). Mapping human epigenomes.* Cell, 155(1), 39–55, *with permission from Elsevier.* (See the color plate.)

"software" layer called the "epigenome." The fundamental units of the epigenome are nucleosomes comprising approximately 147 bp of DNA around a histone octamer. The nucleosomal histones can be chemically modified and exchanged with variants (Zhou, Goren, & Bernstein, 2011). Epigenomics refer to the collection of DNA methylation and covalent modifications of histone proteins.

We can now create unprecedented maps of human and mouse epigenomes as a result of the application of next-generation sequencing (NGS). Powerful, parallel short-read sequencing technologies have proven increasingly high-throughput, fast, accurate, and cost-effective at rates faster than Moore's law (Rivera & Ren, 2013). Overall, the accomplishments and discoveries of the epigenomics field have hinged in large part on the improvements of technologies (Fig. 1).

3. MAPPING DNA METHYLATION

DNA methylation and hydroxymethylation are involved in development, X-chromosome inactivation, cell differentiation, tissue-specific gene expression, imprinting, cancers, and diseases (Jullien & Berger, 2010; Schmitz et al., 2013; Smith & Meissner, 2013). DNA methylation typically occurs at the 5 position of cytosines and plays a crucial role in gene regulation and chromatin remodeling. Most cytosine methylation occurs on cytosines located near guanines called CpG islands. CpG islands are defined as regions that are greater than 500 bp in length with more than 55% GC content. Although cytosine methylation (5mC) is known as a silencing mark repressing genes, cytosine hydroxymethylation (5hmC) has been shown to be an activating mark that promotes gene expression and is a proposed intermediate in the DNA demethylation pathway (Koh & Rao, 2013).

It has been shown that 5mC and/or 5hmC can be used as a diagnostic tool to help identify the effects of nutrition, carcinogens, and environmental factors in relation to diseases (Thomson et al., 2012). What makes these modifications unique is that their impact on gene regulation depends on their locations within the genome. It is therefore important to determine the exact position of the modified bases. Different methods have been developed to map genomic regions associated with DNA methylation. Bisulfite sequencing or whole-genome bisulfite sequencing is a well-established protocol to detect methylated cytosines in genomic DNA (Feil, Charlton, Bird, Walter, & Reik, 1994). Methylated DNA immunoprecipitation sequencing is a commonly used assay to study 5mC or 5hmC modification (Weber et al., 2005). Specific antibodies can be used to study cytosine modifications. If using 5mC-specific antibodies, methylated DNA is isolated from genomic DNA via immunoprecipitation. Anti-5mC antibodies are incubated with fragmented genomic DNA and precipitated, followed by DNA purification and sequencing. Deep sequencing provides greater genome coverage, representing the majority of immunoprecipitated methylated DNA. A more detailed review on DNA methylation assays can be found elsewhere (Koh & Rao, 2013).

4. MAPPING HISTONE MODIFICATIONS BY CHROMATIN IMMUNOPRECIPITATION AND SEQUENCING

Chromosomal DNA is packaged into nucleosomes with DNA wrapped around histone octamers consisting of H2A, H2B, H3, H4 subunits, and their variants. The histone tails and histone proteins are subject to over more than hundred posttranslational modifications (PTMs), and less than a thousand distinct histone isoforms have been discovered in humans (Zhou et al., 2011). Histones can undergo various covalent modifications including acetylation, methylation, phosphorylation, and ubiquitination. Chromatin immunoprecipitation and sequencing (ChIP-seq) enabled the genome-wide identification of histone modifications in an unbiased manner. ChIP-seq allows for detection of DNA–protein interactions and is widely utilized for mapping histone tail modifications and distribution of chromatin-associated proteins (Park, 2009). The technique involves two major steps: (a) cross-linking of a protein to the contacting DNA and (b) subsequent immunoprecipitation of DNA–protein complexes by antibodies directed at a target protein. The pool of fragmented DNA enriched through immunoprecipitation is then sequenced to produce a genome-scale map of local read frequencies, representing both the location and strength of DNA–protein interactions (Park, 2009).

5. LIMITATIONS OF ChIP-seq

Mapping chromatin alterations in cell lines and primary cells have revolutionized our approach toward gene regulation. However, conventional ChIP-seq assays have major limitations. One major limitation of ChIP assays is that they heavily rely on the quality of antibodies. The mass production of more than thousand PTM-specific histone antibodies has greatly facilitated the study of these histone marks and their impact on chromatin function (Perez-Burgos et al., 2004). Indeed, large-scale epigenomics "road map" efforts like the Encyclopedia of DNA Elements (ENCODE) and NIH Epigenomics Roadmap projects rely on these antibodies for genome-wide mapping of histone PTM distribution (Ernst & Kellis, 2010; Kellis et al., 2014). However, recent reports have made some alarming observations regarding the impact of histone PTM antibodies, including off target recognition, strong influence by neighboring PTMs, and an inability to identify the modification state on a particular residue (Rothbart et al., 2015). Consequently, poor antibody choice can lead to misinformed conclusions regarding the location and function of the histone PTM (Baker, 2015). Since our understanding of how histone PTMs contribute to normal biology and human disease depends on the quality of the antibodies, continued rigorous quality control is critical for the accurate data interpretation. Recently, a publicly available and interactive database has been designed (Rothbart et al., 2015). This interactive web portal provides a critical resource to the research community that routinely uses these antibodies as detection reagents for a wide range of applications.

Another limitation of early ChIP-seq procedures is that they require separate IPs that are sensitive to the amount of chromatin input in addition to the quality of the antibody. This compromises the accuracy with which chromatin alterations can be quantitatively evaluated across samples. The lack of quantitative information in ChIP-seq data is a long-standing limitation and can complicate global differences in histone modification levels due to cell-state transitions or genetic mutations in epigenetic regulators frequently observed in cancer (Ryan & Bernstein, 2012). Recent studies have presented strategies for quantitatively comparing ChIP-seq signal intensities by incorporating synthetic histone spike-in controls, but these protocols may not be compatible with low cell numbers (Bonhoure et al., 2014; Grzybowski, Chen, & Ruthenburg, 2015; Orlando et al., 2014).

A major restriction particularly relevant to primary immune cells is that conventional ChIP-seq experiments require large numbers of cells. Various adaptations of ChIP-seq reduced the cell number requirements (Adli, Zhu, & Bernstein, 2010; Goren et al., 2010; Lara-Astiaso et al., 2014). Lara-Astiaso et al. introduced an innovative approach to overcome the limitations posed by trying to do ChIP-seq on small numbers of cells comprising key progenitor populations. The method is called indexing-first ChIP since it relies on indexing before the IP step. In this protocol, barcoding is performed directly on the total cellular chromatin, hence avoiding the low-input enzymatic reactions occurring in conventional ChIP. This study represents an important technological advance that will likely reshape our understanding of chromatin organization in development.

Despite progress toward reducing cell requirements, these upgrades do not address the challenge of quantitative comparison, are low throughput, and have mostly worked only for certain modifications (Adli et al., 2010; Goren et al., 2010; Lara-Astiaso et al., 2014). A more recent study addressed this limitation by both minimizing loss and performing linear amplification of input material, which have individually been demonstrated to increase sensitivity of ChIP experiments (van Galen et al., 2016).

6. MNase-seq FOR NUCLEOSOME POSITIONING

MNase-seq has been develop to map positions of nucleosomes and employs a micrococcal nuclease that preferentially cleaves DNA between nucleosomes, thereby generating mononucleosome preparations of approximately 200-bp DNA with associated core histones. By reading DNA sequences from the free end, the positions of individual nucleosomes can be mapped on a reference genome (Schones et al., 2008). Nucleosome phasings surrounding the transcription start site of genes are identified with notable differences between active and silent genes. In contrast, in other genomic regions where the positioning of nucleosomes is substantially flexible, reproducibly positioned nucleosomes are seen only for a small fraction. Therefore, both DNA sequence-based nucleosome preference and nonnucleosomal factors are possibilities to determine nucleosome organization in mammalian cells (Valouev et al., 2011).

7. DNase-seq AND ATAC-seq FOR DNA ACCESSIBILITY

Similar to MNase, the nuclease DNase I generates double-strand breaks by nicking complementary strands of DNA one strand at a time (Campbell & Jackson, 1980). DNase-seq uses DNase I to cut DNA within chromatin that is open and accessible to the enzyme, either away from the nucleosome-compacted area or not tightly covered by DNA-associated proteins. Conventionally, the positions of frequent cutting are referred to as DNase hypersensitivity sites. Using genomic sequencing of free ends generated by DNase-mediated cutting, researchers have identified hypersensitivity sites across the genome, in particular for distal intergenic regulatory elements with an open-chromosome configuration (Crawford et al., 2004).

The ability of DNase-seq in identifying transcription factor binding sites is strongly dependent on the size of fragments. It is clear from recent studies that for several transcription factors, it is more efficient to use shorter fragments (typically less than 100 bp). However, longer fragments (more than 150 bp) span entire nucleosomes (He et al., 2014; Vierstra, Wang, John, Sandstrom, & Stamatoyannopoulos, 2014) and are less likely to cluster around open chromatin regions where transcription factors bind. DNase I cleavage sites are mostly related to the precise sequence of the three nucleotides on either side of the cleavage site, and this bias has been shown to be strand specific (Lazarovici et al., 2013). Intrinsic DNase I cleavage bias is particularly evident when analyzing a set of sites in aggregate, in which the genomic loci are aligned by the transcription factor consensus motif on DNase I-hypersensitive sites. This issue is not limited to DNase I; it is evident that other nucleases, including MNase (Chung et al., 2010; Gaffney et al., 2012), cyanase, and benzonase (Grontved et al., 2012), cleave DNA in a sequence-specific manner. The Tn5 transposase used in ATAC-seq (Buenrostro, Giresi, Zaba, Chang, & Greenleaf, 2013) is also known to cleave DNA in a sequence-dependent manner.

Despite its popularity, conventional DNase-seq experiments often require millions of cells as input material, averaging out heterogeneity in cellular populations. In many cases, rare and important immune cell types cannot be acquired in amounts sufficient for genome-wide chromatin analyses. The assay of transposase-accessible chromatin, ATAC-seq (Buenrostro et al., 2013), uses hyperactive Tn5 transposase (Goryshin & Reznikoff, 1998) to simultaneously cut and ligate adapters for high-throughput sequencing at regions of increased accessibility. Genome-wide mapping of insertion ends

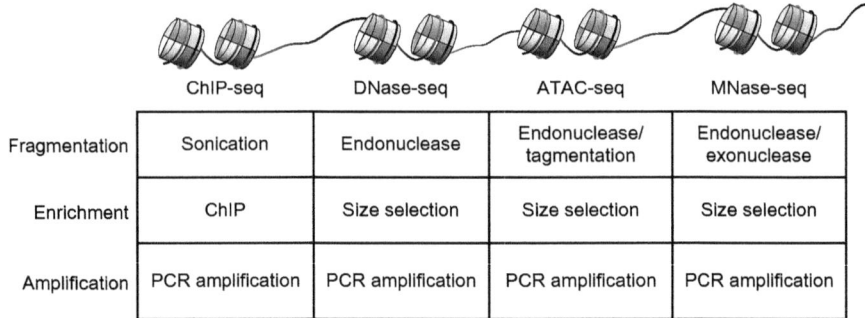

Fig. 2 An overview of ChIP-seq, DNase-seq, ATAC-seq, and MNase-seq. ChIP-seq measures histone modifications, DNase-seq and ATAC-seq map accessible chromatin regions, and MNase-seq is an assay to map positions of nucleosomes.

by high-throughput sequencing allows mapping the chromatin accessibility with a relatively straightforward protocol that can be carried out in hours for a standard sample size of 50,000 cells, instead of weeks compared to DNase I-seq protocols. Because of the paired-end sequencing, ATAC-seq data sets can also be divided into reads that are shorter than the canonical length generally protected by a nucleosome and reads consistent with the approximate length of DNA protected by a nucleosome. It has also been shown that ATAC-seq can provide MNase-seq-like information about the positions of nucleosomes, as well as nucleosome-free regions (Buenrostro, Wu, Chang, & Greenleaf, 2015). Fig. 2 summarizes major technical steps in ChIP-seq, DNase-seq, ATAC-seq, and MNase-seq assays. A more comprehensive overview of these assays and the bias embedded as a result of these steps can be found elsewhere (Meyer & Liu, 2014). Recent adaptations of ATAC-seq and DNase I-seq also allowed for the creation of chromatin accessibility maps in single cells (Buenrostro, Wu, Litzenburger, et al., 2015; Cusanovich et al., 2015; Jin et al., 2015).

8. ANALYSIS OF NGS DATA

All the protocols described earlier involve the sequencing of a library of DNA intended to assay a particular feature of the epigenome. Although these DNA fragments are initially generated using diverse experimental protocols, the upstream analysis converges to a universal set of steps, namely, the quality control, genomic alignment of the sequences, and peak calling. Sequencing machines perform automated base calling as the DNA fragments are sequenced, and millions of DNA fragments can be sequenced simultaneously. These sequences, called reads, do not capture the entire sequence of a DNA fragment, as only the first 50–150 base pairs are sequenced due to technical limitations of current technology. To better understand the computational analysis of data, it is important to keep in mind the origin of the DNA fragments in the library, the type of sequencing being done, and how these two pieces of information relate to the biological question at hand. Software such as Galaxy exists to make the processing of this data as straightforward as possible with simple point-and-click functionality. Users with more bioinformatics backgrounds install and use command-line tools because of their effectiveness, power, and adaptability. In addition, every case is unique, and a "one-size-fits-all" program does not currently exist. These tools require some basic knowledge of Unix command-line and shell scripting, and the user should refer to some basic tutorials on these subjects.

To understand the computational analysis of the sequencing data, we must first review the properties of the sequences being generated. These properties derive from the following experimental design decision: single-end vs paired-end sequencing. In single-end sequencing, one end of the DNA fragment is sequenced up to the number of base pairs (known as the read length) specified by the user. The sequence information for the rest of the DNA fragment is not captured. Although forward and reverse sequence reads may come from copies of the same DNA fragment, the user cannot determine which forward and reverse reads belong to the same fragment. Paired-end sequencing alleviates this limitation. In paired-end sequencing, one end of the DNA fragment is sequenced first, and then that same DNA fragment undergoes another round of sequencing from the opposite end. These reads are "paired" and provide the user with sequence information at both ends of a DNA fragment. The same limitations in read length still apply, and so the center of the DNA fragment may not have been sequenced depending on the fragment's length. Once mapped to a reference genome, the remaining sequence information or the distance between the forward and reverse reads can be inferred. The choice between single-end and paired-end sequencing is usually dictated by cost, sequencing time, and utility. Neither single-end nor paired-end reads contain information regarding the strand from which the DNA fragment was originally derived. Retaining strand information is typically only relevant for RNA sequencing when the user needs to determine if the read comes from the coding or non-coding strand. Several strategies exist to preserve strand information, but in brief, the strand is marked in a way that allows for the complementary strand to be discarded before the library is loaded onto the sequencer so that only the specific strand of interest is sequenced.

Now with an understanding of the sequence information reported by the sequencer, we can move onto quality control. As the sequencing machine calls base pairs, an associated quality score is also reported. These quality scores, called a Phred score, are an estimated probability of error. This score ranges from 1 to 30, with 30 being the highest quality and having a 1 in 1000 probability of being incorrect or, in other words, a 99.9% accuracy. These Phred scores are generated for every base call, and the nucleotide and quality score are each encoded by a single ASCII character. For a given read, the read name, nucleotide sequence, and quality score are reported by the sequencer. All the reads are compiled in a file format known as a "FASTQ" file, which combines the nucleotide sequence (FASTA) with the quality data. These FASTQ files are the springboard for the rest of the downstream analysis. ABI SOLiD is an older sequencing platform that is no longer supported, but the user may run into these files from time to time. ABI SOLiD reads are formatted as two separate files: one that includes the fasta sequence, and another that includes the associated quality sequence.

The data (ie, the FASTQ file) should first be tested against several metrics to ensure quality. Various methods and programs exist, and among the most popular is FASTQC, which tests a sequence file against numerous quality metrics and generates an HTML-formatted report of the results. The reader should refer to the FASTQC website for a complete description of all the tests. The most useful ones to examine include per base sequence quality, sequence duplication, and adapter contamination. These are issues that can typically be remedied by the user with some relatively simple processing. In per base sequence quality, the ends of reads can suffer from decaying quality (this can also happen for the first few nucleotides) resulting in a decline in quality. Additionally, the quality of the second read in a pair can also suffer heavily near the end. One strategy is to trim a predetermined number of bases at the beginning or end of the read (a process known as hard clipping). Another strategy is to trim only the bases that fall below a quality threshold (typically a score of 20), but this strategy is not desirable if low-quality bases are found in the middle of reads. In this case, low-quality bases can be masked (replaced with "N"), and indeed masking may represent the best strategy of the three. Programs such as FASTX-Toolkit can accomplish these trimming or masking steps. A diverse library should contain sequences that only occur once, and high levels of duplication can indicate PCR overamplification or some other enrichment bias. This is especially common if too little starting material is used initially. This means that more rounds of PCR amplification are required to generate enough material for sequencing. Adapters are also typically not present in reads unless DNA fragments are shorter than the read length (such as in the case of sequencing short RNA species) or if the library is contaminated with adapters that failed to properly ligate to DNA fragments. Adapter sequences can be removed using programs such as Trim Galore! or Cutadapt.

Once sequence files are preprocessed to trim/mask low-quality base pairs or remove adapters, the sequences can now be aligned to a reference genome. With enough sequence coverage, genomes can be assembled de novo. De novo assembly is not the focus of this chapter, the reader should refer to other established protocols. The choice of alignment software and the parameters used to align sequence files depends on the experimental design and features being tested. Sequences that are not expected to have high rates of mutations, such as insertions/deletions, substitutions, and splicing, are best aligned with bowtie and bowtie2 (Langmead, Trapnell, Pop, & Salzberg, 2009). This is typically the case for ChIP-seq, ATAC-seq, and DNA-seq libraries. Mutations, such as those that occur in cancer, will need special considerations and are not the focus of this review. Reads from RNA-seq in which splicing occurs are best aligned with STAR (Dobin et al., 2013). Before alignment, an index typically needs to be created for the genome being aligned. The process of generating these indices is specific to the aligner being used, but the general principal involves downloading a reference genome (eg, mm10 or hg19) and an annotation set which includes information about locations and names of genes and other features in the genome. Once sequences are processed and aligned, the output will include reads, their genomic location, and their quality. It is best to separate the aligned reads from the unaligned reads using another program called samtools (Li et al., 2009), and it is not unusual to have a large portion (30–50%) of your reads fail to align. Running FASTQC on this aligned file will show the same quality metrics as mentioned earlier, but usually with some improvement to the results. This is because low-quality sequences, reads contaminated with adapters, or reads with some sequence biases are simply unable to be aligned to the genome. Next, the files need to be coordinate sorted, which is typically done in the aligner—such as with STAR—but bowtie will not coordinate sort the file by default. To accomplish this, the user needs to use PICARD tools (SortSam) to sort the files by coordinates. Once this is accomplished, duplicated sequences (ie, PCR duplicates) can be marked and removed using PICARD MarkDuplicates.

As mentioned earlier, every analysis is unique, and PCR duplicates are not typically removed from RNA-seq data. With this coordinate-sorted bam file, the user can now use a variety of tools and approaches for downstream analysis. Once the initial steps of quality control and alignment have been completed, the downstream analysis diverges depending on the experimental conditions and feature being assayed, and each presents unique challenges for the analysis. For assays we discussed (Fig. 2), after mapping reads to the genome, "peak calling" software is typically used to identify regions of signal enrichment. Many programs have been proposed to perform peak calling among those *macs* is perhaps the most frequently utilized software (Zhang et al., 2008). Detailed guidelines for analysis and quality control of sequencing experiments can be found elsewhere (Landt et al., 2012).

9. PAINTING AN ENHANCER LANDSCAPE

NGS technologies have been used in diverse ways to investigate various aspects of chromatin biology in biological systems including the immune cells. By forming nucleosome structures, chromatin allows binding of transcription factors to *accessible regions*, typically called *regulatory elements*. The most abundant and arguably most important regulatory elements in mammalian cells are called *enhancers*. Enhancers were first described in the 1980s as elements distal to genes, which, in contrast to promoters, affect transcription in an orientation-independent manner (Blom van Assendelft, Hanscombe, Grosveld, & Greaves, 1989).

Pre-NGS studies that relied on sequence conservation suggested an abundance of *cis*-regulatory sequences in noncoding DNA tightly regulating key cytokines and transcription factors. The recent genome-scale methods are identifying genomic locations of putative enhancers complements the conservation-based approaches, as it is specific for cell type, condition, and activation states (Natoli, 2010). Noncoding functional elements are associated with distinctive chromatin structures, which display signature patterns of histone modifications, DNase accessibility, and transcription factor occupancy (Kellis et al., 2014). Active enhancers are evidenced by histone acetylations and DNase-accessible chromatin and binding of sequence-specific transcription factors, coactivators such as EP300, and, often, RNA polymerase II. In contrast, poised enhancers are marked by histone methylation (H3K4me1) (Creyghton et al., 2010; Neph, Vierstra, et al., 2012; Rada-Iglesias et al., 2010; Visel et al., 2009).

 ## 10. CHROMATIN BIOLOGY TO UNDERSTAND GENE REGULATION IN IMMUNE CELLS

The genome-scale enhancer mapping relying on the chromatin signature of enhancers revealed unexpected complexity of enhancer utilization, with around a million enhancers annotated by the ENCODE and NIH Epigenomics Roadmap projects (Creyghton et al., 2010; Neph, Stergachis, et al., 2012; Neph, Vierstra, et al., 2012; Rada-Iglesias et al., 2010; Roadmap Epigenomics et al., 2015; Visel et al., 2009). Studies in immune cells by individual labs also added new dimensions to these international efforts. It became evident that enhancer mapping can teach us more about gene regulation than simply profiling gene expression. Activated B cells and embryonic stem cells both express *Myc*, but utilize distinct sets of enhancers in the *Myc* locus, suggesting a very different mode of regulation (Kieffer-Kwon et al., 2013). $CD4^+$ T helper cell subsets (Th1, Th2, and Th17 cells) also reach to very distinct sets of enhancers for their specialization. In fact, the extent of difference in enhancer elements is an order of magnitude larger than the number of differentially regulated genes (Ciofani et al., 2012; Hawkins et al., 2013; Samstein et al., 2012; Vahedi et al., 2012).

In addition to ontogeny, the tissue microenvironment to which a cell is exposed can have profound effects on the enhancer repertoire. Two parallel studies profiled the expression and chromatin landscape of up to nine populations of mouse macrophages isolated from distinct tissues (Gosselin et al., 2015; Lavin et al., 2015). Tissues included brain microglia, spleen red pulp macrophages, liver Kupffer cells, lung macrophages, large and small peritoneal macrophages, and colonic large and ileal small intestinal macrophages, in addition to thioglycollate-elicited peritoneal macrophages and bone marrow-derived macrophages. The regulatory elements of tissue-resident macrophages, including promoters, active enhancers, and poised enhancers, were mapped using H3K27Ac, H3K4me1, and ATAC-seq (Buenrostro et al., 2013). Superposition of these maps suggests that macrophages utilize distinct enhancer elements based on the tissue they reside in. These tissue-specific enhancers drive divergent programs of gene expression (Gosselin et al., 2015; Lavin et al., 2015). Some poised enhancers are shown to be inherited during development and thus provide ontogenic memory (Gosselin et al., 2015; Lavin et al., 2015). The highest percentages of poised enhancers are at genomic regions present in only one or two macrophage populations, demonstrating complex regulatory mechanisms, which can only be analyzed comprehensively on the level of the chromatin (Lavin et al., 2015).

11. ROLE OF INTRINSIC AND EXTRINSIC SIGNALS ON ENHANCER FORMATION

The discovery that enhancer landscapes are unique raises the question as to what factors are responsible for creating these genomic elements. Studies in macrophages established that lineage-determining transcription factors (LDTFs) exert a pervasive role in controlling the whole repertoire of active regulatory elements (Ghisletti et al., 2010; Heinz et al., 2010). Approximately 80% of PU.1 binding sites were located distal to promoters and associated with p300 binding, the canonical chromatin signature of active enhancers (Ghisletti et al., 2010). Strikingly, most H3K4me1 regions had PU.1 binding in macrophages. The reconstitution of PU.1 expression in PU.1-deficient myeloid progenitors led to sequential nucleosome remodeling and increased H3K4me1 modifications at PU.1 binding sites (Heinz et al., 2010). In addition, PU.1 depletion results in the reduction of H3K4me1 amounts at several tested enhancers (Ghisletti et al., 2010). Together, since PU.1 binding is necessary and sufficient to change the modification state of the neighboring chromatin, it has been concluded that such factors actively shape the epigenome.

However, studies in T helper cells argue strongly against a simple model for the pervasive effect of LDTFs in shaping the enhancer landscape. LDTFs in T cells including T-bet, Gata3, Rorγt, and Foxp3, also called *master regulators*, are both necessary and sufficient for differentiation of each corresponding T-cell subset. Multiple independent studies provided a new and very distinct perspective for the relative roles of various factors in T-cell specialization. In Treg cells, Foxp3 deletion does not affect enhancers annotated by DNase I hypersensitivity (Samstein et al., 2012). Using genomic DNase I footprinting analysis across Tregs (Neph, Vierstra, et al., 2012), it was shown that the pioneer factor Foxo1 seems to set the stage for Foxp3 binding (Samstein et al., 2012). The roles of T-bet and Rorγt on the active enhancer landscape appear to be equally limited (Ciofani et al., 2012; Vahedi et al., 2012). Enhancers including those in the *Ifng* locus were found to be T-bet dependent. Overall though, deletion of T-bet and Rorγt in Th1 and Th17 cells led to limited changes in enhancer landscapes annotated by binding of p300, suggesting different players working on the chromatin in these cell types. Rather, T-bet largely limits p300 recruitment on enhancers of opposite lineages as opposed to primarily establishing the enhancer elements of Th1 cells. Likewise, only a handful of loci including *Il17a*, *Il17f*, and *Il23r* are highly dependent on Rorγt (Ciofani et al., 2012).

The impact of signal-dependent transcription factors (SDTFs) on chromatin landscapes of T cells is not unexpected because a key aspect in the acquisition of distinct T helper cell phenotypes is the cytokine milieu. After encountering diverse microbial pathogens, dendritic cells and other cells of the innate and adaptive immune system produce cytokines, which serve to instruct distinct T-cell fates. The major specifying cytokines exert their effect through signal transducer and activator of transcription (STAT) family proteins.

Recent efforts unraveled the pivotal role of STAT proteins in orchestrating the enhancer repertoire in T cells and macrophages (Ciofani et al., 2012; Ostuni et al., 2013; Samstein et al., 2012; Vahedi et al., 2012). The majority of differentially active enhancers in T helper cells were STAT4 or STAT6 dependent: specifically, in the absence of STAT4 and STAT6 proteins, the recruitment of p300 was disrupted at more than half of enhancers (Vahedi et al., 2012). This was also the case in Th17 cells, in which STAT3 plays a major role in the recruitment of p300 (Ciofani et al., 2012). The interplay between SDTFs and LDTFs is clearly illustrated when the reconstitution of STAT4- and STAT6-deficient cells with LDTFs T-bet and GATA3 failed to recover the active enhancer landscapes annotated by p300 binding (Vahedi et al., 2012). Mapping the chromatin signature of enhancers after the reconstitution of LDTFs in cells lacking SDTFs argues for a primary role of environmental sensors in dictating global landscapes. A similar work in macrophages also revealed that STAT1 and STAT6 play key roles in depositing H3K27Ac and H3K4me1 in response to IFNG and IL-4 (Ostuni et al., 2013). These studies enabled by the mapping the epigenome using NGS suggest a hierarchical model where STATs are the architects of epigenetic landscape and the lineage-specific transcription factors in directing global landscapes.

12. CONCLUSIONS

Immunology along with other fields of biology is undergoing a revolution as a result of NGS technologies. Here, we described NGS methods relevant to mapping chromatin alterations in immune cells. We have also reviewed lessons, we have learnt in immune cells by mapping the chromatin.

The more recent assays mapping the epigenome in single cells have the potential to truly change various paradigms in immunology. These assays are able to provide us a better picture of the immune response in both human and mice. This remarkable success at the same time highlights that biology has irreversibly changed to a data-rich science. To address the challenges of complex data analysis, immunologists require both conceptual understanding and working skill sets of statistical approaches (Spreafico, Mitchell, & Hoffmann, 2015).

ACKNOWLEDGMENTS
This work was supported by NIAID K22AI112570 to G.V. We thank Carol A. Clifton for the adaption of Fig. 1 from Rivera and Ren (2013).

REFERENCES
Adli, M., Zhu, J., & Bernstein, B. E. (2010). Genome-wide chromatin maps derived from limited numbers of hematopoietic progenitors. *Nature Methods, 7*(8), 615–618.
Baker, M. (2015). Reproducibility crisis: Blame it on the antibodies. *Nature, 521*(7552), 274–276.
Blom van Assendelft, G., Hanscombe, O., Grosveld, F., & Greaves, D. R. (1989). The beta-globin dominant control region activates homologous and heterologous promoters in a tissue-specific manner. *Cell, 56*(6), 969–977.
Bonhoure, N., Bounova, G., Bernasconi, D., Praz, V., Lammers, F., Canella, D., et al. (2014). Quantifying ChIP-seq data: A spiking method providing an internal reference for sample-to-sample normalization. *Genome Research, 24*(7), 1157–1168.
Buenrostro, J. D., Giresi, P. G., Zaba, L. C., Chang, H. Y., & Greenleaf, W. J. (2013). Transposition of native chromatin for fast and sensitive epigenomic profiling of open chromatin, DNA-binding proteins and nucleosome position. *Nature Methods, 10*(12), 1213–1218.
Buenrostro, J. D., Wu, B., Chang, H. Y., & Greenleaf, W. J. (2015a). ATAC-seq: A method for assaying chromatin accessibility genome-wide. *Current Protocols in Molecular Biology, 109*, 21 9 1-9.
Buenrostro, J. D., Wu, B., Litzenburger, U. M., Ruff, D., Gonzales, M. L., Snyder, M. P., et al. (2015b). Single-cell chromatin accessibility reveals principles of regulatory variation. *Nature, 523*(7561), 486–490.
Campbell, V. W., & Jackson, D. A. (1980). The effect of divalent cations on the mode of action of DNase I. The initial reaction products produced from covalently closed circular DNA. *The Journal of Biological Chemistry, 255*(8), 3726–3735.
Chung, H. R., Dunkel, I., Heise, F., Linke, C., Krobitsch, S., Ehrenhofer-Murray, A. E., et al. (2010). The effect of micrococcal nuclease digestion on nucleosome positioning data. *PLoS One, 5*(12), e15754.
Ciofani, M., Madar, A., Galan, C., Sellars, M., Mace, K., Pauli, F., et al. (2012). A validated regulatory network for Th17 cell specification. *Cell, 151*(2), 289–303.
Crawford, G. E., Holt, I. E., Mullikin, J. C., Tai, D., Blakesley, R., Bouffard, G., et al. (2004). Identifying gene regulatory elements by genome-wide recovery of DNase hypersensitive sites. *Proceedings of the National Academy of Sciences of the United States of America, 101*(4), 992–997.

Creyghton, M. P., Cheng, A. W., Welstead, G. G., Kooistra, T., Carey, B. W., Steine, E. J., et al. (2010). Histone H3K27ac separates active from poised enhancers and predicts developmental state. *PNAS, 107*(50), 21931–21936.
Cusanovich, D. A., Daza, R., Adey, A., Pliner, H. A., Christiansen, L., Gunderson, K. L., et al. (2015). Epigenetics. Multiplex single-cell profiling of chromatin accessibility by combinatorial cellular indexing. *Science, 348*(6237), 910–914.
Dobin, A., Davis, C. A., Schlesinger, F., Drenkow, J., Zaleski, C., Jha, S., et al. (2013). STAR: Ultrafast universal RNA-seq aligner. *Bioinformatics, 29*(1), 15–21.
Ernst, J., & Kellis, M. (2010). Discovery and characterization of chromatin states for systematic annotation of the human genome. *Nature Biotechnology, 28*(8), 817–825.
Feil, R., Charlton, J., Bird, A. P., Walter, J., & Reik, W. (1994). Methylation analysis on individual chromosomes: Improved protocol for bisulphite genomic sequencing. *Nucleic Acids Research, 22*(4), 695–696.
Gaffney, D. J., McVicker, G., Pai, A. A., Fondufe-Mittendorf, Y. N., Lewellen, N., Michelini, K., et al. (2012). Controls of nucleosome positioning in the human genome. *PLoS Genetics, 8*(11), e1003036.
Ghisletti, S., Barozzi, I., Mietton, F., Polletti, S., De Santa, F., Venturini, E., et al. (2010). Identification and characterization of enhancers controlling the inflammatory gene expression program in macrophages. *Immunity, 32*(3), 317–328.
Goren, A., Ozsolak, F., Shoresh, N., Ku, M., Adli, M., Hart, C., et al. (2010). Chromatin profiling by directly sequencing small quantities of immunoprecipitated DNA. *Nature Methods, 7*(1), 47–49.
Goryshin, I. Y., & Reznikoff, W. S. (1998). Tn5 in vitro transposition. *The Journal of Biological Chemistry, 273*(13), 7367–7374.
Gosselin, D., Link, V. M., Romanoski, C. E., Fonseca, G. J., Eichenfield, D. Z., Spann, N. J., et al. (2015). Environment drives selection and function of enhancers controlling tissue-specific macrophage identities. *Cell, 159*(6), 1327–1340.
Grontved, L., Bandle, R., John, S., Baek, S., Chung, H. J., Liu, Y., et al. (2012). Rapid genome-scale mapping of chromatin accessibility in tissue. *Epigenetics & Chromatin, 5*(1), 10.
Grzybowski, A. T., Chen, Z., & Ruthenburg, A. J. (2015). Calibrating ChIP-seq with nucleosomal internal standards to measure histone modification density genome wide. *Molecular Cell, 58*(5), 886–899.
Hawkins, R. D., Larjo, A., Tripathi, S. K., Wagner, U., Luu, Y., Lonnberg, T., et al. (2013). Global chromatin state analysis reveals lineage-specific enhancers during the initiation of human T helper 1 and T helper 2 cell polarization. *Immunity, 38*(6), 1271–1284.
He, H. H., Meyer, C. A., Hu, S. S., Chen, M. W., Zang, C., Liu, Y., et al. (2014). Refined DNase-seq protocol and data analysis reveals intrinsic bias in transcription factor footprint identification. *Nature Methods, 11*(1), 73–78.
Heinz, S., Benner, C., Spann, N., Bertolino, E., Lin, Y. C., Laslo, P., et al. (2010). Simple combinations of lineage-determining transcription factors prime cis-regulatory elements required for macrophage and B cell identities. *Molecular Cell, 38*(4), 576–589.
Iwasaki, A., & Medzhitov, R. (2015). Control of adaptive immunity by the innate immune system. *Nature Immunology, 16*(4), 343–353.
Jin, W., Tang, Q., Wan, M., Cui, K., Zhang, Y., Ren, G., et al. (2015). Genome-wide detection of DNase I hypersensitive sites in single cells and FFPE tissue samples. *Nature, 528*(7580), 142–146.
Jullien, P. E., & Berger, F. (2010). DNA methylation reprogramming during plant sexual reproduction? *Trends in Genetics, 26*(9), 394–399.
Kellis, M., Wold, B., Snyder, M. P., Bernstein, B. E., Kundaje, A., Marinov, G. K., et al. (2014). Defining functional DNA elements in the human genome. *Proceedings of the National Academy of Sciences of the United States of America, 111*(17), 6131–6138.

Kieffer-Kwon, K. R., Tang, Z., Mathe, E., Qian, J., Sung, M. H., Li, G., et al. (2013). Interactome maps of mouse gene regulatory domains reveal basic principles of transcriptional regulation. *Cell*, *155*(7), 1507–1520.

Koh, K. P., & Rao, A. (2013). DNA methylation and methylcytosine oxidation in cell fate decisions. *Current Opinion in Cell Biology*, *25*(2), 152–161.

Landt, S. G., Marinov, G. K., Kundaje, A., Kheradpour, P., Pauli, F., Batzoglou, S., et al. (2012). ChIP-seq guidelines and practices of the ENCODE and modENCODE consortia. *Genome Research*, *22*(9), 1813–1831.

Langmead, B., Trapnell, C., Pop, M., & Salzberg, S. L. (2009). Ultrafast and memory-efficient alignment of short DNA sequences to the human genome. *Genome Biology*, *10*(3), R25.

Lara-Astiaso, D., Weiner, A., Lorenzo-Vivas, E., Zaretsky, I., Jaitin, D. A., David, E., et al. (2014). Immunogenetics. Chromatin state dynamics during blood formation. *Science*, *345*(6199), 943–949.

Lavin, Y., Winter, D., Blecher-Gonen, R., David, E., Keren-Shaul, H., Merad, M., et al. (2015). Tissue-resident macrophage enhancer landscapes are shaped by the local microenvironment. *Cell*, *159*(6), 1312–1326.

Lazarovici, A., Zhou, T., Shafer, A., Dantas Machado, A. C., Riley, T. R., Sandstrom, R., et al. (2013). Probing DNA shape and methylation state on a genomic scale with DNase I. *Proceedings of the National Academy of Sciences of the United States of America*, *110*(16), 6376–6381.

Li, H., Handsaker, B., Wysoker, A., Fennell, T., Ruan, J., Homer, N., et al. (2009). The sequence alignment/map format and SAMtools. *Bioinformatics*, *25*(16), 2078–2079.

Meyer, C. A., & Liu, X. S. (2014). Identifying and mitigating bias in next-generation sequencing methods for chromatin biology. *Nature Reviews. Genetics*, *15*(11), 709–721.

Natoli, G. (2010). Maintaining cell identity through global control of genomic organization. *Immunity*, *33*(1), 12–24.

Neph, S., Stergachis, A. B., Reynolds, A., Sandstrom, R., Borenstein, E., & Stamatoyannopoulos, J. A. (2012). Circuitry and dynamics of human transcription factor regulatory networks. *Cell*, *150*(6), 1274–1286.

Neph, S., Vierstra, J., Stergachis, A. B., Reynolds, A. P., Haugen, E., Vernot, B., et al. (2012). An expansive human regulatory lexicon encoded in transcription factor footprints. *Nature*, *489*(7414), 83–90.

Orlando, D. A., Chen, M. W., Brown, V. E., Solanki, S., Choi, Y. J., Olson, E. R., et al. (2014). Quantitative ChIP-seq normalization reveals global modulation of the epigenome. *Cell Reports*, *9*(3), 1163–1170.

Ostuni, R., Piccolo, V., Barozzi, I., Polletti, S., Termanini, A., Bonifacio, S., et al. (2013). Latent enhancers activated by stimulation in differentiated cells. *Cell*, *152*(1–2), 157–171.

Park, P. J. (2009). ChIP-seq: Advantages and challenges of a maturing technology. *Nature Reviews. Genetics*, *10*(10), 669–680.

Perez-Burgos, L., Peters, A. H., Opravil, S., Kauer, M., Mechtler, K., & Jenuwein, T. (2004). Generation and characterization of methyl-lysine histone antibodies. *Methods in Enzymology*, *376*, 234–254.

Rada-Iglesias, A., Bajpai, R., Swigut, T., Brugmann, S. A., Flynn, R. A., & Wysocka, J. (2010). A unique chromatin signature uncovers early developmental enhancers in humans. *Nature*, *470*, 279–283.

Rivera, C. M., & Ren, B. (2013). Mapping human epigenomes. *Cell*, *155*(1), 39–55.

Roadmap Epigenomics, C., Kundaje, A., Meuleman, W., Ernst, J., Bilenky, M., Yen, A., et al. (2015). Integrative analysis of 111 reference human epigenomes. *Nature*, *518*(7539), 317–330.

Rothbart, S. B., Dickson, B. M., Raab, J. R., Grzybowski, A. T., Krajewski, K., Guo, A. H., et al. (2015). An interactive database for the assessment of histone antibody specificity. *Molecular Cell*, *59*(3), 502–511.

Ryan, R. J., & Bernstein, B. E. (2012). Molecular biology. Genetic events that shape the cancer epigenome. *Science*, *336*(6088), 1513–1514.

Samstein, R. M., Arvey, A., Josefowicz, S. Z., Peng, X., Reynolds, A., Sandstrom, R., et al. (2012). Foxp3 exploits a pre-existent enhancer landscape for regulatory T cell lineage specification. *Cell*, *151*(1), 153–166.

Schmitz, R. J., He, Y., Valdes-Lopez, O., Khan, S. M., Joshi, T., Urich, M. A., et al. (2013). Epigenome-wide inheritance of cytosine methylation variants in a recombinant inbred population. *Genome Research*, *23*(10), 1663–1674.

Schones, D. E., Cui, K., Cuddapah, S., Roh, T. Y., Barski, A., Wang, Z., et al. (2008). Dynamic regulation of nucleosome positioning in the human genome. *Cell*, *132*(5), 887–898.

Smith, Z. D., & Meissner, A. (2013). DNA methylation: Roles in mammalian development. *Nature Reviews. Genetics*, *14*(3), 204–220.

Spreafico, R., Mitchell, S., & Hoffmann, A. (2015). Training the 21st century immunologist. *Trends in Immunology*, *36*(5), 283–285.

Thomson, J. P., Lempiainen, H., Hackett, J. A., Nestor, C. E., Muller, A., Bolognani, F., et al. (2012). Non-genotoxic carcinogen exposure induces defined changes in the 5-hydroxymethylome. *Genome Biology*, *13*(10), R93.

Vahedi, G., Kanno, Y., Furumoto, Y., Jiang, K., Parker, S. C., Erdos, M. R., et al. (2015). Super-enhancers delineate disease-associated regulatory nodes in T cells. *Nature*, *520*(7548), 558–562.

Vahedi, G., Takahashi, H., Nakayamada, S., Sun, H. W., Sartorelli, V., Kanno, Y., et al. (2012). STATs shape the active enhancer landscape of T cell populations. *Cell*, *151*(5), 981–993.

Valouev, A., Johnson, S. M., Boyd, S. D., Smith, C. L., Fire, A. Z., & Sidow, A. (2011). Determinants of nucleosome organization in primary human cells. *Nature*, *474*(7352), 516–520.

van Galen, P., Viny, A. D., Ram, O., Ryan, R. J., Cotton, M. J., Donohue, L., et al. (2016). A multiplexed system for quantitative comparisons of chromatin landscapes. *Molecular Cell*, *61*(1), 170–180.

Vierstra, J., Wang, H., John, S., Sandstrom, R., & Stamatoyannopoulos, J. A. (2014). Coupling transcription factor occupancy to nucleosome architecture with DNase-FLASH. *Nature Methods*, *11*(1), 66–72.

Visel, A., Blow, M. J., Li, Z., Zhang, T., Akiyama, J. A., Holt, A., et al. (2009). ChIP-seq accurately predicts tissue-specific activity of enhancers. *Nature*, *457*(7231), 854–858.

Weber, M., Davies, J. J., Wittig, D., Oakeley, E. J., Haase, M., Lam, W. L., et al. (2005). Chromosome-wide and promoter-specific analyses identify sites of differential DNA methylation in normal and transformed human cells. *Nature Genetics*, *37*(8), 853–862.

Zhang, Y., Liu, T., Meyer, C. A., Eeckhoute, J., Johnson, D. S., Bernstein, B. E., et al. (2008). Model-based analysis of ChIP-seq (MACS). *Genome Biology*, *9*(9), R137.

Zhou, V. W., Goren, A., & Bernstein, B. E. (2011). Charting histone modifications and the functional organization of mammalian genomes. *Nature Reviews. Genetics*, *12*(1), 7–18.

AUTHOR INDEX

Note: Page numbers followed by "*f*" indicate figures.

A

Abraham, Y., 115, 120–121
Abrassart, D., 189
Abukhdeir, A.M., 248–249
Ackloo, S., 80–100
Adams, C.C., 160–161
Adams, P.D., 313–314, 356–363
Adey, A., 371–372
Adibekian, A., 120–121
Adler, C.H., 249
Adli, M., 370
Aebersold, R., 12–14, 337, 339–341
Agarwal, P., 192–193
Aggarwal, A.K., 106
Agresta, A.M., 83
Aguiar, R.C., 106
Ahn, K., 196–197, 214–215
Ai, N., 246–247
Ait-Si-Ali, S., 108–109
Ajala, O.Z., 54
Aka, J.A., 184–186
Akbarian, S., 349–352
Akiyama, J.A., 376–377
Aksoy, O., 356
Al-Abed, Y., 194–195
Alabert, C., 126–127
Alaimo, P.J., 254–257
Alani, R.M., 110–111
Albaugh, B.N., 317–318
Albitar, M., 246–247, 261–264
Alborn, A.M., 344
Alekseyenko, A.A., 83–84
Alfa, C., 172
Alfaro, J.F., 290–291
Al-Hashimi, H.M., 270–272
Alinakhi, 203–204
Alinger-Scharinger, B., 246, 296
Alkass, K., 342–344
Allahverdi, A., 151–152, 156–157
Allali-Hassani, A., 84–87, 91–92, 98–99, 289, 296–297
Allfrey, V.G., 106

Allis, C.D., 4, 32, 33*f*, 150, 186, 313–314, 332–333, 348*f*, 350
Allison, J.R., 168
Allocco, J.J., 195
Almouzni, G., 356–357
Alpaugh, M., 357–358
Alt, F.W., 190–191
Altman, J., 344
Altucci, L., 203–204
Amador-Noguez, D., 126, 128–129, 132–133
Amengual, J.E., 201–202
Amore, A., 204
An, J., 257–259
Andacht, T., 337
Andersen, J.S., 352
Anderson, G., 26
Andrade, J., 248–249
Andrake, M., 356
Anglin, J.L., 275–276, 281, 286
Anighoro, A., 95
Antczak, C., 287, 295*f*
Antonysamy, S., 86–87, 98–99, 259–261
Aonuma, M., 194
Aoubala, M., 204
Aparicio, O.M., 186, 197–198
Appella, E., 42–43
Arcand, M., 289–290
Ardavanis, A., 273–275
Aris, J.P., 170–171
Arkhammar, P., 187–188
Armstrong, C.M., 186
Arnaudo, A.M., 126, 339–340
Arnesen, T., 108–109
Arnold, K.M., 317–318
Arrowsmith, C.H., 54, 80–100, 246–247
Arruda, P., 80–81
Arvey, A., 377–379
Aspalter, C.-M., 23
Astudillo, Y.M., 290–291
Atta, M., 249–250
Audia, J.E., 81–82, 99

Austin, C., 81–82, 99
Auwerx, J., 190–191, 214
Avalos, J.L., 184–186
Avila, M.A., 253

B
Badeaux, A.I., 150
Badiger, J., 203–204
Bae, Y., 46
Baek, H.J., 248–249
Baek, S., 371
Baell, J., 81–82, 99
Baeza, J., 126–146, 197, 234–236
Bagot, R.C., 333–334, 341, 344–346, 352
Bahirvani, A.G., 42–43
Bailey, D.J., 340–341
Bainbridge, M., 356–357
Baines, A.J., 168–169
Baiocchi, R.A., 247–248
Bajorath, J., 95
Bajpai, R., 376–377
Baker, D.J., 356
Baker, M., 369
Baker, R.W., 32–34, 42, 54
Balabadra, U., 204
Baliban, R.C., 20–24
Ballmer, P., 264–268, 286–287
Bandle, R., 371
Bangham, R., 54
Bannister, A., 106
Bantscheff, M., 106–107, 115, 120–121
Banumathy, G., 356
Bao, Y., 159–160
Baranowski, B., 96–98
Barchas, J., 290–291
Bare, O., 190–191
Barker, J., 202–203
Barozzi, I., 366, 378–379
Barradas, M., 356
Barrett, L.W., 196–197
Barski, A., 248–249, 356–357, 370
Barsyte-Lovejoy, D., 80–100
Barth, T.K., 126–127
Bartkova, J., 248–249
Bartley, S.M., 257–259
Basavapathruni, A., 250, 279–281, 283, 286, 297
Bass, J., 189

Bassi, C., 89
Bastuck, S., 115, 120–121
Batley, S.J., 106
Batzoglou, S., 375–376
Bause, A.S., 193
Baxevanis, A.D., 153
Baylin, S.B., 246–247, 296, 312–313
Bedalov, A., 184–205
Bedford, M.T., 54, 89, 247–248, 273–275, 289, 296–297
Bednarek, M.A., 195
Behbahan, I.S., 127–128
Beli, P., 81
Belinsky, S.A., 246–247
Belton, J.M., 168
Benedik, M.J., 357–358
Benhida, R., 250
Benner, C., 378
Bennett, B.D., 132–133, 143
Bennett, J., 81–82, 99
Bennett, J.M., 246–247, 261–264
Bennett, K.L., 82
Berber, E., 194
Berdasco, M., 126
Berger, A.B., 170–171
Berger, F., 367–368
Berger, S.L., 33f, 312–327
Bergmann, O., 342–344
Bernard, S., 342–344
Bernasconi, D., 369–370
Berndsen, C.E., 106–107
Bernstein, B.E., 366–370, 375–376
Bernstein, N., 261–264, 286–287
Berthelsen, J., 272–273, 289–290
Bertolino, E., 378
Bertrand, P., 250
Bezzi, M., 259–261
Bhanu, N.V., 9–10, 15
Bharathy, N., 42–43
Bhardwaj, R.D., 342–344
Bhatia, R., 192–193
Bhinder, B., 287, 295f
Bhullar, K.S., 214–215
Biggar, K.K., 84–85
Bignell, G., 248–249
Bilenky, M., 356–357, 377
Billington, B.L., 186, 197–198
Bingham, P., 257–259, 286

Bird, A.P., 261–264, 368
Birtwistle, M.R., 332–352
Bissinger, E.-M., 250
Bitterman, K.J., 214–219, 234–236
Bjork-Eriksson, T., 344
Black, J.C., 46
Blackwell, H.E., 197–198, 203–204
Blagg, J., 80–82, 99
Blagoev, B., 120–121, 339–340
Blakesley, R., 371
Blanchard, P., 268–270
Blanco, M.A., 126
Blanco, S., 249
Blander, G., 214
Bleasdale, J.E., 196–197, 214–215
Blecher-Gonen, R., 377–378
Blencowe, B.J., 248–249
Blobel, G., 170–171
Blom van Assendelft, G., 376
Blouin, J., 289–290
Blow, M.J., 376–377
Blum, G., 270–272, 286–287, 295–296, 295f
Bock, I., 54
Boeke, J.D., 184–186, 197–198
Boesche, M., 115, 120–121
Bogan, K.L., 188
Bogyo, M., 106–107
Bolognani, F., 368
Bonaldi, T., 83
Bonanno, J.B., 259–261
Bonday, Z., 259–261
Bondy, S.C., 335
Bonhoure, N., 369–370
Bonifacio, S., 366, 379
Bonin, J., 54
Bonneil, E., 153
Borchardt, R.T., 253
Borchers, C.H., 46–48
Borenstein, E., 377
Boriack-Sjodin, P.A., 84–87, 250, 283, 297
Borra, M.T., 184–186, 195, 214–215
Borroni, E., 264–268, 286–287
Bouchekioua-Bouzaghou, K., 273–275
Bouffard, G., 371
Bounova, G., 369–370
Bountra, C., 81, 99, 246–247
Boyd, S.D., 370

Brace, C.S., 189, 214
Brachmann, C.B., 186, 197–198
Braig, M., 356
Brandli, L., 266–268
Brandt, O., 54
Branscombe, T.L., 89
Braunstein, M., 186
Bravo-San Pedro, J.M., 317–318
Breitkreutz, A., 83–84
Bremang, M., 83
Brenner, C., 188
Brik, A., 151
Brimble, M.A., 126
Britton, L.-M.P., 4–5, 54
Broach, J.R., 186
Bronowska, A.K., 95
Brown, M., 247–248
Brown, P.J., 84–87, 259–261, 287
Brown, R.L., 250, 270–272
Brown, V.E., 369–370
Browne, C.E., 257
Brugmann, S.A., 376–377
Bruice, T.C., 264–266, 270–272
Bruno, J., 190–191
Bruns, A.N., 163
Bryant, B., 126
Bua, D.J., 54
Buc, H., 170–171
Buchholz, B.A., 342–344
Buenrostro, J.D., 371–372, 377–378
Bulfer, S., 290–291
Bunkenborg, J., 352
Bunnage, M.E., 80–81, 99
Burley, S., 80–81
Burlingame, A.L., 25, 106–107, 151–152
Burnett, C., 186–187
Burrell, R.A., 248–249
Busby, S.A., 9–10, 338–339
Busch, R., 336
Bustos, S.P., 89
Butler, A., 248–249
Butler, D., 48–49
Bylebyl, G.R., 151, 153
Byles, V., 193
Byrd, J.C., 247–248
Bystricky, K., 161
Byvoet, P., 335–336

C

Cabal, G.G., 170–171
Cabreiro, F., 186–187
Cacicedo, J.M., 190–191
Cagney, G., 248–249
Cahill, D.J., 95
Cai, G., 281
Cai, L., 313
Cai, X.-C., 246–298
Caldwell, S.D., 214–215
Calvanese, V., 203–204
Campbell, J., 204, 214
Campbell, R.M., 259–261
Campbell, V.W., 371
Campisi, J., 356
Camporez, J.P., 191
Canadien, V., 248–249
Canamero, M., 193
Canella, D., 369–370
Canto, C., 190–191, 214
Cao, B., 34–35
Cao, D., 196, 215
Cao, Q., 248–249
Cao, R., 257–259
Cao, X.-J., 15, 315
Cao, Y., 151–152, 156–157
Caparros, M.-L., 150
Capizzi, J.R., 168–171
Carapeti, M., 106
Carey, B.W., 376–377
Carling, D., 190–191
Carlson, J.E., 106
Carmel, R., 283
Carney, D.P., 196–197, 214–215, 217–219, 221
Caroli, A., 249–250
Caron, M., 289–290
Carpenter, P.B., 248–249, 296
Carr, S.M., 247–248
Carrasco-Legleu, C., 246–247
Carrer, A., 106–107, 313–314
Carsten, A.L., 336
Carvalho, S., 248–249
Casadio, F., 89
Case, A.W., 215–219, 221, 227, 229, 234–236, 240
Castellano, S., 249–250
Caudy, A.A., 128–129, 132–133

Cebrat, M., 110–111
Celano, P., 246–247
Celic, I., 184–186
Cesaroni, M., 89
Cetina, L., 246–247
Chai, X., 184–186
Chakravarthy, S., 159–160
Chalkiadaki, A., 191–193
Chambers, M., 18
Chan, E.Y., 126
Chan, L.C., 273–275
Chang, H.Y., 371–372, 377–378
Chang, S., 86–87, 98–99
Chang, Y.Q., 291–292
Chan-Penebre, E., 84–87
Chao, E.D., 197–198, 203–204
Charlton, J., 368
Chatterjee, C., 150–163
Chavez-Blanco, A., 246–247
Chedin, F., 54
Chekler, E.L., 80–81
Chelbi-Alix, M.K., 153
Chen, B., 203
Chen, D., 189–192
Chen, K.F., 248–249
Chen, L., 248–249
Chen, L.H., 86–87, 98–99
Chen, M.W., 369–371
Chen, P., 275–276, 286
Chen, W., 356
Chen, X., 42–43, 289, 296–297
Chen, Y., 126, 192
Chen, Z., 46, 126, 356, 369–370
Cheng, A.W., 376–377
Cheng, D., 151, 289, 296–297
Cheng, G., 275–276, 286
Cheng, H.L., 190–191
Cheng, J.A., 246–247
Cheng, X., 42–43, 49, 287
Cheng, Z., 341
Cheung, N., 273–275
Cheung, P., 54, 248–249
Cheung, T., 139–140
Chi, P., 313–314
Chiarenza, A., 106, 192
Chicas, A., 356
Chikashige, Y., 170
Childs, B.G., 356

Childs, D., 82
Chim, H., 191
Chin, J.W., 156–157
Chin, S.F., 106
Chmilewski, L.K., 193
Cho, E.C., 247–248, 259–261
Cho, H.S., 249
Cho, J., 46
Choe, N.W., 248–249
Choi, J., 246–247
Choi, Y.J., 369–370
Chopra, V., 204
Chory, E.J., 86–87, 91–92, 96, 281–283, 286
Chou, H.Y., 247–248
Choudhary, C., 134
Chowdhury, S., 184–205
Christiansen, L., 371–372
Chrunyk, B., 196–197, 214–215
Chu, F., 151–152
Chua, K.F., 194
Chuikov, S., 249
Chun, S., 192–193
Chung, H.J., 371
Chung, H.R., 371
Chung, J.H., 214–215
Chung, N., 190–191
Ciofani, M., 366, 377–379
Claridge, S., 261–264, 286–287
Clark, S.J., 246–247
Clarke, I.D., 96–98
Clarke, S.G., 42–43, 89, 247–248, 259–261, 273–275
Clark-Garvey, S., 201–202
Clark-Lewis, I., 150–151
Clasquin, M.F., 128–129, 132–133
Cleary, M.L., 273–275
Clements, A., 184–186
Cliften, P., 214
Coffield, V.M., 248–249, 257
Cohen, H.Y., 193–194, 214–219, 234–236
Cole, J.J., 356, 358–363
Cole, P.A., 46, 106–107, 153–154, 163, 250, 335–336
Colinge, J., 82
Collado, M., 356
Collazo, E., 41, 290–291
Coller, H.A., 126–127

Collins, I., 82
Collins, R.E., 42–43
Comagic, S., 261–264
Comerford, S.A., 313–314
Commerford, S.L., 336
Considine, T., 215–219, 221, 227, 229, 234–236, 240
Contrepois, K., 25
Coon, J.J., 340–341
Copeland, R.A., 246–250, 253, 268, 283–285, 294–295
Corcoran, C.J., 168–171
Cornett, E.M., 32–50, 54–76
Coronel, J., 246–247
Corrales, F.J., 253
Cosentino, C., 127–128
Cosgrove, M.S., 291–292
Costa-Pinheiro, P., 246–247
Cote, J., 33f, 160–161
Cotton, M.J., 370
Courcelles, M., 153
Coutinho, F.J., 97
Couture, J.F., 290–291
Cramer, S.C., 332–333
Cravatt, B.F., 106–107, 118–121
Craven, M.W., 49, 54
Crawford, G.E., 371
Creyghton, M.P., 376–377
Cronkite, E.P., 336
Cross, N.C., 106
Cuddapah, S., 248–249, 356–357, 370
Cui, K., 248–249, 356–357, 370–372
Cui, Y.Q., 290–291
Cunningham, D., 196–197, 214–215
Cuomo, A., 83
Curtis, C., 86–87, 98–99
Curtis, R., 202–203
Cusanovich, D.A., 371–372
Czermin, B., 257–259

D

Daganzo, S.M., 356
Dahiya, R., 246–247
Dai, H., 214–242
Dai, J., 341
Dai, L., 126, 341
Dai, Y., 193

Daigle, S.R., 96, 250, 279–281, 283, 286, 297
Daigo, Y., 249
Dalgliesh, G.L., 248–249
Dalhoff, C., 249–250
Dalla-Favera, R., 192
Dan, C., 82
Daniel, J., 54
Danovi, D., 96–98
Dantas Machado, A.C., 371
Dao, T.T., 217–219
DaRe, J.M., 257
Das, G.D., 344
Dauwerse, H.G., 106
David, E., 370, 377–378
Davie, J.R., 150–151, 156–157
Davies, J.J., 368
Davies, N., 151
Davis, C.A., 375–376
Dawlaty, M.M., 356
Dawling, S., 249
Dawson, P.E., 150–151
Daza, R., 371–372
de Almeida, S.F., 248–249
de Boer, P.A., 248–249
De Carvalho, D.D., 246, 296
De Riva, A., 336
De Santa, F., 378
Deery, M.J., 336
Defossez, P.A., 186–187
DeGrado, W., 151–152
Deguchi, K., 106
Deguchi, T., 290–291
Delacruz, C., 151–152
Delbridge, L.M., 126
Demakos, E.P., 246–247, 261–264
Deng, H.T., 184–186, 291–292
Deng, L., 275–276, 281, 286
Denoyelle, C., 356
Dent, S., 33f
Denu, J.M., 49, 54, 106–107, 126–146, 184–190, 194–195, 197, 214–215, 234–236, 249–250, 317–318
Dequiedt, F., 196
Dessain, S.K., 192
Detrich, H.W., 32
Devkota, K., 272–273, 289–290
Devlin, M.K., 110–111

Dhall, A., 150–163
Dhalluin, C., 106
Dharmarajan, V., 291–292
Dhayalan, A., 42–46, 49, 54, 287
Dhe-Paganon, S., 289–290
Dhingra, V., 337
Dhodary, B., 217–219
Diao, J., 275–276, 286
Dickinson-Anson, H., 344
Dickson, B.M., 4–5, 32–50, 35f, 54–76, 369
Diederich, F., 266–268
Dietmann, S., 249
Dikovskaya, D., 356, 358–363
Dill, B.D., 332–352
DiMaggio, P.A., 20–24, 42–43, 126–127, 314–315, 341
Dimitrova, E., 46, 287
Ding, L., 270–272
Ding, X., 42–43
DiPersio, J., 246–247, 261–264
Dittenhafer-Reed, K.E., 189–190
Dittmann, A., 106–107, 120–121
Dittmar, G., 340
Djondjurov, L.P., 336
Dobbin, M.M., 214
Dobin, A., 375–376
Dobkin, B.H., 332–333
Dokmanovic, M., 151, 153
Dominguez-Sola, D., 106, 192
Dominy, J.E., 191
Domon, B., 337, 340–341
Donahue, G., 81
Dong, A.P., 81, 270–272, 286–287, 289, 295–297, 295f
Donigian, J.R., 184–186
Donohue, L., 370
Dorgan, K.M., 290–291
Dorigo, B., 161
Dormann, H.L., 348f, 350
Dorsey, J.A., 151, 153
Dotan, Z.A., 356
Dou, Z., 81
Dowden, J., 276–279, 286
Dowell, J.A., 126, 128–129
Doyle, B., 259–261
Doyle, T.L., 214–215
Drenkow, J., 375–376
Drewry, D.H., 80–81

Druid, H., 342–344
Druzina, Z., 259–261
Dryja, T.P., 246–247
D'Souza, R.C., 153
Duarte, F.V., 214
Duclos, N., 106
Duenas-Gonzalez, A., 246–247
Dukes, D.F., 112
Dunkel, I., 371
Duns, G., 248–249, 270–272
Duong, T., 170–171
Duquenne, C., 86–87, 96
Durant, S.T., 259–261
Durbin, R., 326
Dutton, P., 151–152
Dyer, P.N., 159–160

E

Easlon, E., 190–191
Eason, M.M., 286
Eaves, G.N., 357–358
Eberhard, D., 115, 120–121
Edayathumangalam, R.S., 159–160
Edelmann, M., 259–261
Edgerly, C., 204
Edwards, A.M., 80–81, 99
Eeckhoute, J., 375–376
Efeyan, A., 356
Egelhofer, T.A., 83–84
Ehrenhofer-Murray, A.E., 371
Ehrlich, M., 246–247
Eichenfield, D.Z., 366, 377–378
Eisenhaber, F., 42–43
Eisenman, R., 153
El Kortbi, M.S., 268–270
Eldeiry, W.S., 246–247
Eliuk, S.M., 25
Ellermann, M., 266–268
Ellis, J.L., 214–242
Elliston, K., 253
Emiliani, S., 194–195
Endris, V., 192
Eom, G.H., 248–249
Eram, M.S., 86, 89, 259–261, 287
Erdjument-Bromage, H., 46–48, 257–259
Erdos, M.R., 366
Eriksson, P.S., 344
Ernst, J., 369, 377

Esmans, E.L., 337
Espejo, A., 54
Esteller, M., 32, 126, 150, 246, 296
Eswaran, J., 257
Evertts, A.G., 126–127, 356
Evjenth, R., 108–109
Ezan, E., 25

F

Fabbri, G., 106, 192
Faber, J., 248–249
Fabre, E., 170–171
Falco, J.P., 246–247
Falk, A., 96–98
Fallahi-Sichani, M., 95
Faller, D.V., 193
Faller, W.J., 356, 358–363
Fan, H., 246–247
Fan, J., 126–146
Fang, J., 46–48
Fang, M.Z., 246–247
Fantes, P., 172
Farmer, R., 202–203
Farrelly, L.A., 332–352
Faulkner, R., 106
Federation, A.J., 86–87, 91–92, 96, 281–283, 286, 289–290
Fedorov, O., 80–81, 257, 286, 296
Feige, J.N., 190–191, 214
Feil, R., 368
Feinberg, A.P., 246–247
Feldman, J.L., 189–190, 197, 234–236
Feldmann, M., 99
Fenaille, F., 25
Feng, Q., 248–249, 257, 275–276
Feng, Y., 287, 289, 296–297
Feng, Z., 248–249
Fennell, T., 375–376
Fenouil, R., 248–249
Ferguson, D., 84–85
Fernandez-Cuesta, L., 106
Ferrari, R., 127–128
Ferreira de Freitas, R., 261
Ferry, J.J., 286
Field, H.I., 247–248
Fields, S., 186–187
Fierz, B., 151–153, 156–159, 163
Filippakopoulos, P., 257

Finley, L.W., 193
Finn, M.G., 112
Finney, G.L., 18
Finnin, M.S., 184–186
Fire, A.Z., 370
Firestein, R., 214
Fischer, A., 214
Fischer, F., 215
Fischle, W., 194–195
Fish, P.V., 81, 246–247
Flis, S., 246, 250, 296
Flores, J.V., 249
Floudas, C.A., 20–24
Flynn, D., 196–197, 214–215
Flynn, R.A., 376–377
Fog, C.K., 89
Fogel, S., 186, 197–198
Folarin, A.A., 96–98
Folco, H.D., 168
Fondufe-Mittendorf, Y.N., 371
Fonseca, G.J., 366, 377–378
Fontecave, M., 249–250
Fourrey, J.L., 268–270
Fraga, M.F., 203–204
Frankel, A., 89
Franken, H., 82
Franzel, B., 215
Freund, C., 49
Frewen, B., 18
Friedhoff, P., 357–358
Friedman Ohana, R., 120–121
Frietze, S., 247–248
Friis, R.M.N., 313
Frisen, J., 342–344
Frost, C., 202–203
Frye, M., 249
Frye, R.A., 189–190, 192
Frye, S.V., 80–81
Fu, Z.F., 337
Fuchs, S.M., 32–34, 36–38, 41–42, 49, 54–55
Fudenberg, G., 168
Fujiwara, T., 253
Fukushima, T., 42–43
Fulco, M., 106–107
Fuller, R.W., 253
Fullwood, M.J., 168
Funabiki, H., 168–170

Furgason, M., 151
Furge, K., 248–249
Furumoto, Y., 366

G

Gaffney, D.J., 371
Gage, F.H., 344
Gagne, D.J., 196–197, 214–215, 217–219
Galan, C., 366, 377–379
Galisson, F., 153
Galletti, P., 253
Galli, G.G., 89
Gallo, M., 97
Galluzzi, L., 317–318
Galuska, S., 195
Ganji, G., 86–87, 96, 286
Garay, J.P., 248–249
Garcia Boy, R., 261–264
Garcia, B.A., 4–28, 32, 42–43, 54, 126–129, 139–140, 150, 314–315, 337–341
Garcia, G.E., 352
Garcia-Trevijano, E.R., 253
Garofalo, R.S., 196–197, 214–215
Garske, A.L., 54
Gatbonton, T., 192–193, 197–201, 203–204
Gauthier, N., 289–290
Gayden, T., 97
Gayther, S.A., 106
Gehlen, L.R., 168
Gelb, M.H., 339–340
Gems, D., 186–187
Genuth, S., 126
George, J., 106
Gerber, S.A., 339–340
Gerhart-Hines, Z., 190–191
German, N., 193
Gerstein, M., 54
Gertler, A.A., 193–194
Gertz, M., 215
Getlik, M., 81
Geze, M., 268–270
Gheyi, T., 259–261
Ghisletti, S., 378
Ghosh, A.K., 268–272
Gil, J., 356
Giles, R.H., 106
Gillet, L.C., 12–14, 340–341
Gimadutdinow, O., 357–358

Giresi, P.G., 371–372, 377–378
Glaser, K.B., 196–197
Gnyszka, A., 246, 250, 296
Goehle, S., 192–193, 200–201
Goelz, S.E., 246–247
Goldman, J.M., 106
Goldstein, B.M., 253
Golshani, A., 248–249
Gomes, A.P., 214–219, 227, 234–236, 240
Gomez-Lopez, G., 193
Gonzales, M.L., 371–372
Gonzales-Cope, M., 4–5, 126–127
Gonzalez-Fierro, A., 246–247
Goodlett, D.R., 337
Goodman, R.H., 187–188
Goren, A., 366–370
Goryshin, I.Y., 371–372
Goss, M., 186–187
Gosselin, D., 366, 377–378
Gossen, M., 340
Gottschling, D.E., 186, 197–199, 203–204
Goya, R., 257–259
Gozani, O., 42–43, 54
Graca, I., 246–247
Gramlich, V., 264–268, 286–287
Grand, R.S., 168
Grandi, P., 106–107, 120–121
Grant, A.J., 253
Graves, T.L., 289–290, 296–297
Gray, J.W., 95
Gray, N.S., 254–257, 291–292
Greaves, D.R., 376
Green, R., 257–259
Greenbaum, D., 106–107
Greenleaf, W.J., 371–372, 377–378
Greenman, C., 248–249
Greer, E.L., 42–43, 313–314
Greger, V., 246–247
Grever, M.R., 247–248
Grimaldi, B., 189
Gronroos, E., 248–249
Grontved, L., 371
Grosse, F., 234–236
Grosso, A.R., 248–249
Grosveld, F., 376
Grove, G.W., 336
Grozinger, C.M., 197–198, 203–204
Grunn, A., 106, 192

Grunstein, M., 42–43, 156–157
Grzybowski, A.T., 4–5, 32–34, 35f, 36–37, 42–43, 46–48, 54, 69, 369–370
Gu, H.B., 83
Gu, W., 192
Guarente, L., 186–187, 190–193
Guasch, L., 108–109, 118–119
Guccione, E., 89
Gucwa, J.S., 118–119
Guerra, C., 356
Guler, G.D., 139–140
Gull, K., 168–169
Gunasekera, A., 196–197
Gunderson, K.L., 371–372
Guo, A.H., 4–5, 32–34, 35f, 36–37, 42–43, 46–48, 54, 69, 369
Guo, A.L., 83
Guo, H., 54
Guo, J., 84–87, 196–197
Gupta, M., 337
Gurley, L.R., 335
Gustin, J.P., 248–249
Gut, P., 312
Gutierrez, J.A., 270–272, 286–287
Gygi, S.P., 190–191, 339–340

H

Ha, T.K., 217–219
Haab, B., 34–35
Haas, W., 190–191
Haase, M., 368
Hackett, J.A., 368
Hadler, M., 289–290
Hagan, I., 168–170
Hahn, M., 357–358
Haider, S., 202–203
Haigis, K.M., 193
Haigis, M.C., 193
Haines, D.R., 253
Hajian, T., 89
Hajjar, K.A., 283
Haj-Yahya, M., 151
Hake, S.B., 348f, 350
Haldbo, S., 20–23
Hall, D.A., 54
Hallows, W.C., 49, 54, 197
Hamamoto, R., 249, 296
Hamashima, H., 83–84

Hamil, R.L., 253
Hamilton, S.R., 246–247
Hamuro, Y., 215, 217–219, 229
Handsaker, B., 375–376
Hanscombe, O., 376
Hansen, T.A., 20–24
Happel, N., 49
Hardin, J.M., 335–336
Hart, C., 370
Hart, G.W., 341
Hattan, S., 339–340
Hauer, C.R., 337
Haugen, E., 376–379
Hawkins, R.D., 377
Hayami, S., 247–249
Hayashi, K., 89
Hayashi, R., 42–43
Hayashi, Y., 203
Hayden, J.E., 168–171
He, C., 106
He, H.H., 371
He, Y., 84–85, 203, 367–368
Head, R., 97
Heard, E., 356
Hearn, S.A., 356
Heatley, S.L., 270–272
Heger, A., 326–327
Heiland, I., 188
Heinke, R., 250
Heinz, S., 326, 378
Heise, F., 371
Heiser, L.M., 95
Helfand, S.L., 186–187, 234–236
Helin, K., 20–24
Heller, R.C., 184–186
Heltweg, B., 192–193, 196, 200–201, 214–215
Hemeon, I., 286–287
Hendricks, C.L., 290–291
Hendriks, I.A., 153
Hennekam, R.C., 106
Henriquez, R., 170–171
Herdewijn, P., 249–250
Hergeth, S., 42–43
Herranz, D., 192–193
Herskowitz, I., 186
Herzog, E.D., 214
Hevel, J.M., 286

Hicks, J.B., 186
Higgins, M., 204
Hirao, M., 189
Hirayama, J., 189
Hirst, M., 356–357
Hisa, J.Y., 247–248
Hixon, J., 202–203, 214–215
Ho, M.C., 259–261, 286–287
Ho, Y., 192–193
Hobson, S., 115, 120–121
Hoehn, M.M., 253
Hoes, I., 337
Hoff, K.G., 184–186
Hoffman, J., 356
Hoffmann, A., 379–380
Hoffmann, J., 356
Hofstra, R.M., 248–249, 270–272
Hole, K., 108–109
Holland, J.C., 246–247, 261–264
Hollema, H., 248–249, 270–272
Hollick, J.J., 204
Holmes, S.G., 186
Holmfeldt, L., 270–272
Holt, A., 376–377
Holt, I.E., 371
Homer, N., 375–376
Honarnejad, S., 95
Hong, V., 112
Hong, W., 276–279, 286
Hong, X., 249
Hopf, C., 106–107, 120–121
Hopping, W., 246–247
Horiuchi, K.Y., 286
Horowitz, S., 270–272
Horsthemke, B., 246–247
Horton, J.R., 291–292
Hou, H., 170
Hou, Z., 246–247
Houten, S.M., 214
Howell, K.E., 339–340
Howitz, K.T., 214–219, 234–236
Hrycyna, C.A., 290–291
Hu, S., 54
Hu, S.S., 371
Huang, B., 257–259, 286
Huang, D., 270–272
Huang, H., 32
Huang, K., 250

Huang, T.W., 127–128
Huang, X.P., 86–87
Huang, Y., 203
Huang, Y.N., 339–340
Huarte, M., 46
Hubbard, B.P., 214–242
Huber, K.V., 82
Hudlebusch, H.R., 248–249
Humphreys, P., 96–98, 249
Hunt, D.F., 9–10, 337
Huntly, B.J., 106
Hussain, S., 249
Huttner, H.B., 342–344
Hwang, Y., 106–109
Hyams, J., 172

I

Ibanez, G., 270–272, 286–287, 290–291, 295–296, 295f
Ignatushchenko, M., 82
Ikawa, M., 248–249
Illig, R., 246, 296
Ilott, N.E., 326–327
Imai, E., 202–203
Imai, S.I., 186, 192, 214
Imakaev, M., 168
Imanishi, M., 83–84
Imhof, A., 257–259, 312
Inglada-Perez, L., 193
Ingvarsdottir, K., 151, 153
Inoue, A., 253
Irvine, W.P., 197–199, 203–204
Isakovic, L., 261–264, 286–287
Ishihama, Y., 137–138, 351
Ishii, H., 126, 150–151, 156–157
Ishikawa, F., 170
Islam, K., 291–292
Issa, J.P.J., 246–247, 261–264
Ito, A., 189–190
Itoh, Y., 202–203
Itou, S., 83–84
Ivanov, G.S., 249
Ivanova, E.C., 336
Iwasaki, A., 366
Iwasaki, O., 168–179
Izrael-Tomasevic, A., 139–140
Izzo, A., 42–43

J

Jackson, D.A., 371
Jackson, D.B., 95
Jackson, M.D., 187–188
Jacob, S.T., 247–248
Jacobsen, D.W., 283
Jaenisch, R., 246–247
Jafari, R., 82
Jagadeeswaran, S., 84–85
Jaitin, D.A., 370
Jakob-Roetne, R., 264–268, 286–287
Jansen, P.W.T.C., 81, 249
Jansson, M., 259–261
Jastrzebski, Z., 246, 250, 296
Jedrychowski, M.P., 191
Jeffery, D.R., 264–266
Jeffries, C.D., 357–358
Jeganathan, K.B., 356
Jeltsch, A., 42–46, 49, 54, 287
Jensen, O.N., 20–24, 126–127
Jenuwein, T., 4, 42–43, 150, 249, 369
Jenuwien, T., 33f
Jeong, S.M., 193
Jha, S., 375–376
Jia, D., 42–43
Jiang, F., 254–257, 291–292
Jiang, H., 196, 215, 281
Jiang, J., 192
Jiang, K., 366
Jiang, L., 54, 106–107
Jiang, Q., 248–249, 257
Jin, F., 126, 341
Jin, J., 250
Jin, L., 250, 253, 279–281, 283, 286, 297
Jin, W., 371–372
Johansen, J.V., 248–249
John, S., 371
Johnson, D.S., 356–357, 375–376
Johnson, J.L., 366–380
Johnson, N.A., 257–259
Johnson, S.M., 370
Johnston, D., 253
Johnston, L.D., 84–87
Johnstone, S.E., 320
Jones, L.H., 80–81
Jones, M.B., 168
Jones, P.A., 246, 296, 312–313
Jones, S., 356–357

Josefowicz, S.Z., 377–379
Joshi, T., 367–368
Juan, C.C., 247–248
Juan, L.J., 160–161
Jullien, P.E., 367–368
Jung, H.R., 20–23
Jung, M., 196, 203–204, 250
Jurczak, M.J., 191
Jurkowska, R., 42–46, 49
Justin, N., 249

K

Kaakkola, S., 264–266
Kadonaga, J.T., 118–119
Kaeberlein, M., 186–187, 214–215
Kaelin, W.G., 312
Kalac, M., 201–202
Kalle, A.M., 203–204
Kallgren, S.P., 170
Kaluzova, M., 189
Kaniskan, H.Ü., 86, 250
Kanno, Y., 366, 377–379
Kantarjian, H.M., 246–247, 261–264
Kanu, N., 248–249
Kapilashrami, K., 246–298
Kapoor, A., 356
Karaman, B., 203–204
Karch, K.R., 4–28
Kareta, M.S., 54
Karlsen, O.A., 108–109
Karschner, E.L., 290–291
Kasper, L.H., 106
Kassel, D.B., 239–240
Katada, S., 312
Katsuoka, Y., 253
Kauer, M., 369
Kaufmann, W., 253
Kautz, A.R., 234–236
Kavanagh, K.L., 48–49
Kazantsev, A.G., 204
Keavey, K., 202–203
Kebede, A.F., 49
Kee, H.J., 248–249
Kelleher, N.L., 106–107, 126–127, 150
Kellis, M., 369, 376
Kelly, R., 259–261
Kelly, T.K., 246, 296

Kenna, M.A., 186
Kennedy, B.K., 186–187, 214–215
Kennedy, S., 259–261, 287
Kent, S.B.H., 150–151
Keogh, M.C., 54
Keren-Shaul, H., 377–378
Kessler, B., 259–261
Khalil, E.M., 108–109
Khan, M.N., 202–203
Khan, S.M., 367–368
Khan, Z., 126, 128–129
Khatri, P., 81, 249
Kheradpour, P., 375–376
Khorasanizadeh, S., 32
Khuong, M.T., 118–119
Kidd, D., 118–119
Kieffer-Kwon, K.R., 377
Kienle, S., 339–340
Kiesslich, T., 246, 296
Killenberg, P.G., 112
Kim, D., 214
Kim, G.W., 184–186
Kim, H., 246–247, 249
Kim, J., 46, 54, 151, 204, 217–219
Kim, J.Y., 248–249
Kim, K., 249
Kim, K.-D., 168–179
Kim, K.H., 217–219
Kim, M., 248–249
Kim, S., 92–93
Kim, S.H., 190–191
Kim, S.M., 248–249
King, R.W., 289, 296–297
Kinoshita, N., 170
Kioi, M., 248–249
Kirkland, K.T., 186–187
Kirkland, T.A., 120–121
Kirmizis, A., 257–259
Kirschner, M.W., 340
Klar, A.J., 186, 197–198
Klauschen, F., 192
Kleer, C.G., 248–249
Klieser, E., 246, 296
Klose, R.J., 46
Klugman, S., 83–84
Knapp, S., 80–81
Knight, J.C., 99
Knobel, S.M., 187–188

Kobayashi, M., 253
Kobayashi, Y., 189
Koch, F., 248–249
Kogure, M., 249
Koh, K.P., 367–368
Kolasinska-Zwierz, P., 83–84
Kondengaden, S., 250
Konishi, H., 248–249
Konze, K.D., 84–87, 96, 250
Kooistra, T., 376–377
Kopcho, J.J., 257
Korboukh, V., 86–87
Korenchuk, S., 86–87, 96, 286
Kornberg, R.D., 32
Kornblith, A.B., 246–247, 261–264
Kouroumalis, E., 273–275
Kouzarides, T., 32, 33f, 150, 246, 248–249
Kozarich, J.W., 106–107
Krajewski, K., 4–5, 32–34, 35f, 36–38, 41–43, 46–49, 54–55, 69, 369
Kratchmarova, I., 120–121, 339–340
Krause, K.L., 357–358
Krautkramer, K.A., 128–129
Krebs, H.A., 187–188
Krishna, M., 54
Krishnan, S., 32–50, 54, 63–64, 75
Kristensen, D.B., 120–121, 339–340
Krivacic, C., 257–259, 286
Krivtsov, A.V., 248–249
Krizhanovsky, V., 356
Krobitsch, S., 371
Kroemer, G., 317–318
Krogan, N.J., 248–249
Ku, M., 370
Kudithipudi, S., 49, 54
Kudo, N., 189–190
Kuhn, H.G., 344
Kuhn, K., 339–340
Kuilman, T., 356
Kulej, K., 12–14
Kundaje, A., 369, 375–377
Kundu, T.K., 108–109
Kung, P.P., 257–259, 286
Kungulovski, G., 54
Kuningas, R., 195
Kuo, A.J., 54, 248–249
Kuplast, K.G., 84–87
Kurash, J.K., 249

Kurchuk, T., 170
Kurtev, M., 190–191
Kurth, M.C., 249
Kusevic, D., 49, 54
Kushida, M., 97
Kustigian, L., 215, 217–219, 221
Kutter, C., 249
Kuzmichev, A., 257–259

L

Labonte, A., 289–290
Labrie, V., 84–87, 91–92
Labuschagne, C.F., 313–314
Lachmi-Weiner, K., 42–43
Ladd, B., 46
LaFrance, L.V., 86–87, 96
Lagouge, M., 190–191, 214
Lahusen, T., 193
Lain, S., 204
Laird, P.W., 246–247, 312–313
Lake, A., 54
Lakshminarasimhan, M., 215
Lam, W.L., 368
Lambert, P.D., 196–197, 214–215, 217–219
Lammers, F., 369–370
Lamming, D.W., 214–219, 234–236
Lan, F., 46, 335–336
Lan, X., 97
Landry, J., 184–186, 194–195
Landt, S.G., 326, 375–376
Landwehrmeyer, G.B., 202–203
Lange, C.A., 248–249
Lange, K.C., 247–249, 268
Lange, V., 340–341
Langer, M.R., 184–186
Langmead, B., 326, 375–376
Lao, U., 193, 201
Lara, E., 203–204
Lara-Astiaso, D., 320, 370
Larjo, A., 377
Larsen, B., 83–84
Larsson, E.A., 82
Laslo, P., 378
Lau, O.D., 108–109
Laurent, G., 193
Lauring, J., 248–249
Lavin, Y., 377–378
Lavu, S., 214–219, 234–236

Lawrence, F., 268–270
Lazarovici, A., 371
Le Romancer, M., 273–275
LeBrasseur, N.K., 356
Leconte, N., 273–275
Lee, B.H., 106
Lee, C.H., 217–219
Lee, D.Y., 42–43
Lee, E.H., 217–219
Lee, J., 89
Lee, J.E., 215, 217–219, 229
Lee, J.M., 249
Lee, J.S., 249
Lee, J.V., 106–107
Lee, K.A., 81, 83–84, 246–247
Lee, K.K., 106–107
Lee, M., 168
Lee, S., 126, 341, 356
Lee, T.H., 151–153, 156–159
Lee, T.I., 320
Lee, Y., 191
Lee, Y.C., 247–248
Leffler, S., 290–291
Lehotzky, A., 203–204
Leko, V., 193
Lemiere, F., 337
Lemieux, M.E., 248–249
Lempiainen, H., 368
Lerin, C., 190–191
Lerner, C., 264–268, 286–287
Lerner, L.M., 253
LeRoy, G., 54, 126–127
Lerrer, B., 193–194
Levin, R.S., 126–127, 314–315, 341
Levin, S., 120–121
Lewellen, N., 371
Li, B., 313
Li, C., 192
Li, F.L., 81, 84–87, 96, 151–152, 156–157, 257, 259–261, 286–287, 296
Li, G., 377
Li, H., 42–43, 192–193, 200–201, 259–261, 326, 375–376
Li, J., 214–219, 227, 234–236, 240
Li, K., 127–128
Li, K.K., 250
Li, L., 192–193, 270–272
Li, S.S., 84–85

Li, Y., 42–43, 248–249, 257
Li, Z., 275–276, 376–377
Lieberman-Aiden, E., 168
Lienard, B.M., 48–49
Liljas, A., 264–266
Lillehaug, J.R., 108–109
Lim, K.B., 239–240
Lima-Fernandes, E., 80–100
Lin, A.W., 356
Lin, H.K., 356
Lin, J., 54
Lin, Q., 254–257, 291–292
Lin, S., 9–10, 15, 54, 128–129, 139–140
Lin, S.J., 186–187, 190–191
Lin, Y., 248–249, 257
Lin, Y.C., 378
Lin, Z.Y., 83–84
Lindsey, G.G., 151
Ling, A.J., 214
Linger, L., 106
Link, V.M., 366, 377–378
Linke, C., 371
Linscheid, M., 273–275
Lipinski, M., 356
Litzenburger, U.M., 371–372
Liu, C.L., 54
Liu, D., 196, 215
Liu, F., 84–87, 91–92, 289, 296–297
Liu, G., 83–84, 247–248
Liu, J., 168, 291–292
Liu, L., 92–93
Liu, M.H., 168
Liu, P.Y., 193
Liu, Q., 272–273, 289–290
Liu, S., 249
Liu, S.C., 81
Liu, T., 193, 375–376
Liu, W., 270–272
Liu, X., 26, 108–109
Liu, X.S., 371–372
Liu, Y., 106–107, 118–119, 371
Liu, Z., 281
Llewellyn, D.B., 261–264, 286–287
Loddenkemper, C., 356
Loe, T.K., 201
Loenen, W.A., 249–250
Loh, C., 214–215
Lohse, B., 272–273, 289–290

Long, L., 151
Lonnberg, T., 377
Lopez-Nieva, P., 203–204
Lorch, Y., 32
Lorenz, K., 192
Lorenzo-Vivas, E., 370
Lovaas, J.D., 193
Low, D., 259–261
Lowary, P.T., 160
Lowe, S.W., 356
Lu, C., 313–314
Lu, H., 246–247
Lu, Q., 214–215
Lu, S.C., 253
Lu, T., 247–249, 268
Lu, W., 128–129, 132–133
Lu, Z., 126
Lua, G.B.J., 151–152, 156–157
Lucio-Eterovic, A.K., 296
Luco, R.F., 248–249
Luger, K., 4, 32, 332–333
Lund, A.H., 89
Lund, P., 187–188
Lund, T., 336
Lundstrom, K., 264–266
Lungu, C., 49
Luo, H., 126, 341
Luo, J., 192
Luo, M., 246–298
Luo, M.K., 291–292
Luo, X., 86
Lupien, M., 247–248
Luu, Y., 377
Lv, K., 268–270
Lyko, F., 250

M

Ma, A., 84–87, 96
Ma, C., 112
Maag, D., 84–87
MacCoss, M.J., 18, 339–340
Mace, K., 366, 377–379
Machado De Oliveira, R., 190–191
Mack, S.C., 97
MacLean, B., 18
Macleod, K., 186, 197–198
MacNevin, C.J., 84–87, 96
MacRae, T.H., 168–169

Madar, A., 366, 377–379
Maddocks, O.D.K., 313–314
Madeo, F., 317–318
Mäder, A.W., 4, 32, 332–333
Maeda, H., 357–358
Maehara, Y., 249
Mahajan, S.S., 201
Mahrouche, L., 153
Mai, A., 203–204, 249–250
Maile, T.M., 139–140
Majer, C.R., 84–87, 96, 253, 279–281, 286, 297
Makalowska, I., 153
Malik, S., 194
Mallika, A., 203–204
Maltby, D., 25
Malureanu, L., 356
Malyukova, A., 193
Mann, C., 25
Mann, M., 134, 137–138, 153, 314, 341, 351
Manning, A.L., 46
Mannisto, P.T., 264–266
Maraver, A., 193
Marchese, J.N., 339–340
Marcheva, B., 189
Marchi, E., 201–202
Marcotte, P.A., 196–197
Margueron, R., 257–259
Marinov, G.K., 369, 375–376
Marmorstein, R., 108–109, 184–186
Marquez, V.E., 253
Martin, B.R., 120–121
Martin, C., 248–249
Martin, J.L., 249–250
Martin, N.I., 268
Martinato, F., 89
Martinet, N., 250
Martinez Molina, D., 82
Martinez, P., 248–249
Martinez-Chantar, M.L., 203–204
Martinez-Garcia, E., 248–249
Martinez-Pastor, B., 127–128
Martini, E.M., 356
Martin-Montalvo, A., 214
Martins, M.G., 246–247
Masjost, B., 264–268, 286–287
Masselot, A., 139–140

Masuno, M., 106
Matevossian, A., 349–352
Mathe, E., 377
Mathieson, T., 82
Mathioudaki, K., 273–275
Mato, J.M., 253
Matson, C., 46, 335–336
Matsui, T., 259–261
Matsui, Y., 89
Matsumoto, K., 357–358
Matsumoto, T., 170
Matthews, D.E., 339–340
Mawson, A., 192
Maze, I., 32, 326–327, 332–352
Mazur, P.K., 81, 249
McBean, J.L., 290–291
McBrian, M.A., 127–128
McBryan, T., 356–357
McCabe, D., 257–259
McCabe, M.T., 86–87, 96, 286
McCammon, M.T., 253
McDonagh, T., 202–203, 214–215
McDonald, S., 336
McDonald, T., 192–193
McDonough, M.A., 48–49
McGinty, R.K., 151, 153, 163
McKnight, S.L., 312
McLeod, M., 172
McMillan, F.M., 249–250
McPherson, H., 253
McVey, M., 186–187
McVicker, G., 371
Meade, S., 204
Mechtler, K., 369
Medzhitov, R., 366
Medzihradszky, K.F., 106–107
Mehta, S., 168
Meier, J.L., 106–121
Meiss, G., 357–358
Meissner, A., 367–368
Melamud, E., 128–129, 132–133, 140–141
Melfi, R., 257–259
Melki, J.R., 246–247
Mellor, K.M., 126
Meloche, S., 153
Merad, M., 377–378
Merino, A., 95
Mernaugh, R.L., 249

Messmer, E., 246–247
Meuleman, W., 377
Mews, P., 312–327
Meyer, C.A., 371–372, 375–376
Meyer, M., 97
Miao, F., 126
Michaloglou, C., 356
Michan, S., 214
Michaud, G.A., 54
Michel, B.Y., 250
Michelini, K., 371
Michon, A.M., 106–107, 120–121
Micklefield, J., 249–250
Mietton, F., 378
Migliori, V., 54, 259–261
Miller, M.D., 357–358
Miller, V.L., 32–34, 42, 54
Milne, J.C., 189–191, 196–197, 214–215, 217–219
Min, D.J., 248–249
Min, J., 54, 184–186, 275–276
Minoda, A., 83–84
Mirsky, A.E., 106
Misteli, T., 168, 248–249
Mitchell, S., 379–380
Miyake, Y., 81
Miyata, N., 202–203
Mizuguchi, T., 168
Mizuno, S., 106
Mizzen, C.A., 150
Moazed, D., 186, 194–195, 197–198, 203–204
Moehlenbrink, J., 247–248
Mohamed, Y.B., 168
Moir, R.D., 214–215
Mok, W.C., 259–261
Mokbel, K., 248–249
Molden, R.C., 15, 339–340
Molina, H., 332–352
Mollah, S., 9–10, 338–339
Montellier, E., 126, 341
Montgomery, D.C., 106–121
Montgomery, M.K., 214
Moore, C.D., 54
Moore, R.J., 25
Moorman, A.F., 248–249
Moret, E.E., 268
Morimoto, J., 203

Morin, R.D., 257–259
Moritz, R.L., 259–261
Morris, B.J., 214
Mortazavi, A., 356–357
Moslehi, J.J., 214
Mostoslavsky, R., 127–128, 190–191
Muhammad, S., 186
Muhsen, U.A., 276–279, 286
Muir, T.W., 150–151, 153–154, 163
Mulhern, D., 83
Muller, A., 368
Müller, A.C., 82
Muller, J., 259–261
Muller, S., 80–81
Mulliez, E., 249–250
Mulligan, P., 46, 335–336
Mullikin, J.C., 371
Mundade, R., 247–249, 268
Mungall, A.J., 257–259
Munro, S., 247–248
Murakami, M., 42–46, 49
Murakami, S., 170
Muratore, T.L., 9–10, 338–339
Murison, A., 97
Murray, J.W., 257
Musselman, C.A., 54
Muthurajan, U.M., 159–160
Myers, R.M., 356–357
Myers, R.W., 195

N
Nady, N., 54
Nagarajan, R., 253
Naito, T., 170
Nakahata, Y., 189
Nakajima, Y., 83–84
Nakakido, M., 249
Nakamura, Y., 249, 296
Nakaseko, Y., 170
Nakatani, Y., 356
Nakayamada, S., 366, 377–379
Napper, A.D., 202–203
Narayanan, K.L., 204
Nare, B., 195
Narita, M., 356
Nasheuer, H.P., 234–236
Natarajan, R., 246, 296
Nathan, D., 151, 153

Natoli, G., 376
Navarro, P., 12–14, 340–341
Nehrbass, U., 170–171
Nelkin, B.D., 246–247
Nelson, D.M., 356, 358–363
Nelson, M., 189
Neph, S., 376–379
Nestor, C.E., 368
Neumann, H., 156–157
Neureiter, D., 246, 296
Newbold, R.F., 248–249
Newcomb, B., 197–198
Newlander, K.A., 86–87, 96
Neylon, E., 201–202
Ng Eaton, E., 192
Ng, S.S., 48–49
Nguyen, G.T., 215
Nguyen, H., 86–87, 98–99
Nguyen, K.T., 257, 286, 296
Nguyen, M.D., 214
Nguyen, P.H., 217–219
Nicklaus, M.C., 108–109, 118–119
Nielsen, M.L., 83
Niki, M., 356
Nikolaev, A.Y., 192
Nikula, K.J., 246–247
Nimura, K., 248–249
Nishida, Y., 134
Niwa, O., 170
Noel, J.P., 290–291
Noh, K.M., 32, 332–334, 341, 344–346, 352
Noland, B.J., 335–336
Noma, K.-I., 168–179
Nordborg, C., 344
Nordlund, P., 82
Noriega, L., 190–191
Norris, G.E., 270–272
North, B.J., 203–204, 214
North, J.A., 163
Nottke, A., 46
Nunez, S., 356

O
Oakeley, E.J., 368
Oberdoerffer, P., 214
O'Brien, C., 332–333
O'Carroll, D., 42–43

O'Connor, O.A., 201–202
Odchimar-Reissig, R., 246–247, 261–264
Ogino, S., 214
Oh, B.K., 246–247
Oh, W.K., 217–219
Ohtani, N., 246–247
Okabe, M., 248–249
Okada, Y., 248–249, 257
Okitsu, Y., 253
Olek, K.M., 82
Olhava, E.J., 96, 250, 279–281, 283, 286, 297
Oliver, S.S., 54
Olsen, J.V., 134
Olson, E.R., 369–370
O'Neill, L.P., 357–358
Ong, C.K., 270–272
Ong, S.-E., 120–121, 314, 339–340
Onishi, A., 54
Opitz, C.A., 188
Oppermann, U., 48–49
Opravil, S., 369
Orlando, D.A., 369–370
Ortiz-Tello, P.A., 41
Osborne, T., 273–275
Osdal, T., 192–193
Osinga, J., 248–249, 270–272
Osterling, D.J., 84–85
Ostuni, R., 366, 379
Ota, I., 248–249
Ott, H.M., 86–87, 96, 286
Ottesen, J.J., 163
Otto, P., 120–121
Oudhoff, M.J., 81
Ow, J.R., 42–43
Owen-Hughes, T.A., 160–161
Oyama, F., 237, 239
Ozbal, C.C., 239–240
Ozsolak, F., 370

P

Pacholec, M., 196–197, 214–215
Pai, A.A., 371
Pal, S., 247–248
Pan, Q., 248–249
Pan, Y.F., 168
Pandey, A., 120–121, 339–340
Pandita, T.K., 192
Pandolfi, P.P., 356
Panning, B., 25, 151–152
Papadokostopoulou, A., 273–275
Pappano, W.N., 84–85
Park, C., 246–247
Park, G.J., 193
Park, H.J., 246–247, 249
Park, J., 217–219
Park, P.J., 368–369
Parker, K., 339–340
Parker, S.C., 366
Parks, L.W., 253, 292
Parl, F.F., 249
Parry, R.V., 276–279, 286
Parton, T., 84–87, 96
Partyka, K., 34–35
Paša-Tolić, L., 25
Pasini, D., 20–23
Pasqualucci, L., 106, 192
Passarge, E., 246–247
Patel, A., 291–292
Patel-Thombre, U., 253
Paterson, A., 126
Patnaik, D., 257
Patra, A., 246–247
Patra, S.K., 246–247
Patricelli, M.P., 106–107
Patterson, G.H., 187–188
Pauli, F., 366, 375–379
Paulini, R., 266–268
Pavletich, N.P., 184–186
Payne-Turner, D., 270–272
Paz, M.I., 352
Pazin, M.J., 150–151, 156–157
Pchelintsev, N.A., 356–357
Peak-Chew, S.Y., 156–157
Pediconi, N., 106–107
Pedro, L., 289–290
Peeper, D.S., 356
Peifer, M., 106
Pellois, J.-P., 153
Peng, C., 126
Peng, X., 377–379
Pereira-Smith, O.M., 248–249
Perez-Burgos, L., 369
Perez-Cardenas, E., 246–247
Perez-Plasencia, C., 246–247
Perfilieva, E., 344

Pesavento, J.J., 150
Peters, A.H., 356, 369
Peterson, A.C., 340–341
Peterson, B.L., 246–247, 261–264
Peterson, C.L., 150–151, 156–157
Peterson, D.A., 344
Petrij, F., 106
Petrossian, T.C., 42–43
Pevzner, P.A., 26
Phalke, S., 259–261
Piao, L.H., 249
Picard, F., 190–191
Piccolo, V., 366, 379
Pichersky, E., 290–291
Pichugina, T., 168
Picotti, P., 340–341
Pierce, K., 193
Pietrocola, F., 317–318
Pignot, M., 273–275
Pike, R.A., 276–279, 286
Pilka, E.S., 48–49
Pillus, L., 184–186, 197–198
Pingoud, A., 357–358
Pinho, A.V., 192
Piper, M.D., 186–187
Pirrotta, V., 257–259
Piston, D.W., 187–188
Planck, J.L., 286
Plazas-Mayorca, M.D., 20–22
Pliner, H.A., 371–372
Pliushchev, M., 84–87
Poirier, M.G., 163
Pollard, S.M., 96–98
Polletti, S., 366, 378–379
Pons, J.F., 202–203
Ponting, C.P., 326–327
Pop, M., 326, 375–376
Popovic, R., 248–249
Portela, A., 32, 150
Posakony, J., 189, 192–193, 200–201
Poustovoitov, M.V., 356
Praz, V., 369–370
Prentki, M., 190–191
Presolski, S.I., 112
Price, N.L., 214
Prinos, P., 80–100
Puigserver, P., 190–191
Puri, A., 356
Purushothaman, I., 333–334, 341, 344–346, 352

Q

Qi, J., 81, 289–290
Qian, J., 377
Qing, Z., 250
Qiu, X., 196, 215
Qiu, X.M., 246–247
Quinn, A.M., 289, 296–297
Quinti, L., 204
Quivy, J.-P., 356

R

Raab, J.R., 4–5, 32–34, 35f, 36–37, 42–43, 46–48, 54, 69, 369
Rabanal, F., 151–152
Rabinowitz, J.D., 128–129, 132–133, 140–141, 143
Racki, L., 151–152
Rada-Iglesias, A., 376–377
Radu, C., 287, 295f
Ragno, R., 249–250
Ragoczy, T., 168
Rai, T.S., 356–363
Raine, K., 270–272
Rajman, L., 214
Rajski, S.R., 273–275
Ralfkiaer, E., 248–249
Ram, O., 370
Ramjan, Z., 54–76
Ramsey, K.M., 189
Rangel-Lopez, E., 246–247
Rank, G., 259–261
Rao, A., 367–368
Rao, V.K., 42–43
Rapaport, J.M., 246–247
Rappsilber, J., 137–138, 351
Rastelli, G., 95
Rathert, P., 42–46, 49, 54, 287
Rauh, D., 215
Ray-Gallet, D., 356–357
Rea, S., 42–43
Reik, W., 368
Reimand, J., 97
Reinberg, D., 42–43, 150, 257–259
Reiter, L., 12–14, 128–129, 340–341

Ren, B., 312–313, 367, 380
Ren, G., 371–372
Rensing, N., 214
Reveron-Gomez, N., 126–127
Reyhanfard, H., 250
Reynoird, N., 81, 249
Reynolds, A.P., 376–379
Reznikoff, W.S., 371–372
Richard, J., 92–93
Richardson, P.L., 196–197
Richmond, R.K., 4, 32, 332–333
Richmond, T.J., 4, 32, 161, 332–333
Richon, V.M., 246–248, 250, 253, 268
Riera, T.V., 215, 217–219, 229
Riley, T.R., 371
Rine, J., 186
Rist, B., 339–340
Rival-Gervier, S., 84–87, 91–92
Rivera, C.M., 312–313, 367, 380
Roadmap Epigenomics, C., 377
Robert-Gero, M., 268–270
Robertson, G., 356–357
Robin-Lespinasse, Y., 273–275
Robinson, E.W., 25
Robledo, M., 193
Rodgers, J.T., 190–191, 214
Roeder, R.G., 151
Roessler, C., 203–204
Rogakou, E.P., 356
Rogina, B., 186–187, 234–236
Roh, T.Y., 248–249, 356–357, 370
Romanoski, C.E., 366, 377–378
Roodi, N., 249
Rooman, I., 192
Rose, A.B., 186
Rosenfeld, C.S., 246–247, 261–264
Roska, R.L., 273–275
Ross, J.R., 290–291
Ross, P.L., 339–340
Rossing, H.H., 248–249
Röst, H., 12–14, 340–341
Rotem, A., 320
Roth, D., 266–268
Roth, J.A., 264–266
Rothbart, S.B., 4–5, 32–50, 54–76, 369
Rottcher, D., 253
Roy, N., 194
Roy, S., 42–43

Royce, T., 54
Ruan, J., 375–376
Ruderman, N.B., 190–191
Rudolph, C., 356
Ruf, A., 264–268, 286–287
Ruff, D., 371–372
Ruminowicz, C., 20–24
Rumpf, T., 203–204
Rush, J., 340
Russell, J.D., 340–341
Russell, R., 96–98
Ruthenburg, A.J., 369–370
Ryan, R.J., 369–370

S

Saavedra, O.M., 261–264, 286–287
Sabari, B.R., 32
Sachchidanand, 203–204
Sadaie, M., 170
Sahar, S., 189
Saitoh, H., 253
Sakabe, K., 341
Sakai, T., 246–247
Salcius, M., 54
Saldanha, A.J., 44f
Salehpour, M., 342–344
Saloura, V., 249, 296
Salzberg, S.L., 326, 375–376
Sammons, M.A., 81
Samstein, R.M., 377–379
Sananbenesi, F., 214
Sanchez-Arevalo Lobo, V.J., 192
Sanchez-Mut, J.V., 246, 296
Sandberg, T., 259–261
Sandstrom, R., 371, 377–379
Sanger, T.D., 332–333
Santini, V., 246–247
Santoni-Rugiu, E., 248–249
Santos, H.A., 356
Sargent, D.F., 4, 32, 332–333
Saris, J.J., 106
Sarkissyan, M., 246, 296
Sartorelli, V., 366, 377–379
Sasse, R., 168–169
Sassone-Corsi, P., 189, 312
Satoh, A., 214
Sauve, A.A., 184–186, 214–215
Savitski, M.M., 106–107, 120–121

Sawada, H., 202–203
Scamps, C., 356
Scarlett, C.J., 193
Schafer, J., 339–340
Schalch, T., 161
Schanzer, J.M., 257
Schapira, M., 81, 246–247, 261
Schenk, S., 196–197, 214–215, 217–219
Schiedel, M., 203–204
Schillinger, D., 253
Schirrmacher, R., 261–264
Schlatter, D., 266–268
Schlegelberger, B., 356
Schlenk, F., 292
Schlesinger, F., 375–376
Schlicker, C., 215
Schmatz, D.M., 195
Schmid, M., 42–43
Schmidt, A., 42–43, 126–127
Schmidt, G., 339–340
Schmidt, K., 86
Schmidt, M.T., 187–188
Schmitt, C.A., 356
Schmitz, R.J., 367–368
Schneider, A., 234–236
Schneider, R., 42–43
Scholz, C., 81
Schones, D.E., 248–249, 326, 356–357, 370
Schramm, V.L., 184–186, 214–215, 270–272, 286–287
Schreiber, S.L., 197–198, 203–204
Schuchlautz, H., 89
Schuhmacher, A.J., 356
Schuler, A.D., 192–193, 200–201
Schultz, P.G., 254–257, 291–292
Schutkowski, M., 215
Schwämmle, V., 20–24
Schwanhausser, B., 340
Schwartz, R.J., 248–249
Schwarz, J., 339–340
Schweitzer, B.I., 54
Schweizer, W.B., 266–268
Scian, M., 201
Scopton, A., 86–87, 91–92, 96, 281–283, 286
Scorilas, A., 273–275
Scott, J.E., 289–290, 296–297
Scotto, L., 201–202

Segel, I.H., 294
Segura-Pacheco, B., 246–247
Seidel, D., 106
Seidenfeld, J., 253
Seitz, V., 257–259
Sekeri-Pataryas, K.E., 356
Seki, T., 82
Selbach, M., 340
Selevsek, N., 12–14, 340–341
Sellars, M., 366, 377–379
Senawong, T., 190–191
Senisterra, G., 89
Sens, K., 184–186
Sentis, S., 273–275
Serrano, M., 193, 356
Settles, B., 49, 54
Severson, T.M., 257–259
Seznec, J., 259–261
Shabanowitz, J., 9–10, 337–339
Shafer, A., 371
Shaffer, D., 356
Shah, S., 106–107
Shahbazian, M.D., 156–157
Shankar, S.R., 42–43
Sheahan, S., 259–261
Shechter, D., 348f, 350
Shen, H., 312–313
Shen, L., 326–327
Shen, R., 248–249
Shen, Y., 26
Shepherd, G.R., 335–336
Sherwin, T., 168–169
Shi, X., 42–43
Shi, Y., 42–43, 46, 150, 313–314, 335–336
Shigematsu, H., 106
Shilo, Y., 153
Shiloh, A., 192
Shim, Y.H., 246–247
Shimazu, T., 249
Shimko, J.C., 163
Shinkai, Y., 42–43, 272–273
Shiratori, H., 248–249
Shogren-Knaak, M.A., 150–151, 156–157, 254–257
Shokat, K.M., 254–257
Shoresh, N., 370
Shulman, N., 18
Shum, D., 287, 295f

Siarheyeva, A., 89
Sidoli, S., 4–28, 126–127
Sidow, A., 370
Siedlecki, P., 261–264
Siethoff, C., 273–275
Sif, S., 247–248
Silver, P.A., 247–248
Silverman, L.R., 246–247, 261–264
Simeoni, S., 249–250
Simithy, J., 12–14
Simon, J.A., 184–205, 248–249
Simon, M., 151–152
Simon, R., 248–249
Simpson, R.J., 259–261
Sims, D., 326–327
Sinclair, D.A., 186–187, 214–242
Singh, V., 270–272
Sinha, A.U., 248–249
Sink, W.D., 153
Sinn, B., 192
Sippl, W., 250
Sirotkin, Y., 26
Slama, J.T., 184–186
Smallegan, M.J., 126, 128–129
Smil, D., 257, 259–261, 286–287, 296
Smith, A.G., 96–98
Smith, B.C., 49, 54, 187–188, 197, 214–215, 249–250
Smith, C.L., 370
Smith, J.J., 196–197, 214–215, 217–219
Smith, J.S., 186, 197–198
Smith, K.M., 42–43
Smith, R.F., 286
Smith, R.W., 234–236
Smith, Z.D., 367–368
Smolikov, S., 46
Sneeringer, C.J., 96, 253, 279–281, 286, 297
Snyder, M., 54
Snyder, M.P., 369, 371–372, 376
Snyder, N.W., 106–107
So, C.W., 273–275
Soccio, R.E., 108–109
Soengas, M.S., 356
Solanki, S., 369–370
Soldi, M., 83
Solomon, M.E., 246–248, 250, 253, 268
Somaiah, N., 356
Somogyvari, M., 186–187

Sondhi, D., 153–154, 163
Sone, K., 249
Song, C., 42–43
Song, J., 96, 279–281, 286, 297
Song, Y., 246–247
Sorger, P.K., 95
Sorum, A.W., 108–109, 118–121
Sos, M.L., 106
Soshnev, A.A., 333
Sousa, E.J., 246–247
Souza, A.L., 193
Spadotto, V., 83
Spalding, K.L., 342–344
Spang, R., 193
Spann, N.J., 366, 377–378
Spannhoff, A., 291–292
Spector, D.L., 168, 356
Spiegelman, B.M., 190–191
Spreafico, R., 379–380
Sprenger, R.R., 20–23
Sripathy, S., 201
Staerk, D., 272–273, 289–290
Stallcup, M.R., 42–43
Stamatoyannopoulos, J.A., 371, 377
Staples, O.D., 204
Starai, V.J., 186
Stark, C., 83–84
Stebbins, J., 184–186
Stec, I., 248–249
Steegborn, C., 215
Steen, H., 120–121, 339–340
Stein, R.L., 215, 217–219, 221, 257
Steine, E.J., 376–377
Stemman, O., 340
Stenoien, D.L., 25
Stenzinger, A., 192
Stephens, P., 270–272
Stergachis, A.B., 376–379
Sterner, D.E., 151, 153
Sternglanz, R., 184–186, 194–195
Strahl, B.D., 4–5, 32–50, 54–55, 63–64, 69, 75, 150
Strathern, J.N., 186
Strehle, A., 214
Stricker, S., 96–98
Struck, A.W., 249–250
Stugiewicz, M., 83
Su, F., 192

Su, L.J., 247–248
Su, T., 127–128
Suarez, D.P., 86–87, 96
Sudbery, I., 326–327
Suenkel, B., 215
Suga, H., 203
Sugawara, K., 237, 239
Sugimoto, K., 42–43
Suh-Lailam, B.B., 286
Sullivan, S., 153
Sun, H.W., 366, 377–379
Sun, J.-M., 150–151, 156–157
Sun, Y.T., 193, 246–247
Sun, Z.W., 42–43
Sung, M.H., 377
Sur, M., 332–333
Sussmuth, S.D., 202–203
Sutter, B.M., 313
Sutton, A., 184–186
Suzuki, T., 202–203
Svensson, L.A., 264–266
Swanson, M.S., 289, 296–297
Sweatt, J.D., 342
Sweet, S.M., 248–249
Sweetman, G.M., 82
Sweis, R.F., 84–87
Swierczynski, S., 246, 296
Swigut, T., 376–377
Sylvestersen, K.B., 83
Syrjanen, S., 178
Szewczyk, M.M., 80–100, 259–261, 287

T

Tachibana, M., 42–43, 272–273
Tafrov, S.T., 184–186
Tagami, H., 356
Tai, D., 371
Taja-Chayeb, L., 246–247
Takahashi, H., 366, 377–379
Takawa, M., 249
Talukdar, P., 203–204
Tamaru, H., 42–43
Tamas, R., 42–46, 49, 54
Tan, C.S., 82
Tan, M., 126, 341
Tan, Y.T., 259–261
Tanaka, A., 168–169
Tanaka, S., 194

Taneja, N., 168
Taneja, R., 42–43
Tang, Q., 371–372
Tang, Y., 192, 356
Tang, Z., 377
Tanizawa, H., 168–171
Tanner, J., 357–358
Tanner, K.G., 184–186, 194–195
Tanny, J.C., 194–195
Tarpey, P., 270–272
Taskinen, J., 264–266
Tatar, M., 234–236
Tate, S., 12–14, 340–341
Tatlock, J.H., 81, 257–259, 286
Taylor, D.M., 204
Taylor, P.M., 153
Tchkonia, T., 356
Telling, A., 168
Tempel, W., 86–87, 91–92, 96, 257, 281–283, 286, 296
Tempst, P., 46–48, 257–259
Tenhunen, J., 264–266
Termanini, A., 366, 379
Teyssier, C., 42–43
Thastrup, O., 187–188
Theisen, J.W., 118–119
Thelen, J.N., 189–190
Thelen, J.P., 151
Therkelsen, C.A., 96, 250, 279–281, 283, 286, 297
Thomas, P.M., 126–127, 248–249
Thompson, A., 273–275, 339–340
Thompson, C., 86–87, 96, 286
Thompson, D.H., 290–291
Thompson, M.L., 249–250
Thompson, P.R., 106–111, 273–275
Thomson, J.P., 368
Thorpe, K., 106
Tian, X.R., 86–87, 96
Tian, Z., 25
Tilgmann, C., 264–266
Tindell, C., 139–140
Ting, Y.S., 26
Tinsley, J., 96–98
Tissenbaum, H.A., 186–187
Tögel, I., 82
Toguchida, J., 246–247
Tolić, N., 25

Tomazela, D.M., 18
Tominaga, K., 248–249
Tomishima, Y., 83–84
Tomlins, S.A., 248–249
Tong, A., 248–249
Topark-Ngarm, A., 190–191
Torres-Ferreira, J., 246–247
Toyokawa, G., 247–249
Tran, T.L., 217–219
Trapnell, C., 326, 375–376
Treilleux, I., 273–275
Trievel, R.C., 32–50, 54, 63–64, 75, 270–272, 290–291
Trifonov, V., 106, 192
Trindler, C., 266–268
Tripathi, S.K., 377
Tripepi, G., 202–203
Trojanowski, J.Q., 332–333
Trojer, P., 42–43
Tropberger, P., 42–43
Trotman, L.C., 356
Trout, K.L., 153
Tsai, L.H., 214
Tsai, P., 168
Tsai, Y.S., 26
Tseng, C.K., 253
Tsukada, Y., 46–48
Tsunoda, T., 247–248
Tu, B.P., 313
Tully, S.E., 120–121
Turberfield, A.H., 46
Turecek, F., 339–340
Turner, B.M., 357–358
Tzavaras, N., 333–334, 341, 344–346, 352

U

Ueberheide, B.M., 9–10, 338–339
Ueda, K., 249
Ugarkar, B.G., 257
Ulmanen, I., 264–266
Unoki, M., 247–248
Upadhyay, A.K., 291–292
Ura, K., 248–249
Urdinguio, R.G., 246, 296
Urich, M.A., 367–368
Utley, R.T., 160–161
Uyeda, H.T., 120–121
Uzawa, S., 168–170

V

Vadgama, J.V., 246, 296
Vahedi, G., 366–380
Valdes-Lopez, O., 367–368
Valentini, S., 186–187
Valouev, A., 370
Van Aller, G.S., 86–87, 96, 286
van Berkum, N.L., 168
van de Sluis, B., 356
van den Berg, E., 248–249, 270–272
van der Horst, C.M., 356
van Deursen, J.M., 356
Van Dongen, W., 337
van Duivenbode, I., 248–249, 270–272
van Galen, P., 320, 370
van Haeringen, A., 248–249
van Haren, M., 268
Van Lint, C., 194–195
van Ommen, G.J., 248–249
Van Onckelen, H., 337
Van Rechem, C., 46
van Ree, J.H., 356
van Tuyn, J., 356–357
van Ufford, L.Q., 268
Vanhoutte, K., 337
Vanner, R.J., 97
Varambally, S., 248–249
Varela, I., 270–272
Vassilev, A., 108–109
Vastag, L., 140–141
Vaughan, R.M., 32–50, 54, 63–64, 75
Vaziri, H., 192
Vedadi, M., 81, 84–87, 91–92
Vempati, S., 248–249
Venneti, S., 106–107
Venturini, E., 378
Verdin, E., 134, 188, 194–196, 312
Verlaan-de Vries, M., 153
Verma, S.K., 86–87, 96
Vernot, B., 376–379
Vertegaal, A.C., 153
Vicidomini, C., 249–250
Vickers, R., 96–98
Vidgren, J., 264–266
Vieira, F.Q., 246–247
Vierstra, J., 371, 376–379
Villanova, L., 194
Villeneuve, L.M., 246, 296

Vincent, P.C., 246–247
Viny, A.D., 370
Visel, A., 376–377
Vizioli, M.G., 356, 358–363
Vogelstein, B., 246–247
Vollor, L., 204
Vought, V.E., 291–292
Vousden, K.H., 313–314
Vredeveld, L.C., 356

W

Wagner, E.J., 248–249
Wagner, E.K., 54
Wagner, S.A., 81
Wagner, U., 377
Walker, J.M., 128–129
Walsh, C.P., 286
Walter, J., 368
Walters, R.A., 335
Wan, M., 371–372
Wang, C., 120–121
Wang, C.H., 290–291
Wang, D., 106–107
Wang, F., 257
Wang, G.G., 313–314
Wang, H., 257–259, 371
Wang, J., 170, 193, 203
Wang, L., 106–111, 192–193, 247–248, 257–259
Wang, M., 196, 215
Wang, M.W., 272–273, 289–290
Wang, Q., 248–249
Wang, S., 34–35
Wang, X., 9–10, 15, 249, 356
Wang, Y., 83, 170, 246–247
Wang, Z., 192–193, 248–249, 341, 356–357, 370
Warbrick, E., 172
Ward, S.G., 276–279, 286
Warnecke, P., 246–247
Warren, M.E., 46–48
Warth, A., 192
Washburn, M.P., 337
Washizuka, K., 83–84
Wasney, G.A., 289, 296–297
Watt, F.M., 249
Wauters, E., 192
Weber, M., 368

Wei, H., 247–249, 268
Wei, S., 106–107, 151–153, 156–159
Wei, T.Y., 247–248
Weichert, W., 192
Weinberg, R.A., 192
Weiner, A., 370
Weinert, B.T., 81, 134
Weinhold, E., 249–250, 273–275
Weinstein, E.J., 289, 296–297
Weirich, S., 49, 54
Weiss, T., 42–43
Weisshart, K., 234–236
Wellen, K.E., 313–314, 317–318
Weller, C.E., 150–163
Welstead, G.G., 376–377
Wenderski, W., 333–334, 341, 344–346, 352
Werner, T., 82
Wernimont, A.K., 86–87, 91–92, 96, 281–283, 286
West, L.E., 42–43
Westman, E.A., 214–215
Westphal, C.H., 196–197, 214–215, 217–219
Westphall, M.S., 340–341
Westwood, N.J., 204
Whetstine, J.R., 46, 335–336
White, C.L., 159–160
Wickramasinghe, P., 168
Widom, J., 160
Wiesner, J.B., 257
Wiessler, M., 261–264
Wigle, T.J., 84–87, 279–281, 286
Wijshake, T., 356
Wilczek, C., 259–261
Wilkinson, A.W., 81, 249
Williams, L., 168
Williams-Ashman, H.G., 253
Williamson, B., 339–340
Williamson, D.H., 187–188
Willis, I.M., 214–215
Wilson, J.R., 249
Wingreen, N.S., 126–127
Winkler, F.K., 264–268, 286–287
Winston, S., 337
Winter, D., 377–378
Wittig, D., 368
Wojcik, E.J., 92–93

Wolberger, C., 184–186
Wold, B., 356–357, 369, 376
Wolters, D.A., 337
Wong, L.S., 249–250
Wood, J.G., 214–219, 234–236
Woodcock, C.L., 151–153, 156–159
Wooderchak, W.L., 290–291
Woodman, B., 204
Woodroofe, C.C., 120–121
Woods, A., 168–169
Wooten, J., 250
Workman, J.L., 106–107, 160–161
Workman, P., 82
Wozniak, D.F., 214
Wright, A.T., 106–107
Wright, T.J., 248–249
Wu, B., 371–372
Wu, C., 192
Wu, C.C., 339–340
Wu, G., 270–272
Wu, H., 42–43, 270–272, 286–287, 295–296, 295f
Wu, J., 192, 287, 289, 296–297
Wu, Q.X., 246–247, 249
Wu, W., 249
Wu, X., 20–24, 126
Wu, Y., 341
Wu, Y.Y., 246, 296
Wynn, D.P., 290–291
Wysocka, J., 376–377
Wysoker, A., 375–376

X

Xia, L., 54, 257–259
Xiang, Z., 204
Xiao, B., 249
Xiao, C., 193
Xiao, Y., 249
Xie, N., 287, 289, 296–297
Xie, Z., 54, 341
Xing, L., 259–261
Xiong, G., 249
Xiong, J., 42–43
Xiong, Y., 86–87
Xu, B., 126–127, 192
Xu, H., 168
Xu, N., 193, 196–197

Xu, R.M., 184–186, 196, 215, 275–276
Xu, X., 289–290
Xynopoulos, D., 273–275

Y

Yadav, N., 89, 289, 296–297
Yamane, K., 46
Yamatsuta, K., 202–203
Yan, L., 203
Yanagida, M., 168–170
Yancheva, N.Y., 336
Yandell, D.W., 246–247
Yang, B., 153
Yang, C., 289, 296–297
Yang, J.L., 217–219
Yang, J.S., 126, 341
Yang, M.H., 193
Yang, N., 196, 215
Yang, R., 151–152, 156–157
Yang, X., 86
Yang, X.J., 184–186
Yang, Y., 247–248
Yao, Y., 275–276, 281, 286
Yates, J.R., 337, 339–340
Ye, X., 356
Yen, A., 377
Yen, R.W.C., 246–247
Yesselman, J.D., 270–272
Yi Goh, X., 248–249
Yi, J.S., 289–290
Yokoyama, Y., 168–169
Yonemura, K., 357–358
Yonezawa, M., 42–43
York, A., 42–43
Yoshida, K., 89
Yoshida, M., 189–190
Yoshikawa, K., 96–98
Yoshimatsu, M., 247–248
Yoshino, J., 189
You, L., 249
Young, N.L., 20–24
Young, R.A., 320
Yu, H.Q., 291–292
Yu, W., 86–87, 91–92, 96, 257, 281–283, 286, 296
Yu, X., 54
Yu, Y., 42–43

Yuan, J., 143
Yuan, Z.-F., 9–10, 15
Yusufzai, T., 118–119

Z

Zaba, L.C., 371–372, 377–378
Zaleski, C., 375–376
Zamdborg, L., 248–249
Zang, C., 371
Zang, W., 203
Zaretsky, I., 370
Zee, B.M., 4–5, 42–43, 126–127, 248–249, 314–315, 317, 341
Zehnder, L., 257–259, 286
Zeissler, U., 42–43
Zeng, H., 81, 270–272, 286–287, 295–296, 295f
Zeng, L., 106
Zeng, T., 356–357
Zhan, L., 261–264, 286–287
Zhan, X., 192
Zhang, C., 246–247
Zhang, D., 46, 106–107
Zhang, J.Q., 42–43, 83–84, 246–247, 249–250, 270–272
Zhang, L., 126
Zhang, M.Q., 356
Zhang, Q., 42–43, 187–188
Zhang, R., 356
Zhang, T., 376–377
Zhang, X., 42–46, 49, 151–152, 156–157, 270–272, 291–292
Zhang, X.D., 193
Zhang, Y., 33f, 46–48, 248–249, 275–276, 289–290, 296–297, 326, 371–372, 375–376
Zhao, H., 246–247
Zhao, K., 108–109, 184–186, 326, 356–357
Zhao, P., 34–35
Zhao, Q., 259–261
Zhao, R., 25
Zhao, W., 192
Zhao, Y., 32, 192, 356–357
Zhao, Z.J., 246–247
Zheng, H., 42–43
Zheng, L., 192
Zheng, S., 247–248
Zheng, W., 203
Zheng, W.H., 270–272, 286–287, 291–292, 295–296, 295f
Zheng, Y., 110–111, 126–127
Zheng, Y.G., 151–152, 163, 250, 287, 289, 296–297
Zheng, Y.-J., 264–266
Zhengtian, Y., 86
Zhou, F., 54
Zhou, J., 83, 249
Zhou, L., 249
Zhou, M.M., 106
Zhou, T., 371
Zhou, V.W., 366–369
Zhou, Z., 170
Zhou, Z.S., 290–291
Zhu, B., 42–43
Zhu, H., 54
Zhu, J., 81, 370
Zhu, L., 193
Zhu, L.J., 326
Zhu, X., 54, 97
Ziegler, M., 108–109
Zielenkiewicz, P., 261–264
Zovkic, I.B., 342
Zurcher, G., 264–268, 286–287
Zurita-Lopez, C.I., 259–261
Zweidler, A., 336
Zweygarth, E., 253

SUBJECT INDEX

Note: Page numbers followed by "*f*" indicate figures, and "*t*" indicate tables.

A

Accelerator mass spectrometry (AMS)
 description, 342–345
 nucleosomal turnover in human brain with, 346, 347*f*
Acetylated histone H3 analogs
 fractions, 157–159
 generation, 157–159, 158*f*
 pET3a vector, 157–159
 semisynthesis design, 156–157
Acetylation
 dysregulation, 192
 histone
  data analysis and kinetic modeling, 140–145
  labeling and methods, 128–132
  sample analysis, 132–140
 lysine Acetyltransferase, 106
Acetyl-SILAC, 317–318. *See also* Histone modifications
Activity-based capture
 applications, 120–121
 enrichment, 115–116
 materials, 114–115
Activity-based protein profiling (ABPP), 106–107
Allosteric activation, 215
AMS. *See* Accelerator mass spectrometry (AMS)
Antibodies
 KMT assays with G9a
  equipment and reagents, 38
  procedures, 39
 methylation-specific, 83–84
 PTMs, 36–37
Antibody bead preparation, 359
Antigen-antibody reactions, 176
Anti-NPC protein antibody, 171*f*
Arginine methylation, 83–84
ArrayNinja, 34–35, 55–56
 benchmarking against ImageQuant TL
  homogeneous noise, 69–70
 inhomogeneous noise, 70–71
 local noise correction, 71–72
 screenshot of, 69, 70*f*
 data analysis features
  microarray image preparation, 65–66
  step-by-step quantification, 66, 68*f*
 default quantification, 72
 development, 75–76
 features, 75
 limitations and assumptions, 74–75
 local noise thresholding, 72–73
 Morph checkbox, 73–74
 nonlocal noise thresholding, 72
 phpMyAdmin, 58, 58*f*
 planning tool
  input fields, 63
  screenshots, 62*f*
  starting, 61
 requirement, 56–57
 source plates, 56
  population of, 64
  slide layout and, 64
 starting, 57
 tables
  empty table template, 60
  printer settings table, 60
  source plate, importing, 61
  source plate table, 59–60
  SubstrateTable, 59
  table management interface, 57, 58*f*
  values, importing, 60
 variables, 56
 variegated spot morphology, 73–74, 73*f*
 VirtualBox, 57
Assisted allosteric activation, 215
ATAC-seq, 371–372, 372*f*

B

Benzonase, 357–358
 digestion, 361*f*, 362–363
 endonuclease, 357–358

Biology
  chemical probes in, 80–81
  sirtuin, 200–201
Biomarker assay, 82–83
  exogenous PMT cell-based, 87–89
  for PMTs, 84–86
  for PRDM9, 89–91, 90f
Bomb pulse labeling
  analytical considerations, 343f, 345–347
  coupled to accelerator mass spectrometry, 342–345, 343f
Bottom-up mass spectrometry
  data analysis
    abundance calculation, 15–18
    DDA and DIA experiments, 15
    histones, 14–15
    software-based peak area extraction, 15–18
  derivatization and digestion, 10–11
  desalting, 11–12
  histone proteins, 9–10
  materials and buffer recipes, 10
  online RP-HPLC and acquisition, 12–14, 14t
  trypsin, 9–10
  workflows, 5f
Brain
  current proteomic methods, 337–342
  histone turnover in, 334–336
  human postmortem, 342–347
Bridged-urea STAC preparation, 229–234, 229f

C

Cell-active chemical inhibitors, 80–81
Cell apoptosis, in GBM cell, 98–99, 98f
Cell-based assay, 80–81
  biomarker, exogenous PMT, 87–89
  epigenetics, 91–92
  overexpression, 91–92
Cell culture, 359–362
Cell lysis, 88
Cellular inhibition, of PMTs, 83
Cellular responses, to chemical probes, 93–94, 94f
Cellular senescence, 356
Cellular Thermal Stability Assays (CETSA), 82

Centromeres, visualizing clusters of, 170, 170f
CETSA. See Cellular Thermal Stability Assays (CETSA)
Chemical acetylation-labeling reagents, 134–135, 136t
Chemical inhibitors, cell-active, 80–81
Chemical probes
  active in cells, 82–83
  assay readout choice, 83–84
  in biology, 80–81
  in cell-based experiments, 81–92
  cellular responses to, 93–94, 94f
  high-quality, 81
  phenotypic assays using, 93–94
  requirements for, 81–82
ChIP. See Chromatin immunoprecipitation (ChIP)
Chromatin, 332–333
  isolation from neurons, 349
  preparation, 348–352
Chromatin biochemistry tools, 36–37
Chromatin biology, 377–378
Chromatin dynamics, 341
Chromatin immunoprecipitation (ChIP), 83–84
  conventional, 357–358
  protocol, 356–357
Chromatin immunoprecipitation sequencing (ChIP-seq), 356–357, 359–362. See also Histone modifications
  antibody choice, 320–321
  bead preparation, 322
  cell harvest and cross-linking, 321
  cell lysis, 322
  computational pipeline, 326–327
  DNA purification, 324–325
  elution, 324
  epigenetic landscape by, 320
  library preparation, 325–326
  limitations, 369–370
  reverse cross-linking, 324
  setting Up the IP, 323
  washes, 323–324
Chromatin interaction analysis by paired-end tag sequencing (ChIA-PET), 168

Chromosomal DNA, 368–369
Chromosome conformation capture (3C), 168
Chronic myeloid leukemia (CML), 192–193
Collision-induced dissociation (CID), 6
Co-overexpression assay, enzyme/substrate, 89–91
Cyclooxygenase IV (CoxIV), 118–119
Cysteines, 335

D

Data-dependent acquisition (DDA), 12, 15
Data-independent acquisition (DIA), 12–15
Deacetylation
  on fluorophore-tagged peptides, 214–215
  of native peptides, 215
  of natural amino acid, 215
  of peptide, 236f
  reaction, 236–240
Demethylases, 33f
Designing and synthesizing inhibitors
  pan-inhibitors, 250–253, 252f
  target-selective inhibitors
    5′-alkoxy SAM analogs, 264–266
    5′-alkyl/alkenyl SAM analogs, 266–268
    5′-alkylthio SAM analogs, 254–264
    5′-aminoalkyl SAM analogs, 268–273
    5′-amino SAM analogs, 273–283
    complementary approaches, 253–254
    SAM, 253–254
DMSO, 92–93
DNA, 4
  methylation, 367–368
  primer synthesis and gene sequencing, 152–153
  purification, 324–325
DNA methyltransferases (DNMTs), 246–247
DNA–protein complex, 332–333
DNase-seq, 371–372, 372f

E

Electron transfer dissociation (ETD), 6
Electrospray ionization mass spectrometry (ESI-MS), 152–153
Enhancers, 376
Enzyme overexpression assay, 89

Enzyme specificity profiling
  G9a KMT activity, 42–46
  JMJD2A KDM activity, 46–49
Enzyme/substrate Co-overexpression assay, 89–91
Epigenetic biomarker changes, 95–96
Epigenetic chemical probes, 99–100
Epigenetic enzymes, cellular substrates of, 83
Epigenetic regulators, 32
Epigenetics, 81
  cell-based assays, 91–92
  defined, 81
  targets, 96
Epigenome, 366–367, 367f
EpiProfile, 15–18, 17f
Erasers, 33f
Eukaryotic gene transcription, 332
Eukaryotic genomes, 32
Evaluating inhibitors
  assay formats, 283, 285f
  FP-based detection methods, 289–290
  MS-based detection methods, 291–292
  radiometric assays
    balanced conditions, 283–285
    filter paper-based scintillation assay, 286–287
    methyltransferases and different assay formats, 285f
    scintillation proximity assay, 287–288
  radiometric filter paper assay
    materials, 292
    potency of inhibitors in vitro, 294–295
    preliminary test, 293
    steady-state kinetic parameters measurement, 293–294
    SAH formation, 290–291
Exogenous PMT cell-based biomarker assay, 87–89
Extracted ion chromatogram (XIC), 12–14

F

FASTQ, 374
Fission yeast cells, fixation of, 175
Fixation, 176
Fluorophore-tagged peptides, deacetylation on, 214–215

G

G9a/GLP inhibitor effects, 92f
G9a, KMT assays with
  equipment and reagents, 38
    antibody detection, 38
    radioisotope detection, 38
  high-throughput profiling, 42–46
  optimization, 34–35, 35f
  procedure, 38
    antibody detection, 39
    radioisotope detection, 39
  solution-based assays, 40
  substrate specificity, 44f
GenePix Associated List (GAL) file, 54–56
Gene positioning, 170–171, 171f
Glioblastoma (GBM) stem cell, 96–99, 98f

H

H3.3
  associated nucleosomal turnover, 334f
  histone variant proteins, 333–334
  peptide labeling, 343f
  synthesis and accumulation, 347f
HDACs. See Histone deacetylases (HDACs)
Hi-C. See Chromosome conformation capture (3C)
High-energy collision dissociation (HCD), 6
HIRA, 356–358
His-tagged SIRT1, 216, 234–236
Histone
  extraction from cells
    acid extraction, 8–9
    cell harvest, 7
    materials and buffer recipes, 6–7
    nuclei isolation, 7–8
  in human postmortem brain, bomb pulse labeling
    analytical considerations, 343f, 345–347
    coupled to accelerator mass spectrometry, 342–345, 343f
  octamers, 159–160
  offline fractionation
    histone variant purification, 18–19, 19f
    materials and buffer recipes, 18
  proteins, 4
  PTMs, 4–5, 150
  sumoylation, 153

Histone acetylation
  data analysis and kinetic modeling
    small-molecule data, 140–141
    stoichiometry, 141–143
    turnover rate, 143–145
  HDAC, 126–127
  mass spectrometry, 126
  PTMs, 126–127
  sample analysis
    digesting histones into single amino acids, 133
    histone digest by HPLC–MS, 133–134
    metabolite label analysis, 132–133
    site-specific acetylation-labeling kinetics, 139–140
    stoichiometry, 134–139
  sample preparation
    extracting histones, 131–132
    general experimental design, 128–129, 129f
    isotopically labeled precursors, 129–131
    metabolites extract, 131
Histone deacetylases (HDACs), 126–127
Histone-DNA complexes
  characterization of, 32–34
  histone code, 32
  posttranslational modifications on, 32
  regulation of lysine methylation, 33f
Histone modifications
  by chromatin immunoprecipitation and sequencing, 368–369
  genome-wide mapping, ChIP
    bead preparation, 322
    cell harvest and cross-linking, 321
    cell lysis, 322
    computational pipeline, 326–327
    DNA purification, 324–325
    elution, 324
    epigenomic analyses by, 320
    library preparation, 325–326
    materials and buffer recipes, 320–321
    reverse cross-linking, 324
    setting Up the IP, 323
    washes, 323–324
  turnover dynamics analysis, SILAC, 312–313
    acetyl-SILAC, 317–318
    methylation, 315–317

methyl-SILAC, 315
SILAC-mass spec methodology, 314, 318–320
Histone peptides, 351
Histone/protein mark detection, 86–87
Histone proteins, 348f
Histone PTM analysis
  EpiProfile, 15–18, 17f
  MS, 5–6
Histone PTM-specific antibodies
  antibodies, 36–37
  custom print formats design, 34–35
  KDM assays with JMJD2A, 40–42, 46–49
  KMT assays with G9a, 38–40, 42–46
  radioisotopes, 36
Histone turnover, 333–334, 334f
  in brain, 334–336
    acquisition types, 340–341
    label free vs. stable isotope incorporation, 339–340
    mass spectrometry study, 337–338, 341–342
    sample preparation considerations, 338–339
  mass spectrometry analysis of, 348–352
Histone variants
  mass spectrometry analysis of, 348–352
  proteins, 333
HPLC method, 22
Human brain, nucleosomal turnover in, 346, 347f
Human postmortem brain, bomb pulse labeling
  analytical considerations, 343f, 345–347
  coupled to accelerator mass spectrometry, 342–345, 343f
Hybridization, 177

I

IF-FISH approach
  analysis, 169f
  centromeres, 170, 170f
  gene associations, detection and confirmation, 168–169
  gene positioning, to nuclear architecture, 170–171, 171f
  protocol
    antigen-antibody reactions, 176
    cells preparation for microscopy, 177
    fission yeast cells, 175
    fixation, 176
    fluorescent labeling of templates, 174–175
    hybridization, 177
    permeabilization of cells, 175–176
    preparation, 174
  supply
    buffers, 173
    equipment, 171–172
    materials, 172–173
  telomeres, 170, 170f
ImageMagick, 65
ImageQuant TL, ArrayNinja
  benchmarking
    homogeneous noise, 69–70
    inhomogeneous noise, 70–71
    local noise correction, 71–72
    screenshot of, 69, 70f
Imidazo[1,2-b]thiazole STAC preparation, 219–221, 219f
Imidazo[4,5-c]pyridine STAC preparation, 227–228, 227f
Immune responses
  ChIP-seq limitations, 369–370
  chromatin immunoprecipitation and sequencing, 368–369
  DNA methylation mapping, 367–368
  DNase-seq and ATAC-seq, 371–372, 372f
  enhancer landscape
    active enhancers, 376
    definition, 376
    intrinsic and extrinsic signals, roles, 378–379
    Pre-NGS studies, 376
  epigenome, 366–367, 367f
  gene regulation, 377–378
  MNase-seq, 370, 372f
  multitiered design, 366
  NGS data analysis
    ABI SOLiD, 374
    computational analysis, 373–374
    De novo assembly, 375–376

Immune responses (*Continued*)
 DNA fragments, 372–373
 FASTQ, 374
 "peak calling" software, 375–376
 Phred scores, 374
Immunoblotting, 88
Inhibitor enabled discovery, 92
 inhibitor handling and inhibitor libraries, 92–93
 on-target phenotypic effects, 94–99
 phenotypic assays using chemical probes, 93–94
Inhibitors
 cell-active chemical, 80–81
 chemical, cell-active, 80–81
 effects, 92*f*
 sirtuins
  discover, 204
  identification, 197–200
  indole-containing compounds, 202–203
  mechanism-based deacetylase, 203
  N-ε-trifluoroacetyl-lysine pseudosubstrate compound, 203
  nicotinamide, 201–202, 202*f*, 214–215
  splitomicin, 200–201
  synthetic small-molecule, 203–204
  yeast and mammalian, 201–202

J

JMJD2A, KDM assays with
 equipment and reagents, 40
 high-throughput profiling, 46–49
 procedure, 41
 substrate specificity, 47*f*

K

KDM assays with JMJD2A
 equipment and reagents, 40
 high-throughput profiling, 46–49
 procedure, 41
 substrate specificity, 47*f*
KMT assays with G9a
 equipment and reagents, 38
  antibody detection, 38
  radioisotope detection, 38
 high-throughput profiling, 42–46
 optimization, 34–35, 35*f*
 procedure, 38
  antibody detection, 39
  radioisotope detection, 39
 solution-based assays, 40
 substrate specificity, 44*f*

L

LC–MS/MS
 analysis for histone acetylation-labeling kinetics, 140
 data analysis, 139
 sample preparation for, 139–140
Lysine acetyltransferase (KAT)
 ABPP, 106–107
 activity-based capture
  applications, 120–121
  enrichment, 115–116
  materials, 114–115
 central role, 106
 competitive chemical proteomic profiling, 108–109, 110*f*
 future applications and directions, 120–121
 KAT–ligand interactions
  competitive chemoproteomic analyses, 118–119
  competitive profiling, 114
  CoxIV immunoblot, 118–119
  elution and Western blot, 117
  enrichment, 115–116
  immunoblotting methods, 117
  materials, 114–115
  LC-MS/MS, 118
 proteomes preparation
  cell lysate, 111
  materials, 111
 quantitative gel densitometry, 118–119
 regulation, 106–107, 107*f*
 study, 106–107, 108*f*
 synthesis of capture probes
  alkyne-modified bisubstrate inhibitors, 111–112
  biotinylated bisubstrate inhibitors, 110–113
  materials, 111
Lysine methylation, 82–84
Lysis, 359–362

Subject Index

M

Mass spectrometry (MS)
  acquisition method, 22
  approaches, 5–6, 5f
  bottom-up, 6
  peptide cleanup prior analysis
    desalting, 138–139
    materials, 137
    solvents, 137
  top-down, 5–6
Metabolic disorders, 190–191
Metabolic isotope labeling, 340
Methylation, 83–85
  arginine, 83–84
  lysine, 82–84
  steady-state levels of, 91–92
Methylation-specific antibodies, 83–84
Methyl-SILAC. See also Histone modifications
  heavy-labeled medium preparation, 315–316
  histone methylation, 315, 316f
  mammalian cell culture, 316–317
  peptide quantification and kinetics modeling, 317
Methylthioadenosine (MTA), 253
Methyltransferases, 33f, 36
  COMT, 249
  designing and synthesizing inhibitors
    pan-inhibitors, 250–253, 252f
    target-selective inhibitors, 253–283, 284f
  DNMTs, 246–247
  evaluating inhibitors
    assay formats, 283, 285f
    FP-based detection methods, 289–290
    MS-based detection methods, 291–292
    radiometric assays, 283–288, 285f
    radiometric filter paper assay, 288f, 292–295, 295f
    SAH formation, 290–291
  NSD2, 248–249
  PKMT, 248–249
  PRMT7, 247–248
  transmethylation reactions catalyzed by, 249, 251f
Microarrays, 54–55

Micrococcal nuclease (MNase) digestion, 357–358
Middle-down mass spectrometry
  data analysis
    identification and quantification, 22–24
    interplay score, 23
    variable modifications, 23
  digestion, 21
  drawback, 20
  endoproteinase GluC, 19–20
  histone
    PTMs, 19–20
    tail peptides, 20
  materials and buffer recipes, 20–21
  WCX-HILIC, 21–22
  workflows, 5f
MNase-seq, 370, 372f
Mononucleosomes (MNs)
  147 bp 601 DNA generation, 160
  177 bp repeat 601 DNA preparation, 161
  generation of, 160f
  histone octamers, 159–160
  nucleosome arrays, generation, 161–163, 162f
  reconstitution, 160–161

N

NAD^+ levels, 189
NAM. See Nicotinamide (NAM)
National Cancer Institute's Developmental Therapeutics Program's, 199
Native chemical ligation (NCL), 150–151
Neuroepigenetics, 332–333, 342, 347
Neuronal histone peptides, mass spectrometry analysis, 351
Neurons
  acid extraction of histones from, 350
  chromatin isolation from, 349
Neuroplasticity, 332–334, 334f
NGS technologies, 376
Nicotinamide (NAM)
  production, 236–239, 236f
  standard curve, 237
Nuclear architecture, gene positioning to, 170–171, 171f
Nuclear membrane
  anti-NPC protein antibody, 171f
  sad1, 170

Nucleolus, 170–171
Nucleosomal turnover, 346, 347f
Nucleosome arrays, generation, 161–163, 162f

O

Offline fractionation, histone species
  histone variant purification, 18–19, 19f
  materials and buffer recipes, 18
On-target phenotypic effects, 94–99
Open-air bomb testing, 345
Overexpression assay
  cell-based, 91–92
  enzyme, 89

P

Pan-sirtuin activators, 214–215
Phenotypic assay, 93–94
PMTs. *See* Protein methyltransferases (PMTs)
PNC1-OPT assay
  assay of SIRT1 activators using, 236–239, 236f
  STACs using, 216–217
Polycomb repressor complex 2 (PRC2), 248–249
Posttranslational modifications (PTMs), 4, 32, 332–333
  DNA topology and chromatin structure, 246
  histone, 150
  histone acetylation, 126–127
PRDM9, biomarker assay for, 89–91, 90f
Protein
  acylation, 184
  chemical steps, 184–186, 185f
  deacylases, 184–186
  histone variant, 333
  synthesis, 335
Protein arginine methyltransferases (PRMTs), 247–248
Protein lysine methyltransferases (PKMTs), 247–249
Protein methyltransferases (PMTs), 81, 86, 247–248
  biomarker assays for, 84–91
  cell-based biomarker assay, 87–89
  cellular inhibition of, 83

Proteomic methods, histone turnover in brain
  acquisition types, 340–341
  label free *vs.* stable isotope incorporation, 339–340
  mass spectrometry study, 337–338, 341–342
  sample preparation considerations, 338–339
PTMs. *See* Posttranslational modifications (PTMs)

Q

QuikChange Mutagenesis, 157–159

R

Radioactive labeling, 336
Radioactive tracers, 335–336
Radioisotopes
  KMT assays with G9a
    equipment and reagents, 38
    procedures, 39
  PTMs, 36
RapidFire mass spectrometry assay, 217
RapidFire O-Ac-ADPR detection assay, 239–240
Recombinant His-tagged SIRT1, 216, 234–236
Regulatory elements, 376
Response biomarker, 94–95
Reversed-phase HPLC (RP-HPLC), 152–153

S

S-adenosyl-L-methionine (SAM), 246–247
SAH, 253
SAHF. *See* Senescence-associated heterochromatic foci (SAHF)
Scintillation proximity assay (SPA), 195
SDS-PAGE, 88
Senescence
  cellular, 356
  chromatin changes in, 356–357
  establishment of, 356
  long-term, 356
Senescence-associated heterochromatic foci (SAHF), 356

Signal-dependent transcription factors (SDTFs), 379
Signal transducer and activator of transcription (STAT), 379
SILAC. See Stable isotope labeling using amino acids in cell culture (SILAC)
SILAC-mass spec methodology, 314, 318–320
(+)-Sinefungin (SNF), 253
SIRT1, 192–193
　activating compounds, synthesis, 215–234
　activators
　　RapidFire mass spectrometry assay, 217
　　using PNC1-OPT assay, 216–217, 236f
　assay of, 216–217
　deregulation of, 214
　expression and purification of recombinant His-tagged, 216, 234–236
　overexpression of, 214
SIRT1 activators (STACs), 214–215
　assay of, 237, 240
　using PNC1-OPT assay, 216–217, 236–239
　using RapidFire mass spectrometry assay, 217
　using RapidFire O-Ac-ADPR detection assay, 239–240
Sirtuins
　activity assays
　　acetylated peptide, 194–195
　　charcoalbinding assay, 195
　　enzyme-linked deacetylase assays, 196
　　fluorescence quenching, 196–197
　　immiscible organic phase, 194–195
　　majority, 197
　　nicotinamide, 197
　　SPA, 195
　and cancer, 191–194
　cellular NAD^+ levels regulation, 189
　functions, 189–190
　inhibitors
　　discover, 204
　　identification, 197–200
　　indole-containing compounds, 202–203
　　mechanism-based deacetylase, 203
　　N-ε-trifluoroacetyl-lysine pseudosubstrate compound, 203
　　nicotinamide, 201–202, 202f, 214–215
　　splitomicin, 200–201
　　synthetic small-molecule, 203–204
　　yeast and mammalian, 201–202
　and metabolic disorders, 190–191
　and metabolism, 187–188
S-labeled methionines, 335
Small ubiquitin-like modifier protein (SUMO), 150–151, 155–156
S-methionine, 335
Sonication, 356–357
Splitomicin, 197–198, 198f, 200–201
Stable isotope labeling using amino acids in cell culture (SILAC), 312–313, 337, 339–342, 343f. See also Histone modifications
　acetyl-SILAC, 317–318
　vs. ChIP-seq, 313–314
　limitations, mass spec, 318–320
　methylation, 315–317
　methyl-SILAC, 315
　quantitative proteomics, 314
STACs. See SIRT1 activators (STACs)
Stoichiometry
　data analysis and kinetic modeling, 141–143
　histone acetylation, sample analysis, 134–139
Sumoylated histone H4 analogs
　generation, 156, 157f
　recombinant K12C preparation, 154–155
　recombinant SUMO-3-aminoethanethiol preparation, 155–156
　semisynthesis design, 153–154

T

"Target hunting" assays, 93–94
Target-selective inhibitors
　5′-alkoxy SAM analogs, 264–266
　5′-alkyl/alkenyl SAM analogs, 266–268
　5′-alkylthio SAM analogs, 254–264
　5′-aminoalkyl SAM analogs, 268–273
　5′-amino SAM analogs, 273–283
　complementary approaches, 253–254

Target-selective inhibitors (*Continued*)
  general synthetic strategies to prepare, 254, 255f
  SAM, 253–254
Target validation, 80–81, 98–99
Telomeres, 170, 170f
Thialysine analogs, generation, 159
Thiazolopyridine STAC preparation, 221–227, 222f
Top-down mass spectrometry
  advantage, 24–25
  data analysis, 24–28
  fractionation, 25
  materials and buffer recipes, 25
  using direct infusion, 25–26
  workflows, 5f
Trifluoroacetic acid (TFA), 152–153

W

Weak cation exchange hydrophilic interaction liquid chromatography (WCX-HILIC), 20–22
Western blot analysis, 88
Writers, 33f

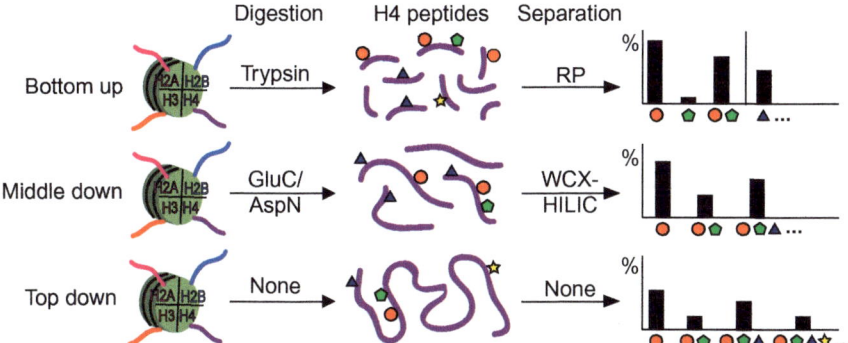

K.R. Karch et al., Fig. 1 Workflows for bottom-up, middle-down, and top-down histone PTM analysis by high-resolution tandem MS. In bottom-up MS, the relative abundances and PTM cooccurrences can be monitored for PTMs contained within a single tryptic peptide. Longer peptides are generated in middle-down MS, allowing for better connectivity than bottom-up MS. In top-down MS, full connectivity is preserved, allowing for identification of complete protein isoforms.

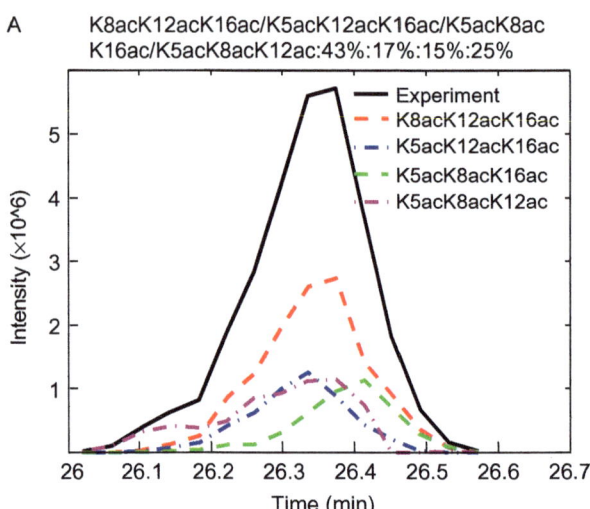

	1,Sample 1			2,Sample2		
Peptide	RT(min)	Area	Ratio	RT(min)	Area	Ratio
KSTGGKAPR(H3 9-17)						
unmod	22.57	8.58E+08	0.245241	22.22	4.21E+08	0.248842
K9me1	25.6	1.04E+09	0.296224	25.29	3.75E+08	0.221391
K9me2	14.42	5.76E+08	0.164828	15.09	3.75E+08	0.221797
K9me3	14.36	3.45E+08	0.098683	14.97	2.44E+08	0.144006
K9ac	20.83	5.84E+07	0.016689	20.85	1.67E+07	0.009887
K14ac	20.94	1.71E+08	0.0488	20.97	6.25E+07	0.036953
K9me1K14ac	24.09	1.93E+08	0.055091	23.99	6.23E+07	0.036792
K9me2K14ac	13.76	1.85E+08	0.052866	13.47	9.81E+07	0.057953
K9me3K14ac	13.58	6.87E+07	0.01964	13.29	3.52E+07	0.020818
K9acK14ac	19.26	6.78E+06	0.001938	19.27	2.64E+06	0.001561

K.R. Karch et al., Fig. 3 EpiProfile allows for quantification of histone PTMs, including isobaric species. (A) Fragment ion XICs for H4 4–17 peptide containing three acetyl groups. These fragment ion XICs are used for quantification because the parent ion XICs overlap. (B) Example of EpiProfile quantification output from "histone_ratios.xls" file. The retention time, area under the XIC and relative abundance (ratio) for each peptide in each sample are listed.

E.M. Cornett et al., Fig. 1 The dynamic regulation of lysine methylation on histone H3. Shown are major sites of methylation (me), acetylation (ac), and phosphorylation (p) on the N-terminal tail domain of histone H3. Known writers (methyltransferases; KMTs) and erasers (demethylases; KDMs) of lysine methylation are clustered by major histone substrate residue(s). Methylation products and substrates (mono-, me1; di-, me2; tri-, me3) of KMT and KDM reactions, respectively, are listed. Enzyme identification reflects both conventional and generic (Allis et al., 2007) nomenclature.

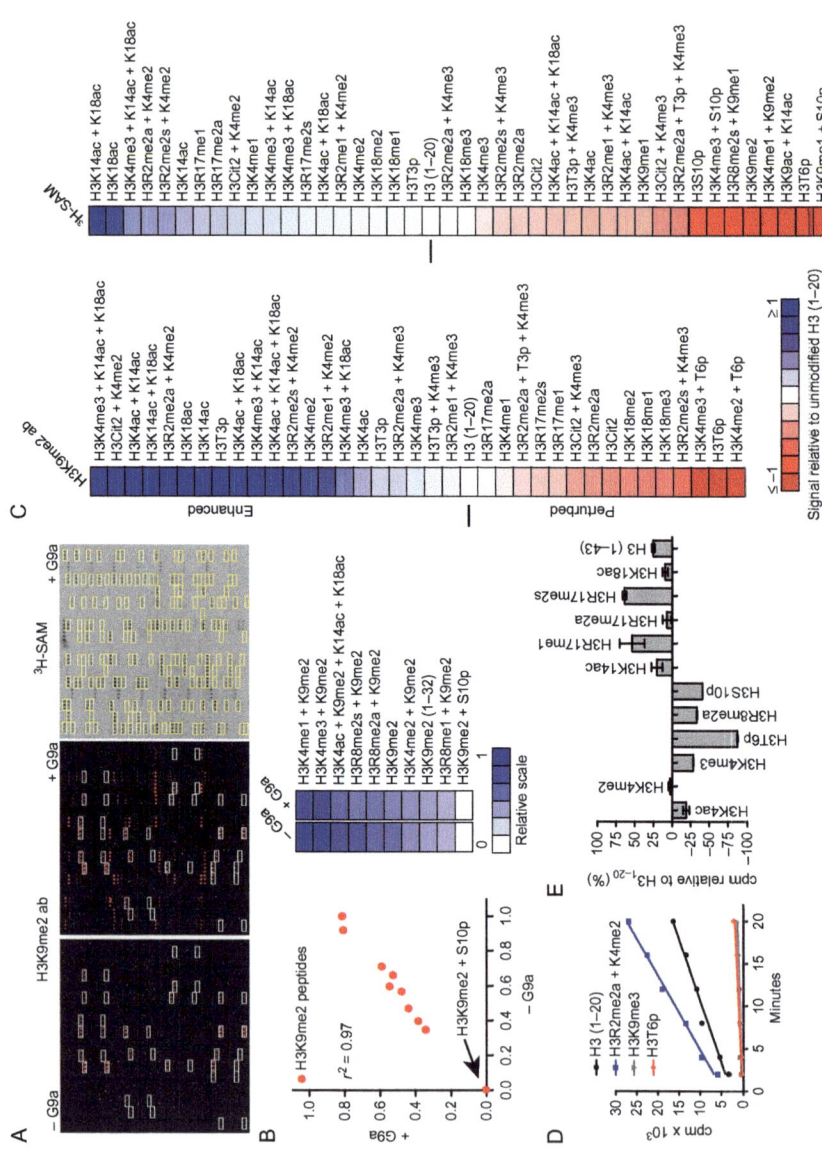

E.M. Cornett et al., Fig. 3 See legend on opposite page.

E.M. Cornett et al., Fig. 3 Analysis of G9a histone substrate specificity and the combinatorial PTM influence on KMT activity. G9a histone substrate specificity was profiled using the described antibody and radioisotopic detection strategies on large-format histone peptide microarrays. (A) Representative images of arrays detected with an H3K9me2 antibody (Abcam #1220) following hybridization in the absence or presence of 1 μM G9a for 2 h (*left*). *White boxes* demarcate peptides detected by H3K9me2 antibody in the absence of G9a. Image of autoradiography film exposed for 1.5 months following a 2-h array assay with 1 μM G9a and 5 μCi ^{3}H-SAM (*right*). For comparison, *yellow boxes* demarcate peptides detected with H3K9me2 antibody in the presence of G9a. (B) Scatter plot (*left*) and heat map (*right*) of H3K9me2 peptides detected by the above-mentioned H3K9me2 antibody in the absence or presence of G9a. Correlation coefficient (r^2) was calculated by linear regression analysis using GraphPad Prism v6. For heat maps, relative signal intensities are plotted using JavaTreeView (Saldanha, 2004) from 0 (*white*, no binding) to 1 (*blue*, strong binding). (C) Heat maps depicting the effects of combinatorial PTMs on the enzymatic activity of G9a from panel (A). Enhanced (1, *blue*) and occluded (−1, *red*) effects are depicted. Peptide signal intensities are presented relative to H3$_{(1-20)}$ (0, *white*) following detection with the above-mentioned H3K9me2 antibody (*left*) or autoradiography (*right*). (D) In-solution ^{3}H-SAM filter-binding assays monitoring G9a activity as a function of time on the listed histone peptide substrates. Data points are presented as counts per minute (cpm), each from three independent measurements. (E) In-solution ^{3}H-SAM filter-binding assays monitoring G9a activity following a 10-min incubation with the listed histone peptide substrates. Data points are presented as cpm relative to an H3$_{(1-20)}$ substrate. Error bars represent ±S.E.M. from three independent experiments.

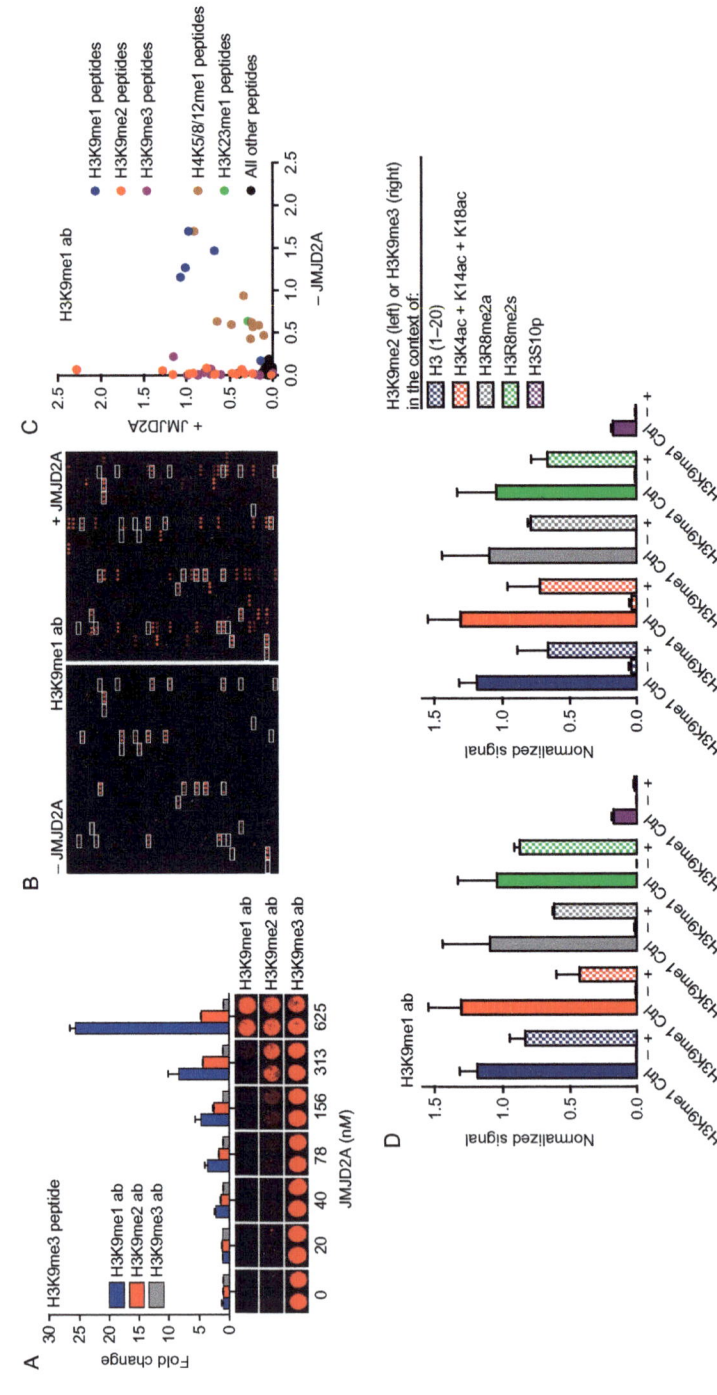

E.M. Cornett et al., Fig. 4 See legend on opposite page.

E.M. Cornett et al., **Fig. 4** Analysis of JMJD2A histone substrate specificity and the combinatorial PTM influence on KDM activity. (A) Optimization of JMJD2A demethylase activity on H3$_{(1-20)}$K9me3 peptides in a 48-well microarray (see Fig. 2A). The indicated orders of H3K9 methylation were detected by antibody hybridization (H3K9me1, EpiCypher #13-0014; H3K9me2, Active Motif #39239; H3K9me3, Active Motif #39161) following 18 h incubation with the indicated concentrations of JMJD2A. Fold change is expressed relative to signal intensity in the absence of enzyme. Error bars represent ±S.E.M. (B) Representative images of arrays detected with an H3K9me1 antibody (EpiCypher #13-0014) following hybridization in the absence or presence of 313 nM JMJD2A for 18 h. *White boxes* demarcate peptides detected by H3K9me1 antibody in the absence of JMJD2A. (C) Scatter plot of all peptides on the large array detected with the above-mentioned H3K9me1 antibody following hybridization in the absence or presence of JMJD2A. Signal intensities are normalized to IgG control spots. (D) Bar graphs depicting the effects of combinatorial PTMs on the enzymatic activity of JMJD2A from panel (B). Signal intensities from H3K9me1 antibody detection are normalized to IgG control spots. Shown are signals for the indicated peptides that also contain H3K9me2 (*left*) or H3K9me3 (*right*), both in the absence (−) and presence (+) of JMJD2A. To control for H3K9me1 antibody specificity in the context of these additional PTMs, normalized signal intensities from these peptides that also contain H3K9me1 (H3K9me1 Ctrl) are plotted as a reference.

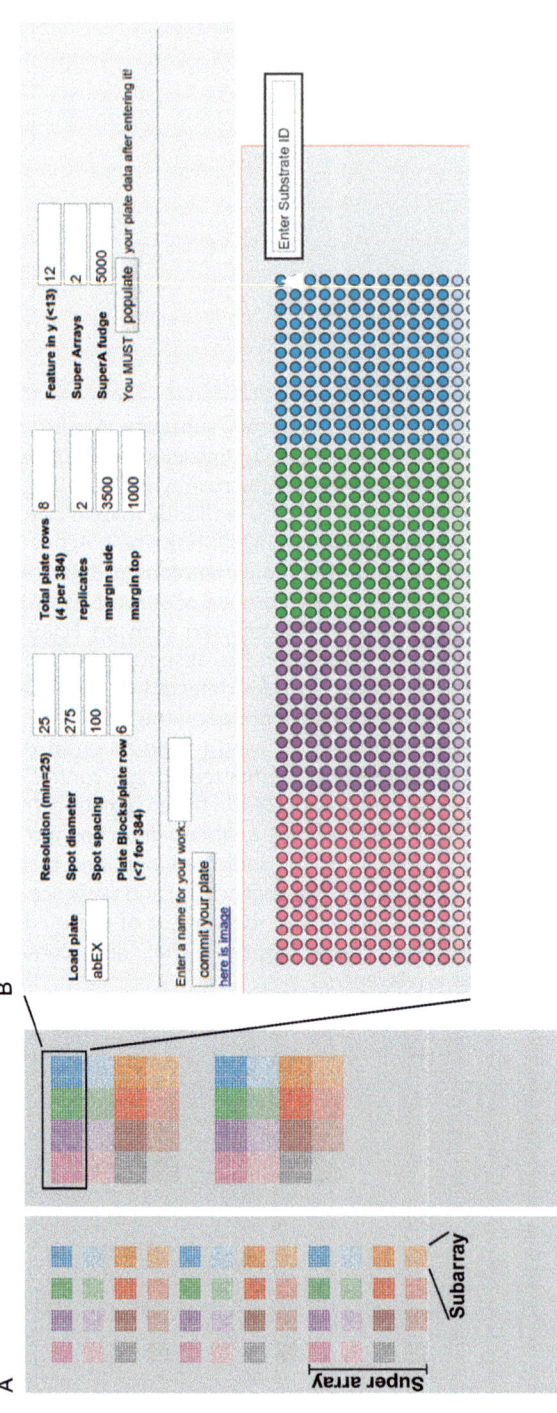

B.M. Dickson et al., Fig. 2 Screenshots of the ArrayNinja planning tool. (A) A microarray layout consisting of 48 subarrays and three super arrays (*left*), and a microarray layout of two super arrays (*right*). (B) The ArrayNinja planning interface for designing array layouts, populating source plate tables, and naming experiments.

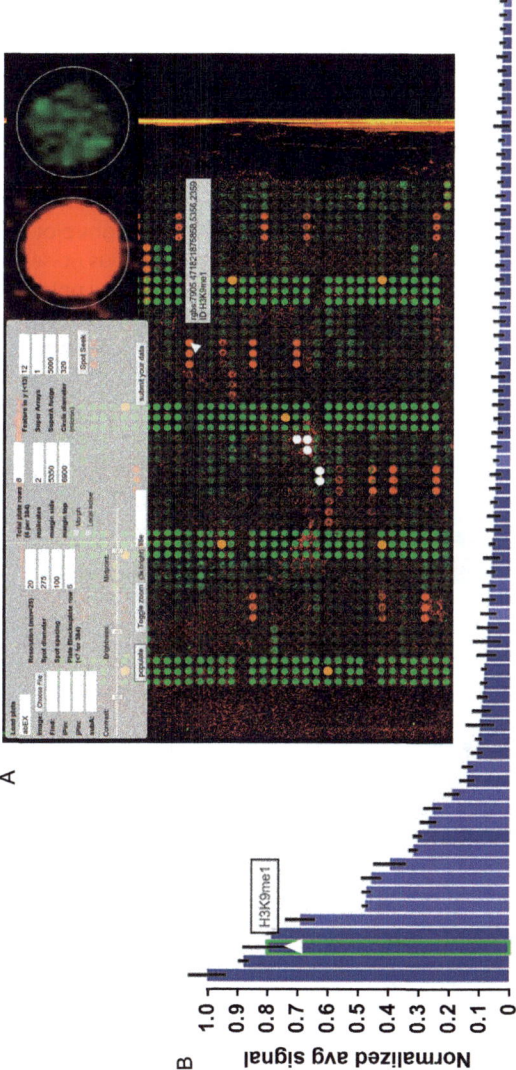

B.M. Dickson et al., Fig. 3 Data analysis with ArrayNinja. (A) Screenshot of the ArrayNinja quantification interface. Interactive bar graphs (B) and tables (C) displaying quantified array results are automatically generated after clicking the "submit your data" button.

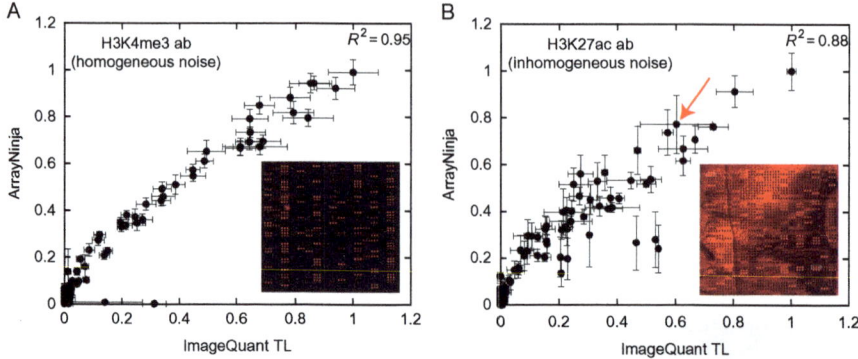

B.M. Dickson et al., Fig. 4 Benchmarking ArrayNinja against ImageQuant TL. (A) Scatter plot comparing quantification methods of arrays with homogeneous noise between ImageQuant TL using local noise thresholding and ArrayNinja using nonlocal noise thresholding. Arrays were hybridized with an H3K4me3 antibody (Epicypher #13-0004). Correlation coefficient (R^2) was calculated by linear regression using gnuplot. (B) Scatter plot comparing quantification methods of arrays with inhomogeneous noise between ImageQuant TL using local noise thresholding and ArrayNinja using local noise thresholding. Arrays were hybridized with an H3K27ac antibody (Diagenode #C15410196). *Red arrow* points to a quantized data point from average intensity of spots marked by *white boxes*. R^2 was calculated as above.

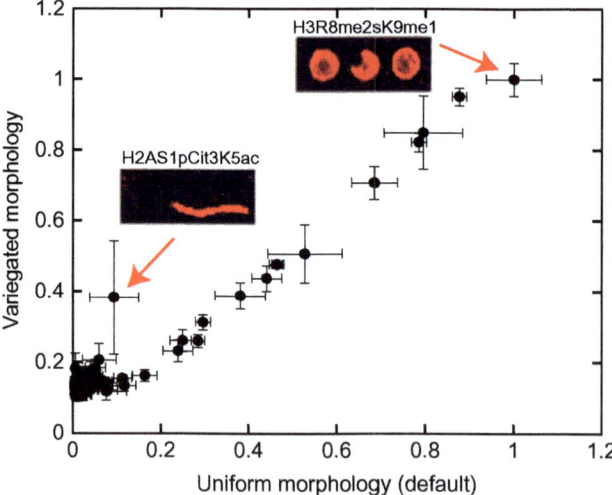

B.M. Dickson et al., Fig. 5 Comparison of whole-spot and variegated quantification methods with ArrayNinja. Scatter plot comparison of whole-spot and variegated morphology quantification methods. Inset images illustrate how spot morphology impacts correlation between the two methods.

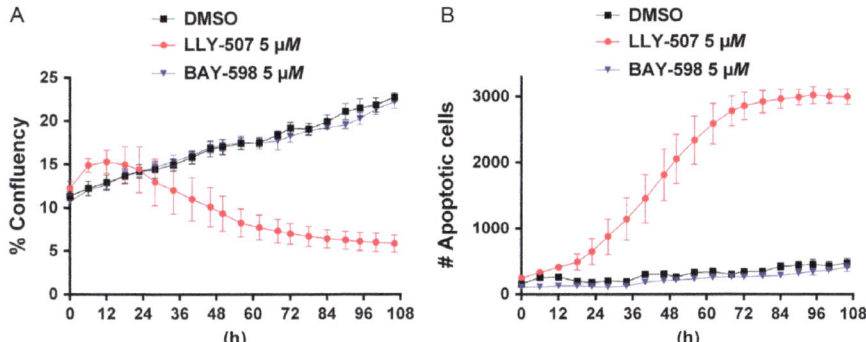

D. Barsyte-Lovejoy et al., Fig. 6 LLY-507 impairs cell growth and induces rapid cell apoptosis in GBM primary cells. (A) Cell confluence and (B) Apoptosis measurements of GBM511 cells treated with the indicated probes at 5 μM over 10 days. Note the striking differences in the phenotypes observed with the two SMYD2 probes, LLY-507 and BAY-598 probes.

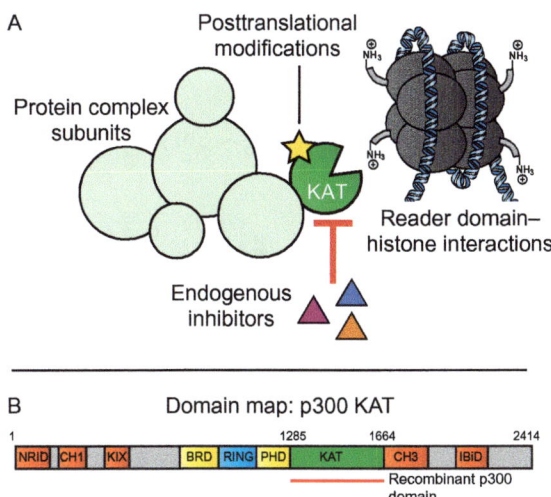

D.C. Montgomery and J.L. Meier, Fig. 1 Differences in the regulation of lysine acetyltransferase (KAT) activity in biochemical and cellular contexts. (A) Cellular acetylation is regulated by a multitude of different KAT activities, which are present as members of multiprotein complexes and regulated by expression level, substrate availability, endogenous metabolic inhibitors, and reversible posttranslational modifications. (B) KATs are large, multidomain enzymes whose native activity is challenging to reconstitute in vitro. Shown is the domain architecture of the prototypical KAT p300. Underlined in *red* is the commonly studied, commercially available p300 KAT domain.

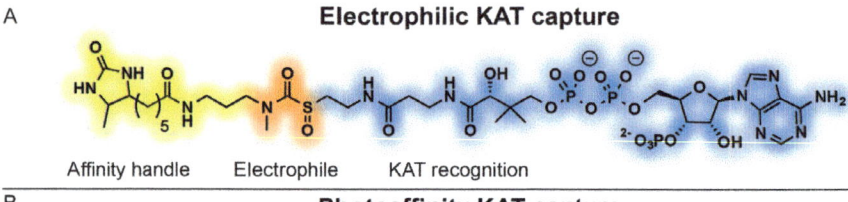

D.C. Montgomery and J.L. Meier, Fig. 2 Overview of activity-based probes for studying cellular KAT activity. (A) KATs can be labeled covalently with sulfoxide thiocarbamates, which transfer a reactive acyl group to the KAT active sites. This was the first method shown to be capable of detecting KAT active sites in cell lysates. (B) KATs can be labeled covalently using high-affinity bisubstrate inhibitors containing clickable photoaffinity groups. This approach allows stringent washing, but is limited by the low yield of the photocrosslinking step. (C) KATs can be enriched via a noncovalent capture method using bead-based capture. This approach enables low-abundance nuclear KATs to be studied and is the primary method described herein.

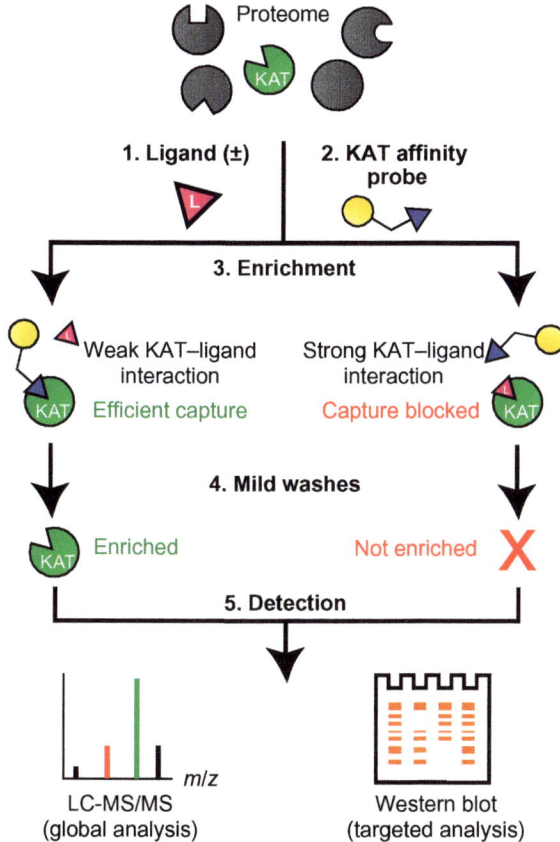

D.C. Montgomery and J.L. Meier, Fig. 3 Analyzing KAT–ligand interactions by competitive chemical proteomic profiling. In a two-step procedure, KAT-containing proteomes are incubated with a ligand of interest, and then these proteomes are mixed with a streptavidin-immobilized KAT capture probe. Ligands that interact strongly with KATs block affinity capture. Analysis of treated and untreated samples by LC-MS/MS or western blot enables quantitative rank-ordering of how KATs interact with ligands of interest.

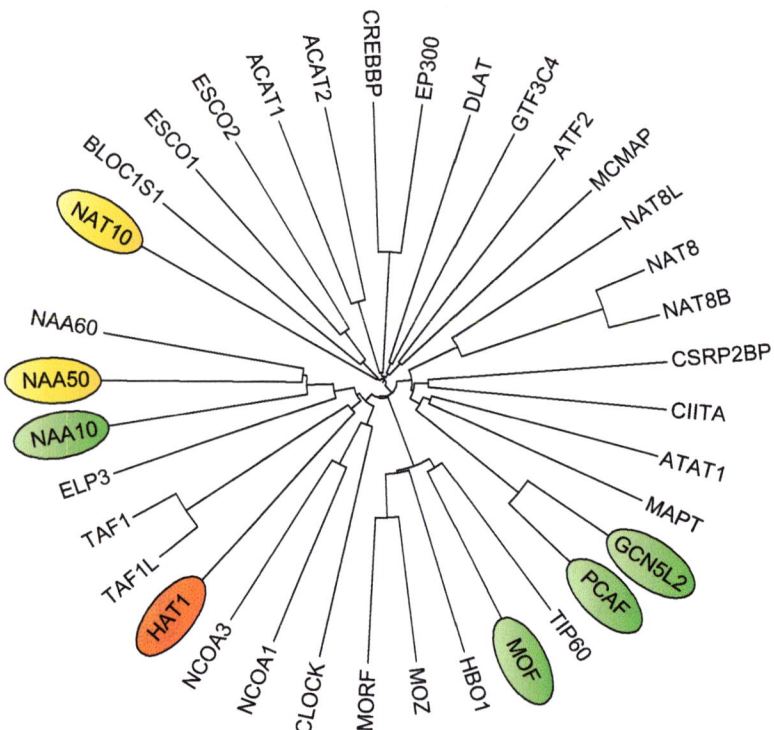

D.C. Montgomery and J.L. Meier, Fig. 4 Phylogenetic tree of putative and validated KAT enzymes (http://apps.thesgc.org/resources/phylogenetic_trees/). *Circled* enzymes represent KATs whose activity has been monitored in an endogenous context using chemical proteomics. *Orange circle*, KAT sulfoxide thiocarbamate probe; *yellow circles*, KAT-clickable photoaffinity bisubstrate probe; *green circles*, KAT-biotin bisubstrate probe.

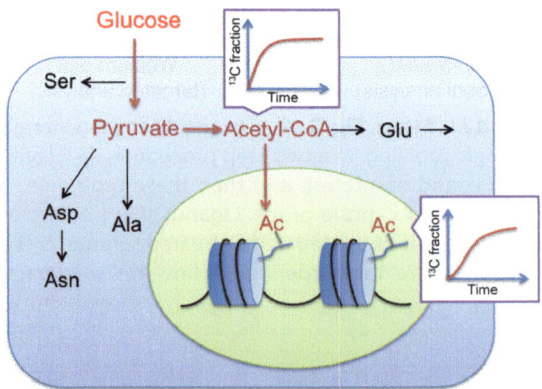

J. Fan et al., Fig. 2 Quantifying histone acetylation rate by metabolic labeling from ^{13}C-glucose. Labeling from glucose is first incorporated into acetyl-CoA by glucose metabolism and then incorporated onto histones by HATs. Glucose also labels various amino acids at the same time.

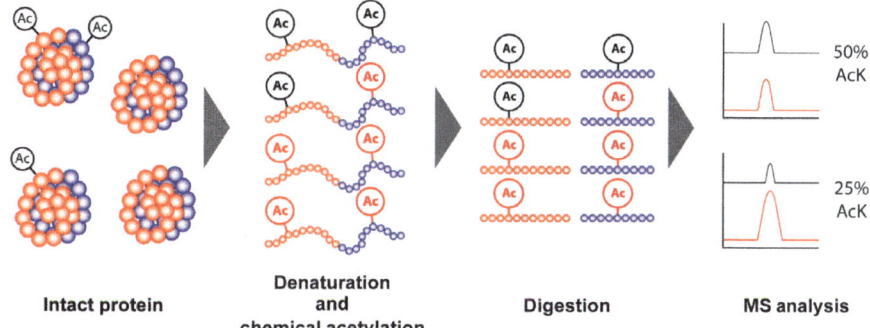

J. Fan et al., Fig. 3 General scheme for determining acetylation stoichiometry. A protein population is denatured and chemically acetylated using isotopic acetic anhydride followed by trypsin digestion. The heavy and light acetyl peptides can be measured by high-resolution mass spectrometry.

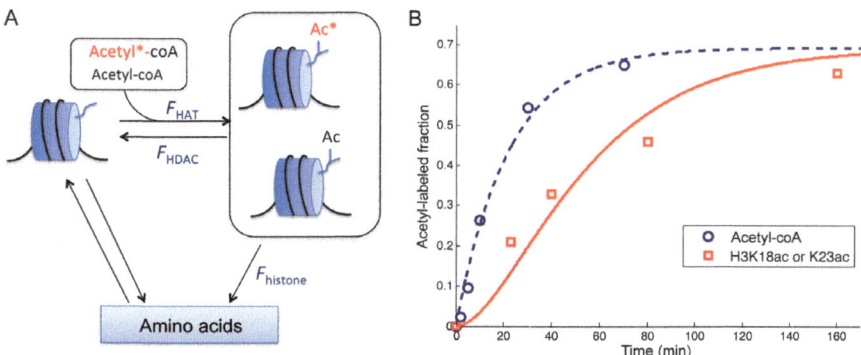

J. Fan et al., Fig. 6 Quantifying histone acetylation rate with metabolic labeling, using model presented in (A). The isotopic acetyl group (*red with* *) is incorporated onto histone lysine sites by HATs and removed by HDACs or turnover of old acetylated histones. Example of acetyl-CoA and monoacetylated H3 peptide $K_{18}QLATK_{23}AAR$ labeling kinetics is shown in (B). *Circles and squares* are experimental data and *lines* are fitting result.

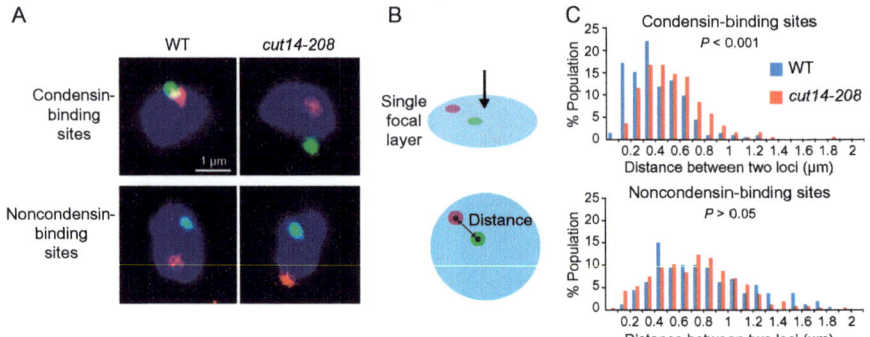

K.-D. Kim et al., Fig. 1 Analysis on FISH foci reflecting positions of two gene loci. (A) The two paired gene loci were visualized in wild-type and *cut14-208* condensin mutant cells. Because the *cut14-208* mutation is temperature sensitive, wild-type and mutant cells were cultured at the restrictive temperature (36°C) for 1 h. The FISH foci (*green* and *red*) representing the two loci bound by condensin were positioned nearby in wild-type cells but not in the mutant (*top*), whereas the two control loci (noncondensin-binding sites) were consistently separated (*bottom*). DAPI signals are shown in *blue*. (B) Two FISH foci are present on a single focal layer (*top*), and the distance between centers of the two FISH foci is defined as a physical distance between the two loci (*bottom*). (C) Distributions of distances between the two gene loci in wild-type and *cut14-208* mutant cells.

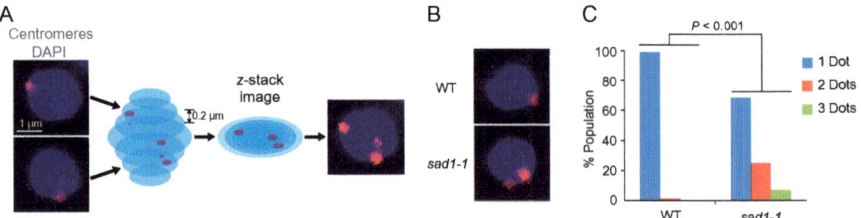

K.-D. Kim et al., Fig. 2 Analysis on centromeric clusters. (A) FISH visualization of centromeric foci (*red*). Centromeric clusters are captured by a *z*-stack projection. DAPI signals are shown in *blue*. (B) Centromeric clusters in the wild-type and *sad1-1* mutant. Since the *sad1-1* is a temperature-dependent mutation, wild-type and mutant cells were cultured at 36°C for 1 h, and subjected to FISH analysis. (C) Numbers of centromeric clusters in wild-type and mutant cells.

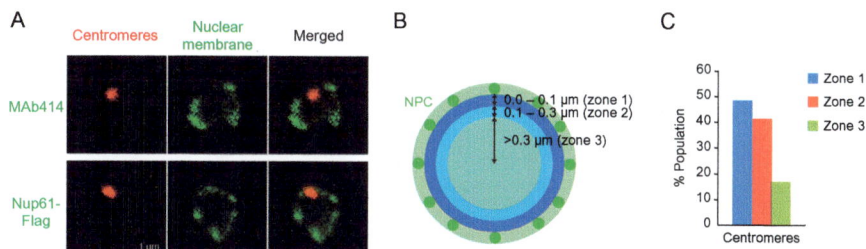

K.-D. Kim et al., Fig. 3 Gene positioning relative to the nuclear architecture. (A) Anti-NPC protein antibody (MAb414) was used to visualize the nuclear membrane (*top*). Alternatively, a NPC subunit (Nup61) fused to a Flag epitope was expressed from the endogenous locus with its own promoter, and Nup61-Flag proteins were visualized by an IF approach (*bottom*). In the same cells, centromeres were covisualized by FISH (*red*). (B) NPC signals are used to trace the nuclear membrane, and the nucleus is divided into the three zones based on distance from the membrane. (C) The distance between the centromeric signal and the nuclear membrane was measured in more than 100 cells, and assigned to one of the nuclear zones.

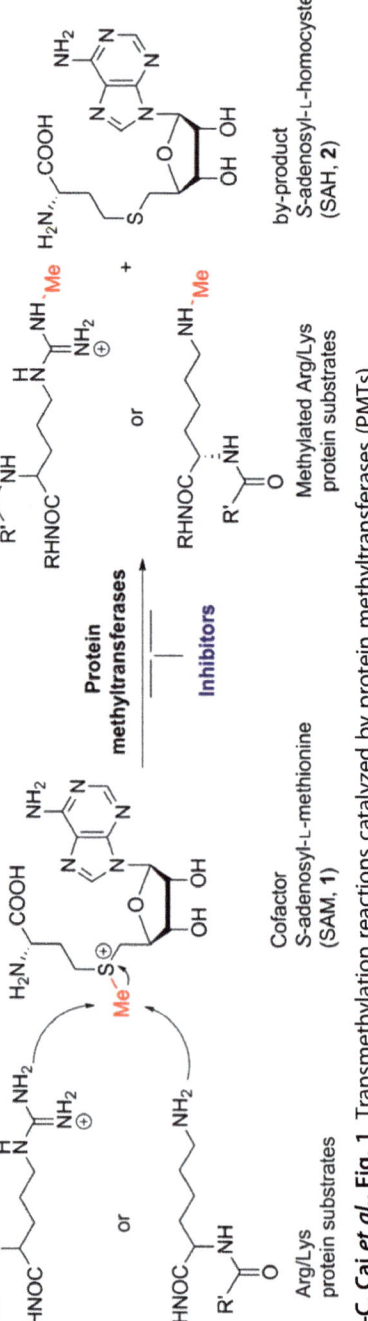

X.-C. Cai et al., **Fig. 1** Transmethylation reactions catalyzed by protein methyltransferases (PMTs).

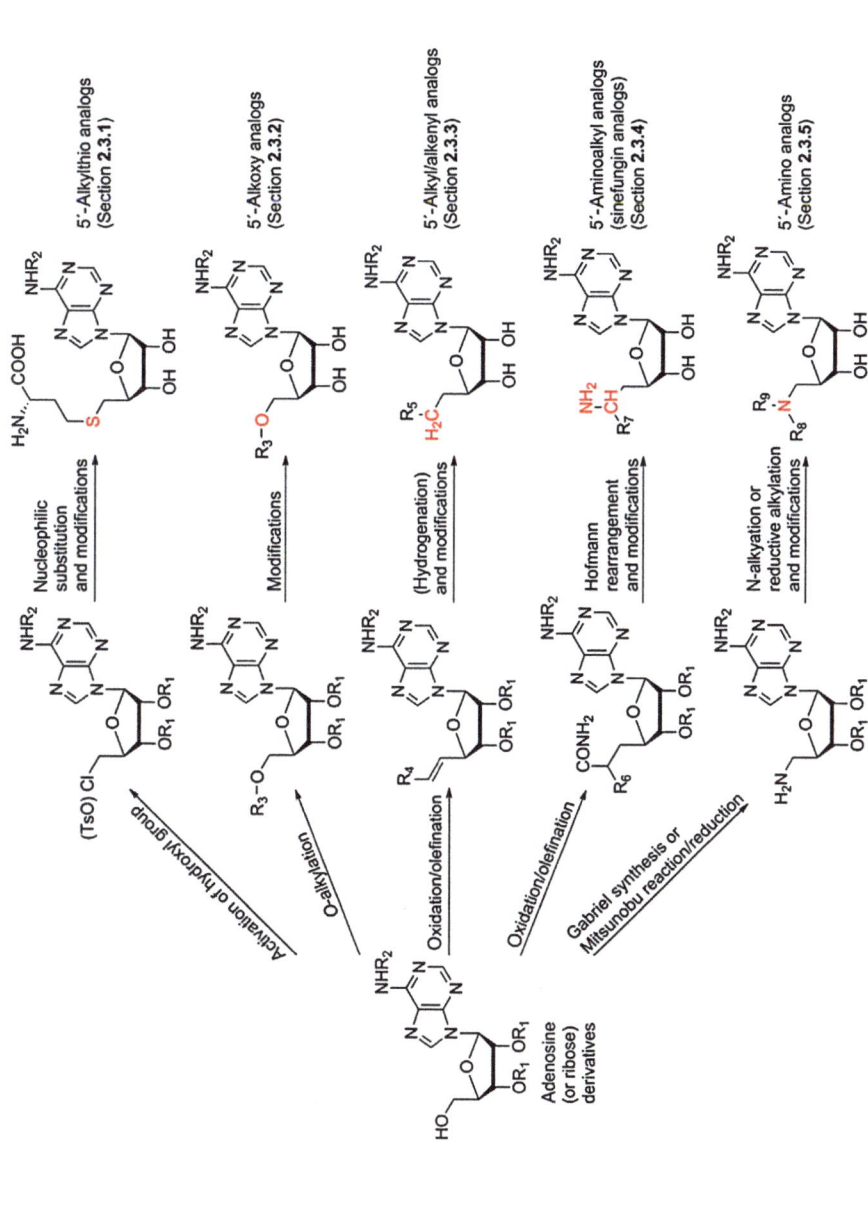

X.-C. Cai et al., Fig. 3 General synthetic strategies to prepare SAM-based methyltransferase inhibitors.

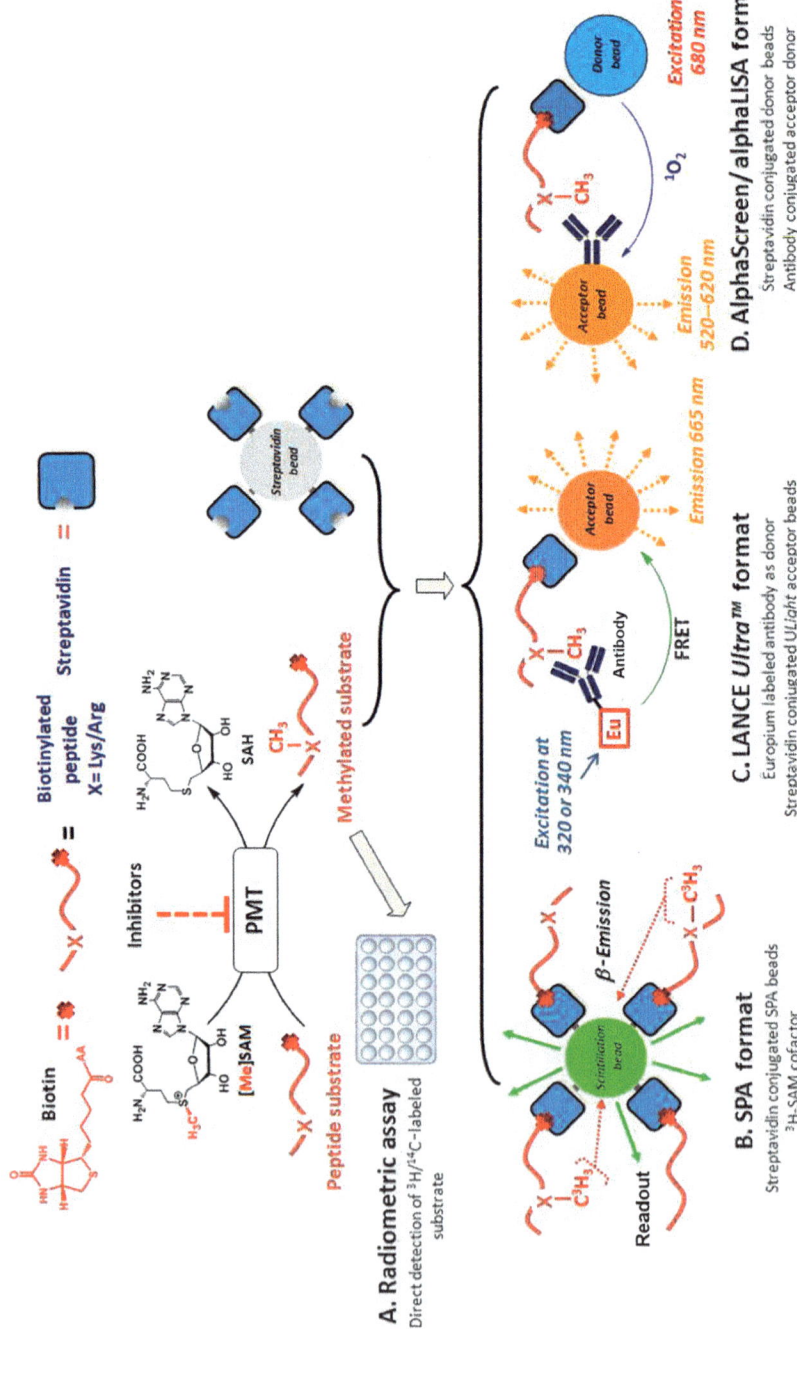

X.-C. Cai et al., Fig. 21 See legend on opposite page.

X.-C. Cai et al., Fig. 21 Assay formats to evaluate inhibitors targeting methyltransferases. (A) Radiometric filter paper/plate assay is based on the transfer of the radiolabeled methyl group from [^{3}H]-SAM to the substrate. β-Emission from the radiolabel to the scintillant fluid results in luminescence signal as a readout. (B) Scintillation proximity assay is a variant of the radiometric assay featured by the capture of the [^{3}H]-labeled substrate on streptavidin-conjugated scintillant beads. (C) LANCE *Ultra*™ involves the capture of the biotinylated peptide on streptavidin-coated acceptor beads. The methylated peptide is recognized by lanthanide (Eu)-labeled antibodies. Förster resonance energy transfer between the lanthanide donor ($\lambda_{ex} = 320-340$ nm) and the acceptor beads gives a fluorescence emission at 665 nm. (D) AlphaScreen(AlphaLISA) format involves the capture of the methylated, biotin-tagged peptide product on streptavidin-coated donor beads. The methylated peptide is recognized by antimethyllysine antibody-conjugated on acceptor beads. Excitation of the donor beads at 680 nm results in emission of a singlet oxygen ($^1O_2$), which excites the acceptor beads causing an emission at 520–620 nm (AlphaScreen) or at 615 nm (AlphaLISA). *Adapted from Luo, M. (2012). Current chemical biology approaches to interrogate protein methyltransferases. ACS Chemical Biology, 7 (3), 443–463.*

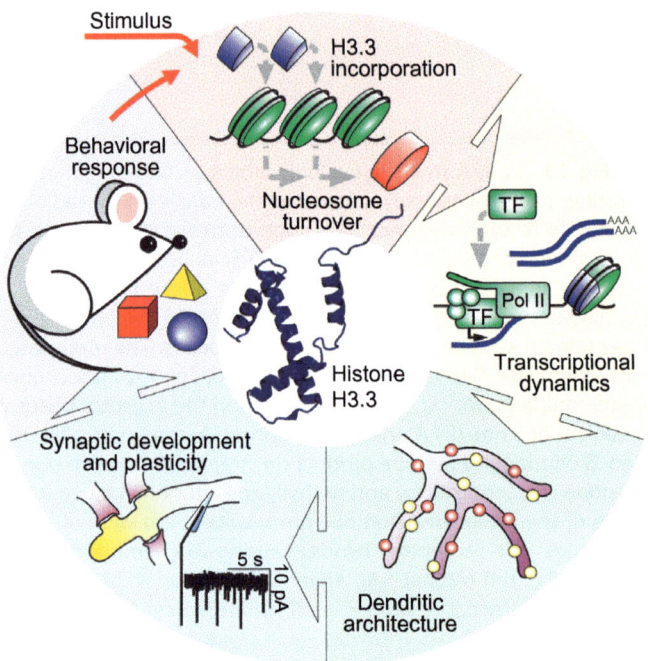

L.A. Farrelly et al., Fig. 1 Histone turnover is a critical mediator of neurological plasticity. In the CNS, during periods of heightened neuronal activity and cellular plasticity, H3.3-associated nucleosomal turnover is increased to allow for activity-dependent transcriptional plasticity that results in alterations in dendritic architecture, synaptic connectivity, and behavioral plasticity (eg, cognition).

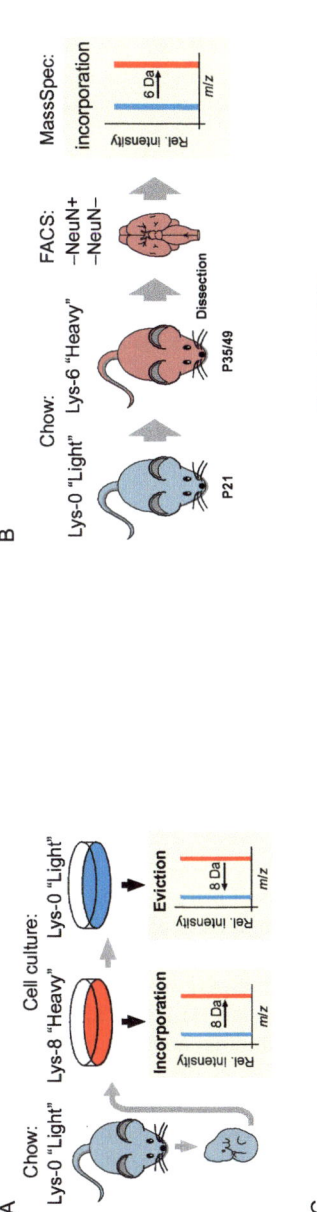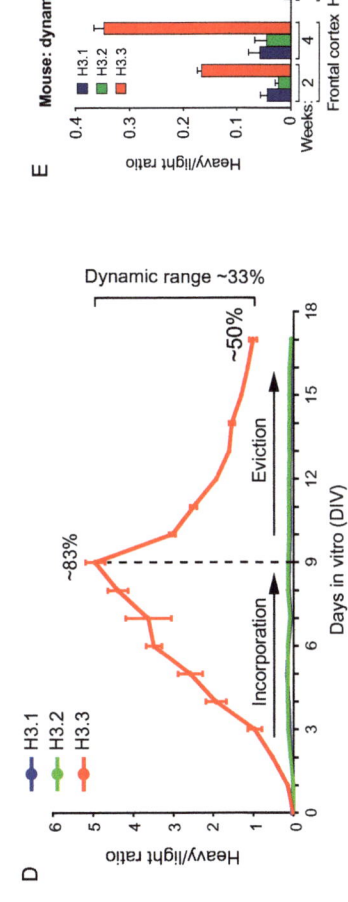

L.A. Farrelly *et al.*, Fig. 2 See legend on next page.

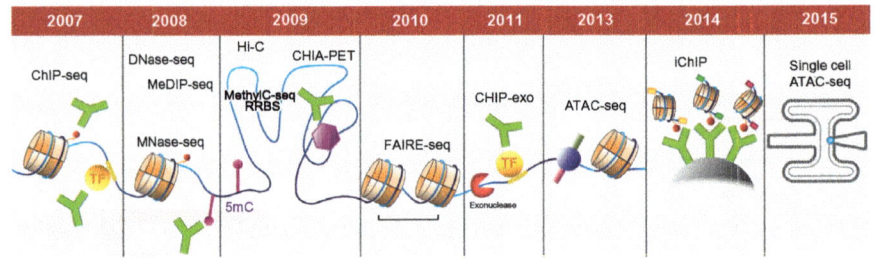

J.L. Johnson and G. Vahedi, Fig. 1 Timeline of sequencing-based technologies for mapping epigenomes. *Adapted from Rivera, C. M., & Ren, B. (2013). Mapping human epigenomes.* Cell, *155(1), 39–55, with permission from Elsevier.*

L.A. Farrelly et al., Fig. 2 Mass spectrometry-based assessments of histone turnover in neurons using SILAC. (A) Schematic of SILAC to assess chromatin-associated H3.x incorporation and eviction from cultured neurons. (B) Schematic describing the SILAC mouse model to assess chromatin-associated H3.x incorporation in adult neurons. (C) Amino acid sequences for H3.x proteins highlighting differences between H3.1 vs H3.2 and H3.3 (H3.3, *red*). Putative target peptides for mass spectrometric analysis in the N-terminal tail (*yellow*) or histone core (*blue*) are indicated. Target peptides in *blue* are used for all subsequent analyses. (D) SILAC time course of H3.x chromatin incorporation and eviction in mouse embryonic neurons over the course of 17 d in vitro. Percentages reflect H3.3 peptide labeling by SILAC. (E) SILAC LC–MS/MS analysis of H3.1/2 vs H3.3 in NeuN+ mouse chromatin from multiple brain structures after 2 or 4 weeks of feeding on a heavy lysine (6 Da) diet. *Panels (D) and (E) displayed with permission from Maze, I., Wenderski, W., Noh, K. M., Bagot, R. C., Tzavaras, N., Purushothaman, I., et al. (2015). Critical role of histone turnover in neuronal transcription and plasticity.* Neuron, *87(1), 77–94.*

Edwards Brothers Malloy
Ann Arbor MI. USA
July 18, 2016